GENETICS – RESEARCH AND ISSUES SERIES

BACTERIAL DNA, DNA POLYMERASE AND DNA HELICASES

GENETICS – RESEARCH AND ISSUES SERIES

Sex Chromosomes: Genetics, Abnormalities, and Disorders
Cynthia N. Weingarten and Sally E. Jefferson (Editors)
2009. ISBN: 978-1-60741-304-2

Genetic Diversity
Conner L. Mahoney and Douglas A. Springer (Editors)
2009. ISBN: 978-1-60741-176-5

The Human Genome: Features, Variations and Genetic Disorders
Akio Matsumoto and Mai Nakano (Editors)
2009. 978-1-60741-695-1

Bacterial DNA, DNA Polymerase and DNA Helicases
Walter D. Knudsen and Sam S. Bruns (Editors)
2009. ISBN: 978-1-60741-094-2

GENETICS – RESEARCH AND ISSUES SERIES

BACTERIAL DNA, DNA POLYMERASE AND DNA HELICASES

WALTER D. KNUDSEN
AND
SAM S. BRUNS
EDITORS

Nova Science Publishers, Inc.
New York

Copyright © 2010 by Nova Science Publishers, Inc.

All rights reserved. No part of this book may be reproduced, stored in a retrieval system or transmitted in any form or by any means: electronic, electrostatic, magnetic, tape, mechanical photocopying, recording or otherwise without the written permission of the Publisher.

For permission to use material from this book please contact us:
Telephone 631-231-7269; Fax 631-231-8175
Web Site: http://www.novapublishers.com

NOTICE TO THE READER

The Publisher has taken reasonable care in the preparation of this book, but makes no expressed or implied warranty of any kind and assumes no responsibility for any errors or omissions. No liability is assumed for incidental or consequential damages in connection with or arising out of information contained in this book. The Publisher shall not be liable for any special, consequential, or exemplary damages resulting, in whole or in part, from the readers' use of, or reliance upon, this material. Any parts of this book based on government reports are so indicated and copyright is claimed for those parts to the extent applicable to compilations of such works.

Independent verification should be sought for any data, advice or recommendations contained in this book. In addition, no responsibility is assumed by the publisher for any injury and/or damage to persons or property arising from any methods, products, instructions, ideas or otherwise contained in this publication.

This publication is designed to provide accurate and authoritative information with regard to the subject matter covered herein. It is sold with the clear understanding that the Publisher is not engaged in rendering legal or any other professional services. If legal or any other expert assistance is required, the services of a competent person should be sought. FROM A DECLARATION OF PARTICIPANTS JOINTLY ADOPTED BY A COMMITTEE OF THE AMERICAN BAR ASSOCIATION AND A COMMITTEE OF PUBLISHERS.

Library of Congress Cataloging-in-Publication Data

Bacterial DNA, DNA polymerase, and DNA helicases / [edited by] Walter D. Knudsen and Sam S. Bruns.
   p. ; cm.
 Includes bibliographical references and index.
 ISBN 978-1-60741-094-2 (hardcover)
 1. Bacterial genetics. 2. DNA polymerases. 3. DNA helicases. I. Knudsen, Walter D. II. Bruns, Sam S.
 [DNLM: 1. Bacteria--genetics. 2. DNA Helicases--physiology. 3. DNA-Directed DNA Polymerase--physiology. QW 51 B13045 2009]
 QH434.B333 2009
 572.8'293--dc22
                           2009018833

Published by Nova Science Publishers, Inc. ✦ *New York*

Contents

Preface		vii
Research and Review Studies		1
Chapter I	Codon and Codon Pairs Usage in Bacteria *Boris I. Bachvarov, Kiril T. Kirilov and Ivan G. Ivanov*	3
Chapter II	Maillard Reaction and Spontaneous Mutagenesis in *Escherichia coli* *Roumyana Mironova, Yordan Handzhiyski, Toshimitsu Niwa, Alfredo Berzal-Herranz, Kirill A. Datsenko, Barry L. Wanner and Ivan Ivanov*	51
Chapter III	Nucleoid Architecture and Dynamics in Bacteria *Ryosuke L. Ohniwa, Kazuya Morikawa, Toshiko Ohta, Chieko Wada and Kunio Takeyasu*	91
Chapter IV	Genome Reduction in *Bacillus subtilis* and Enhanced Productivities of Recombinant Proteins *Yasushi Kageyama, Takuya Morimoto, Katsutoshi Ara, Katsuya Ozaki and Naotake Ogasawara*	119
Chapter V	Molecular Detection and Characterization of Food and Waterborne Viruses *Alain Houde, Kirsten Mattison, Pierre Ward and Daniel Plante*	137
Chapter VI	hTERT in Cancer Chemotherapy: A Novel Target of Histone Deacetylase Inhibitors *Jun Murakami, Jun-ichi Asaumi, Hidetsugu Tsujigiwa, Masao Yamada, Susumu Kokeguchi, Hitoshi Nagatsuka, Tatsuo Yamamoto and You-Jin Lee*	187
Chapter VII	Dynamics of DNA Polymerase *Ping Xie*	225

Chapter VIII	Inhibiting Viral DNA Polymerase for HBV Therapy *Min Jiang and Yuanhao Li*	263
Chapter IX	Modeling DNA Translocation and Unwinding by Helicase RecG *Ping Xie*	283
Chapter X	*De Novo* DNA Synthesis by DNA Polymerase *Xingguo Liang, Tomohiro Kato and Hiroyuki Asanuma*	299
Chapter XI	Bacteriophage ϕ29 DNA Polymerase: An Outstanding Replicase *Miguel de Vega and Margarita Salas*	329
Chapter XII	Bacterial Replicative DNA Helicases *Subhasis B. Biswas, and Jessica J. Clark, Deepa S. Kurpad and Esther E. Biswas*	353
Chapter XIII	Effect of Substrate Traps on Hepatitis C Virus NS3 Helicase Catalyzed DNA Unwinding: Evidence for Enzyme Catalyzed Strand Exchange *Ryan S. Rypma, Angela M. I. Lam and David N. Frick*	389

Short Communication

	Discovery of Orphan Helicases and Deorphanization by Genome-Wide Analyses in Two Model Organisms, *S. Cerevisiae* and *C. Elegans* *Toshihiko Eki and Fumio Hanaoka*	411
Index		433

Preface

Bacteria are a large group of unicellular microorganisms. Most bacteria have a single circular chromosome that can range in size. DNA is organized into chromosomes and, in organisms other than bacteria, it is found only in the cell nucleus. Deoxyribonucleic acid(DNA)is a nucleic acid that contains the genetic instructions used in the development and functioning of all known living organisms and some viruses. Their main role is the long-term storage of information. DNA is often compared to a set of blueprints or a code, since it contains the instructions needed to construct other components of cells. This book presents current research on bacterial DNA, DNA polymerase and DNA helicases.

Chapter I - The new generation of fast and low-cost DNA sequencing methods led to a rapid accumulation of genomic data in the DNA databases after the year 2000. Hundreds of bacterial genomes have been sequenced during the last decade, however, in rare cases they were profoundly analyzed. Apparently it is much easier today to generate raw DNA sequencing data rather than to decipher their biological meaning. This study aims to shed light on the relationship between codon usage and gene expression in 158 prokaryotes belonging to the two major classes: *Archaea* and *Bacteria*. *Archaea* includes 2 families (*Crenarchaeota* and *Euryarchaeota*) and *Bacteria* 13 families (*Actinobacteria, Aquificae, Bacteroidetes, Chlamydiae, Chlorobi, Cyanobacteria, Deinococcus-Thermus, Firmicutes, Fusobacteria, Planctomycetes, Proteobacteria, Spirochaetes* and *Thermotogae*) and the total number of genera was 9 for *Archaea* and 22 for *Bacteria*. In order to study the link between codon usage and gene expression four local DNA databases were created containing respectively: a) all protein coding sequences of each organism; b) highly expressed genes; c) ribosomal proteins genes and d) low expressed genes. These databases were used to compare codon usage patterns of: a) different strains of one bacterial species; b) different subsets of genes in one species and c) different bacterial genomes. They were also employed to identify missing and atypical codons as well as to search for correlation between codon usage and the formal prokaryotic taxonomy. The obtained results indicate that the formal taxonomy of many bacterial species does not fit their codon usage pattern (as predicted by the "Genome Theory"). On the other hand there are obvious similarities in the codon usage pattern of distant (according to the formal taxonomy) organisms. The latter suggests that either the Genome Theory or the official taxonomy of microorganisms need revisions. Taking into consideration that at least two tRNAs occupy the two (A and P) functional ribosomal sites during the translation elongation step, the authors have also studied the effect of neighbor codons (codon pairs) on gene expression. Aiming to identify preferential and rare codon pairs, the authors have developed own criteria for determining their frequency of occurrence and for evaluation of their possible effect on translation. This approach was applied to analyze the distribution of all (3904) codon pairs in the *Escherichia coli* genome and the authors found that their frequency of occurrence varied from zero to 4913 times. The predicted effect of some (particularly 3'-terminal) codon pairs on gene expression is confirmed experimentally.

Chapter II - DNA is a fragile organic molecule of finite chemical stability, capable of interacting with a variety of chemicals. In the context of the living cell this DNA reactivity is manifested as mutability. In addition to the chemical mutability, the perpetuation of DNA via imperfect replication is another source of DNA changeability. It is worth mentioning that DNA repair is error-tolerant and always leaves behind some lesions exist. Were repair mechanisms perfect, then the door of evolution would be locked. Risky rather than benign at individual level, mutations are the living things' long term strategy for survival and evolution. The creative power of mutations can explain why organisms have evolved special mechanisms to enhance mutagenesis under certain conditions. Hypermutation in bacteria under stress (SOS response) as well as hypermutation of immunoglobulin genes in response to antigen challenge in vertebrates are key examples.

The present study is focused on spontaneous mutagenesis in *Escherichia coli* K-12.

Spontaneous mutations are the net result of DNA damage and repair under normal physiological conditions. Bacterial DNA is loosely packed with proteins and occupies almost the entire cellular space. Therefore, it is extremely vulnerable to chemical attack, especially during replication and transcription, when the two DNA strands are separated. In the same arena, the cytoplasm, and at the same time, play also many small but highly reactive compounds. Although the collision of the latter with DNA seems to be unavoidable, to date the impact of ordinary cellular metabolites on spontaneous mutagenesis has not been systematically studied.

In year 2012 we shall celebrate the 100 anniversary of the discovery of the carbonyl-amine reaction by the French chemist Louis Camille Maillard (Maillard, 1912). Here, the authors provide evidence that the Maillard reaction, referred to as non-enzymatic glycosylation or glycation in the literature, is an important endogenous source of spontaneous mutations in *E. coli*. Experimental data allowed us to estimate that glycation affects on average one per 10^5 to 10^4 nucleotides in the *E. coli* chromosome. Based on experimental evidence, the authors postulate the existence in *E. coli* of a novel DNA repair enzyme, DNA amadoriase or DNA deglycase, especially designed to combat glycation induced spontaneous mutagenesis in *E. coli*.

Chapter III - Both prokaryotes and eukaryotes store their genomic DNA in an environment, which balances the physical properties of double-stranded DNA with the mechanical effects of DNA-protein interactions. The origin of genome packing remains a mystery; it seems to go back to the very beginning of life itself. Since then, the principle of higher-order genome construction and architecture has been maintained and shared among three domains of living things. The genomes are stored as flexible higher-order structures that can shuttle between relatively active and inactive states. The major differences reside in the structural protein components of the architecture. Bacteria and Eukaria utilize HU and histones as the most fundamental structural proteins, respectively, forming the nucleoid in bacterial cell and the nucleosomes in the nucleus. On the other hand Archaea employ either HU or histones depending on their phylogenetic origins. Nevertheless, the hierarchies of nucleoid fibers look extremely similar among different species in different domains. Focusing upon the mechanism of bacterial genome folding, recent structural investigations have revealed a step-wise folding of the genome DNA; from 10 nm to 30 nm fibers, and to 80 nm and further condensed beaded structures, depending upon the growth conditions and differential contribution of nascent single-stranded RNA. The nucleoid condensation is mainly brought about by another nucleoid protein, Dps, which occurs widely in the bacterial domain, but not in Archaea or Eukarya. The regulatory mechanisms of the *dps* gene are greatly different among bacterial species. In ☐-Proteobacteria such as *Escherichia coli*, the *dps* gene is governed by two regulatory systems, IHF-☐s and OxyR, and is induced towards the stationary phase and under oxidative stress. The DNA topology control by *E. coli* Fis also participates in the regulation of the nucleoid condensation. In contrast, in Firmicutes such as *Staphylococcus aureus*, the gene is mainly controlled by oxidative stress-response system, PerR, and the Dps expression is directly coupled with the nucleoid condensation. Here the key features of the nucleoid dynamics will be discussed from the biochemical and structural biological points of views.

Chapter IV - *Bacillus subtilis*, which is one of the most important host microorganisms for large-scale industrial production of useful proteins, has a genome of 4.2 Mb with approximately 4106 protein-coding genes. Some of these genes are expected to be unnecessary for industrial production of proteins under controlled conditions and may be wasteful with regard to energy consumption. The authors attempted to reduce the genome size of *B. subtilis* by deleting unnecessary regions of the genome to allow the construction of simplified host cells as a platform for the further development of novel genetic systems with increased productivity.

First, the authors generated the strain MGB469 with deletion of all prophage (SPβ and PBSX) and prophage-like (pro1-7 and skin) sequences, with the exception of pro7, as well as two large operons that produce secondary metabolites (*pks* and *pps*). These cells showed normal growth, but no beneficial effects were observed with regard to recombinant protein production from plasmids carrying the corresponding genes. Second, the authors constructed several multiple-deletion mutants containing additional deletions in the MGB469 genome, resulting in total genome size reductions of 0.78 to 0.99 Mb. In most of the multiple-deletion series, extensive deletion mutants showed no beneficial improvements in traits as host strains. The strain MG1M with a total genome size reduction of 0.99 Mb showed unstable phenotypes with regard to growth rate, cell morphology, and recombinant protein productivity after successive culture, making it inappropriate for further studies. In addition, strain MGB943 derived from another lineage with genome reduction of 0.94 Mb showed reduced recombinant cellulase productivity. Finally, they generated another multiple-deletion series including the mutant MGB874 with a total genome deletion of 0.87 Mb. In comparison to wild-type cells, the metabolic network of the mutant strain was reorganized after entry into the transition state due to the synergistic effects of multiple deletions. Moreover, the levels of production of extracellular cellulase and protease from transformed plasmids carrying the corresponding genes were markedly increased.

Our results demonstrated the effectiveness of a synthetic genomic approach with reduction of genome size to generate novel and useful bacteria for industrial uses.

Chapter V - Enterically transmitted viruses such as norovirus (NoV), rotavirus (RV), hepatitis A virus (HAV) and hepatitis E virus (HEV) are currently recognized as a major cause of food and waterborne illness in humans worldwide. Most of these viruses cannot be cultured *in vitro*, are stable in the environment for long periods of time, are generally infectious at low doses (only a few viral particles are needed to trigger illness) and would be usually present in low numbers in contaminated food and water samples. There are also increasing concerns about the possible zoonotic transmission or the emergence of new recombinant strains from animal enteric viruses closely related to human pathogenic strains. Until recently, the viral contamination of food and water remained undetected due to a lack of appropriate detection methods. Electronic microscopy (EM) and ELISA assays commonly used on stool suspensions were rather insensitive for food and water applications. Due to their increased sensitivity, conventional and now real-time quantitative reverse transcription PCR (qRT-PCR) assays were developed and widely used for the detection and identification of the nucleic acid of these challenging viruses. Along with microarrays, these new approaches have also allowed significant advances in the molecular typing of these organisms. Rapid, sensitive and reliable molecular detection and typing methods will have a significant impact in the mitigation of outbreaks and could contribute in the prevention of food/waterborne transmission. This review will provide recent trends and advances in the field of DNA polymerase-based detection and characterization methods of these 4 main food and waterborne viruses. It will focus on the use of different conventional and qRT-PCR methods and the microarray technology.

Chapter VI - Chromatin structure plays an important role in the regulation of gene transcription. Chromatin structure can be modified by various post-translational modifications, including histone acetylation, phosphorylation, methylation and ribosylation. Among those modifications, histone acetylation/deacetylation is the most important mechanism for regulating transcription and is regulated by a group of enzymes known as histone acetyltransferases/histone deacetylases (HDACs).

Recently, HDAC inhibitors have been shown to be a novel and promising new class of anti-cancer agent that can regulate the transcription of genes by disrupting the balance of acetylation/deacetylation in particular regions of chromatin. A number of HDAC inhibitors are currently in phase I and II clinical trials against a variety of cancers. Although some promising candidates have been identified (e.g., $p21^{WAF1}$ and c-Myc), the precise molecular targets remain uncertain. In this article, the authors focus on one of the DNA polymerases, telomerase, as a new candidate molecular target for HDAC inhibitors. Telomerase is composed primarily of the catalytic subunit (hTERT) and the RNA template (hTERC), and its activity correlates with levels of hTERT mRNA. hTERT expression is apparently governed by complicated regulatory pathways. Based on recent studies, the hTERT gene is likely to be targeted by histone acetylation/deacetylation.

Chapter VII - Replicative DNA polymerase (DNAP) is an enzyme that synthesizes a new DNA strand on a single-stranded template with a high processivity and a high fidelity. The high fidelity is mainly realized via a mechanism of proofreading that is performed at the exonuclease active site spatially separate from the polymerase active site. Here the authors present a detailed account of our proposed model for the processive nucleotide incorporation and switching transition of DNA between the polymerase and exonuclease active sites by DNAP. Based on the model, the authors present detailed theoretical studies on its dynamics. The moving time of DNAP to next site after a correct incorporation and that after an incorrect incorporation are analytically studied. The experimentally measured dependence of polymerization rate on tension applied to the template is well simulated. The transfer rates of DNA from the polymerase to exonuclease active sites after a correct incorporation and after incorrect incorporations as well as the transfer rate from the exonuclease to polymerase active sites are analytically studied. Moreover, the backward motion of DNAP when large tensions are applied to the template is also analytically studied. The theoretical results are in good agreement with available experimental data. Some predicted results are presented.

Chapter VIII – Hepatitis B virus infection is a major public health problem worldwide especially in East Asia. The viral infection, if persistent, may lead to chronic hepatitis, cirrhosis, and hepatocellular carcinoma. In1998, FDA approved GlascoSmithKline's Epivir-HBV (Lamivudine), a small molecule drug that inhibits HBV DNA polymerase and interferes with viral replication. Since then, a few similar drugs have been approved and many candidates are currently at preclinical and clinical development. The use of small molecular drugs that target HBV DNA polymerase is a milestone in the treatment of chronic hepatitis B. In this review, the authors will focus on HBV DNA polymerase and discuss the preclinical and clinical development process for the HBV polymerase inhibitors. The DNA polymerase mutants associated with drug resistance will also be discussed. The mechanism of the drug resistance and further understanding of these DNA polymerase inhibitors may help the determination of better clinical regimen for HBV therapy and patient care.

Chapter IX - RecG is a DNA helicase involved in the repair of damage at a replication fork by catalyzing the reversal of the fork to create a Holliday junction. Here, based on previous structural and biochemical studies a model is presented on how RecG catalyzes the processive translocation and unwinding of DNA so as to realize the fork reversal. In the model, a power stroke induces the DNA translocation and unwinding. Moreover, in order to effectively unwind DNA of a helical structure, it is assumed that the residues (i.e., the long α helix) connecting the wedge domain and the helicase domains in which the dsDNA-binding loop is located behave elastically along the torsional direction. Thus, accompanying the power stroke, the wedge domain is forced elastically to rotate relative to the dsDNA-binding loop that binds dsDNA strongly. Using the model, the calculated DNA translocation size per ATP hydrolysis is consistent with previous experimental data. Moreover, it is showed that RecG has a strong preference for negatively supercoiled DNA, i.e., RecG shows a much higher ATPase activity and longer processivity when interacting with the negatively supercoiled DNA than with the relaxed and/or positively supercoiled DNAs. This is also in agreement with the previous experimental data. In addition, the rotational rate or the rotational angle of RecG relative to DNA per ATP hydrolysis and the torsional elastic coefficient of the residues connecting the wedge domain and the helicase domains are predicted.

Chapter X - In this chapter, the *de novo* DNA synthesis by DNA polymerase in the absence of any added template and/or primer is described. As early as the 1960s, Kornberg et al. reported that DNA polymerase I could *de novo* synthesize DNA, although it was later pointed out that the contaminated DNA or RNA may provide the seeds for DNA synthesis and the contaminated transferase that can polymerize dNDPs might play the role of template-independent DNA synthesis. The synthesized DNA polymers were characterized as homopolymer pairs such as poly(dA)/poly(dT) and poly(dI)/poly(dC), or the alternating copolymers poly(dA-dT), poly(dG-dC) and poly(dI-dC). In late 1990s, after a long silent time, Ogata et al. found that the highly purified thermophilic DNA polymerase could carry out *de novo* DNA synthesis at higher temperatures (> 65°C) and the synthesized DNA polymers were composed of repetitive palindromic sequences such as $(TACATGTA)_n$, and $(TAAT)_n$. Surprisingly, Liang et al. reported that thermophilic DNA polymerases could carry out *de novo* DNA synthesis even at room temperature. Basically, *de novo* DNA synthesis is considered to occur in two stages, although the detail mechanism is not clear. In the initial stage, dNTPs are polymerized to short DNA oligos, and then the selected repetitive sequences are elongated to long DNA in the elongation stage through self-priming and primer extension. More interestingly, the *de novo* DNA synthesis can be greatly accelerated by adding endonuclease such as restriction enzymes, which digest DNA at their recognition sites. A Cut-grow mechanism was proposed to explain both the paradox of endonuclease and the extremely high efficiency: the synthesized long DNA is cut to shorter seeds for the elongation of the next cycle so that DNA can be exponentially amplified. The possible biological and evolutionary roles of *de novo* DNA synthesis are also discussed.

Chapter XI - Due to the limited processivity of replicative DNA polymerases (replicases), as well as to their incapacity to unwind parental duplex DNA to allow replication fork progression, their replication efficiency depends on the functional assistance of accessory proteins as processivity factors and helicases. In addition, the inability of DNA polymerases to start de novo DNA synthesis requires the use of a short RNA/DNA molecule to provide the 3'-OH group required to initiate DNA replication. This requisite for a primer creates a dilemma to replicate the ends of linear genomes: once the last primer for the lagging strand synthesis is removed, a portion of ssDNA at the end of the genome will remain uncopied. Bacteriophage ϕ29 has overcome these issues by means of the unique catalytic features of an outstanding enzyme, the ϕ29 DNA polymerase. This replicase belongs to the family B (eukaryotic-type) of DNA-dependent DNA polymerases and has served as model to understand the enzymology of these polymerases. As most of the family B members, ϕ29 DNA polymerase contains both 3'-5' exonuclease and polymerization activities residing in two structurally independent domains. During two decades, site-directed mutagenesis studies of individual residues contained in regions of high amino acid similarity have provided the functional insights of this enzyme, extrapolative to other family B members. However, ϕ29 DNA polymerase is endowed with two distinctive features: high processivity and strand displacement capacity that allow it to replicate the viral genome from a single binding event, without requiring the assistance of unwinding and processivity factors. Recent crystallographic resolution of the structure of the apo and binary/ternary complexes of ϕ29 DNA polymerase, together with the biochemical studies of site-directed mutants, have given insights into the structural basis responsible for the coordination of the processive polymerization and strand displacement. In addition, such structures have provided the mechanism of translocation of family B DNA polymerases. Another difference with respect to the rest of replicases is the ability of φ29 DNA polymerase to use a protein (terminal protein, TP) as primer, circumventing the end replication problem. Recent resolution of the structure of the φ29 DNA polymerase/TP heterodimer, together with the biochemical analysis of chimerical DNA polymerases and TPs, have given the clues of the specificity of the interaction between both proteins, suggesting a model for the transition from initiation to elongation. The authors will also discuss how the basic research on the φ29 DNA polymerase properties have led to the development of DNA amplification technologies based on this outstanding enzyme.

Chapter XII - Replicative DNA helicases are energy-transducing enzymes, which act as the engine of the cellular replication apparatus and unwind duplex DNA in the replication fork. These enzymes are true multifunctional enzymes that are responsible for organization and execution of DNA replication from initiation to termination. The most well studied of them all is DnaB helicase of *Escherichia coli*, which is the prototype of this class of DNA helicases. Early genetic studies of the *dnaB* gene defined important roles of DnaB protein in the elongation stage of DNA replication. DnaB protein acts as the organizer of the replisome by engaging in multiple protein-protein and protein-DNA interactions during the initiation, elongation, and termination stages of DNA replication. DnaB protein and its orthologs have homo-hexameric structure in solution and in the ssDNA-bound state. This homo-hexameric subunit structure allows for the formation of a ring-structure; ssDNA passes through the ring's central hole. Formation of this unique protein-ssDNA complex is a key element in this enzyme's ability to carry out its unique functions. Due to its stable ring structure, *E. coli* DnaB helicase requires the assistance of DnaC protein in order to form the initial DnaB•ssDNA complex. It forms a DnaB•DnaC•ATP complex that attenuates its nucleotidase activity but helps in binding ssDNA. DnaC protein is released upon slow ATP hydrolysis leading to organization of the replication proteins in the fork and replisome formation. DnaB protein engineers the assembly of the replisome through its specialized interaction with DnaG primase and DNA polymerase III holoenzyme. DnaB•ssDNA complex formation stimulates its ATPase activity of DnaB hexamer several fold. The energy of ATP hydrolysis powers translocation of DnaB protein on ssDNA in a $5'\to3'$ direction and helps unwind duplex DNA during translocation. The rapid and processive DNA unwinding also requires direct participation of a topoisomerase to remove the DnaB-generated super-twists without which the DNA unwinding and the replication fork would likely stall. In this review article, the authors summarize the structural and functional properties of the *E. coli* replicative DNA helicase and compare it to analogous replicative helicases from other gram-negative and gram-positive bacteria.

Chapter XIII - The helicase encoded by the hepatitis C virus (HCV) is shown in this chapter to catalyze homologous DNA strand exchange. Single-stranded DNA oligonucleotides complementary to either the short or long strand of a partially duplex DNA substrate affected the rate and extent of unwinding catalyzed by either the full-length HCV NS3 protein fused to a portion of HCV NS4A, or a truncated NS3 protein lacking the protease domain. The oligonucleotides did not, however, sequester HCV helicase and prevent it from separating the original duplex after a single binding cycle. Furthermore, when DNA oligonucleotides were pre-incubated with HCV helicase, they did not prevent subsequent duplex separation, indicating that the enzyme catalyzes strand exchange. The protease portion of NS3 was not needed for this strand exchange. Fluorescent DNA substrates were further used to directly monitor both this ssDNA assisted unwinding and homologous strand exchange, and the effect of a protein trap (poly(U) RNA) on HCV helicase-catalyzed strand exchange was examined. The results demonstrate that HCV helicase can simultaneously bind at least three DNA strands and imply that HCV NS3 helicase could play an important role not only in viral RNA unwinding, but also in the folding of the HCV genome.

Short Communication - Eukaryotic DNA helicases are members of the helicase superfamily together with two other functional classes, RNA helicases and chromatin remodeling ATPases. Members of this superfamily play crucial roles in various nucleic acid- and chromatin-mediated cellular processes such as DNA replication, repair and recombination, pre-mRNA splicing, ribosome biogenesis, RNA interference, and chromatin remodeling. Since these reactions are essential for the maintenance, expression, and regulation of genetic information in the chromosome, dysfunctional helicase genes may lead to genetic diseases such as cancer. Indeed, genetic mutations of the human RecQ-like BLM helicase and WRN DNA helicase result in Bloom syndrome and Werner syndrome, which are characterized by the early development of various cancers and premature aging, respectively. Most helicases share conserved amino acid sequence motifs; their corresponding genes are classified into five superfamilies (SF1-SF5) based on the occurrence and characteristics of conserved motifs.

The genome sequences of two representative eukaryotes, the budding yeast *Saccharomyces cerevisiae* and the nematode *Caenorhabditis elegans* were determined in 1996 and 1998, respectively. Subsequently, high-throughput genome-wide analyses have been systematically performed to clarify the cellular roles of 6000 and 19000 genes discovered in each genome, including gene expression profiling, proteome analyses, comprehensive analyses of loss-of-function phenotypes, and genetic interaction analyses. A huge amount of functional data from these studies has been analyzed and deposited in public databases such as the *Saccharomyces* Genome Database (SGD) and the WormBase. However, a few decades after the initial sequencing, a large number of genes remain functionally-unknown (i.e., orphan) in both organisms; for instance, over 1000 genes are uncharacterized even in yeast. A large number of helicase-like genes have been found in the sequenced eukaryotic genomes, including many genes encoding orphan helicase-related proteins. Because of the biological importance of the helicase family, a comprehensive analysis of the functions of helicase family members in two model organisms, *S. cerevisiae* and *C. elegans*, has been performed. In this Short Communication, the authors briefly summarize the results from studies of helicase-like proteins in both organisms and describe the effectiveness and limitations of a genome-wide analysis of orphan family members.

Research and Review Studies

In: Bacterial DNA, DNA Polimerase and DNA Helicases
Editors: W. D. Knutsen and S. S. Bruns

ISBN 978-1-60741-094-2
© 2009 Nova Science Publishers, Inc.

Chapter I

Codon and Codon Pairs Usage in Bacteria

Boris I. Bachvarov, Kiril T. Kirilov and Ivan G. Ivanov[*]

Institute of Molecular Biology,
Bulgarian Academy of Sciences, Bulgaria

Abstract

The new generation of fast and low-cost DNA sequencing methods led to a rapid accumulation of genomic data in the DNA databases after the year 2000. Hundreds of bacterial genomes have been sequenced during the last decade, however, in rare cases they were profoundly analyzed. Apparently it is much easier today to generate raw DNA sequencing data rather than to decipher their biological meaning. This study aims to shed light on the relationship between codon usage and gene expression in 158 prokaryotes belonging to the two major classes: *Archaea* and *Bacteria*. *Archaea* includes 2 families (*Crenarchaeota* and *Euryarchaeota*) and *Bacteria* 13 families (*Actinobacteria, Aquificae, Bacteroidetes, Chlamydiae, Chlorobi, Cyanobacteria, Deinococcus-Thermus, Firmicutes, Fusobacteria, Planctomycetes, Proteobacteria, Spirochaetes* and *Thermotogae*) and the total number of genera was 9 for *Archaea* and 22 for *Bacteria*. In order to study the link between codon usage and gene expression four local DNA databases were created containing respectively: a) all protein coding sequences of each organism; b) highly expressed genes; c) ribosomal proteins genes and d) low expressed genes. These databases were used to compare codon usage patterns of: a) different strains of one bacterial species; b) different subsets of genes in one species and c) different bacterial genomes. They were also employed to identify missing and atypical codons as well as to search for correlation between codon usage and the formal prokaryotic taxonomy. The obtained results indicate that the formal taxonomy of many bacterial species does not fit their codon usage pattern (as predicted by the "Genome Theory"). On the other hand there are obvious similarities in the codon usage pattern of distant (according to the formal taxonomy) organisms. The latter suggests that either the

---

[*] Corresponding author: e-mail: iivanov@bio21.bas.bg

Genome Theory or the official taxonomy of microorganisms need revisions. Taking into consideration that at least two tRNAs occupy the two (A and P) functional ribosomal sites during the translation elongation step, we have also studied the effect of neighbor codons (codon pairs) on gene expression. Aiming to identify preferential and rare codon pairs, we have developed own criteria for determining their frequency of occurrence and for evaluation of their possible effect on translation. This approach was applied to analyze the distribution of all (3904) codon pairs in the *Escherichia coli* genome and we found that their frequency of occurrence varied from zero to 4913 times. The predicted effect of some (particularly 3'-terminal) codon pairs on gene expression is confirmed experimentally.

1. The Genetic Code

1.1. Introduction

Genetic information is coded by the four nitrogen bases, adenine (A), guanine (G), cytosine (C) and thymine/uracil (T/U) ordered consecutively in the polynucleotide chains of DNA and RNA. This information is decoded by the help of transfer RNAs (tRNAs) endowed with dual function. They recognize specific combinations of nucleotides in messenger RNA (mRNA) and at the same time carry activated α-aminoacids to be incorporated into the polypeptide chain growing on the surface of the translating ribosomes.

Following discovery of the DNA double helix, deciphering the genetic code was one of the greatest challenges of the last century. The main attributes of the genetic code are postulated in the early 60th and can be summarized as follows: a) the genetic code is triplet, i.e. three nucleotides (codon) encode one aminoacid; b) it is not interrupted by gaps or commas; c) the genetic code is unidirectional; d) it is universal; e) the genetic code is degenerated.

Braking of the genetic code, however, did not reveal immediately the meanings of the codons. Before the era of DNA sequencing, the genetic code was deciphered by two types of fundamental studies: a) translation of synthetic polyribonucleotides in bacterial lysates devoid of mRNAs and b) studying the capability of synthetic trinucleotides with known primary structure to cause specific binding of labeled α-aminoacyl-tRNAs to active ribosomes.

The first approach was employed by Marshall Nirenberg and Heinrich Mattei (Nirenberg et al. 1963). They demonstrated that poly(U) promoted synthesis of polyphenylalanine, whereas the poly(A) and poly(C) directed synthesis of polylysine and polyproline. From these observations they concluded that the triplets UUU, AAA and CCC coded for phenylalanine, lysine and proline respectively.

The second strategy was employed by Marshall Nirenberg and Philip Leder (Nirenberg and Leder 1964). Testing all combinations of trinucleotides they assigned the meaning of all possible triplets and proved that 61 of 64 codons represented aminoacids and the other three codons terminated protein synthesis. Thus the standard genetic code (see Table 1) was completely deciphered by 1966.

Table 1. Standard genetic code

First \ Second	U		C		A		G		Third
U	UUU	Phe	UCU	Ser	UAU	Tyr	UGU	Cys	U
	UUC	Phe	UCC	Ser	UAC	Tyr	UGC	Cys	C
	UUA	Leu	UCA	Ser	UAA	Stop	UGA	Stop	A
	UUG	Leu	UCG	Ser	UAG	Stop	UGG	Trp	G
C	CUU	Leu	CCU	Pro	CAU	His	CGU	Arg	U
	CUC	Leu	CCC	Pro	CAC	His	CGC	Arg	C
	CUA	Leu	CCA	Pro	CAA	Gln	CGA	Arg	A
	CUG	Leu	CCG	Pro	CAG	Gln	CGG	Arg	G
A	AUU	Ile	ACU	Thr	AAU	Asn	AGU	Ser	U
	AUC	Ile	ACC	Thr	AAC	Asn	AGC	Ser	C
	AUA	Ile	ACA	Thr	AAA	Lys	AGA	Arg	A
	AUG	Met (start)	ACG	Thr	AAG	Lys	AGG	Arg	G
G	GUU	Val	GCU	Ala	GAU	Asp	GGU	Gly	U
	GUC	Val	GCC	Ala	GAC	Asp	GGC	Gly	C
	GUA	Val	GCA	Ala	GAA	Glu	GGA	Gly	A
	GUG	Val	GCG	Ala	GAG	Glu	GGG	Gly	G

1.2. Codons with Dual Function

Several codons serve dual function. The triplet AUG codes for methionine in internal position and plays the role of initiation codon when placed at the beginning of the coding sequence. The latter function in prokaryotes is also executed by the triplets GUG and UUG, coding (as internal codons) for valine and leucine respectively. In eukaryotes the alternative initiation codons are used less frequently.

Another codon with dual function is the termination codon UGA. Occasionally it is recognized in both prokaryotes and eukaryotes as selenocysteine codon (Stadtman 1996). In *E. coli* the incorporation of selenocysteine is controlled by four genes, *selA*, *selB*, *selC* and *selD* (Jiang et al. 2002). The product of *selC* is a suppressor tRNA^{Ser} bearing the anticodon UCA, whereas the products of *selA* and *selD* genes are required for conversion of this tRNA into selenocysteine tRNA. The recognition of UGA as selenocysteine codon and therefore the incorporation of selenocysteine in the growing polypeptide chain is dependent on the product of the *selB* gene (a protein homologous to the elongation factor Tu).

1.3. The Wobble Hypothesis

The wobble hypothesis was formulated by Frances Crick in 1966 (Crick 1966a; Jiang et al. 2002) to explain the degeneracy of the genetic code. He proposed that the codon-anticodon interaction includes a strict Watson-Crick base-pairing between the first two codon bases and a less specific (wobble) pairing between the bases in the third position.

Table 2. The wobble interactions

Crick's wobble rule		Modified wobble rule	
1st anticodon	3rd codon	1st anticodon	3rd codon
U	A G	U	U C A G
		xo^{5}U	U A G
		xm^{5}Um, Um, xm^{5}U	A G
		xm^{5}s^{2}U	A (G)
		Gψ(1st, 2nd)	U C A
C	G	C	G
		Cm	(A) G
		f^{5}C	A G
		U	A G
		L	A
A	U	A	U C G (A)
		I	U C (A)
		I	U C A
G	U C	G	U C
		G	U C
		Q	U C
		m^{7}G	U C A G

Modified from Knight at all (Knight et al. 2001c) (f, formyl; I, inosine; L, lysidine; m, methyl; N, anynucleotide; Q, queosine; s, thiol-substituted; xo^{5}U, hydroxymethyluridine derivative; xm^{5}U, methyluridine derivative; Ψ=pseudouridine.)

This hypothesis is in accordance with the finding that the 5' nucleotides in the anticodons are usually modified. The modified bases, like inosine, dihydrouridine, pseudouridine, methylated nucleotides, etc., are capable of interacting with more than one nucleotide in the third codon position (see Table 2). Inosine for instance can form pairs with the four bases in RNA, i.e. this position is wobbling and specifies the number of codons recognized by one tRNA. F. Crick predicted that the 61 sense codons can be translated by 32 tRNAs (31 for the 20 aminoacids and one for initiation of translation). The recent genomic studies showed that except for mitochondria, the number of tRNAs in all other organisms vary from 40 to 50.

Table 3. List of prokaryotic organisms

class	Family	Genus		Acces.		All genes	Low expression genes	High expression genes	Ribosome genes
1	2	3	4	5	6	7	8	9	10
Archaea	Crenarchaeota	Thermoprotei	1	NC_000854	Aeropyrum pernix	1840	23	125	66
			2	NC_002754	Sulfolobus solfataricus	2977	44	157	69
			3	NC_003106	Sulfolobus tokodaii	2826	33	131	67
			4	NC_003364	Pyrobaculum aerophilum	2605	37	136	70
	Euryarchaeota	Archaeoglobi	5	NC_000917	Archaeoglobus fulgidus	2420	62	122	61
		Halobacteria	6	NC_002607	Halobacterium sp. NRC-1	2075	39	118	57
		Methanobacteria	7	NC_000916	Methanobacterium thermoautotrophicum	1873	46	124	61
		Methanococci	8	NC_000909	Methanococcus jannaschii	1729	25	127	66
			9	NC_005791	Methanococcus maripaludis	1722	43	125	62
		Methanomicrobia	10	NC_003901	Methanosarcina mazei Goe1	3371	107	157	65
			11	NC_003552	Methanosarcina acetivorans	4540	96	142	65
		Methanopyri	12	NC_003551	Methanopyrus kandleri AV19	1687	64	154	66
		Thermococci	13	NC_000961	Pyrococcus horikoshii	1801	16	119	61
			14	NC_003413	Pyrococcus furiosus	2065	52	138	67
			15	NC_000868	Pyrococcus abyssi	1769	43	128	64
		Thermoplasmata	16	NC_005877	Picrophilus torridus DSM 9790	1535	49	145	64
			17	NC_002578	Thermoplasma acidophilum	1482	30	123	58
			18	NC_002689	Thermoplasma volcanium	1499	69	142	59
Bacteria	Actinobacteria	Actinobacteridae	19	NC_004369	Corynebacterium efficiens YS-314	2950	114	133	59
			20	NC_003450	Corynebacterium glutamicum ATCC 13032	2993	180	137	58
			21	NC_002935	Corynebacterium diphtheriae	2272	106	135	59
			22	NC_002945	Mycobacterium bovis subsp. bovis AF2122/97	3920	205	156	62
			23	NC_002755	Mycobacterium tuberculosis CDC1551	4187	145	147	60
			24	NC_002677	Mycobacterium leprae	1605	52	94	53
			25	NC_004572+ NC_004551	Tropheryma whipplei	796 +/- 17.7	26 +/- 2.1	109 +/- 0.7	55 +/- 2.1
				NC_004572	Tropheryma whipplei Twist	808	27	108	56
				NC_004551	Tropheryma whipplei TW08/27	783	24	109	53
			26	NC_003155	Streptomyces avermitilis MA-4680	7575	730	264	96
			27	NC_003888	Streptomyces coelicolor A3(2) chromosome	7512	691	191	68
			28	NC_004307	Bifidobacterium longum	1729	109	119	58
	Aquificae	Aquificales	29	NC_000918	Aquifex aeolicus	1529	36	131	59
	Bacteroidetes	Bacteroides(class)	30	NC_004663	Bacteroides thetaiotaomicron VPI-5482	4778	231	414	61
			31	NC_002950	Porphyromonas gingivalis W83	1909	63	143	57
	Chlamydiae	Chlamydiales	32	NC_002620	Chlamydia muridarum	904	21	122	54
			33	NC_000117	Chlamydia trachomatis	895	18	124	57
			34	NC_005043+ NC_000922+ NC_002491+ NC_002179	Chlamydophila pneumoniae	1087 +/- 30.7	24 +/- 3.3	139 +/- 13.9	55 +/- 1.2
				NC_005043	Chlamydophila pneumoniae	1113	28	156	56

Table 3. (Continued)

					TW-183				
				NC_000922	Chlamydophila pneumoniae CWL029	1054	23	132	54
				NC_002491	Chlamydophila pneumoniae J138	1069	23	143	54
				NC_002179	Chlamydophila pneumoniae AR39	1112	20	124	56
			35	NC_003361	Chlamydophila caviae GPIC	998	22	141	54
			36	NC_005861	Parachlamydia sp. UWE25	2031	28	139	61
	Chlorobi	Chlorobia	37	NC_002932	Chlorobium tepidum TLS	2252	44	145	58
	Cyanobacteria		38	NC_005125	Gloeobacter violaceus	4430	160	154	60
		Chroococcales	39	NC_005070	Synechococcus sp. WH8102	2517	54	143	62
			40	NC_000911	Synechocystis PCC6803	3 167	51	146	57
			41	NC_004113	Thermosynechococcus elongatus BP-1 chromosome	2475	57	122	56
		Nostocales	42	NC_003272+ NC_003276	Nostoc sp. PCC 7120+Nostoc sp. PCC 7120 plasmid pCC7120alpha	5751	245	145	64
		Prochlorophytes		NC_005071+ NC_005072+ NC_005042	Prochlorococcus marinus CCMP1378	1953 +/- 283.3	47 +/- 12.7	149 +/- 4.4	65 +/- 6.4
			43	NC_005071	Prochlorococcus marinus MIT9313	2265	62	151	72
			44	NC_005072	Prochlorococcus marinus CCMP1378	1712	39	144	62
			45	NC_005042	Prochlorococcus marinus CCMP1375	1882	41	152	60
	Deinococcus-Thermus	Deinococci	46	NC_001263	Deinococcus radiodurans chromosome 1	2629	85	144	61
			47	NC_005835	Thermus thermophilus HB27 chromosome	1982	70	143	58
	Firmicutes	Bacillales	48	NC_003909+ NC_004722	Bacillus cereus	5419 +/- 260.9	354 +/- 26.2	209 +/- 21.9	90 +/- 18.4
				NC_003909	Bacillus cereus ATCC 10987	5603	335	193	77
				NC_004722	Bacillus cereus ATCC14579	5234	372	224	103
			49	NC_003997+ NC_007530	Bacillus anthracis Ames	5282 +/- 41.7	286 +/- 18.4	181 +/- 1.4	69 +/- 2.8
				NC_003997	Bacillus anthracis Ames	5311	273	180	67
				NC_007530	Bacillus anthracis str. Ames 0581	5252	299	182	71
			50	NC_002570	Bacillus halodurans	4066	228	159	62
			51	NC_000964	Bacillus subtilis	4112	240	155	59
			52	NC_004193	Oceanobacillus iheyensis	3500	166	140	66
			53	NC_003212	Listeria innocua chromosome	2968	199	142	64
			54	NC_002973+ NC_003210	Listeria monocytogenes	2834 +/- 17.7	189 +/- 24.7	140 +/- 1.4	67 +/- 3.5
				NC_002973	Listeria monocytogenes str. 4b F2365	2821	171	139	69
				NC_003210	Listeria monocytogenes	2846	206	141	64
			55	NC_002745+ NC_002758+ NC_003923	Staphylococcus aureus	2647 +/- 61.3	79 +/- 46.0	116 +/- 10.4	60 +/- 2.6
				NC_002745	Staphylococcus aureus N315 chromosome	2594	132	128	63
				NC_002758	Staphylococcus aureus Mu50 chromosome	2714	59	111	59
				NC_003923	Staphylococcus aureus MW2 chromosome	2632	47	109	58
			56	NC_004461	Staphylococcus epidermidis ATCC 12228	2419	103	130	66
		Clostridia	57	NC_004557	Clostridium tetani E88	2373	143	144	46

			58	NC_003030	Clostridium acetobutylicum ATCC824	3672	254	181	64
			59	NC_003366	Clostridium perfringens	2660	126	120	64
			60	NC_003869	Thermoanaerobacter tengcongensis	2588	134	169	64
			61	NC_004668	Enterococcus faecalis V583	3113	211	144	69
			62	NC_004567	Lactobacillus plantarum	3009	273	128	67
			63	NC_005362	Lactobacillus johnsonii NCC 533	1821	47	122	59
			64	NC_002662	Lactococcus lactis	2267	144	124	60
			65	NC_002737+ NC_003485+ NC_004070+ NC_004606	Streptococcus pyogenes	2069 +/- 36.1	109 +/- 9.8	121 +/- 3.0	60 +/- 2.6
		Lactobacillales		NC_002737	Streptococcus pyogenes M1 GAS	1697	104	118	58
				NC_003485	Streptococcus pyogenes MGAS8232	1845	101	120	57
				NC_004070	Streptococcus pyogenes MGAS315	1865	123	125	63
				NC_004606	Streptococcus pyogenes SSI-1	1861	107	122	60
			66	NC_004116+ NC_004368	Streptococcus agalactiae	1817 +/- 80.5	69 +/- 84.9	114 +/- 24.0	62 +/- 9.2
				NC_004116	Streptococcus agalactiae 2603	2124	129	131	68
				NC_004368	Streptococcus agalactiae NEM316	2094	9	97	55
			67	NC_004350	Streptococcus mutans UA159	1960	135	127	56
			68	NC_003028+ NC_003098	Streptococcus pneumonia	2109 +/- 21.2	88 +/- 27.6	127 +/- 2.8	65 +/- 2.8
				NC_003028	Streptococcus pneumonia TIGR4	2094	107	129	67
				NC_003098	Streptococcus pneumonia R6	2043	68	125	63
			69	NC_005303	Onion yellows phytoplasma	754	25	127	55
			70	NC_002771	Mycoplasma pulmonis	782	12	107	57
			71	NC_000908	Mycoplasma genitalium	484	11	106	54
		Mollicutes	72	NC_004432	Mycoplasma penetrans	1037	24	103	51
			73	NC_000912	Mycoplasma pneumoniae	689	12	102	56
			74	NC_005364	Mycoplasma mycoides	1016	21	83	53
			75	NC_006908	Mycoplasma mobile 163K	633	24	111	58
			76	NC_002162	Ureaplasma urealyticum	614	9	99	55
Fusobacteria	F usobacterales		77	NC_003454	Fusobacterium nucleatum	2067	86	152	61
Planctomycetes	Planctomycetacia		78	NC_005027	Pirellula sp.	7325	205	217	65
Proteobacteria	Alphaproteobacteria		79	NC_002696	Caulobacter crescentus	3737	221	169	64
			80	NC_004463	Bradyrhizobium japonicum	8317	529	148	59
			81	NC_005296	Rhodopseudomonas palustris CGA009	4814	341	227	61
			82	NC_003317+ NC_003318	Brucella melitensis chromosome I+Brucella melitensis chromosome II	3198	213	180	70
			83	NC_004311+ NC_004310	Brucella suis chromosome II+Brucella suis chromosome I	3264	173	158	59
			84	NC_002678	Mesorhizobium loti	6746	473	193	58
			85	NC_003305+ NC_003304	Agrobacterium tumefaciens str. C58 (U. Washington) linear chromosome+Agrobacterium tumefaciens str. C58 (U. Washington) circular chromosome	4661	354	159	61
			86	NC_003047	Sinorhizobium meliloti 1021	3341	205	145	61

Table 3. (Continued)

					chromosome				
			87	NC_003103	Rickettsia conorii Malish 7	1374	10	90	56
			88	NC_000963	Rickettsia prowazekii	835	12	129	57
			89	NC_002978	Wolbachia endosymbiont of Drosophila melanogaster	1195	25	130	60
		Betaproteobacteria	90	NC_002929	Bordetella pertussis	3436	277	177	67
			91	NC_002927	Bordetella bronchiseptica	4994	459	198	64
			92	NC_002928	Bordetella parapertussis	4185	381	193	64
			93	NC_003295	Ralstonia solanacearum chromosome	3440	265	176	60
			94	NC_005085	Chromobacterium violaceum ATCC 12472	4407	305	204	66
			95	NC_003112+ NC_003116	Neisseria meningitidis	2072 +/- 9.9	56 +/- 4.2	153 +/- 2.1	62 +/- 1.4
		Betaproteobacteria		NC_003112	Neisseria meningitidis MC58	2079	53	154	63
				NC_003116	Neisseria meningitidis Z2491	2065	59	151	61
			96	NC_004757	Nitrosomonas europaea	2461	75	141	53
		Deltaproteobacteria	97	NC_005363	Bdellovibrio bacteriovorus	3587	150	169	63
			98	NC_002937	Desulfovibrio vulgaris subsp. vulgaris str. Hildenborough	3379	172	171	65
			99	NC_002939	Geobacter sulfurreducens	3445	191	205	63
		Epsilonproteobacteria	100	NC_002163	Campylobacter jejuni	1634	41	125	58
			101	NC_000915+ NC_000921	Helicobacter pylori	1534 +/- 60.1	26 +/- 0.7	165 +/- 0.7	56 +/- 0.7
				NC_000915	Helicobacter pylori 26695	1576	26	165	55
				NC_000921	Helicobacter pylori J99	1491	25	164	56
			102	NC_004917	Helicobacter hepaticus	1875	20	125	56
			103	NC_005090	Wolinella succinogenes	2044	68	131	54
		Gammaproteobacteria	104	NC_004347	Shewanella oneidensis MR-1 chromosome	4324	235	203	68
			105	NC_004344	Wigglesworthia brevipalpis	654	16	135	57
			106	NC_005061	Blochmannia floridanus	583	17	126	59
			107	NC_002528+ NC_004061+ NC_004545	Buchnera	538 +/- 30.7	11 +/- 1.5	121 +/- 2.1	59 +/- 1.7
				NC_002528	Buchnera sp. APS	564	11	123	58
				NC_004061	Buchnera aphidicola Sg	545	12	119	58
				NC_004545	Buchnera aphidicola	504	9	122	61
			108	NC_000913+ NC_002655+ NC_002695+ NC_004431	Escherichia coli	5059 +/- 522.4	299 +/- 52.4	234 +/- 24.1	68 +/- 2.9
				NC_000913	Escherichia coli K12	4279	327	238	66
				NC_002655	Escherichia coli O157:H7 EDL933	5324	345	256	66
				NC_002695	Escherichia coli O157:H7	5253	299	243	72
				NC_004431	Escherichia coli CFT073	5379	226	200	67
			109	NC_005126	Photorhabdus luminescens	2972	68	92	24
			110	NC_003197+ NC_003198+ NC_004631	Salmonella typhimurium	4390 +/- 64.2	319 +/- 58.1	250 +/- 72.8	69 +/- 0.6
				NC_003197	Salmonella typhimurium LT2 chromosome	4451	386	334	70
				NC_003198	Salmonella typhi chromosome	4395	288	209	69
				NC_004631	Salmonella typhi Ty2	4323	283	207	69
			111	NC_004741+ NC_004337	Shigella flexneri 2a	4124 +/- 79.3	252 +/- 9.9	201 +/- 6.4	64 +/- 0.7

				NC_004741	Shigella flexneri 2a 2457T	4068	259	196	63
				NC_004337	Shigella flexneri 2a	4180	245	205	64
			112	NC_003143+ NC_004088+ NC_005810	Yersinia pestis biovar	3957 +/- 115.6	232 +/- 23.7	197 +/- 9.6	67 +/- 6.1
				NC_003143	Yersinia pestis CO92 chromosome	3885	219	186	68
				NC_004088	Yersinia pestis KIM chromosome	4090	217	201	60
				NC_005810	Yersinia pestis biovar Mediaevails	3895	259	204	72
			113	NC_002971	Coxiella burnetii	2009	44	137	59
			114	NC_002940	Haemophilus ducreyi 35000HP	1717	53	153	64
			115	NC_000907	Haemophilus influenzae	1714	64	145	60
			116	NC_004578	Pseudomonas syringae	5471	378	214	71
			117	NC_002947	Pseudomonas putida KT2440	5350	463	252	64
			118	NC_002516	Pseudomonas aeruginosa	5567	466	203	69
			119	NC_004459+ NC_004460+ NC_005439+ NC_005140	Vibrio vulnificus	4746 +/- 295.6	325 +/- 7.8	248 +/- 31.8	71 +/- 6.4
				NC_004459+ NC_004460	Vibrio vulnificus CMCP6 chromosome I+Vibrio vulnificus CMCP6 chromosome II	4537	330	270	75
				NC_005140+ NC_005139	Vibrio vulnificus YJ016 chromosome II+Vibrio vulnificus YJ016 chromosome I	4955	319	225	66
			120	NC_004605+ NC_004603	Vibrio parahaemolyticus RIMD 2210633 chromosome 2+Vibrio parahaemolyticus RIMD 2210633 chromosome 1	4832	308	229	71
			121	NC_002505+ NC_002506	Vibrio cholerae chromosome 1+Vibrio cholerae chromosome 2	3835	223	186	70
			122	NC_003919	Xanthomonas axonopodis pv. citri str. 306	4312	248	197	64
			123	NC_003902	Xanthomonas campestris pv. campestris str. ATCC 33913	4181	253	198	63
			124	NC_002488	Xylella fastidiosa chromosome	2766	79	149	60
			125	NC_004556	Xylella fastidiosa Temecula1 plasmid pXFPD1.3	2034	84	157	62
			126	NC_004342+ NC_005823	Leptospira intrerrogans	3877 +/- 683.1	124 +/- 22.6	155 +/- 7.1	61 +/- 2.1
				NC_004342	Leptospira intrerrogans I	4360	140	160	62
	Spirochaetes	S pirochaetales		NC_005823	Leptospira interrogans serovar Copenhageni, chromosome I	3394	108	150	59
			127	NC_001318	Borrelia burgdorferi chromosome	851	20	121	55
			128	NC_002967	Treponema denticola ATCC 35405	2767	77	153	59
			129	NC_000919	Treponema pallidum	1036	23	132	56
	Thermotogae	T hermotogales	130	NC_000853	Thermotoga maritima	1858	67	120	55

*The gray rows include average data about the strains with the indicated accession numbers.

1.4. Genetic Code Evolution

The theories for the origin of genetic code might be classified as "early" and "late". The early theories predict that the genetic code has emerged before the pre-biotic macromolecules (nucleic acids and proteins) whereas the late theories assume that it is a product of late events. According to the early theories, the ancient aminoacid and nucleotides have interacted directly, whereas the late theories assume that the interactions between these two (chemically distinct) classes of compounds has been mediated by specific adapters.

Crick (Crick 1966b; Crick et al. 1976) postulate that the evolution of genetic code has proceeded in a way to minimize the effect of eventual harmful mutations. After optimization of the genetic code it is conserved and established as a standard genetic code (about 3 billion years ago). This fact is explained by the lethality of any changes affecting the meaning of the codons, which is an obstacle for their inheritance.

The most popular of the late theories (Lahav 1993) presume that the genetic code is emerged at the time of the "RNA world", i.e. when RNA has played multiple roles (structural, adapter, catalytic, etc.). According to another theory (Woese 1967), the evolution of the genetic code has followed the improvement of the translational apparatus to improve the recognition of aminoacids with close physicochemical properties. This theory is supported by the fact that the hydrophobicity of α-aminoacids usually correlates with that of the base in the second codon position and also that the codons corresponding to α-aminoacids with similar physicochemical properties are closely related. For instance, the codons of the five most hydrophobic α-aminoacids Phe, Leu, Ile, Met and Val, carry U as a second base and those of the six most hydrophilic α-aminoacids His, Gln, Asn, Lys, Asp and Glu, have A in the second position. Codons with C as a second base code for α-aminoacids whose hydrophobicity/hydrophylicity is in between compared to that of the α-aminoacids coded by U and A containing codons. No correlation is found between the codons bearing G in second position and the hydrophobicity/hydrophylicity of the corresponding α-aminoacids.

1.5. Genetic Code Is Universal

Universality of the genetic code means that the 64 codons have the same meanings in all Earth organisms. In principle this seems to be true, however, there are a number of exceptions from this classiucal rule.

The most striking deviations from the standard genetic code were found in mitochondria. The genome of animal mitochondria contains 10-20 protein coding genes, two for rRNAs and 22 for tRNAs (Knight et al. 2001a). This means that 22 tRNAs are enough to decode all mitochondrial sense codons, which can be explained by the wobble hypothesis (Blanc and Davidson 2003). The first base in the anticodon of the mitochondrial tRNAs always recognizes either two (A or G/U or C) or four bases at third codon position. Therefore, one anticodon can read two or four different mitochondrial codons. This is the reason why the stop-codon UGA (in the standard genetic code) is read as tryptophan and the isoleucine codon AUG as methionine in the mitochondria. The main deviations from the standard genetic code are summarized in the review of (Knight et al. 2001b).

The deviations in the genetic code in mitochondria and also in some cellular parasites are regarded as results of late evolutionary events affecting the anticodons of the mitochondrial tRNAs. For instance, the arginine codons are decoded in the *Echinodermata* mitochondria by serine tRNA containing methylated G (m^{7}GCU) in the first anticodon position (Schneider and de Groot 1991). The methylated G may interact with any one of the four nucleotides in third codon position and therefore can participate in decoding of the four arginine codons AGA, AGC, AGG и AGU.

The deviations from the standard genetic code might be related also with structural alterations affecting other domains (rather than the anticodon) of the tRNAs (Murgola 1985). Such changes could alter their spatial structure and the interaction of tRNA with the corresponding aminoacyl-tRNA synthetase. Mutations in the aminoacyl-tRNA synthetases genes is another potential source of wrong reading and therefore could be a driving force for the evolution of the genetic code in cellular organelles and intracellular parasites (Arnez and Moras 1997).

In some cases the spurious deviation from the standard genetic code is due to RNA editing leading to the transition C→U (Cermakian et al. 1998). In *Leishmania tarentolae* the tRNAs are synthesized in the nucleus and transferred to the mitochondria. The anticodon of the tryptophan tRNA is CAA and it decodes the tryptophan codon UGG in the cytoplasm. In mitochondria however, the CAA is edited to UCA. The latter can base pair both UGG and the stop-codon UGA and due to this the stop-codon UGA is decoded in mitochondria as tryptophan (Alfonzo et al. 1999).

The most popular hypotheses explaining the deviations from the standard genetic code are summarized in the review of (Massey et al. 2003).

According to the *codon capture* hypothesis, some codons are eliminated by the evolution because of the conversion of the duplex GC into AU (Osawa and Jukes 1988; Osawa and Jukes 1989; Osawa et al. 1992). If the direction of mutation fluctuations is change, the eliminated codons will be restored, however, they will be recognized now by tRNAs belonging to other isoacceptor groups and therefore will be translated with new meanings.

The *ambiguous intermediate* hypothesis assumes that some mutations in tRNA could change its structure in a way to promote decoding of more than a single aminoacid (Schultz and Yarus 1994; Schultz and Yarus 1996). In case that this change is favorable, it will be fixed in the population and the corresponding codon will obtain new meaning preserving its original primary structure.

According to the *genome streamlining* hypothesis the driving force for the deviations of the genetic code in organelles and intracellular parasites is the minimization of their genome (Andersson and Kurland 1995; Andersson and Kurland 1998).

1.6. Genetic Code Is Degenerated

The fact that 61 sense codons codes for 20 aminoacids means that the genetic code is degenerated. Codons with the same meaning (coding for the same aminoacid or serving as termination codons) are denoted *synonymous*. Therefore, the 64 codons can be classified by their meaning/function in 21 different groups of which 20 codes for aminoacids and one function as terminators of translation.

The difference between the synonymous codons in the groups consisting of two codons is always transition (purine→purine or pyrimidine→pyrimidine) in third position. In the groups containing four codons the third base could be any one of the four (A, G, C or U) and the groups of six codons represent compilations of the groups of four and two codons.

Although the synonymous codons encode the same aminoacid, this does not mean that they are equally used. Three decades ago it was shown that the usage of synonymous codons may vary not only between species but also between the genes of one operon (Gouy and Gautier 1982). Ames и Hartmann (Ames and P.Hartmann 1963) assume that the codon bias is a mechanism for fine regulation of gene expression. Clarke (Clarke 1970) postulates that the usage of synonymous codons is not random and probably is related with the pool of the decoding tRNAs. This hypothesis is supported by the early genomic data showing that the codon usage in bacteriophage MS2 differs from that of the host bacterium. Fiers et al (Fiers et al. 1976; Jiang et al. 2002) concluded that this difference was due to the selection of fast translating codons in the MS2 genome allowing a higher elongation rate during translation of the phage mRNAs. Fitch (Fitch 1976) found predomination of C (compared to U) in the third position of the synonymous codons used in MS2, which favors specific rather than nonspecific (wobble) interactions. In the phage ΦX174 the preference is given to the codons carrying T (U) compared to A or G in third position (Sanger et al. 1977). Post et al. (Post et al. 1979) observed a difference between the codon composition of the ribosomal protein genes in *E. coli* and that of the low expressed *lacI* gene.

The post genomic era opened new thoroughfares for studying the evolution of genetic code, codon usage and its role in the regulation/modulation of gene expression. The new generations of fast and low-cost DNA sequencing methods (Shendure et al. 2005) led to a rapid accumulation of genomic data in the databases of both prokaryotes and eukaryotes after the year 2000. For prokaryotes such data can be found at DDBJ (http://www.ddbj.nig.ac.jp/), EMBL (http://www.ebi.ac.uk/embl/) and GenBank (http://www.ncbi.nlm.nih.gov/Genbank/index.html). Although thousands of small genomes (bacterial, viral, mitochondrial, chloroplasts, etc.) have already been sequenced, they are not profoundly analyzed (Field et al. 2006). Apparently, it is much easier today to generate raw sequencing data rather than to decipher their biological meaning (Binnewies et al. 2006). Hundreds of prokariotic and thousands of viral and organel genomes have already been sequenced and this volumious information allowed application of advanced statistical methods for generating high quality data for the usage of synonymous codons in different organisms as well as for the biological significance of the degeneration of the standard genetic code.

Grantham et al. (Grantham et al. 1980a; Grantham et al. 1980b; Grantham et al. 1981) showed that grouping the organisms by the type of their codon usage corresponds to the formal taxonomy. According to this concept (known also as *Genome theory*), the type of codon usage, i.e. the frequency of occurrence of the different synonymous codons in the genome, is specific for each organism. The genome theory is supported by Ikemura and co-authors (Aota and Ikemura 1986; Aota et al. 1988). The species specificity of codon usage is employed sometimes in evolutionary studies for dating gene transfer. At the early stages of evolution the codon usage pattern of the transferred genes is close to that of the donor organism. In the timecourse of evolution however, the codon usage pattern is changed to get closer to that of the recipient genome. This process of adaptation is denoted *amelioration* (improvement) of the codon usage (Lawrence and Ochman 1997). Sometimes it is possible to calculate the rate of amelioration and therefore the time of gene transfer. Lawrence and Ochman (Lawrence and Ochman 1998) conclude that 755 (of 4288) genes of *E. coli* have been transferred to the genome in 234 independent events after its deviation from *Salmonella*. In *Bacillus subtilis* 13% of the genes have different codon usage pattern and probably are of foreign origin (Moszer 1998; Moszer et al. 1999). Inserted genes are found in the genome of *Termotoga maritima* (Nelson et al. 1999) as well as in many other organisms. Among the numerous factors resulting in establishment of species specificity of codon usage it is noteworthy mentioning *the efficiency of translation, GC content, reparation of mutations, stability of mRNA*, etc.

Gouy and Gautier (Gouy and Gautier 1982) investigated 83 *E. coli* genes and observed a clear cut correlation between the type of codon usage and level of gene expression. They found that highly expressed genes prefer codons forming with the corresponding anticodons moderately stable base-pairs. These codons are denoted *optimal*.

Ikemura (Ikemura 1985) found correlation between codon usage and the pool of tRNAs in *E. coli, S. typhimurium* and *S. cerevisiae*. He assumed that the concentration of tRNA was a limitation factor during the translation of mRNAs and therefore could be an additional tool for regulation of gene expression. The cytoplasmic concentration of tRNAs is different for the different organisms and probably, varies in the life cycle of one organism/cell (Dong et al. 1996; Berg and Kurland 1997). It is reported that the concentration of tRNAs whose genes are located in rRNA operons is higher compared to the products of the dispersed genes and they usually decode preferential codons (Dong et al. 1996; Komine et al. 1990). The biosynthesis of these (*main*) tRNAs is synchronized with that of the rRNAs, which is related to the bacterial growth phase (Dong et al. 1996; Grantham et al. 1980b). Those codons decoded by the *main* tRNAs are translated 3-6 times faster compared to the corresponding synonymous codons (Sorensen et al. 1989).

1.7. Codon Usage and GC Content

It seems probable that the extent of degeneration of the genetic code and the species specificity of codon usage are related with the genomic GC content. All codons with G and C in the first two positions are members of four codon groups. This might mean that the codon-anticodon interaction on account of these two bases only is strong enough to ensure a specific recognition without the contribution of the third base (Knight et al. 2001c). When the first two positions of the codon are occupied by A or U, the interaction between the triplets is weak and the correct recognition depends on the impact of the third base. It is shown that the frequency of errors related with the translation of the asparagines codon AAU is much higher in comparison with the synonymous codon AAC (Akashi 1994).

The GC content of the prokaryotic genomes varies from 25% to over 75% (Lobry and Sueoka 2002) and from 7% to 95% between the genes of the genome (Wan et al. 2004). The local GC content of the codons is designated as GC1, GC2 and GC3 and refers to the first, second and third position respectively. The GC content might vary also between the two strands of DNA.

As a rule, the leading strands contains more G than C and more T than A (Wan et al. 2004). It is found that in *E. coli* the GC content is about 5% higher in the genes staying closer to the origin of replication (Deschavanne and Filipski 1995) and is slightly higher (58%) in the genes coding for longer polypeptides (over 800 aminoacids) compared to those of the smaller proteins (below 300 aminoacids). It seems likely that the GC content plays role in the bias of both synonymous codons and aminoacids (Ermolaeva 2001). Lynn et al. (Lynn et al. 2002) observed correlation between GC content, codon usage and living temperature of 32 bacterial and 8 arheal species.

1.8. Methods for Estimation of Codon Usage

The methods for evaluation of codon usage can be classified as: a) Criteria employing the zero hypothesis for usage of synonymous codons and b) Criteria based on the comparison of the observed values for usage of synonymous codon versus those of the preferential codons (Comeron and Aguade 1998). The first group of methods includes: χ^2 analysis (Shields et al. 1988; Akashi and Schaeffer 1997; Wernegreen and Moran 1999); determination of the effective number of codons (Nc) (Wright 1990); codon bias index (CBI) (Morton 1993; Morton 1994) and the intrinsic codon bias index (ICDI) (Freire-Picos et al. 1994). The second group embraces: the codon adaptation index (CAI) (Sharp and Li 1987), determination of the frequency of optimal codons (Fop) (Ikemura 1981) and determination of codon preference in genes with different expression (Karlin et al. 1998; Karlin et al. 2001). Since the most common of all methods is the codon adaptation index (CAI), it will be described in details.

1.8.1. Codon Adaptation Index (CAI)

The CAI method is developed by Sharp and Li (Sharp and Li 1987) for measuring the preference of synonymous codons and also for rough estimation of the efficiency of gene expression. It calculates the bias of individual synonymous codons for each aminoacid in high expressed genes, assuming that these codons are favorable for the fast translation. To this end the observed (real) number of codons is divided by the expected (probable) number for each synonymous codon:

$$RSCU_{ij} = \frac{X_{ij}}{\frac{1}{n_i}\sum_{j=1}^{n_i} X_{ij}} \quad (1)$$

where X_{ij} is the number of the j codon of the i aminoacid, and n_i is the number of the synonymous codons for the i aminoacid.

The CAI for a certain gene is obtained by dividing the RSCU by its maximal value for the corresponding codon group.

$$CAI = \frac{CAIobs}{CAImax} \quad (2)$$

where

$$CAIobs = (\prod_{j=1}^{L} RSCU_j)^{1/L} \quad (3)$$

$$CAImax = (\prod_{j=1}^{L} RSCU_j \max)^{1/L} \quad (4)$$

where RSCU$_j$ is the RSCU value for the j codon encoding the i aminoacid; RSCU$_{jmax}$ is the maximal value of RSCU for the group of synonymous codons coding for the i aminoacid and L is the number of all codons in the investigated gene.

Because the CAI is a measure for the deviation of the observed (real) codon composition of a gene from its optimal codon composition, it can be used also as criterion for prediction of the efficiency of gene expression (Carbone et al. 2003). That is, the expected gene expression is higher when CAI tends to 1.

Besides its wide application, the CAI has some disadvantages. It reflects the degree of preference but does not count the nature of that preference. CAI cannot be used to asses the likely compatibility between a gene and the candidate host (Gustafsson et al. 2004). Further CAI doesn't report on for the codons distribution along the gene, codon context and the presence of unfavorable secondary structures in mRNA that could interfere with the efficiency of translation. For instance, although the level of expression of all ribosomal protein genes is the same, the CAI of the genes located in the leading DNA strand is usually higher compared to those of the lagging strand.

1.8.2. χ^2- Analysis

χ^2- Analysis is based on the comparison between the observed and expected frequency of occurrence of codons assuming that the codon usage reflects the local base composition only (Wernegreen and Moran 1999). This method is precise and does not need preliminary information about the codon preference.

1.9. Codon Usage and Gene Expression

Varenne et al. (Varenne et al. 1984) first explained the difference in gene expression of two human genes (hIFN-α1 and hIFN-β) in *E. coli* by the difference in the organization of four (extremely rare) arginine codons AGG in their genes. The four arginine codons are scattered in the hIFN-β gene whereas they are organized in two clusters (tandems) at positions 11-12 and 163-164 respectively in the hIFN-α1 gene. The yield of recombinant protein obtained from identical expression vectors is 15% for hIFN-β and less than 1% for hIFN-α1. To explain this phenomenon the authors launched the so called *Concentration hypothesis*, assuming that the occupation of the two (A and P) ribosomal sites by one and the same rare isoacceptor tRNA decreases drastically its cytoplasmic pool, thus making the elongation step dependent on the concentration of the corresponding tRNA.

To check this hypothesis, Rosenberg et al. (Rosenberg et al. 1993) investigated the influence of tandems of AGG codons on the expression of the phage T7 gene *9*. They found that the yield of protein depended on both the number of AGG codons and their location in the gene. When the AGG codons were placed at the beginning of the gene (after the 13th codon) the yield of protein decreased, however, it increased when the tandems of rare ARG codons were situated after the 223rd codon.

2. Codon Usage in Prokaryotes

Taking advantage of the progress of prokaryotic genomics, we have studied the codon usage in 158 prokaryotic genomes. Such information is partly available at http://www.kazusa.or.jp/codon/; http://www.cbs.dtu.dk/services/GenomeAtlas/, etc., however, it usually refers to the average (typical for the whole genome) codon usage of an organism. Besides this information, we provide also data concerning codon usage in three subsets of genes: highly expressed (denoted *high*) genes, ribosomal protein (*rib*) genes and low expressed (*low*) genes. The differential codon usage data thus obtained were compared firstly, between the three gene classes in the frame of one genome and secondly, between different species and taxons. The latter sheds light on the evolution of genetic code and allow checking the validity of the *genome theory* (Grantham et al. 1980b) over a wide spectrum of organisms belonging to different taxa.

2.1. Databases and Methodology

2.1.1. Databases

The protein coding genes (NC_*.ffn files) and the respective NC_*.asn and NC_*.ptt files describing the genes were downloaded from the NCBI FTP database at ftp://ftp.ncbi.nih.gov/genomes/Bacteria (where NC_* is the relevant accession number of the corresponding chromosome/genome). Using a script written in Visual Basic Studio (VBS), all protein coding sequences of each bacterium were further fragmented into n*.txt files, where n is the number of protein coding genes. Using a script written in Visual Basic for Applications (VBA), all fragmented files were renamed using the designation of the corresponding protein (adopted from the *.ptt files). The taxonomy of the prokaryotic organisms was determined from the lineage data of each NC_*.asn file. A local database was created in which the plasmids sequences (if available in main database) were neglecting and the chromosomes (if more than one per genome) were united. Some species in the local database were represented by more than one strain. Their sequences were compared and if the genome difference was negligible they were further considered as a single genome. In case of substantial difference, their genomes were analyzed separately. Thus the number of individual organisms/genomes was reduced from 158 to 130. They all are listed according to their accession numbers in Table 3.

2.1.2. Gene Classes

Three gene classes (see above) were selected by a script written in Perl. This program searches for keywords and copies the corresponding files from the local database. The group of highly expressed genes (*high*) includes the genes of ribosomal proteins (the latter were also analyzed as a separate type of genes denoted *rib*), protein participating in the synthesis of tRNA, chaperons, transcription and translation factors, outer membrane proteins, etc. The group of low expressed genes (*low*) is represented by the genes of regulatory proteins, helicases, polymerases, etc. The exact number of genes in each category for each organism is presented in Table 3. More information about the three subsets of genes can be found at www.bio21.bas.bg/imb/codonusage.

2.1.3. Codon Counting

Codons were identified and counted for each gene using a script written in Perl. The same script was also used to identify and count atypical initiation codons as well as stop-codons converted into sense codons.

2.1.4. Calculation of Codon Usage

For each gene subgroup (*i*) codon usage *(CU$_{ij}$)* is calculated as follows:

$$CU_{ij} = \frac{N_{count}(C_{ij})}{N_{count}(A_{ik})} * 100, \tag{5}$$

where $N_{count}(C_{ij})$ is the observed number of codon C_j in the subgroup of genes *i*, and $N_{count}(A_{ik})$ is the observed number of appearance of the amino acid A_k in the subgroup of genes *i* coded by the codon C_j. To include termination codons, the three stop-codons (TAA, TGA and TAG) were considered as a separate (21st) codon group.

2.1.5. Comparison of Codon Usage between Species

Common standard deviation is calculated as follows:

$$SD(CU_{ij}) = \sqrt{\frac{\sum (CU_{ijn} - \overline{CU_{ijn}})^2}{n-1}}, \tag{6}$$

where CU_{ijn} is the codon usage value of codon *(C$_j$)* in the subgroup of genes *i* in the strain *n*. $\overline{CU_{ijn}}$ represents the average value of CU_{ij} for all (*n*) strains of the same species.

Standard deviation per codon for each subgroup of genes *i* is calculated as follows:

$$SD_i = \sum_{j=1}^{64} SD(CU_{ij}) \Big/ 64 \tag{7}$$

2.1.6. Difference Analysis

The difference $Dif(i_1 : i_2)$ is defined as:

$$Dif(i_1 : i_2) = (CU_{i_1 j} - CU_{i_2 j}), \tag{8}$$

where $CU_{i_1 j}$ and $CU_{i_2 j}$, are as in equation 5. To compare $Dif(i_1 : i_2)$ between different species the latter were normalized for both positive and negative values.

Normalization for positive values:

$$\overline{Dif}_+(i_1:i_2) = \frac{Dif_+(i_1:i_2)}{\sum_{1}^{X} Dif_+(i_1:i_2)} * 100 \qquad (9)$$

Normalization for negative values:

$$\overline{Dif}_-(i_1:i_2) = \frac{Dif_-(i_1:i_2)}{\sum_{1}^{Y} Dif_-(i_1:i_2)} * 100 \qquad (10)$$

In both equations X and Y are the numbers of positive and negative $Dif(i_1:i_2)$ respectively.

2.1.7. Correspondence Analysis

A common distance is calculated for each codon and for each aminoacid as follows:
For codons:

$$Dist(i_1:i_2) = \sum_{1}^{64} |Dif(i_1:i_2)| \qquad (11)$$

For aminoacids (AA):

$$Dist_{AA}(i_1:i_2) = \sum_{1}^{X} |Dif(i_1:i_2)|, \qquad (12)$$

where X is the number of synonymous codons per amino acid. The *Dist* values were further normalized as follows:

$$Dist_{\overline{AA}}(i_1:i_2) = \frac{\sum_{1}^{X} \lfloor Dif_{AA}(i_1:i_2) \rfloor}{\sum_{1}^{21}\sum_{1}^{X} \lfloor Dif_{AA}(i_1:i_2) \rfloor} * 100. \qquad (13)$$

2.1.8. Correlation Analysis

Correlation analysis was carried out to estimate the relationship between two data sets scaled to be independent. The population correlation calculation returns the covariance of two data sets divided by the product of their standard deviation following the equation:

$$C(X,Y) = \frac{\sum (x-\bar{x})(y-\bar{y})}{\sqrt{\sum (x-\bar{x})^2 \sum (y-\bar{y})^2}}, \qquad (14)$$

where x and y are the sample means average for the codon usage distribution in the first (array X) and in the second (array Y) genome. Correlation C(X,Y) analysis was performed for all possible combinations of organisms thus producing a two dimensional (130*130) matrix of correlations (for details see the tables at www.bio21.bas.bg/imb/codonusage). The values of C(X,Y) vary from +1 for highly correlated organisms to -1 for highly distinguished organisms. The average means for each gene class (C_{av}) is obtained by dividing the sum of the entire matrix by 130^2.

2.1.9. Presentation of Results (Color Figures)

Because of the difficulty to present the immense information in numbers, the results are presented in color tables where the numbers are substituted by colors using a program written in VBA.

2.2. Results and Discussion

A local database was created by downloading bacterial DNA sequences from the NCBI FTP database. Using common techniques the latter was fragmented and organized according to the formal bacterial taxonomy (see Methods). The species in the database belong to two major classes of prokaryotes, *Archaea* and *Bacteria* (Table 3). *Archaea* includes 2 families (*Crenarchaeota* and *Euryarchaeota*) and *Bacteria* 13 families (*Actinobacteria, Aquificae, Bacteroidetes, Chlamydiae, Chlorobi, Cyanobacteria, Deinococcus-Thermus, Firmicutes, Fusobacteria, Planctomycetes, Proteobacteria, Spirochaetes* and *Thermotogae*). The total number of genera was 9 for *Archaea* and 22 for *Bacteria* respectively. Some species in the genomic database were represented by more than one strain. The local genomic database (containing originally 158 individual genomes and further reduced to 130 genomes for codon usage analysis) was used to derive four auxiliary databases containing: a) a full set of protein coding genes (denoted *all*); highly expressed genes (*high*); ribosomal protein genes (*rib*) and low expressed genes (*low*). Although the ribosomal protein genes belong to the group of highly expressed genes, they were analyzed separately because of their overexpression in all prokaryotes.

The four derivative local databases (*all, high, rib* and *low*) were explored to compare the codon usage pattern of: a) different strains; b) different subsets of genes in one species and c) different bacterial genomes. Finally, they were also used to identify missing and atypical codons in the protein coding genes.

2.2.1. Codon Usage in Different Strains of Bacterial Species

Nineteen species in our local database were represented by more than one bacterial strain as follows: 11 species were represented by 2 strains, 5 species by 3 strains and 3 species by 4 strains (Table 4).

Table 4. Bacterial species represented by more than one species

N	Name	Accession number	n	all N of genes	all StDev per number of genes value	all %	all StDev per codon %	leg N of genes	leg StDev per number of genes value	leg %	leg StDev per codon %	all high N of genes	all high StDev per number of genes value	all high %	all high StDev per codon %	ryb N of genes	ryb StDev per number of genes value	ryb %	ryb StDev per codon %
1	Tropheryma whipplei	NC_004572+NC_004551	2	796	17.7	2.2	0.15	26	2.1	8.3	0.50	109	0.7	0.7	0.29	55	2.1	3.9	0.71
2	Chlamydophila pneumoniae	NC_005043+NC_000922+NC_002491+NC_002179	4	1087	30.7	2.8	0.07	24	3.3	14.1	1.00	139	13.9	10.0	0.72	55	1.2	2.1	1.13
3	Bacillus cereus	NC_004722+NC_003909	2	5419	260.9	4.8	0.24	354	26.2	7.4	0.61	209	21.9	10.5	0.68	90	18.4	20.4	1.25
4	Bacillus anthracis Ames	NC_003997+NC_007530	2	5282	41.7	0.8	0.02	286	18.4	6.4	0.96	181	1.4	0.8	1.00	69	2.8	4.1	2.28
5	Listeria monocytogenes	NC_002973+NC_003210	2	2834	17.7	0.6	0.13	189	24.7	13.1	1.58	140	1.4	1.0	0.79	67	3.5	5.3	2.13
6	Staphylococcus aureus	NC_002745+NC_002758+NC_003923	3	2647	61.3	2.3	0.14	79	46.0	58.0	0.89	116	10.4	9.0	0.57	60	2.6	4.4	0.77
7	Streptococcus agalactiae	NC_004368+NC_004116	2	1817	80.5	4.4	0.20	69	84.9	123.0	1.97	114	24.0	21.1	1.16	62	9.2	14.9	3.26
8	Streptococcus pneumonia	NC_003098+NC_003028	2	2109	21.2	1.0	0.20	88	27.6	31.5	1.32	127	2.8	2.2	0.30	65	2.8	4.4	1.26
9	Streptococcus pyogenes	NC_003485+NC_004606+NC_004070+NC_002737	4	2069	36.1	1.7	0.20	109	9.8	9.0	0.57	121	3.0	2.5	0.40	60	2.6	4.4	1.52
10	Neisseria meningitidis	NC_003116+NC_003112	2	2072	9.9	0.5	0.30	56	4.2	7.6	0.83	153	2.1	1.4	0.43	62	1.4	2.3	0.55
11	Helicobacter pylori	NC_000915+NC_000921	2	1534	60.1	3.9	0.30	26	0.7	2.8	1.00	165	0.7	0.4	0.60	56	0.7	1.3	0.79
12	Buchnera	NC_004061+NC_004545+NC_002528	3	538	30.7	5.7	1.37	11	1.5	14.3	2.77	121	2.1	1.7	1.68	59	1.7	2.9	2.35
13	Escherichia coli	NC_004431+NC_000913+NC_002655+NC_002695	4	5059	522.4	10.3	0.49	299	52.4	17.5	0.99	234	24.1	10.3	0.98	68	2.9	4.2	2.54
14	Salmonella typhimurium	NC_003197+NC_004631+NC_003198	3	4390	64.2	1.5	0.14	319	58.1	18.2	0.26	250	72.8	29.1	1.19	69	0.6	0.8	0.93
15	Shigella flexneri 2a	NC_004741+NC_004337	2	4124	79.2	1.9	0.23	252	9.9	3.9	0.22	201	6.4	3.2	0.36	64	0.7	1.1	1.01
16	Yersinia pestis biovar	NC_005810+NC_003143+NC_004088	3	3957	115.6	2.9	0.11	232	23.7	10.2	0.64	197	9.6	4.9	0.95	67	6.1	9.2	3.13
17	Vibrio vulnificus	NC_004459+NC_004460+NC_005140+NC_005139	2	4746	295.6	6.2	0.26	325	7.8	2.4	1.00	248	31.8	12.9	0.69	71	6.4	9.0	1.40
18	Leptospira intrerrogans	NC_004342+NC_005823	2	3877	683.1	17.6	0.16	124	22.6	18.2	0.49	155	7.1	4.6	0.79	61	2.1	3.5	1.34
	Average		18	3020	134.9	3.96	0.26	159	23.55	20.34	0.98	165	13.13	7.01	0.75	64	3.77	5.46	1.57
					283.3	14.5	10.5		12.7	26.9	10.4		4.4	2.9	11.6		6.4	9.9	12.4
	Prochlorococcus marinus	NC_005072+NC_005071+NC_005042		1953				47				149				65			
	Prochlorococcus marinus CCMP1378	NC_005072	3	1712				39				144				62			
	Prochlorococcus marinus MIT9313	NC_005071		2265				62				151				72			
	Prochlorococcus marinus CCMP1375	NC_005042		1882				41				152				60			

*The last column shows the standard deviation per codon as calculated by equation 7.

To evaluate the differences between the genomes of the corresponding strains we determined firstly the number of protein coding genes and secondly, the codon usage pattern of the four gene subsets *all, high, rib* and *low*. As seen in Figure 1 and Table 4, except for the strains of *Prochlorococcus marinus*, the genomic differences between the rests of the strains were negligible, which allowed to be considered as one species genome. Due to the great difference between the *Prochlorococcus marinus* strains, however, they were further analyzed separately.

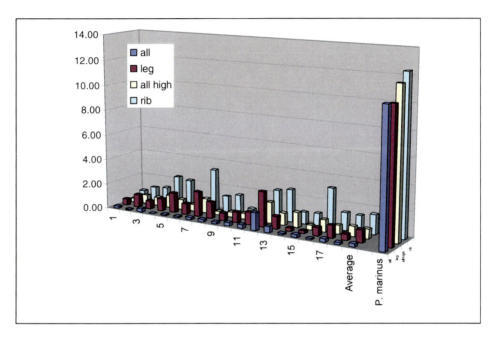

Figure 1. Codon usage in bacterial species represented by more than one strain. X axis: species name; Y axis: codon usage standard deviation as calculated by equation (7); Z axis: codon usage standard deviation in different gene subsets.

2.2.2. Comparative Analysis of Codon Usage in Different Subsets of Protein Coding Genes

The codon usage pattern in the three subsets of protein coding genes, *all, high, rib* and *low* was determined as described above. In Figure 1 the preferential codons are marked in green, rare in red and the color darkness in all columns corresponds to the deviation of the calculated codon usage values from their statistical means. The latter differed for the different codons and depended on the number of synonymous codons in the respective codon group. For example, for the groups consisting of two synonymous codons the statistical mean is 50%; for three codons it is 33.33%; for four it is 25% and for six synonymous codons it is 16.67%. The coincidences between calculated and statistical means are marked in white (the numerical meanings of the data presented in Figure 2 can be found at www.bio21.bas.bg/imb/codonusage).

Figure 2A represents the deviations of codon usage values for each one of the 64 codons for the *all* genes in the 130 bacterial genomes.

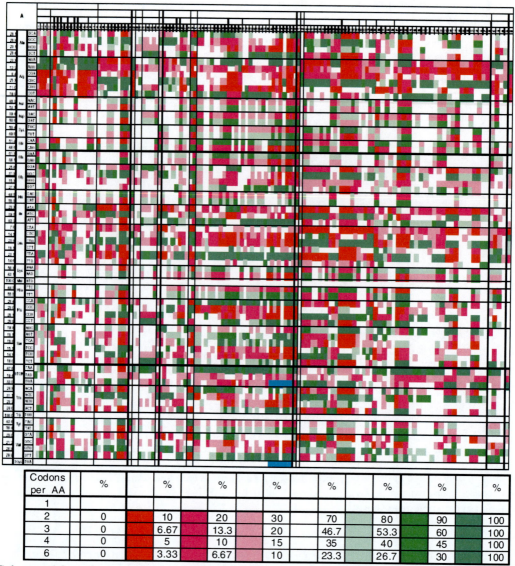

*Colors used for codon usage in dependence of number of codons per AA

Figure 2. Codon usage in prokaryotes. A) Whole set of genes (*"all"* genes). The first four rows are the same as in Table 1 and represent the formal taxonomy of the organism. The differences are summarized for each codon, which is normalized to the corresponding aminoacid and is presented in percents. First column shows the relative differences in codon usage for each aminoacid summarized for all organisms; second and third columns contain amino acids and codons ordered in ascending order, with selenocysteine added at the end. The codon usage distribution is presented in colors as it is described in the table below. Turquoise blue denotes the selenocysteine codon TGA and the arrows on the top show the place of the *E. coli* strains. The numerical data as well as the color figures Figure 2B, 2C and 2D (not presented in this paper) can be found at www.bio21.bas.bg/imb/codonusage.

Similar figures (Figure 2B, 2C and 2D) were also prepared for the three subsets (*all, high, rib* and *low*) of genes. To save printing space, however, these figures are not enclosed but they can be found at www.bio21.bas.bg/imb/codonusage. All these figures (Figure 2A,

2B, 2C and 2D) indicate undoubtedly that the codon usage pattern is specific for the different bacterial species and varies between taxons. An example of an extreme deviation is the conversion of the stop-codons (TGA) in *Molicutes* (family *Firmicutes*) into selenocysteine codon (indicated in the figure by turquoise blue).

2.2.3. Correspondence Analysis

To compare the codon usage pattern of the three subsets of genes (*all, high, rib* and *low*) with that of the *all* genes correspondence analysis was carried out as described in Methods. Figure 3A indicates that the codon usage pattern of the low expressed genes (*low*) is close to that of the whole genome (*all* genes). In numbers, the overall difference (distance) between the *all* and *low* genes is $Dist(all:low) = 114\%$ or less than 2% per codon. For the *high* and *rib* subsets these differences are 218% and 408% respectively. Figure 3A shows also that the differences between *all* and *rib* genes vary between the taxa, which is well visible in *Firmicutes, Proteobacteria, Actinobacteria* and *Euryarchaeota* and is less remarkable in *Crenarchaeota, Cyanobacteria* and *Spirochaetes*.

The data presented in Figure 2A and Figure 2D (see for numerical meanings www.bio21.bas.bg/imb/codonusage) were further processed to identify the aminoacids (Figure 3B) and/or codons (Figure 3C) responsible for the differences shown in Figure 3A. To this end distances ($Dist_{\overline{AA}}(all:rib)$) were calculated for all aminoacids and the results are represented in Figure 3B. The differences between codon usages of *all* and *rib* genes ($Dif(all:rib)$ was further normalized for the positive ($\overline{Dif}_{+}(i_1:i_2)$) and negative ($\overline{Dif}_{-}(i_1:i_2)$) means respectively (Figure 3C). As seen in Figure 3B, there were no striking differences in codon usage between the *all* and *rib* genes when calculations were made on the basis of 21 codon groups (i.e. on the total number of codon groups independently of the number of synonymous codons in the group). Taking into consideration that the number of synonymous codons varies from 1 to 6 per group, however, it is more correct to compare the differences between the groups encoded by equal number of synonymous codons. That is why the aminoacids were further classified on the basis of the number of synonymous codons (i.e. 6, 4, 3 and 2 codons respectively).

2.2.4. Aminoacids Coded by Six Codons

There are three amino acids coded by 6 codons: Arg, Leu and Ser.

Figure 2A shows that among the six synonymous Arg codons (AGA, AGG, CGA, CGC, CGG and CGT) *Archaea* prefer AGA and AGG (except for the species NC_002607 where both AGA and AGG are rare codons like in *Enterobacteriaceae*). Unlike *Archaea*, *Actinobacteria, Cyanobacteria* and *Proteobacteria* use mainly CGC, *Proteobacteria* (like *Firmicutes*) prefer CGC and CGT and *Spirochaetes, Firmicutes* and *Proteobacteria* prefer AGA. It is seen from Figure 3B that the deviation in codon usage between the *all* and *rib* genes for the six Arg codons is 8.6%. Our comparative analysis shows (Figure 3C) that the CGT is preferential for *all* genes of almost all prokaryotes. In *Enterobacteriales* AGA is extremely rare for *all* but preferential for the *rib* genes.

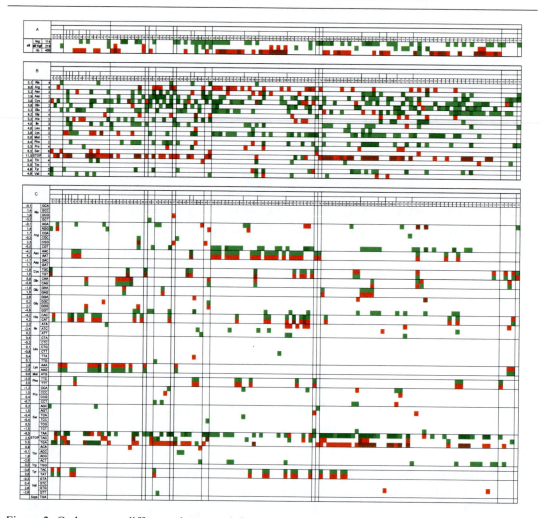

Figure 3. Codon usage difference between deferent gene subsets. The first four rows are as in Table 1 and represent the name and order of the organisms according to the formal taxonomy. A) The differences in codon usage are summarized for each organism calculating the distance (Dist) between the gene classes studied. The average value of this distance (Distav) is presented for each gene set. The values over 600% are indicated in brown, over 400% in red, below 100% in green and below 50% in dark green. B) Difference in codon usage between the set of *all* and *rib* genes. The distances are calculated for each aminoacid [$Dist_{\overline{AA}}(all:rib)$], normalized to the corresponding organism and presented in percents. Values exceeding 15% are marked in brown, higher than 10% in red, the lower than 2% and 1% are colored in green and dark green respectively. The number of synonymous codons is shown on the right side of each aminoacid and the relative distance in codon usage for each aminoacid summarized for all organisms is presented on the left side. C) Difference in codon usage between the *all* and *rib* genes [$Dif(all:rib)$]. Further normalization for the positive and another for the negative values was made as [$\overline{Dif}_{+/-}(all:rib)$]. Differences over 7.81 % are marked in red, over 12.5% in brown, lower than minus7.81% in green and lower than minus 12.5 % in dark green. The differences in codon usage for each aminoacid summarized for all organisms are shown on the left side of the corresponding aminoacid.

Leucine is coded by CTA, CTC, CTG, CTT, TTA and TTG. *Proteobacteria* use mainly CTG and also TTA and CTC, whereas *Firmicutes* prefer TTA (Fig. 2A). Compared to the Arg codons, the codon usage differences (4.8%) for the Leu codons between the *rib* and *all* genes (Figure 3B) are more randomly distributed (Figure 3C).

Serine is coded by AGC, AGT, TCA, TCC, TCG and TCT and almost all of them are more or less randomly used in the different prokaryotic species. There are some exceptions such as *Chlamydiae* using preferentially TCT; *Firmicutes*, use mostly TCT and TCA and *Proteobacteria*, prefere AGC, TCT and TCG (Figure 2A). The codon usage difference between *high* and *all* genes in the case of Ser subgroup is 6.3%. This difference is due mainly to AGT and TCT, which are preferential and avoided respectively in the *rib* genes (Figure 3C).

2.2.5. Aminoacids Coded by Four Codons

Aminoacids coded by 4 codons are Ala, Gly, Pro, Thr and Val. The most significant differences in codon usage between *all* and *rib* genes were registered for Gly (6.2%), Thr (5.4%) and Pro (5.3%) (Figure 3B).

Alanine is coded by GCA, GCC, GCG and GCT. The data presented in Figure 2A show that unlike *Archaea* (where the four codons are randomly used), in bacteria each family has its own codon preference. For example, *Actinobacteria*, *Cyanobacteria* and *Deinococcus-Thermus* prefer GCC; *Chlamydiae* и *Firmicutes* prefer GCT and *Proteobacteria* use both GCC and GCT.

The four synonymous codons of Gly (GGA, GGC, GGG and GGT) are almost randomly used in *Archaea*, whereas in most bacteria the preferential codon is GGT except for *Actinobacteria*, where the preferential codon is GGC and *Proteobacteria* using both GGC and GGT. Proline is coded by CCA, CCG, CCT and CCC. Our data show that a part of *Archaea* prefer CCA, *Actinobacteria* - CCG, *Chlamydiae* - CCT, *Firmicutes* - CCA followed by CCT, *Proteobacteria* - CCC, *Epsilonproteobacteria* – CCG and a part of *Gammaproteobacteria* prefer CCA. Among the four synonymous codons of Thr (ACA, ACG, ACC and ACT) *Actinobacteria* and *Cyanobacteria* prefer ACC, *Firmicutes* - ACT and *Proteobacteria* use both ACC and ACT.

Valine is coded by GTA, GTG, GTC and GTT. The data presented in Fig. 2A illustrate that most bacteria prefer GTT, except for *Deinococcus-Thermus* and a part of *Cyanobacteria*, *Beta* and *Epsilon proteobacteria,* where the preferential codon is GTG. Some species of *Alphaproteobacteria* and *Actinobacteria* use mostly GTC.

2.2.6. Aminoacids Coded by Three Codons

Isoleucine is the only aminoacid coded by three codons (ATA, ATC and ATT). As seen in Figure 2A, *Proteobacteria, Actinobacteria* and *Deinococcus-Thermus* use mainly ATC, *Chlamydiae, Spirochaetes* and *Firmicutes* (especially *Molicutes*) - ATT and *Archaea* - ATA.

2.2.7. Aminoacids Coded by Two Codons

Eight aminoacids are coded by two codons: Asp, Asn, Cys, Glu, Gln, His, Lys and Tyr. More significant differences in codon usage between *all* and *rib* genes among these aminoacids is registered for three of them: Asn (5.2%), His (5.2%) and Tyr (4.9%) (Fig. 3B). Our data show that in the cases when T or C is in third position, the preferential is usually the C containing codon. We also found that the two synonymous codons of Lys are randomly used in most of the prokaryotes except for *Archaea* where the preferential codon is AAG.

2.2.8. Stop-Codons

The most remarkable difference (11.3%) in codon usage between the *all* and *rib* genes is observed in the group of the termination codons (TAA, TAG and TGA; see Fig. 3B). It is due to the strong preference of TAA by most of the ribosome genes. Prokaryotes use mainly TAA, although some species of *Archaea, Actinobacteria, Cyanobacteria* and *Thermotogae* prefer TGA (Figure 2A). As already mentioned, in *Molicutes* one of the stop-codons (TGA) is converted into selenocysteine codon.

2.2.9. Missing and Atypical Initiation Codons

Our data show that some codons are missing in one or another subset of genes (Table 5). The most missing codons were found in the *rib* genes, which are (probably) related with the small number of genes in this group as well as with their overexpression in all bacterial species. Interesting results were obtained with the initiation codons. Although the canonical initiation codon is ATG, the translation of some genes is initiated by alternative codons such

Table 5. Missing codons in different gene subsets

\"Low\" genes			\"High\" genes			\"Rib\" genes		
N	AA	Codon	N	AA	Codon	N	AA	Codon
1	Arg	CGC	2	Arg	CGC	1	Arg	AGG
6	Arg	CGG	1	Arg	CGG	2	Arg	CGA
1	Pro	CCG				4	Arg	CGC
3	STOP	TAG				9	Arg	CGG
1	STOP	TGA				1	Arg	CGT
						2	ile	ATA
						7	Leu	TTA
						3	Leu	CTA
						1	Leu	CTC
						1	Leu	CTG
						1	Pro	CCC
						1	Pro	CCG
						3	Ser	TCC
						1	Ser	TCA
						7	STOP	TAG
						2	STOP	TGA
						1	Trp	TGG

N- number of organisms in which the corresponding codon is missing

as GTG and TTG (Kozak 1999). The results presented in Fig. 4 indicate that 99.86% of the initiation codons used in prokaryotes are structured as NTG. Of this number 77.73% belong to ATG, 13.12% to GTG, 8.88% to TTG and 0.13% to CTG. As seen from the figure, the alternative initiation codons are not randomly distributed between the taxonomic groups. For example, in some families of *Archaea* and *Bacteria* (*Actinobacteria*, *Planctomycetes* and *Thermotogae*) GTG and TTG are found more frequently used compared to other taxa.

Our data show that in 0.14% of all cases (i.e. out of 100%) the translation is initiated by other (rather than NTG) codons. The consensus structure of the initiation codons in these cases is ATN, where in 0.13% the third base (N) is different from G.

2.2.10. Codon Usage, Genome Theory and Prokaryotic Taxonomy

As already mentioned, the *genome theory* postulates (Grantham et al. 1980b; Grantham et al. 1981) that grouping the genomes on the bases of their type of codon usage correlates with their formal taxonomy. Our correlation analyses (Figure 5) shows that the three gene subsets (*all, high, rib* and *low*) differ in extent of correlation.

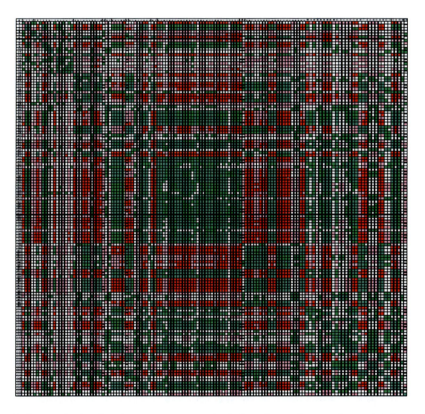

Figure 5. Correlation analysis of codon usage pattern in 130 prokaryotic organisms. (Each organism is compared to all 130 organisms). A) Whole set of genes (*all* genes). The first four rows are the same as in Table 1. The correlation values are presented in colors as follows: very high correlation (C > 0.9) – green; high correlation (C > 0.8) – bright green; neutral correlation (C > 0.5 and C < 0.8) – no color (white); weak correlation (C < 0.5) – red; very weak or negative correlation (C < 0.2) – dark red. The numerical data as well as the color figures 5B, 5C and 5D, representing the correlation analysis data for the *low*, *high* and *rib* genes are not presented in this paper and can be found at www.bio21.bas.bg/imb/codonusage.

The highest correlation was observed at the *all* genes (C_{av} = 0.595) and this value was close to that of the *low* genes (C_{av} = 0.594). The C_{av} value, however, was lower for the *high* genes (C_{av} = 0.560) and particularly for the *rib* genes (C_{av} = 0.554). As seen from the figure, except for few taxa, the codon usage pattern correlates in general with the formal taxonomy (as in the NCBI FTP database). One exception is the species of *Firmicutes* and *Chlamydiae*, fitting perfectly our correlation data, whereas their codon usage pattern resembled that of *Spirochaetales*. An illustration of the opposite case was *Proteobacteries*, where the codon usage pattern differs even between the species of one genus (*Alphaproteobacteria*).

Based on these results we conclude that the formal taxonomy of many bacterial species do not fit their codon usage pattern, as predicted by the *genome theory*. On the other hand there are obvious similarities in codon usage between distant (according to the formal taxonomy) organisms (e.g., the codon usage pattern of *Rickettsiales* is closer to that of *Firmicutes* rather than to *Alphaproteobacteria*). These observations suggest that either *Genome theory* or the formal taxonomy of microorganisms might need revision in the light of the new genomic data and their bioinformatic analysis.

3. Codon Pairs jn *Escherichia Coli*

Bossi and Roth (Bossi and Ruth 1980) were the first to conclude that the intermolecular interactions between the tRNAs situated in the A and P ribosomal sites affect the efficiency of translation. The tRNAs belonging to one isoacceptor group have different spatial structure and therefore it is logic to assume that the combinations between tRNAs on the ribosome surface during translation as well as between the tRNAs and the two translation termination factors (RF1 and RF2) are not equally favorable for the efficiency of translation (Smith and Yarus 1989). The combination of two non- (or less) compatible isoacceptor tRNAs would alter the local rate of translation and this could be another mechanism for regulation/modulation of gene expression (Kittle 2006; Hatfield and Gutman 1993). Because the combinations of tRNAs and tRNA-RF1/RF2 are pre-determined by the combinations of codons (codon pairs), their significance for the efficiency of translation can be estimated indirectly by studying the distribution of codon pairs along the gene or even along the whole genome.

The finding of Bossi and Roth (Bossi and Ruth 1980) that the codon pairs are not randomly used was later confirmed by Gutman and Hatfield (Gutman and Hatfield 1989). Their statistical analysis based on 237 protein coding genes of *E. coli* showed that some codon combinations (of the total 61^2 = 3721) were overrepresented and others were underrepresented with respect to their theoretically predicted means. They found also a correlation between codon pairs usage and efficiency of translation. After the genome of *E. coli* was sequenced (Blattner et al. 1997) in full, we (Boycheva et al., 2003) determined the distribution of all codon pairs in this genome and proved their non-random usage.

In this chapter we present our results from the statistical analysis on occurrence of all combination of codons in pairs in the entire genome of *E. coli* including 4290 open reading frames (ORFs) and their correlation with gene expressen.

3.1. Databases and Methodology

3.1.1. Databases
The full sets of 4290 open reading frames (ORFs) and the subset of 2658 protein coding sequences in the *E. coli* genome were obtained from the Kyoto Encyclopedia of Genes and Genomes (Release 16.0, October 2000) at www.tokyo-center.genome.ad.jp/kegg/kegg.html).

3.1.2. Methodology
The ORFs and protein-coding sequences were analyzed by a program written in Perl. It divides the string of codon pairs into two frames, thus mimicking simultaneous codon-anticodon interactions of two adjacent tRNAs on the elongating ribosome. A row of codon pairs for a coding sequence begins with an initiator codon (Ai) and the codon next to it (A_1). It continues with the second and the third (A_1, A_2), the third and the fourth (A_2, A_3) codon, *etc.*, and finishes with the combination A_N: A_{STOP} (see Figure 6).

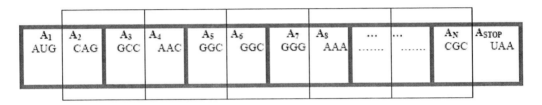

Figure 6. Grouping of codons into codon pairs. Each codon from the coding sequence is assigned "A". The numeric identifier shows the codon position after the initiation codon (Ai). The program divides the coding sequence into the following codon pairs: Ai: A1; A1:A2; A2:A3; A3:A4; A4:A5; AN:ASTOP.

For each of the 3904 codon pairs, our program estimates the following parameters:

A. *Observed Number of Occurrence* (N_{OBS}). This is the real number of occurrence of a individual codon pair (N_{OBS}) in the full set of analysed nucleotide sequences. The set of 4290 ORFs in the genome of *E. coli* consists in 1358854 codon pairs and the smaller subset of 2658 protein-coding sequences includes 906166 codon pairs.

B. *Expected Number of Occurrence* (N_{EXP}). The N_{EXP} is defined as:

$$N_{exp} = (P_A * P_B) * N_{TOT} \qquad (15)$$

where P_A and P_B are the probabilities for occurrence of the individual codons A and B in the codon pair and N_{TOT} is the total number of codons in the analyzed nucleotide sequence. The P_A and P_B are defined as:

$$P_A = \frac{N_{OBS(A)}}{N_{TOT}} \quad \text{and} \quad P_B = \frac{N_{OBS(B)}}{N_{TOT}} \qquad (16)$$

where $N_{OBS(A)}$ and $N_{OBS(B)}$ are the observed number of occurrences of two codons (A and B) in the codon pair.

C. *Expected Random Deviation* (D_{EXP}).

$$D_{EXP} = \sqrt{N_{TOT} * (P_A * P_B) * (1 - P_A * P_B)} \qquad (17)$$

D. *Normalized offset value* (r). To measure the difference between the observed (N_{OBS}) and expected (N_{EXP}) number of occurrence of a codon pair in the analyzed nucleotide sequence, a normalized offset value r was defined as follows: If the product ($P_A P_B$) declines and r raises to the second power, the result is *chi-square*. So, the r and chi-square values are asymptotically the same (Eq. 19).

$$r = \left(\frac{N_{OBS} - N_{EXP}}{D_{EXP}} \right) \qquad (18)$$

$$r^2 = \left(\frac{N_{OBS} - N_{EXP}}{\sqrt{N_{TOT} * (P_A * P_B) * (1 - P_A * P_B)}} \right)^2 \cong \chi^2 = \frac{(N_{OBS} - N_{EXP})^2}{N_{EXP}} \qquad (19)$$

Compared to the *chi-square* test giving only positive meanings, the normalized offset depends on the deviation of the observed frequency from the value of the expected one, *i.e.* when $N_{OBS} > N_{EXP}$, r is positive and conversely, when $N_{OBS} < N_{EXP}$, r is negative. The normalized offset r value is much more informative and convenient to use.

E. The Δ_{REG} value. The Δ_{REG} value is defined as a difference between the r_{high} and r_{low} values for each codon pair (Eq. 20):

$$\Delta_{REG} = r_{high} - r_{low} \qquad (20)$$

where r_{high} is the r value for the codon pair in highly expressed genes and r_{low} is the r value for the same codon pair in poorly expressed genes.

3.2. Results

3.2.1. Statistical Analysis

Distribution of all ($64^2 = 4096$) codon pairs was determined in both the full set of 4290 open reading frames (ORFs) and the subset of 2658 protein coding sequences in the *E. coli* genome. From this number the combinations stop:stop and stop:sense codons (*i.e.* 192 pairs) were subtracted and therefore we considered the distribution and frequency of occurrence of the remaining 3904 codon pairs. The latter were divided into five groups: *missing pairs* and pairs occurring in the entire *E. coli* genome *1-200, 200-600, 600-1000* and *1000-4913 times*. The pairs belonging to the latter four groups are shown in Figure 7. As seen from the figure, almost 50 % of all possible codon combinations appeared less than 200 times, while 6% occured more than 1000 times. The most frequently used pair was the CUG:GCG (N_{OBS} =

4913 times). The group of missing codon pairs includes 19 members. As seen in Table 6, except for two pairs (CCU:AGG and ACU:AGA), all other missing pairs represented combinations between sense and stop-codons. Surprisingly, the type of stop-codon in the missing pairs was biased. As shown in the table, except for one codon pair (ACU:UGA, in which the stop codon was UGA) the rests of the missing pairs contained the stop-codon UAG.

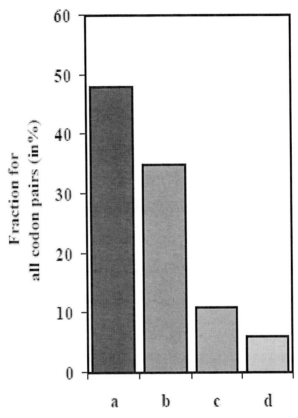

Figure 7. Frequency of occurrence of codon pairs in the *E.coli* genome. The codon pairs found in the *E. coli* genome are divided into four groups depending on their frequency of occurrence: a, b, c and d corresponds to codon pairs occurring *1-200, 200-600, 600-1000* and *1000-4913 times* in the entire *E. coli* genome.

The data presented below indicate that most of the sense codons in the missing codon combinations are rare codons. However, two of them ACC and GGC (coding for Thr and Gly respectively) are quite frequent, representing 2.3% and 3.0% of the total number of all *E. coli* codons. To evaluate whether or not the observed frequency of occurrence of a codon pair is higher or lower than expected, we applied the **r** index (see Methods). By definition **r** is positive for the overrepresented and negative for the underrepresented codon pairs. The variations in sign and value of **r** are well illustrated with the combinations between sense and stop-codons (see Table 7).

Table 6. Missing codon pairs in the *E.coli* genome

Codon pairs*	
(Phe:Stop)	<u>UUU</u>:UAG
(Ser:Stop)	**UCU**:UAG
(Ser:Stop)	**UCC**:UAG
(Tyr:Stop)	UAC:UAG
(Cys:Stop)	**UGC**:UAG
(Leu:Stop)	CUC:UAG
(Leu:Stop)	**CUA**:UAG
(Pro:Stop)	**CCU**:UAG
(Pro:Stop)	CCC:UAG
(Pro:Stop)	**CCA**:UAG
(His:Stop)	CAU:UAG
(Thr:Stop)	**ACU**:UAG
(Thr:Stop)	*ACC*:UAG
(Thr:Stop)	ACG:UAG
(Val:Stop)	GUC:UAG
(Gly:Stop)	*GGC*:UAG
(Thr:Stop)	ACU:UGA
(Thr:Arg)	ACU:**AGA**
(Pro:Arg)	**CCU:AGG**

*Underlined codons are missing in the protein coding genes; rare codons are bolded; frequently used codons are given in italics and the rest of the codons are moderately used.

The data presented above indicate that most of the sense codons in the missing codon combinations are rare codons. However, two of them ACC and GGC (coding for Thr and Gly respectively) are quite frequent, representing 2.3% and 3.0% of the total number of all *E. coli* codons. To evaluate whether or not the observed frequency of occurrence of a codon pair is higher or lower than expected, we applied the **r** index (see Methods). By definition **r** is positive for the overrepresented and negative for the underrepresented codon pairs. The variations in sign and value of **r** are well illustrated with the combinations between sense and stop-codons (see Table 7). It is noteworthy mentioning that the extremely rare Arg codons AGG and AGA were highly biased as penultimate codons (positive **r** values), whereas the most frequently used CUG codon was avoided at this position (negative **r** values). In some cases, the **r** value was found to depend on the type of stop-codon combined with a sense codon. For example, when the preferential Ala codon GCC is combined with the most frequently used stop-codon UAA, the **r** value is negative, however, if this codon is combined with the less frequently used stop-codon UGA, the **r** value is positive. Taking into consideration that the two stop-codons UAA and UGA are recognized by the same release factor RF2, we are tempted to speculate that the combination GCC:UGA favors the interaction of RF2 with the UGA codon.

Our results support the literature data showing that the C-terminal amino acids in the *E. coli* proteins are biased (Brown et al. 1990; Buckingham 1994; Sato et al. 2001). A list of the most over- and underrepresented C-terminal amino acids is shown in Table 8. To classify the C-terminal amino acids, we have applied the sum of normalized offset **r** values for over- or underrepresented synonymous codons situated before the stop-codons. As seen in Table 8, all Lys and Arg codons (except for the Arg codon CUG) in combination with the three stop-codons were overrepresented whereas the threonine codons were always underrepresented.

Table 7. Normalized offset r values for the sense codons preceding stop codons in all *E.coli* ORFs

Codon (Amino acid)	Normalized r value		Codon:UAG
	Codon:UAA	Codon:UGA	
A)			
AGG(Arg)	+6.4	+5.2	+2.5
AGA(Arg)	+4.7	+3.9	+2.8
GGA(Gly)	+3.5	+1.3	+2.7
B)			
ACC(Thr)	-7.4	-3.4	-3.8
GGC(Gly)	-5.0	-1.7	-2.8
CUG(Leu)	-4.2	-6.7	-1.7
C)			
CCA(Pro)	-1.4	+2.6	-1.1
AGU(Ser)	+5.2	-1.2	+1.3
UCC(Ser)	-3.8	+8.9	-1.1
CGU(Arg)	+6.6	-1.6	+1.2
UUC(Phe)	-3.9	+9.1	-2.5
GCC(Ala)	-7.1	+10.6	-2.5

Overrepresented (A) and underrepresented (B) sense codons preceding stop-codons; C) Sense codons with changeable **r** value depending on the nature of the stop-codon. Designations as in Table 6.

Table 8. Normalized offset values (r) for synonymous penultimate codons

Amino acid	Codon pairs	(r)	(r)	Amino acid	Codon pairs	(r)	(r)	Amino acid	Codon pairs	(r)	(r)
Lys	AAG:UAA	+16.9	+32.0	Ala	GCC:UGA	+10.6	+16.5	Arg	CGA:UAG	+4.5	+14.2
	AAA:UAA	+15.1			GCA:UGA	+5.9			CGG:UAG	+3.2	
Arg	CGA:UAA	+7.2	+26.9	Ser	UCC:UGA	+8.9	+16.6		AGA:UAG	+2.8	
	CGU:UAA	+6.6			UCA:UGA	+7.7			AGG:UAG	+2.5	
	AGG:UAA	+6.4			AGG:UGA	+5.2			CGU:UAG	+1.2	
	AGA:UAA	+4.7		Arg	CGA:UGA	+4.0	+14.3	Ile	AUA:UAG	+7.2	+7.2
	CGC:UAA	+2.0			AGA:UGA	+3.9		Lys	AAA:UAG	+1.5	+6.8
His	CAC:UAA	+5.8	+13.4		CGG:UGA	+1.2			AAG:UAG	+5.3	
	CAU:UAA	+7.6		Lys	AAG:UGA	+3.0	+5.8	Gly	GGC:UAG	-2.8	-2.8
Thr	ACC:UAA	-7.4	-19.2		AAA:UGA	+2.8		Ala	GCC:UAG	-2.5	-2.5
	ACG:UAA	-5.1		Thr	ACC:UGA	-3.4	-11.3	Pro	CCG:UAG	-1.3	-2.4
	ACU:UAA	-3.9			ACG:UGA	-3.1			CCA:UAG	-1.1	
	ACA:UAA	-2.8			ACU:UGA	-2.8		Thr	ACG:UAG	-1.2	-2.3
Leu	CUG:UAA	-4.2	-12.1		ACA:UGA	-2.0			ACU:UAG	-1.1	
	UUA:UAA	-3.5		Ser	UCG:UGA	-2.2	-7.0				
	CUC:UAA	-2.9			UCU:UGA	-2.0					
	UUG:UAA	-1.5			AGC:UGA	-1.6					
Val	GUG:UAA	-4.8	-11.0		AGU:UGA	-1.2					
	GUC:UAA	-4.7		Leu	CUG:UGA	-6.7	-6.7				
	GUA:UAA	-2.5									

To investigate the hypothetical relationship between codon pairs usage and gene expression, we have analyzed two subsets of genes: 203 highly expressed (*high*) and 176 poorly expressed (*low*) genes. The group of *high* genes includes genes of ribosomal proteins, outer membrane proteins and proteins involved in the carbohydrate metabolism. The *low* subset of genes is chosen among those coding for regulatory proteins.

We have useed in our analysis the quantity Δ_{REG} (see Methods), which has positive values for the codon pairs overrepresented in the *high* genes ($r_{high} > +1$) and at the same time are underrepresented ($r_{low} < -1$) in the *low* group of genes. Conversely, Δ_{REG} is negative for the overrepresented codon pairs in *low* genes ($r_{low} > +1$) and underrepresented in the *high* genes ($r_{high} < -1$). As shown in Figure 8, the pairs with negative Δ_{REG} value were used twice more frequently in the *low* than in the *high* genes. It is evident from this data that the *low* genes tend to favor codon combinations with negative Δ_{REG} values, whereas the codon pairs with positive Δ_{REG} values are typical for the *high* group of genes. This observation suggests that the use of codon pairs might be related with the modulation of translation. Thus, the codon pairs with negative Δ_{REG} values is expected to have attenuating (pausing) effect and conversely, those with positive Δ_{REG} values might have enhancing effect on translation (see Table 9).

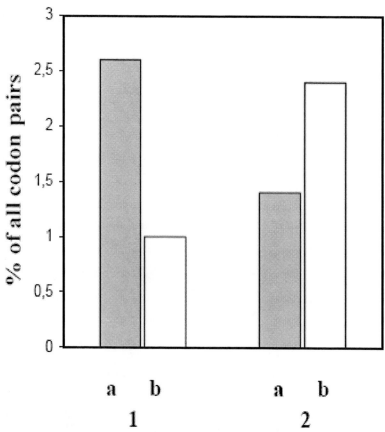

Figure 8. Codon pairs with negative (a) and positive (b) ΔREG values in *low* (1) and *high* (2) subsets of genes. The bar height (ordinate) represents the fraction of the total number of codon pairs in the corresponding group of genes (57842 codon pairs in the group of *low* and 83017 pairs in the group of *high* genes).

Table 9. Codon pairs with predicted modulatory effect on translation

Codon pair*	tRNAs and their fractions out of total tRNA (%)**	Amino acids	N_{OBS} inLXG	N_{EXP} inLXG	N_{OBS} inHXG	N_{EXP} inHXG	r high	r low	Δ_{REG}
A *CUG*:GAG	Leu1,3:Glu2	Leu:Glu	158	99.7	96	120.7	-2.3	+5.8	-8.1
UUU:UUC	Phe:Phe	Phe:Phe	40	24.8	12	38.8	-4.3	+3.0	-7.3
CAG:GAG	Gln2:Glu2	Gln:Glu	94	63.8	38	55.7	-2.4	+3.8	-6.2
UUA:CUG	Leu5:leu1,3	Leu:Leu	113	82.3	27	43.1	-2.5	+3.4	-5.8
GGU:GUA	tRNA1B:tRNA1	Gly:Val	21	11.7	39	60.4	-2.8	+2.8	-5.2
GCU:GGU	tRNA1B:tRNA3	Ala:Gly	31	20.3	78	103.8	-2.5	+2.4	-4.9
B CAA:GAG	Gln1:Glu2	Gln:Glu	19	38.4	28	16.6	+2.8	-3.1	+5.9
AUG:UGU	Metm:Cys	Met:Cys	3	9.3	20	9.8	+3.3	-2.1	+5.4
GAC:GUA	Asp1:Val1	Asp:Val	7	15.5	68	46.9	+3.1	-2.2	+5.3
GGU:CUG	Gly3:Leu1,3	Gly:Leu	57	75.9	315	265.9	+3.0	-2.2	+5.2
CCA:AGC	Pro3:Ser3	Pro:Ser	6	14.6	20	10.5	+2.9	-2.3	+5.2

*Designations as in Table 6.
** The numbers correspond to the spot of individual electrophoretic components fractionated in two-dimensional polyacrylamide gel electrophoresis.
(A) Hypothetically pausing codon pairs; (B) hypothetically non-pausing codon pairs; LXG: *low* expressed genes; HXG: highly (*high*) expressed genes.

Table 10. Effect of 3' terminal codon pairs on the expression of the chloramphenicol acetyltransferase (*cat*) gene in *E. coli* LE392

Codon pairs*	[a]N_{OBS}	[b]N_{EXP}^1	[c]N_{EXP}	[d]r^1	[e]r	[f]mRNA (%)	[g]Protein (%)	[h]Protein/ RNA
i) CCC:UGA	21	7.9	6.9	+4.7	+4.9	96	74	0.8
ii) CCC:UAA	6	17.0	14.8	−2.7	−2.3	88	84	0.9
ii) CCU:UGA	2	6.1	8.8	−1.7	−2.3	88	52	0.6
iii) CCU:UAA	17	13.3	18.9	+1.0	+0.9	88	90	1.0
iii) CGC:UGA	28	32.2	27.6	−0.7	−0.1	64	32	0.5
iii) GCG:UAA**	99	88.9	90.7	+1.1	+0.9	100	100	1.0
iv) CCC:UAG	0	2.06	2.0	−1.5	−1.7	25	10	0.4
iv) CCU:UAG	0	1.6	2.3	−1.2	−1.5	100	17	0.2
iv) CCU:AGG:UAA	15	15.8	3.4	−0.2	+6.4	107	138	1.3
iv) CCU:AGG:UAG	0	1.6	2.3	−1.2	−1.5	70	40	0.6

[a] Observed frequency of occurrence in the *E. coli* genome.
[b] Expected frequency of occurrence computed when the amino acid bias is eliminated.
[c] Expected frequency of occurrence computed when the amino acid bias is taken into consideration.
[d] Normalized offset value defined as a difference between N_{OBS} and N_{EXP}^1.
[e] Normalized offset value defined as a difference between N_{OBS} and N_{EXP}.
[f] Yield of mRNA related to that of the wild type *cat* gene taken as 100%.
[g] Yield of protein.
[h] Yield of protein normalized to the yield of mRNA.
*: *i)* overrepresented codon pairs; *ii)* underrepresented codon pairs; *iii)* neutral codon pairs and *iv)* missing codon pairs.
**The sense: stop codon pair at the 3' terminus of the wild type *cat* gene.

Furthermore, one can speculate that the clusters of codon pairs with negative Δ_{REG} values might cause ribosome pausing to allow a proper protein folding. Analysis of the codon pairs composition, however, showed that the hypothetical translation efficiency (pausing or non-pausing) of a codon pair might not correlate with the frequency of occurrence of the corresponding individual codons in the pair. For instance, the pausing codon pair CUG: ACG (Leu:Thr) is composed of two frequently used codons. One of them (CUG) is the most frequently used codon (5,28% of all codons) in the entire *E. coli* genome. On the contrary, most of the pausing codon pairs contain at least one preferential codon (Table 9).

To examine the relationship between codon pairs usage and concentration of the decoding tRNAs, we used the tRNA-codon recognition pattern from the database of Dong (Dong et al. 1996). In this pattern each of the 46 *E.coli* tRNAs is assigned a numeric identifier corresponding to its electrophoretic spot on a two-dimensional polyacrylamide gel. This analysis shows that the hypothetical translation efficiency of the codon pairs does not correlate with the concentration of the corresponding tRNAs. For instance, the pausing codon pairs GCU: GGU (Ala:Gly) and GGU: GUA (Gly:Val) are recognized by the following abundant tRNAs: $tRNA_{1B}^{Ala}$ (5,04% of total tRNAs), $tRNA_3^{Gly}$ (6,76 %) and $tRNA_1^{Val}$ (5,96%) (see Table 9). Moreover, we found that one half of all 65 pausing codon pairs are recognized by abundant tRNAs. The precise analysis of the relationship between the composition of pausing and non-pausing codon pairs and the concentration of the cognate tRNAs is complicated because of the fact that some codons are translated by more than one tRNA.

As seen in Table 9, the two codon pairs CUG:ACG (*pausing*) and CUG:CCA (*non-pausing*) contain CUG, which is translated by two tRNA species ($tRNA_1^{Leu}$ and $tRNA_2^{Leu}$) characterized with a sevenfold difference in their abundance (6,94% and 1,03% respectively). A *pausing* and a *non-pausing* codon pair could contain the same individual codons but arranged in a reverse way (Table 10). For example, the codon pair CAC:ACC is identified as a *non-pausing*, whereas the inverse pair ACC:CAC is *pausing*. The fact that the individual codons in both pairs are recognized by the same tRNA species means that the efficiency (may be the rate) of translation depends not only on the concentration of the decoding tRNAs but also on the way they are ordered in the A and P ribosomal sites.

3.2.2. Experimental Studies

The analysis of distribution of codon pairs in the *E. coli* genome revealed four types of 3' terminal codon pairs depending on their frequency of occurrence: *overrepresented*, *moderately represented*, *underrepresented* and *missing* pairs. Based on this we have postulated the existence of a correlation between the statistically favored sense:stop codon pairs and the translation termination efficiency. To examine this prediction, representatives of the four group of codon pairs were inserted in front of the chloramphenicol acetyltransferase (CAT) gene and their effect on gene expression was evaluated by the yield of CAT protein and mRNA. The missing pairs were represented by one sense:sense (CCU:AGG, coding for Pro:Arg) and two sense:stop (CCU:UAG and CCC:UAG) codon pairs, the overrepresented codon pair was CCC:UGA (coding for Pro:Stop), the underrepresented were CCC:UAA and CCU:UGA and the neutral (native of the *cat* gene) codon pairs were CCU:UAA and

CGC:UGA respectively. As shown in Figure 9, all codon pairs were substituted for the natural UAA stop-codon.

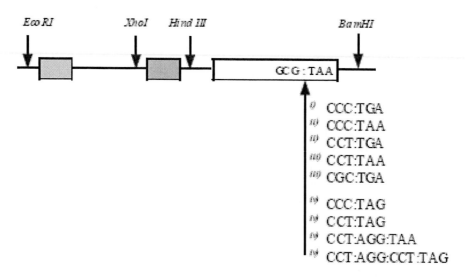

Figure 9. Cloning strategy and design of the derivative *cat* genes. The derivative (by PCR) *cat* genes are cloned in a pBR322 based expression plasmid carrying a strong synthetic constitutive promoter (P1) and a strong SD sequence (R9). The new codon pairs (bolded) are substituted for the native TAA stop codon. (Designations as in Table 10).

This figure shows also that the missing sense:sense codon pair was combined once with the preferential UAA stop-codon and secondly with the missing sense:stop codon pair CCU:UAG. The latter led to a unique construct bearing at the 3'-end two tandems of missing codon pairs (CCU:AGG:CCU:UAG). Additionally, the series of 8 *cat* gene constructs was designed to shed also light on the effect of the last (C-terminal) amino acid on the efficiency of *cat* gene expression. In the wild type CAT protein the ultimate amino acid is Ala (coded by GCG:UAA), whereas in the new constructs the last one is either Pro (CCC:UAG, CCU:UAG, CCC:UAA, CCU:UAA, CCC:UGA and CCU:UGA) or Arg (AGG:UAA and CGC:UGA). Statistically, it is possible to assess weather the sense:stop codon pair bias reflects the amino acid bias or it is a result of the interaction between the last aminoacyl-tRNA and corresponding release factor. In this case r^1 was close to r (see above) for the combinations of the two Pro codons CCC or CCU with all stop-codons (see Table 11), which means that the bias to the codon pair CCC:UGA (r^1 = +4.7 r = +4.9) was not related with that of the ultimate aminoacid (Pro). The prepared series of *cat* gene constructs allowed estimating also the effect of the combinations of all stop-codons (UAA, UGA and UAG) with the two Pro codons CCC and CCU on *cat* gene expression.

The new gene constructs were expressed in *E. coli* LE392 cells under a strong constitutive promoter (P1) and a consensus (AAGGAGGU) SD ribosome-binding site. The high level of expression in this system allowed visualizing the effect of most of the newly introduced modifications by simple polyacrylamide-SDS gel electrophoresis (see Figure 10). Quantitative expression data, however, we obtained by ELISA using a CAT specific monoclonal antibody. Bearing in mind that sometimes small changes in gene structure might interfere with transcription, the level of CAT mRNA was measured too. Thus the obtained data made it possible to evaluate the effect of the 3'terminal alterations on the efficiency of both translation and transcription.

Table 11. Amino acids corresponding to selected *pausing* codon pairs and their *non-pausing* inverse counterparts

Codon pair*	tRNA	Amino acids	Δ_{REG}
ACC:CAC	Thr1,3:His	Thr:His	-3.2
CAC:ACC	His:Thr1,3	His:Thr	+4.0
UUA:AAG	Leu5:Lys	Leu:Lys	-3.7
AAG:UUA	Lys:Leu5	Lys:Leu	+4.2
CGA:GUU	Arg2:Val1,2A,2B	Arg:Val	-2.5
GUU:CGU	Val1,2A,2B:Arg2	Val:Arg	+2.7
CAU:GCU	His:Ala1B	His:Ala	-2.7
GCG:CAC	Ala1B:His	Ala:His	+2.5
CGC:GGC	Arg2:Gly3	Arg:Gly	-3.0
GGU:CGU	Gly3:Arg2	Gly:Arg	+5.4

*The *pausing* codon pairs are bolded.

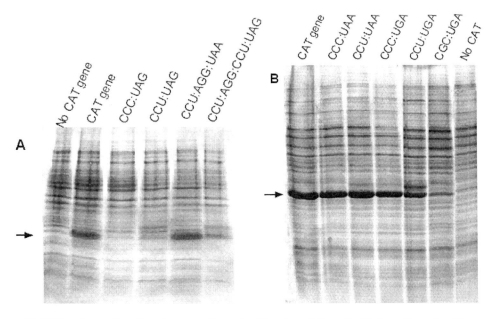

Figure 10. SDS polyacrylamide gel electrophoresis of lysates of *E. coli* cells transformed with *cat* gene expressing plasmids. A: *cat* gene variants carrying missing codon pairs; B: *cat* gene variants bearing overrepresented, underrepresented and neutral sense:stop codon pairs (see legend to Fig. 9). The CAT protein position is marked by arrows.

Based on the results presented in Figure 11 and Table 11, we concluded that the missing codon pairs CCU:UAG and CCC:UAG (both coding for Pro:Stop) had a strong suppressing effect on *cat* gene expression, thus decreasing the yield of CAT to 17% and 10% respectively compared to the native 3' terminal codon pair GCG:UAA (Ala:Stop). In order to check, whether this effect is due to the substitution of the C-terminal amino acid Pro for Ala, the two Pro codons CCU and CCC were combined with the stop-codons UAA and UGA. Our data indicateed that the suppressive effect of the two new codon pairs CCU:UAA and CCC:UAA was negligible, whereas the combinations with the synonymous stop-codon UGA was not. The yield of protein definitely depended on the type of stop-codon used and decreased gradually in the order UAA>UGA>UAG.

In order to clarify whether the variations in the yield of CAT were related with changes in the efficiency of transcription, the yield of CAT mRNA was measured too. The results showed that the cellular content of CAT mRNA for all combinations of the two Pro codons CCU and CCC with the three stop-codons (except for the pair CCC:UAG) was close to that of the wild-type construct. This means that not the insertion of Pro itself but the combinations of Pro codons with the three stop-codons predetermine the efficiency of CAT mRNA translation. Therefore we concluded that the missing codon pairs CCU:UAG and CCC:UAG had the strongest attenuating effect on translation. The latter could be explained by the inefficient termination of translation or by ribosome pausing (Bjornsson and Isaksson 1996; Jiang et al. 2002). Pausing might provoke formation of a line of translating ribosomes resulting in suppression of gene expression (Lesnik et al. 2000; Wolin and Walter 1988).

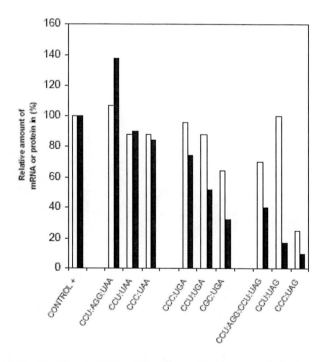

Figure 11. Relative yield of CAT and CAT mRNA in *E. coli* cells transformed with *cat* gene expressing plasmids. The yields of protein (black bars) and mRNA (white bars) are related to those obtained with the wild type *cat* gene.

Another conclusion drawn from these results is that the insertion of the missing sense:sense codon pair CCU:AGG (Pro:Arg) in front of the termination codon has an enhancing effect on translation. As seen from Figure 11 and Table 11, the yield of protein obtained from this construct wass even higher than that of the control. The same enhancing effect was observed also when the same codon pair was inserted in front of the missing pair CCU:UAG. The latter led to more than two-fold increase in the yield of CAT (from 17% to 40%) in comparison with the referent construct.

3.3. Conclusion

The relationship between translation efficiency and codon pairs bias was first investigated by Gudman and Hatfield (1989) on the bases of 237 *E. coli* genes. According to this study the highly expressed genes are enriched with underrepresented pairs and the poorly expressed genes in overrepresented codon pairs. Based on this observation they concluded that the overrepresented codon pairs were translated slower compared to the underrepresented pairs. Our analysis however, shows that the poorly expressed genes are enriched in codon pairs, which are overrepresented in the poorly expressed genes and in the same time are underrepresented in the highly expressed genes. Consequently, these codon pairs, denoted as *pausing*, might be translated slower in comparison with the overrepresented in the highly expressed genes, which are at the same time underrepresented in the poorly expressed genes (the latter are denoted as *non-pausing*).

It is shown in an *E. coli* assay system (Irwin et al. 1995) that the overrepresented codon pair ACG:CUG (Thr-Leu) inhibits translation compared to the synonymous underrepresented codon pair ACC:CUG coding for the same dipeptide. Although this effect was predicted (Gutman and Hatfield 1989) it has not been confirmed in a T7 assay system (Cheng and Goldman 2001). According to the data presented here, the ACC:CUG pair is underrepresented in both highly and poorly expressed genes (r_{low}= -4.8 and r_{high}= -5.2) and the synonymous ACG:CUG codon pair is overrepresented in both groups of genes (r_{low}= +10.0 and r_{high}= +11.2). This means that according to the criteria accepted in this study the two codon pairs should not affect the translation efficiency in *E. coli*.

Another example concerns the codon pairs AAG:UUA and UUA:AAG, being *non-pausing* and *pausing* respectively (Table 10). Both pairs are composed of the same synonymous Arg and Leu codons and therefore are decoded by the same tRNAs. The only difference is that the two decoding tRNAs are situated in a reverse order in the A and P sites during translation. Therefore it seems reasonable to assume that the specific structure of the individual tRNAs in combination with the specific architecture of the ribosome decoding center serve as a driving force for the evolution of genetic code led to the establishment of the present codon pairs bias in the *E. coli* genome.

The relationship between translation termination and codon pairs bias (sense:stop codons) is based on the following molecular interactions: tRNA:RF, mRNA:rRNA and tRNA:tRNA. If one of these interactions is not optimal for the termination, a frameshifting or nonsense codon translation could occur (Pedersen and Curran 1991; Poole et al. 1995; Stormo et al. 1986). The efficiency of translation termination depends on: *(i)* the nature of nucleotide following the stop-codon (called also *extended stop-codon*) (Major et al. 1996; Poole et al. 1995); *(ii)* the type of stop-codon and *(iii)* the nature of the penultimate sense codon. Based on our results, we are tempted to speculate that a correlation does exist between the statistically favored sense:stop codon combinations and translation termination efficiency. For instance, we found that the arginine codons are highly biased as penultimate and might express positive effect on the translation termination efficiency. This conclusion is supported by the study of Gursky and Beabealashvilli (Gursky and Beabealashvilli 1994) showing that the introduction of rare arginine codons before the stop-codon enhances gene expression. The variations in both **r** values (see Table 7) can be explained by the nature of the different release factors participating in the translation termination step. For instance, RF1 recognizes UAA and UAG, whereas RF2 recognizes the UAA and UGA codons (Scolnick and Caskey 1969). Therefore it can be speculated that the preference to RF1 or RF2 for decoding the UAA might depend on the nature of the penultimate codon.

The results presented in this study indicate that the frequency of occurrence of codon pairs in the *E.coli* genome vary from zero to 4913 times, which is an indication for selection pressure on the fixation of codon combinations during the evolution of genetic code. Our results show also that the translation efficiency is much more dependent on the type of codon pairs rather than the individual synonymous codons (rare or frequently used) in the coding sequence. Generally, we assume that the frequency of occurrence of the codon pairs located at the 3'-terminus of the genes have a well documented effect on translation in *E. coli* and therefore on the yield of protein. As a rule, this effect is better demonstrated with the missing sense:stop rather than with the sense:sense codon pairs.

References

Akashi H. 1994. Synonymous codon usage in *Drosophila melanogaster*: natural selection and translational accuracy. *Genetics* 136:927-935.

Akashi H, Schaeffer SW. 1997. Natural selection and the frequency distributions of "silent" DNA polymorphism in *Drosophila*. *Genetics* 146:295-307.

Alfonzo JD, Blanc V, Estevez AM, Rubio MA, Simpson L. 1999. C to U editing of the anticodon of imported mitochondrial tRNA(Trp) allows decoding of the UGA stop codon in *Leishmania tarentolae*. EMBO J 18:7056-7062.

Ames, B and P.Hartmann. 1963.The histidine operon. *Cold Spring Harbor Symposium Quantitative Biology* 28, 348-356.

Andersson SG, Kurland CG. 1995. Genomic evolution drives the evolution of the translation system. *Biochem. Cell Biol.* 73:775-787.

Andersson SG, Kurland CG. 1998. Reductive evolution of resident genomes. *Trends Microbiol.* 6:263-268.

Aota S, Ikemura T. 1986. Diversity in G + C content at the third position of codons in vertebrate genes and its cause. *Nucleic. Acids Res.* 14:6345-6355.

Aota S, Gojobori T, Ishibashi F, Maruyama T, Ikemura T. 1988. Codon usage tabulated from the GenBank Genetic Sequence Data. *Nucleic Acids Res.* 16 Suppl:r315-r402.

Arnez JG, Moras D. 1997. Structural and functional considerations of the aminoacylation reaction. *Trends Biochem. Sci.* 22:211-216.

Berg OG, Kurland CG. 1997. Growth rate-optimised tRNA abundance and codon usage. *J. Mol. Biol.* 270:544-550.

Binnewies TT, Motro Y, Hallin PF, Lund O, Dunn D, La T, Hampson DJ, Bellgard M, Wassenaar TM, Ussery DW. 2006. Ten years of bacterial genome sequencing: comparative-genomics-based discoveries. *Funct. Integr. Genomics* 6:165-185.

Bjornsson A, Isaksson LA. 1996. Accumulation of a mRNA decay intermediate by ribosomal pausing at a stop codon. *Nucleic. Acids Res.* 24:1753-1757.

Blanc V, Davidson NO. 2003. C-to-U RNA editing: mechanisms leading to genetic diversity. *J. Biol. Chem.* 278:1395-1398.

Blattner FR, Plunkett G, III, Bloch CA, Perna NT, Burland V, Riley M, Collado-Vides J, Glasner JD, Rode CK, Mayhew GF, Gregor J, Davis NW, Kirkpatrick HA, Goeden MA, Rose DJ, Mau B, Shao Y. 1997. The complete genome sequence of *Escherichia coli* K-12. *Science* 277:1453-1474.

Bossi L, Ruth JR. 1980. The influence of codon context on genetic code translation. *Nature* 286:123-127.

Boycheva, S., Shkodrov, G. and Ivanov, I. Codon pairs in the genome of *Escherichia coli*. *Bioinformatics* 29 (2003) 987-998.

Brown CM, Stockwell PA, Trotman CN, Tate WP. 1990. The signal for the termination of protein synthesis in procaryotes. *Nucleic. Acids Res.* 18:2079-2086.

Buckingham RH. 1994. Codon context and protein synthesis: enhancements of the genetic code. *Biochimie* 76:351-354.

Carbone A, Zinovyev A, Kepes F. 2003. Codon adaptation index as a measure of dominating codon bias. *Bioinformatics* 19:2005-2015.

Cermakian N, McClain WH, Cedergren R. 1998. tRNA nucleotide 47: an evolutionary enigma. *RNA* 4:928-936.

Cheng L, Goldman E. 2001. Absence of effect of varying Thr-Leu codon pairs on protein synthesis in a T7 system. *Biochemistry* 40:6102-6106.

Clarke B. 1970. Darwinian evolution of proteins. *Science* 168:1009-1011.

Comeron JM, Aguade M. 1998. An evaluation of measures of synonymous codon usage bias. *J. Mol. Evol.* 47:268-274.

Crick FH. 1966a. Codon--anticodon pairing: the wobble hypothesis. *J. Mol. Biol.* 19:548-555.

Crick FH. 1966b. The genetic code--yesterday, today, and tomorrow. *Cold Spring Harb Symp. Quant. Biol.* 31:1-9.

Crick FH, Brenner S, Klug A, Pieczenik G. 1976. A speculation on the origin of protein synthesis. *Orig. Life* 7:389-397.

Deschavanne P, Filipski J. 1995. Correlation of GC content with replication timing and repair mechanisms in weakly expressed *E.coli* genes. *Nucleic Acids Res.* 23:1350-1353.

Dong H, Nilsson L, Kurland CG. 1996. Co-variation of tRNA abundance and codon usage in Escherichia coli at different growth rates. *J. Mol. Biol.* 260:649-663.

Ermolaeva MD. 2001. Synonymous codon usage in bacteria. *Curr. Issues Mol. Biol.* 3:91-97.

Field D, Wilson G, van der GC. 2006. How do we compare hundreds of bacterial genomes? *Curr. Opin. Microbiol.* 9:499-504.

Fiers W, Contreras R, Duerinck F, Haegeman G, Iserentant D, Merregaert J, Min JW, Molemans F, Raeymaekers A, Van den BA, Volckaert G, Ysebaert M. 1976. Complete nucleotide sequence of bacteriophage MS2 RNA: primary and secondary structure of the replicase gene. *Nature* 260:500-507.

Fitch WM. 1976. Is there selection against wobble in codon-anticodon pairing? *Science* 194:1173-1174.

Freire-Picos MA, Gonzalez-Siso MI, Rodriguez-Belmonte E, Rodriguez-Torres AM, Ramil E, Cerdan ME. 1994. Codon usage in *Kluyveromyces lactis* and in yeast cytochrome c-encoding genes. *Gene* 139:43-49.

Gouy M, Gautier C. 1982. Codon usage in bacteria: correlation with gene expressivity. *Nucleic. Acids Res.* 10:7055-7074.

Grantham R, Gautier C, Gouy M. 1980a. Codon frequencies in 119 individual genes confirm consistent choices of degenerate bases according to genome type. *Nucleic. Acids Res.* 8:1893-1912.

Grantham R, Gautier C, Gouy M, Mercier R, Pave A. 1980b. Codon catalog usage and the genome hypothesis. *Nucleic. Acids Res.* 8:r49-r62.

Grantham R, Gautier C, Gouy M, Jacobzone M, Mercier R. 1981. Codon catalog usage is a genome strategy modulated for gene expressivity. *Nucleic. Acids Res.* 9:r43-r74.

Gursky YG, Beabealashvilli RS. 1994. The increase in gene expression induced by introduction of rare codons into the C terminus of the template. *Gene* 148:15-21.

Gustafsson C, Govindarajan S, Minshull J. 2004. Codon bias and heterologous protein expression. *Trends Biotechnol.* 22:346-353.

Gutman GA, Hatfield GW. 1989. Nonrandom utilization of codon pairs in *Escherichia coli*. *Proc. Natl. Acad. Sci. USA* 86:3699-3703.

Hatfield, G. W. and Gutman, G. A. Codon Pair Utilization Bias in Bacteria, Yeast and Mammals. *Transfer RNA in Protein Synthesis*. 1993. Ref Type: Thesis/Dissertation.

Ikemura T. 1981. Correlation between the abundance of Escherichia coli transfer RNAs and the occurrence of the respective codons in its protein genes: a proposal for a synonymous codon choice that is optimal for the *E. coli* translational system. *J. Mol. Biol.* 151:389-409.

Ikemura T. 1985. Codon usage and tRNA content in unicellular and multicellular organisms. *Mol. Biol. Evol.* 2:13-34.

Irwin B, Heck JD, Hatfield GW. 1995. Codon pair utilization biases influence translational elongation step times. *J. Biol. Chem.* 270:22801-22806.

Jiang ZH, Mu Y, Li WJ, Yan GL, Luo GM. 2002. The progress in mechanism of selenoprotein biosynthesis. *Sheng Wu Hua Xue Yu Sheng Wu Wu Li Xue Bao* (Shanghai) 34:395-399.

Karlin S, Mrazek J, Campbell AM. 1998. Codon usages in different gene classes of the *Escherichia coli* genome. Mol Microbiol 29:1341-1355.

Karlin S, Mrazek J, Campbell A, Kaiser D. 2001. Characterizations of highly expressed genes of four fast-growing bacteria. *J. Bacteriol.* 183:5025-5040.

Kittle, J. D. Radical Changes in the Engineering of Synthetic Genes for Protein Expression. *BioPharm International*, 12-18. 2006. Ref Type: Thesis/Dissertation

Knight RD, Landweber LF, Yarus M. 2001a. How mitochondria redefine the code. *J. Mol. Evol.* 53:299-313.

Knight RD, Freeland SJ, Landweber LF. 2001b. Rewiring the keyboard: evolvability of the genetic code. *Nat. Rev. Genet.* 2:49-58.

Knight RD, Freeland SJ, Landweber LF. 2001c. A simple model based on mutation and selection explains trends in codon and amino-acid usage and GC composition within and across genomes. *Genome Biol* 2:RESEARCH0010.

Komine Y, Adachi T, Inokuchi H, Ozeki H. 1990. Genomic organization and physical mapping of the transfer RNA genes in *Escherichia coli* K12. *J. Mol. Biol.* 212:579-598.

Kozak M. 1999. Initiation of translation in prokaryotes and eukaryotes. *Gene* 234:187-208.

Lahav N. 1993. The RNA-world and co-evolution hypotheses and the origin of life: implications, research strategies and perspectives. *Orig. Life Evol. Biosph.* 23:329-344.

Lawrence JG, Ochman H. 1997. Amelioration of bacterial genomes: rates of change and exchange. *J. Mol. Evol.* 44:383-397.

Lawrence JG, Ochman H. 1998. Molecular archaeology of the *Escherichia coli* genome. *Proc. Natl. Acad. Sci. USA* 95:9413-9417.

Lesnik T, Solomovici J, Deana A, Ehrlich R, Reiss C. 2000. Ribosome traffic in *E. coli* and regulation of gene expression. *J. Theor. Biol.* 202:175-185.

Lobry JR, Sueoka N. 2002. Asymmetric directional mutation pressures in bacteria. *Genome Biol.* 3:RESEARCH0058.

Lynn DJ, Singer GA, Hickey DA. 2002. Synonymous codon usage is subject to selection in thermophilic bacteria. *Nucleic. Acids Res.* 30:4272-4277.

Major LL, Poole ES, Dalphin ME, Mannering SA, Tate WP. 1996. Is the in-frame termination signal of the *Escherichia coli* release factor-2 frameshift site weakened by a particularly poor context? *Nucleic. Acids Res.* 24:2673-2678.

Massey SE, Moura G, Beltrao P, Almeida R, Garey JR, Tuite MF, Santos MA. 2003. Comparative evolutionary genomics unveils the molecular mechanism of reassignment of the CTG codon in *Candida* spp. *Genome Res.* 13:544-557.

Morton BR. 1993. Chloroplast DNA codon use: evidence for selection at the psb A locus based on tRNA availability. *J. Mol. Evol.* 37:273-280.

Morton BR. 1994. Codon use and the rate of divergence of land plant chloroplast genes. *Mol. Biol. Evol.* 11:231-238.

Moszer I. 1998. The complete genome of *Bacillus subtilis*: from sequence annotation to data management and analysis. *FEBS Lett* 430:28-36.

Moszer I, Rocha EP, Danchin A. 1999. Codon usage and lateral gene transfer in *Bacillus subtilis*. *Curr. Opin. Microbiol.* 2:524-528.

Murgola EJ. 1985. tRNA, suppression, and the code. *Annu. Rev. Genet* 19:57-80.

Nelson KE, Clayton RA, Gill SR, Gwinn ML, Dodson RJ, Haft DH, Hickey EK, Peterson JD, Nelson WC, Ketchum KA, McDonald L, Utterback TR, Malek JA, Linher KD, Garrett MM, Stewart AM, Cotton MD, Pratt MS, Phillips CA, Richardson D, Heidelberg J, Sutton GG, Fleischmann RD, Eisen JA, White O, Salzberg SL, Smith HO, Venter JC,

Fraser CM. 1999. Evidence for lateral gene transfer between *Archaea* and bacteria from genome sequence of *Thermotoga maritima. Nature* 399:323-329.

Nirenberg MW, Matthaei JH, Jones OW, Martin RG, Barondes SH. 1963. Approximation of genetic code via cell-free protein synthesis directed by template RNA. *Fed. Proc.* 22:55-61.

Nirenberg M, Leder P. 1964. Rna codewords and protein synthesis. The effect of trinucleotides upon the binding of srna to ribosomes. *Science* 145:1399-1407.

Osawa S, Jukes TH. 1988. Evolution of the genetic code as affected by anticodon content. *Trends Genet* 4:191-198.

Osawa S, Jukes TH. 1989. Codon reassignment (codon capture) in evolution. *J. Mol. Evol.* 28:271-278.

Osawa S, Jukes TH, Watanabe K, Muto A. 1992. Recent evidence for evolution of the genetic code. *Microbiol. Rev.* 56:229-264.

Pedersen WT, Curran JF. 1991. Effects of the nucleotide 3' to an amber codon on ribosomal selection rates of suppressor tRNA and release factor-1. *J. Mol. Biol.* 219:231-241.

Poole ES, Brown CM, Tate WP. 1995a. The identity of the base following the stop codon determines the efficiency of in vivo translational termination in *Escherichia coli. EMBO J.* 14:151-158.

Post LE, Strycharz GD, Nomura M, Lewis H, Dennis PP. 1979. Nucleotide sequence of the ribosomal protein gene cluster adjacent to the gene for RNA polymerase subunit beta in Escherichia coli. *Proc. Natl. Acad. Sci. USA* 76:1697-1701.

Rosenberg AH, Goldman E, Dunn JJ, Studier FW, Zubay G. 1993. Effects of consecutive AGG codons on translation in Escherichia coli, demonstrated with a versatile codon test system. *J. Bacteriol.* 175:716-722.

Sanger F, Air GM, Barrell BG, Brown NL, Coulson AR, Fiddes CA, Hutchison CA, Slocombe PM, Smith M. 1977. Nucleotide sequence of bacteriophage ΦX174 DNA. *Nature* 265:687-695.

Sato T, Terabe M, Watanabe H, Gojobori T, Hori-Takemoto C, Miura K. 2001. Codon and base biases after the initiation codon of the open reading frames in the *Escherichia coli* genome and their influence on the translation efficiency. *J. Biochem.* 129:851-860.

Schneider SU, de Groot EJ. 1991. Sequences of two rbcS cDNA clones of *Batophora oerstedii*: structural and evolutionary considerations. *Curr. Genet.* 20:173-175.

Schultz DW, Yarus M. 1994. Transfer RNA mutation and the malleability of the genetic code. *J. Mol. Biol.* 235:1377-1380.

Schultz DW, Yarus M. 1996. On malleability in the genetic code. *J. Mol. Evol.* 42:597-601.

Scolnick EM, Caskey CT. 1969. Peptide chain termination. V. The role of release factors in mRNA terminator codon recognition. *Proc. Natl. Acad. Sci. USA* 64:1235-1241.

Sharp PM, Li WH. 1987. The codon Adaptation Index--a measure of directional synonymous codon usage bias, and its potential applications. *Nucleic. Acids Res.* 15:1281-1295.

Shendure J, Porreca GJ, Reppas NB, Lin X, McCutcheon JP, Rosenbaum AM, Wang MD, Zhang K, Mitra RD, Church GM. 2005. Accurate multiplex polony sequencing of an evolved bacterial genome. *Science* 309:1728-1732.

Shields DC, Sharp PM, Higgins DG, Wright F. 1988. "Silent" sites in *Drosophila* genes are not neutral: evidence of selection among synonymous codons. *Mol. Biol. Evol.* 5:704-716.

Smith D, Yarus M. 1989. tRNA-tRNA interactions within cellular ribosomes. *Proc. Natl. Acad. Sci. USA* 86:4397-4401.

Sorensen MA, Kurland CG, Pedersen S. 1989. Codon usage determines translation rate in *Escherichia coli. J. Mol. Biol.* 207:365-377.

Stadtman TC. 1996. Selenocysteine. *Annu. Rev. Biochem.* 65:83-100.

Stormo GD, Schneider TD, Gold L. 1986. Quantitative analysis of the relationship between nucleotide sequence and functional activity. *Nucleic. Acids Res.* 14:6661-6679.

Varenne S, Buc J, Lloubes R, Lazdunski C. 1984. Translation is a non-uniform process. Effect of tRNA availability on the rate of elongation of nascent polypeptide chains. *J. Mol. Biol.* 180:549-576.

Wan XF, Xu D, Kleinhofs A, Zhou J. 2004. Quantitative relationship between synonymous codon usage bias and GC composition across unicellular genomes. *BMC Evol. Biol.* 4:19.

Wernegreen JJ, Moran NA. 1999. Evidence for genetic drift in endosymbionts (*Buchnera*): analyses of protein-coding genes. *Mol. Biol. Evol.* 16:83-97.

Woese, C. R. The Genetic Code. The Molecular Basis for Genetic Expression. New York: *Harper and Row.* 1967. Ref Type: Generic

Wolin SL, Walter P. 1988. Ribosome pausing and stacking during translation of a eukaryotic mRNA. *EMBO J.* 7:3559-3569.

Wright F. 1990. The 'effective number of codons' used in a gene. *Gene* 87:23-29.

Reviewed by:

Dr. Ashkan Golshany, PhD, Associate Professor of Molecular Genetics, College of Natural Sciences, Department of Biology, Carleton University, Nesbitt Building, Room 314; 1125 Colonel By Drive; Ottawa, ON, K1S 5B6, Canada. E-mail: ashkan_golshani@carleton.ca. Tel.: (613) 520-2600 ext 1006. Fax: (613) 520-3539.

In: Bacterial DNA, DNA Polimerase and DNA Helicases
Editors: W. D. Knutsen and S. S. Bruns

ISBN 978-1-60741-094-2
© 2009 Nova Science Publishers, Inc.

Chapter II

Maillard Reaction and Spontaneous Mutagenesis in *Escherichia coli*

Roumyana Mironova[1], Yordan Handzhiyski[1], Toshimitsu Niwa[2], Alfredo Berzal-Herranz[3], Kirill A. Datsenko[4], Barry L. Wanner[4] and Ivan Ivanov[1]

[1]Department of Gene Regulations, Institute of Molecular Biology,
Bulgarian Academy of Sciences, 1113 Sofia, Bulgaria
[2]Department of Clinical Preventive Medicine,
Nagoya University School of Medicine, Nagoya, Japan
[3]Instituto de Parasitología y Biomedicina 'López-Neyra', CSIC,
Parque Tecnológico de Ciencias de la Salud, Avda del Conocimiento s/n,
18100 Armilla, Granada, Spain
[4]Department of Biological Sciences, Purdue University,
West Lafayette, IN, USA

Abstract

DNA is a fragile organic molecule of finite chemical stability, capable of interacting with a variety of chemicals. In the context of the living cell this DNA reactivity is manifested as mutability. In addition to the chemical mutability, the perpetuation of DNA via imperfect replication is another source of DNA changeability. It is worth mentioning that DNA repair is error-tolerant and always leaves behind some lesions exist. Were repair mechanisms perfect, then the door of evolution would be locked. Risky rather than benign at individual level, mutations are the living things' long term strategy for survival and evolution. The creative power of mutations can explain why organisms have evolved special mechanisms to enhance mutagenesis under certain conditions. Hypermutation in bacteria under stress (SOS response) as well as hypermutation of immunoglobulin genes in response to antigen challenge in vertebrates are key examples.

The present study is focused on spontaneous mutagenesis in *Escherichia coli* K-12.

Spontaneous mutations are the net result of DNA damage and repair under normal physiological conditions. Bacterial DNA is loosely packed with proteins and occupies almost the entire cellular space. Therefore, it is extremely vulnerable to chemical attack, especially during replication and transcription, when the two DNA strands are separated. In the same arena, the cytoplasm, and at the same time, play also many small but highly reactive compounds. Although the collision of the latter with DNA seems to be unavoidable, to date the impact of ordinary cellular metabolites on spontaneous mutagenesis has not been systematically studied.

In year 2012 we shall celebrate the 100 anniversary of the discovery of the carbonyl-amine reaction by the French chemist Louis Camille Maillard (Maillard, 1912). Here, we provide evidence that the Maillard reaction, referred to as non-enzymatic glycosylation or glycation in the literature, is an important endogenous source of spontaneous mutations in *E. coli*. Experimental data allowed us to estimate that glycation affects on average one per 10^5 to 10^4 nucleotides in the *E. coli* chromosome. Based on experimental evidence, we postulate the existence in *E. coli* of a novel DNA repair enzyme, DNA amadoriase or DNA deglycase, especially designed to combat glycation induced spontaneous mutagenesis in *E. coli*.

Abbreviations

AGE	Advanced Glycation Endproduct
NBT	Nitroblue Tetrazolium
Rif	Rifampicin
OD	Optical Density
CA	carboxyalkyl
CM	carboxymethyl
CML Nε-	(carboxymethyl)lysine
ELISA	Enzyme Linked Immunosorbent Assay
LB	Luria-Bertani
NER	Nucleotide Excision Repair
BER	Base Excision Repair
DSB	Double Strand Break
AG	aminoguanidine
ASA	Acetyl Salicylic Acid,
PN	Pyridoxine
PM	Pyridoxamine
PLP	Pyridoxal 5'-Phosphate,
EGA	Early Glycation Adduct
SRR	Spontaneous Mutation Rate
ccc	covalently closed circular
TRIS	trishydroxymethylamino-methane
IPTG	Isopropyl β-D-1-thiogalactopyranoside
X-gal	5-bromo-4-chloro-3-indolyl-Я-D-galactoside
cfu	colony forming units
SDS-PAGE	Sodium Dodecyl Sulfate Polyacrylamide Gel Electrophoresis
8-oxo-dGTP	8-hydroxydeoxyguanosine
LC-MS/MS	Tandem mass spectrometry coupled with liquid chromatography,

SD	Standard Deviation
AP	Amadori Product
KO	Knock Out

Introduction

Our current knowledge about the reasons, underlying spontaneous mutagenesis, is far from being complete. The marked advance in our understanding of how the DNA-dependent DNA polymerases work let us know that errors made during DNA replication are kept at a low level due to the proofreading activity of these enzymes. The latter, together with the post-replication mismatch repair, account for an extraordinary fidelity of DNA replication – approximately one wrong base *per* billion nucleotides. On the other side, bacteria mutate at a relatively high rate of about one mutation *per* 10^8 to 10^6 cells *per* generation, depending on the particular gene target.

The relative contribution of replication errors to the overall DNA mutability is a controversial issue (Whittle and Johnston, 2006), but it is obvious that other, replication-independent factors also play an important role. In the absence of an external mutagenic assault, the factors impacting mutagenesis should be endogenous by nature. We would like to emphasize that the border we usually draw between ordinary cellular metabolites and endogenous mutagens is provisory. Water and oxygen are life indispensable constituents, but they represent a significant source of oxidative and hydrolytic lesions in DNA. Actually, H_2O and O_2 are the only "endogenous mutagens" studied in detail so far.

According to Barnes and Lindahl (2004) "Living organisms dependent on water and oxygen for their existence face the major challenge of faithfully maintaining their genetic material under a constant attack from spontaneous hydrolysis and active oxygen species and from other intracellular metabolites that can modify DNA bases". In addition, Pfeifer (2006) suggested "that mCpG transitions are not caused simply by spontaneous deamination of 5-methylcytosine in double-stranded DNA but by other processes including, for example, mCpG-specific base modification by endogenous or exogenous mutagens..." Indeed, by reaction of the cellular metabolite glyoxal with 5-methyl deoxycytidine, the deamination product deoxythymidine is formed (Kasai et al., 1998). This chapter aims at adding to our understanding of spontaneous mutagenesis more knowledge about "other processes" and "other intracellular metabolites" that can modify DNA under normal physiological conditions.

Living organisms do depend for their existence not only on water and oxygen, but also on carbon and energy sources. At the beginning of the past century the French chemist Louis Camille Maillard conducted research on peptide synthesis. At that time, when the mechanisms of protein biosynthesis were still a mystery, his far goal was to understand the natural way of amino acid polymerization under mild physiological conditions. For this purpose L. C. Maillard used D-glucose, a widespread sugar in biologic systems, as a soft condensing agent. Thus he realized that glucose, through its aldehydic functionality, is capable of reacting with amino acids (Maillard, 1912). In this way, L. C. Maillard did not only make a contribution to basic organic chemistry but ingeniously predicted that "The consequences of these facts appear... interesting in various fields of Science: not only in human physiology and pathology..."

The significance of the Maillard reaction, known also as non-enzymatic browning reaction or carbonyl-amine reaction, for food chemistry was recognized soon. However, it took many years for biologists and physicians to grasp the physiological role of the Maillard reaction. Studies in the late 1960s revealed the existence of an "abnormal fast-moving hemoglobin band" in diabetic patients during routine electrophoretic screening for hemoglobin variants (Rahbar, 1968; Rahbar, 2005). In the same year, it has been shown that the fast-moving haemoglobin subfraction HbA1c can be prepared *in vitro* by incubation of hemoglobin with glucose in the absence of any enzyme catalysts (Bookchin and Gallop, 1968). Later on, in the scientific lexicon were introduced the terms "non-enzymatic glycosylation" and "glycation" in order to distinguish the Maillard reaction, proceeding in biologic systems, form the enzymatic glycosylation, which is a quite different, genetically programmed process. In the early 1980s, it has been hypothesized that glycation plays an important role in the pathogenesis of diabetic complications and aging (Monnier et al., 1981; Monnier and Cerami, 1981). Subsequently, this hypothesis found a plethora of experimental support, and many reviews have been dedicated to the link between glycation, diabetes and aging (Brownlee, 1995; Thorpe and Baynes, 1996; Baynes, 2000; Ulrich and Cerami, 2001; Baynes, 2002; Monnier and Sell, 2006; Nass et al., 2007; Ahmed and Thornalley, 2007). Although the carbonyl-amine reaction was discovered a century ago, its chemistry remains a still growing avenue of research. Two well defined stages can be distinguished in the Maillard reaction – early and advanced. The early stage includes the reversible formation of Schiff bases between the carbonyl and amino groups of the reactants, followed by a rearrangement of the Schiff bases to significantly more stable aldoamines (Amadori products) or ketoamines (Heyns products). The early stage of the Maillard reaction is relatively well understood (Kuhn andWeygand, 1937; Hodge, 1953). However, the same does not hold true for the chemical transformations of the Amadori and Heyns products into the so called Advanced Glycation End Products (AGEs) (Cerami, 1986). Generally, these are enolisation, oxidation, dehydration and fragmentation reactions, taking place in the advanced stage of the Maillard reaction. AGEs are stable covalent adducts that, when accumulating in biologic amines, may significantly impair their structure and physiological function. The chemical structure of a number of AGEs formed *in vitro* as well as in human plasma and tissues has been determined (Nakayama et al., 1980; Pongor et al., 1984; Ahmed *et al.*, 1986; Sell and Monnier, 1989; Ahmed and Thornalley, 2003; Thornalley et al., 2003; Thorpe and Baynes, 2003). It is believed, however, that AGEs identified so far, represent only a minor fraction of all AGEs produced in the human body.

AGEs are formed in a sophisticated interplay between the two stages of the Maillard reaction. Under anaerobic conditions, AGEs are derived directly from the Amadori product. However, under aerobic conditions, oxidative degradation of the Amadori products takes place, leading to the generation of α-oxoaldehydes (Anet, 1960; Kato, 1960) that are much more reactive than monosaccharides. This "interruption" between the early and the advanced stage is what we call the intermediate step of the Maillard reaction. In turn, α-oxoaldehydes, which are small, diffusible and highly reactive species, again attack amino compounds thus enhancing the initial chemical burden. The formation of AGEs, either directly from the

Amadori product or through its degradation intermediates, is known as the classical, or the Hodge pathway (Hodge, 1953). The Schiff base, formed early in the Maillard reaction, may also undergo non-enzymatic fragmentation to α-oxoaldehydes, which initiate another chain of chemical transformations known as the Namiki pathway (Hayashi and Namiki, 1980). In addition, under physiological conditions free monosaccharides undergo a slow oxidative degradation (autoxidation) to H_2O_2 and the corresponding α-oxoaldeydes, which are precursors for AGE formation in the Wolff pathway, also recognized as a separate reaction chain in the Maillard chemistry (Thornalley et al., 1984; Wolff and Dean, 1987; Thornalley et al., 1999). The picture becomes increasingly colorful, pasting therein the carbonyl products released during lipid peroxidation, which also react with amines to form AGE-like adducts called Advanced Lipoxidation End Products (ALEs) (Fu et al., 1996; Baynes and Thorpe, 2000). It is not always possible to specify the exact metabolite, causing the formation of a particular advanced end product *in vivo*. For example, Nε-(carboxymethyl)lysine, considered for a long time a typical glycation marker, has been shown to arise also in incubation mixtures of proteins with polyunsaturated fatty acids (Fu et al., 1996). Therefore, Khalifah et al. (2005) suggested to use the term "Advanced Metabolic End Products" (AMEs) for those modifications that arise independently of direct glucose and glycated protein breakdown.

The Maillard reaction may have serious implications for biologic systems, mainly because of the involvement of life essential amines in this reaction. We are tempted to speculate that the basic molecules of live, proteins and DNA are acid derivatives because evolution of life has been tightly linked to oxidation as the main route for energy release from carbon sources. Acids are less reactive with oxygen than alcohols and carbonyls, and hence, better protected from an unsustainable structural damage due to oxidation. However, being amines, proteins and DNA are vulnerable to carbonyl attack. The utilization of glucose as an energy source in living systems could be considered another evolutionary trick, directed against the carbonyl-amine reaction.

Under physiological conditions glucose persists predominantly in cyclic form and because of that, among the known sugars, it is the least reactive with amines. In plants and most insects glucose is delivered to cells in the form of non-reducing disaccharides, sucrose and trehalose respectively, whereas in animals' blood glucose concentration is kept low (~5 mM) *via* stringent hormonal control. In this way the extracellular milieu is defended against hyperglycemic stress. Indeed, there is experimental evidence that the concentration of AGEs is higher in cellular proteins than in plasma proteins (Ahmed and Thornalley, 2003). It is unlikely that glucose is directly involved in a Maillard reaction within cells. During glycolysis there is an outburst of more reactive intermediates such as glucose 6-phosphate, fructose 6-phosphate, and glyceraldehyde 3-phosphate. These, together with α-oxoaldehydes (glyoxal, methylglyoxal, 3-deoxyglucosone *etc*.), form a highly reactive intracellular pool of carbonyls that may account for enhanced glycation inside cells.

In eukaryotes, DNA is insulated from the cytoplasm by a phospholipid membrane and additionally wrapped with nucleophilic histones. It seems that there is a serious barrier for the carbonyl pool to reach DNA. Similar thoughts, perhaps, delayed the studies on DNA glycation.

The first evidence for DNA glycation and its possible physiological implications was presented by Bucala et al., (1984). Thereafter, glycation of nucleotides and DNA has been investigated mainly *in vitro* (Vaca et al., 1994; Seidel and Pischetsrieder, 1998; Kasai et al., 1998). Despite data indicating that DNA glycation may have mutagenic/carcinogenic effect (Lee andCerami, 1991; Bucala et al., 1993; Pushkarsky et al., 1997), glycation of DNA under physiological conditions *in vivo* has not been demonstrated for a long time. In this regard, the experimental proof for the presence of lipid peroxidation adducts in DNA came first. Adducts of malonaldehyde, the major product of lipid peroxidation, have been observed in DNA from total white blood cells of healthy individuals held on different fat-based diets (Fang et al., 1996).

In contrast to eukaryotic DNA, the bacterial chromosome is directly sunk in the cytosol and lacks a tight protein envelope, which makes it more vulnerable to chemical attack. With this in mind, we undertook relevant investigations and found that the chromosome of *Escherichia coli* K-12 accumulates both Amadori products and AGEs under normal growth conditions *i. e.* in the absence of a mutagenic challenge (Mironova et al., 2005).

In the same study we presented data pointing to a link between DNA glycation and spontaneous mutagenesis in *E. coli*. At the 9th International Symposium of the Maillard Reaction, P. Thornalley and co-workers reported the quantification (by electrospray ionization LC-MS/MS) of two AGEs formed by glycation of deoxyguanosine with glyoxal and methylglyoxal (Fleming et al., 2007). The detection of similar adducts in DNA by another approach, a ^{32}P-postlabelling assay, has been reported previously (Vaca et al., 1994). However, P. Thornalley and co-workers presented the first conclusive evidence for the formation of glycation adducts in eukaryotic DNA (of human mononuclear leukocytes) *in vivo*. In some of the DNA samples, the content of the oxidative DNA biomarker 8-hydroxydeoxyguanosine (8-oxo-dGTP) was found to be lower than that of the two guanosine AGEs. Thus, the work of P. Thoranlley and co-workers demonstrated that even the eukaryotic DNA, which is shielded by nucleosomes and higher order chromatin structures, is susceptible to glycation under physiological conditions *in vivo* and that damage of DNA, caused by glycation, is at least comparable to, if not greater than, that caused by oxidation.

In the present chapter, we report new experimental data in favor of the link between DNA glycation and spontaneous mutagenesis in *E. coli*. We found that the level of DNA glycation is strain-dependent and correlates with the spontaneous mutation rate. An *E. coli* strain, severely impaired in key repair mechanisms, accumulated substantial amounts of glycation adducts in its chromosome.

In consistency with others (Murata-Kamiya et al., 1998; Murata-Kamiya et al., 1999), we proposed that among known repair mechanisms, the nucleotide excision repair (NER) is most likely to account for repair of glycated DNA. Furthermore, we detected in *E. coli* a novel enzyme activity, which seems to be especially designed for removal of Amadori products from DNA. We called this putative DNA repair enzyme DNA amadoriase or DNA deglycase.

In addition, acetyl salicylic acid and pyridoxal 5'-phosphate were found to exert opposite effects on DNA glycation and mutability. Whilst inhibiting DNA glycation, both these compounds led to an increase in the spontaneous mutation rate.

Experimental Procedures

Bacterial Strains and Growth Media

The following *E. coli* K12 strains were used in this study: 33W1485 (wild type), AB1157 (F$^-$ thr-1 leu-6 proA2 his-4 argE3 thi-1 lacY1 galK2 ara-14 xyl-5 mtl-1 tsx-33 rspL31 supE44), 2410 (el4$^-$ (mcrA) D(mcrCB$^-$ hsdSMR$^-$ mrr) 171 sbcC recB recJ uvrC umuC::Tn5 (kan^r) lac gyrA96 relA1 lr{F' proAB lacIqZDM15 Tn10 (tet^r)} Su$^-$), XL1-Blue (recA1 endA1 gyrA96 thy-1 hsdR17 supE44 relA1 lacF' proAB lacI^{q}ZΔM15 Tn10 tet^r) and BW28357 (rrnB DlacZ478 hsdR514 D(araBAD)567 D(rhaBAD)568 rph$^+$). Strains 33W1485, AB1157 and 2410 were purchased form the National Bank for Industrial Microorganisms and Cell Cultures (Sofia, Bulgaria). Strains BW37935 and BW37942, isogenic to BW28357, with either full (ΔfrlB) or partial (ΔfrlB') deletion of the frlB gene respectively, were constructed by one-step inactivation of the chromosomal gene as described elsewhere (Datsenko and Wanner, 2000). The partial deletion ΔfrlB' preserves a potential translation initiator in the 3'-coding region of the frlB gene. The rich composition of the LB medium was described elsewhere (Sambrook et al., 1989) and M9G medium was minimal M9 medium (Sambrook et al., 1989) containing 0.1% glucose as a carbon source. This medium was supplemented with 0.5 μg/ml thiamine and the appropriate amino acids at a final concentration of 40 μg/ml. The minimal salt agar (1.5%) for selection of the Arg$^+$ prototrophic mutants was prepared with the same medium devoid of arginine.

Isolation of *E. coli* Chromosomal DNA

E. coli cells corresponding to 125 OD$_{590}$ were collected by centrifugation and resuspended in 8 ml of 25 mM Tris HCl pH 8, 10 mM EDTA, 0.14 M NaCl and 50 mM glucose. SDS and Proteinase K were added to final concentrations of 1% and 250 μg/ml respectively and the suspension (final volume 10 ml) was incubated for 40 min at 37°C. After two phenol extractions DNA was ethanol precipitated and dissolved in 1 ml of 10 mM Tris HCl pH 8, 1 mM EDTA (TE 10-1) containing 100 μg RNase A and 20 units RNase T1. Following 30 min incubation at 37°C, Proteinase K was added to a final concentration 100 μg/ml and the incubation was continued for additional 30 min. The samples were phenol extracted, ethanol precipitated and DNA dissolved in TE 10-1 to a final concentration of 5 mg/ml. DNA was sonicated to an average fragment size below 1300 bp and its quality was estimated by the A$_{260}$/A$_{280}$ spectral ratio (2 ± 0.1).

Determination of Early Glycation Products in DNA

Early glycation products in DNA were determined by the nitroblue tetrazolium (NBT) reduction assay (Johnson et al., 1982) after some modification as previously described (Mironova et al., 2005). Briefly, the DNA samples (100 μl) were mixed with 1 ml 100 mM sodium carbonate buffer (pH 10.8) containing 0.25 mM NBT, and incubated for 5 hours at 37°C. Absorbance at 525 nm was read against distilled water, and the content of early glycation products was determined using an extinction coefficient of 12 640 cm^{-1} M^{-1} for monoformazan, which is formed upon reduction of NBT by early glycation adducts.

Fluorescence Spectroscopy

Fluorescence measurements were carried out on a Shimadzu model RF-5000 fluorescence spectrophotometer. DNA samples at an approximate concentration of 5 mg/ml were sonicated using ultrasonic homogenizer (Model CP 50, Cole-Parmer Instrument Co.) three times for 1 min (50% duty cycle at 80% power). In the comparative analyses, the DNA samples were serially diluted and the fluorescence at λ_{ex}=360 nm/λ_{em}=440 nm was measured. The emission intensity was plotted *versus* the concentration of DNA and the graph was used for exact calculation of the fluorescence related to 5 mg/ml DNA. The Raman fluorescence of the solvent (λ_{ex}=370 nm/λ_{em}=420 nm) was subtracted from the DNA fluorescence spectra because of its interference especially at low DNA concentrations.

Competitive ELISA

Glycated bovine serum albumin (AGE-BSA) was prepared by incubating 10 mg/ml BSA with 0.5 M glucose in a water solution for three months at 37°C. Microtiter plates were coated over night at 37°C with 100 µl *per* well AGE-BSA solution at a concentration of 10 µg/ml in 50 mM sodium carbonate (pH 9.6). The coupling solution was then discarded and replaced with 100 µl 2% BSA in PBS. After blocking for 3 h at 37°C, the plates were washed twice with 0.05% Tween 80 in PBS and once with PBS. DNA samples or the AGE-BSA standard, dissolved in PBS (50 µl *per* well), were mixed with 50 µl horseradish peroxidase labeled anti-CML antibody (Niwa et al., 1996), diluted 1:30 in assay buffer (0.25% BSA in PBS). Competition reaction was carried out for 3 h at 37°C and after washing as above, color was developed with 100 µl *o*-phenylenediamine dissolved at 0.15 mg/ml in 0.2 M K$_2$HPO$_4$citrate buffer pH 6, 0.02% H$_2$O$_2$.

The reaction was stopped with 100 µl 0.8 M H$_2$SO$_4$, and the absorption at 490 nm was read using an ELISA plate reader. Data were processed by 4P logistic curve fit (Mironova et al., 2005). The immunoreactivity of DNA to the anti-CML antibody was expressed in AGE-BSA equivalents (eq), where 1 eq equals to the immunoreactivity of 1 µg AGE-BSA.

Mutational Analyses

To estimate the frequency of spontaneous loss of rifampicin (Rif) resistance, overnight *E. coli* cultures, grown from a single colony, were used to inoculate fresh LB medium at an initial optical density of 0.1 OD$_{590}$. The cultures were then incubated at 37°C until OD$_{590}$ = 2.0 and cells from 1 ml culture were collected by centrifugation. The cell pellets were resuspended in 100 μl LB and spread on LB agar dishes, containing 100 μg/ml Rif. In parallel, serial dilutions of the starting cell cultures were made on LB plates to be used for counting the viable cell number.

The spontaneous reversion of *E. coli* AB1157 to arginine prototrophy was determined as follows: Overnight LB cultures (prepared as above) were used to inoculate fresh LB medium at a starting concentration of 0.1 OD_{590}. The cultures were grown at 37°C to a stationary phase and cells were collected by centrifugation. The cell pellet was washed twice with non-supplemented M9G medium and resuspended in the same medium at approximately 5 OD_{590}/ml. Samples of 200 μl were spread on selective M9G plates (no arginine) and the number of Arg^+ colonies was scored after 3 days of growth at 37°C. To estimate the number of viable cells, serial dilutions of the concentrated cell stock were made on LB plates.

In an additional experiment, designed to explore the effect of a long term bacterial cultivation in thiamine supplemented medium, a single *E. coli* AB1157 colony was inoculated into M9G medium and cultivated overnight at 37°C. This culture was used to inoculate LB medium containing 15 mM thiamine as well as control cultures without thiamine at an initial optical density of 0.1 OD_{590}. Then, continuous subculturing was performed encompassing seven passages over 12 hours each. Individual passages were followed by reinoculation of fresh media at 0.1 OD_{590} and spreading of stationary phase culture aliquots onto LB plates for counting the viable cells. At the end of the culturing period, the cells were washed twice with non-supplemented M9G medium, resuspended in the same medium and plated onto four selective minimal salt agar dishes (2 OD590 *per* dish). The revertants were scored after 3 days of growth at 37°C and the reversion frequencies were expressed as a number of revertants *per* 10^8 cells *per* cell generation. The number of generations in each individual passage was calculated using the equation:

$$N = 1.4 \times 10^7 \times 2^n,$$

where *n* is the number of generations, N is the number of cells *per* ml culture at the end of the passage and the coefficient 1.4×10^7 determined in preliminary experiments, accounts for the number of cells *per* ml starting culture (0.1 OD_{590}). Two experiments on separate cultures were carried out each in duplicate and the standard deviations (SD) were calculated.

Cloning of the frlB Gene

A 5'-primer containing the translational initiator codon of the *frlB* gene, flanked by a *Hind*III restriction site (5'-CCCAAGCTT*ATG*TTGGATATTGATAAAAGCACCGT-3'), and a 3'-primer, containing the stop codon, followed by a *Bam*HI site (5'-CGCGGATCC*TTA*ATA-TTCCACCAGACCACCGTAA-3'), were used to PCR-amplify genomic DNA from the *E. coli* strain DH1. PCR was performed with *Taq* polymerase under the following conditions: denaturation for 2 min at 95°C and 5 cycles of annealing for 30 s at 56°C, primer extension for 1 min at 72°C and heating for 30 s at 95°C. Bulk DNA was synthesized within 30 cycles, each consisting of denaturing for 30 s at 95°C, annealing for 30 s at 70°C and polymerization for 1 min at 72°C. One additional cycle of 1 min at 72°C was included at the end of the PCR for competition of DNA synthesis. The PCR-product was hydrolyzed with *Hind*III and *Bam*HI and ligated to the plasmid pJP₁R₉ (Ivanov et al., 1987) treated with the same restriction endonucleases. The resulting construct was transformed into the *E. coli* XL1-Blue and checked by sequencing.

Amadoriase Assay

The FrlB enzyme activity was measured in the reverse reaction through the conversation of glucose 6-phosphate and N-α-t-Boc-N-L-lysine to N-α-t-Boc-N-ε-fructoselysine 6-phosphate. The reaction mixture contained 25 mM HEPES, 0.1 mM EGTA, 5 mM MgCl₂, 0.15 M glucose 6-phosphate, 0.3 M N-α-t-Boc-N-L-lysine and 100 µg/ml total *E. coli* protein from cells overexpressing the *frlB* gene. The final volume of the reaction mixture was 200 µl. Incubation was carried out at 37°C for 2 hours. For measurement of the reaction product, N-α-t-Boc-N-ε-fructoselysine 6-phosphate, the reaction mixture (100 µl) was directly mixed with 1 ml 100 mM sodium carbonate buffer (pH 10.8) containing 0.25 mM NBT and the absorbance at 525 nm was measured against distilled water over 20 minutes. The enzyme activity was expressed as a change in the absorbance at 525 nm *per* minute (ΔA_{525}/min).

Results and Discussion

The Genetic Background Impacts the Level of Chromosomal DNA Glycation

Glycation is a chemical reaction obeying the mass action law and thus depending on the concentration of the reactants. During exponential growth in rich media, *E. coli* divides rapidly and multiplies its single chromosome to several copies *per* cell. At the same time, under aerobic conditions, the burst of reactive glycolyic intermediates and their oxidation derivatives creates an excellent milieu for the Maillard reaction to take off. Early glycation adducts and AGEs accumulate gradually in the bacterial chromosome, approaching a maximum in stationary phase cells. It is noteworthy that AGEs appear in the *E. coli* DNA in a very short time. We observed a substantial reactivity of the *E. coli* chromosome to an anti-CML antibody after 12 hours of growth (Mironova et al., 2005). The immunoreactivity of DNA to this antibody was indicative of modifications with carboxyalkyl (CA), most probably with carboxymethyl (CM), residues. It is unlikely that carboxyalkyl moieties are formed on DNA in the classical Hodge pathway (Hodge, 1953). In this pathway, AGEs arise slowly in the time course of days and even months. CM adducts, however, can be formed in DNA by reaction of guanine with the highly reactive α-oxoaldehyde glyoxal (Kasai et al., 1998), which is released early in the Maillard reaction due to either glucose or Schiff base oxidation. Therefore, it seems more likely that the anti-CML antibody binds to DNA adducts, which are derived in the Namiki and/or Wolff pathways (Hayashi and Namiki, 1980; Thornalley et al., 1984; Thornalley et al., 1999). In addition, the *E. coli* DNA accumulates fluorescent adducts (Bucala et al., 1984; Mironova et al., 2005), similar to that found in glycated and oxidized proteins. To the best of our knowledge, to date, the chemical structure of fluorescent nucleotide AGEs has not been reported.

In the present study, for estimation of the extent of DNA glycation, we used three approaches, described elsewhere (Mironova et al., 2005). Briefly, these were the nitroblue tetrazolium (NBT) reduction assay for measurement of early glycation products, ELISA for evaluation of the immunoreactivity of DNA to an anti-CML antibody and fluorescence spectroscopy for quantification of fluorescent adducts in DNA.

In a previous study we have found that DNA, isolated from two *E. coli* strains, contains varying amounts of glycation adducts (Mrionova et al., 2005). Here, to further evaluate the impact of the bacterial genotype on the level of DNA glycation, we investigated three *E. coli* stains with quite different genetic characteristics. *E. coli* 33W1485 is a wild type *E. coli* K-12 strain with no mutations reported. The *E. coli* strain AB1157 is defective in the catabolism of several sugars - lactose (*lacY1*), galactose (*galK2*), arabinose (*ara*) and xylose (*xyl*). The third strain, *E. coli* 2410, is impaired in key DNA repair mechanisms, including nucleotide excision repair (*uvrC*), mismatchrepair (*recJ*), DNA translesion synthesis (*umuC*) and recombination repair (*recB*). The three *E. coli* strains were grown in LB medium for 12 hours, chromosomal DNA was then extracted and analyzed for the level of glycation. The chromosome of the wild type strain 33W1485 proved to be the least glycated one, whereas strain AB1157 accumulated slightly more glycaion adducts in its DNA (Figure 1).

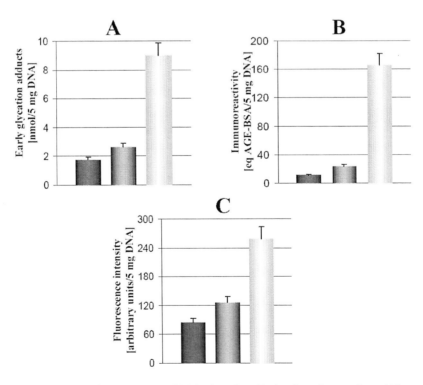

Figure 1. Impact of the *E. coli* genotype on DNA glycation. Early glycation products (A), immunoreactivity to an anti-CML antibody (B) and fluorescent adducts (C) characteristic for chromosomal DNA of *E. coli* strains 33W1485 (■), AB1157 (■) and 2410 (□). Data are mean ±SD (n=3).

Possibly, sugars that can not be metabolized accumulate within AB1157 cells and either directly or through oxidation derivatives additionally modify DNA. Remarkably, the chromosome of the repair deficient strain *E. coli* 2410 was significantly affected by glycation, which implies that glycation adducts in DNA are subjected to repair. Glycated DNA should be repaired because, according to our calculations based on the NBT reduction assay, roughly one *per* 10^5 to 10^4 nucleotides in the *E. coli* chromosome is involved in early glycation. This is a quite high magnitude of DNA damage, comparable to that imposed on DNA by base tautomerization – a problem, basically solved by the proofreading activity of DNA polymerases.

There Is a Correlation between DNA Glycation and Spontaneous Mutagenesis

If glycation contributes to DNA mutability, there should be a link between the level of DNA glycation and the spontaneous mutation rate. With this in mind, we went on determining the frequency of spontaneous mutations in the *E. coli* strains 33W1485, AB1157 and 2410, because they had demonstrated varying levels of DNA glycation. The three *E. coli* strains lack a common genetic marker appropriate for comparative mutational analyses. Suitable in such cases are the dominant forward mutations, which appear independently of the particular genetic background.

An example for such mutations is the acquisition of resistance to the antibiotic rifampicin (Rif) (Hartmann et al., 1967). Wild type *E. coli* strains are sensitive to this antibiotic (Rif^S), which binds the catalytic β-subunit of the RNA polymerase (McClure and Cech, 1978). However, due to spontaneous mutation in the RNAβ-subunit gene, *E. coli* may become resistant to rifampicin (Rif^R) (Jin and Gross, 1988). When we studied the spontaneous mutations in the *E. coli* strains 33W1485, AB1157 and 2410 to Rif^R phenotype, we observed a correlation between the mutation rate and the extent of DNA glycation (Figure 1 and Figure 2). The repair deficient strain 2410, with the highest amount of glycation adducts in its chromosome, mutated to Rif^R phenotype twenty times more frequently than the wild type strain (Figure 2). Strain 2410 is impaired in several repair mechanisms, but among them, the NER-deficiency is manly responsible for the enhanced mutagenesis under normal, non-stressful conditions. This holds true, because:

1. The mismatch repair is only weakly affected in strain 2410 (*recJ*). The RecJ exonuclease is responsible for removal of nearly half the base mismatches, the other half being removed by ExoI or ExoX. Moreover, the RecJ exonuclease is dispensable and its disfunction can be compensated by another exonuclease, ExoVII (Li, 2008).
2. The UmuC protein is a subunit of the *E. coli* DNA polymerase Pol V, belonging to the Y-family of error-prone DNA polymerases (Goodman, 2002). These polymerases, which overcome replication stalls in severely damaged DNA, are part of the SOS response, which is activated only when bacteria are subjected to DNA

damage. The UmuC⁻ phenotype is not expressed in strain 2410, because we grew this strain under normal, non-stressful conditions.

3. The RecB polypeptide, a subunit of the RecBCD enzyme, plays a key role in homologous recombination (Michel et al., 2007). The repair role of this enzyme under physiological conditions, however, is negligible, because the recombination repair, called also double-strand break (DSB) repair, is induced only by double-strand breaks in DNA. In vegetative cells as well as in the absence of an environmental insult, double-strand breaks appear rarely in the *E. coli* DNA.

All the above considerations imply that the high mutation rate of the *E. coli* strain 2410 to Rif resistance should be attributed mainly to the UvrC⁻ phenotype. The UvrC protein (endonuclease) belongs to the NER machinery, which thus seems to be involved in repair of glycated DNA. NER operates at sites of bulky chemical modifications in DNA like those introduced by glycation. Our results are also in conformity with other investigations, indicating that NER may be involved in both the repair and the fixation of glyoxal- and methylglyoxal-induced mutations in the chromosomal *lacI* gene of *E. coli* (Murata-Kamiya et al., 1998; Murata-Kamiya et al., 1999). The new data, we report here, are in favor of the link between glycation and spontaneous mutagenesis established by us in a previous study (Mironova et al., 2005). In this latter study, we found that supplementation of the *E. coli* growth medium with the glycation inhibitor thiamine (vitamin B1) results in suppression of both DNA glycation and mutability. We now expanded these investigations by including five other compounds with anti-glycation activity and studying their impact on *E. coli* DNA glycation and mutability.

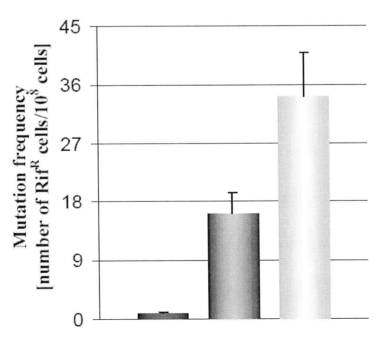

Figure 2. Spontaneous mutation rate to Rif resistance of *E. coli* strains 33W1485 (■), AB1157 (■) and 2410 (□). Data are mean ± SD (n=5).

Suppression of DNA Glycation May Result in Enhanced Mutagenesis

The pathological consequences of glycation in diabetic and elder humans have promoted an active search for compounds that can counteract glycation under physiological conditions *in vivo* (Monnier, 2003). Many chemicals have been tested to date, however, their effect on DNA glycation and mutability has been overlooked. Here, we provide data obtained with six widely used glycation inhibitors including aminoguanidine (AG), acetyl salicylic acid (ASA) (aspirin), thiamine (vitamin B1), and the vitamin B6 vitamers, pyridoxine (PN), pyridoxamine (PM), and pyridoxal 5'-phosphate (PLP). These compounds were added as supplements to the *E. coli* growth medium, and their interference with DNA glycation and spontaneous mutagenesis was studied.

The *E. coli* strain AB1157 was cultured to a stationary phase in LB medium either in the presence or absence of a given inhibitor and the level of chromosomal DNA glycation was measured as above. In the figures in this subheading, for conciseness, we present data only about the early glycation adducts in DNA. We would like to emphasize, however, that the immunoreactivity of DNA to the anti-CML antibody as well as the amount of DNA fluorophores followed the trend of the early glycation adducts.

In parallel, the *E. coli* strain AB1157 was also investigated for the spontaneous reversion rate to Arg$^+$ phenotype. The arginine auxotrophy of this strain (*argE3*) is due to mutation in the chromosomal N-acetylornithinase gene. The Gln-338 codon CAA in this gene is converted to nonsense ochre (TAA) triplet (Kato and Shinoura, 1997; Meinnel et al., 1992). The reversion to arginine prototrophy in the N-acetylornithinase gene could be due to either base substitution at A:T base pairs in the ochre codon (back mutations) or to G:C →A:T transition and A:T →T:A, G:C→T:A transversions in several tRNA genes creating suppressor mutations (Sledziewska-Gojska et al., 1992). Another line of experimental evidence indicates that treatment of *E. coli* with glyoxal induces point mutations in the *lacI* gene by all types of base-pair substitutions with G:C→A:T transitions and G:C→T:A transversions being predominant (Murata-Kamiya et al., 1997). Because glycation by endogenous glyoxal may have similar consequences, the Arg$^+$ reversion appears appropriate for testing the effect of glycation inhibitors on DNA mutability. The values for the spontaneous reversion rate (SRR), as well as those for the content of early glycation adducts (EGA) in DNA, were normalized to that of control samples (no inhibitors). The reader may refer to the text in the figure legends, where the corresponding absolute values (EGAc and SRRc) for the control samples are given. Efficient suppression of glycation is usually achieved by applying the inhibitors at high concentrations, with a magnitude in the millimolar (mM) range (Booth et al., 1997; Khalifah et al. (2005). Aminoguanidine and acetyl salicylic acid are exogenous compounds to *E. coli*, whereas under physiological conditions the concentration of vitamins is kept low, within the micromolar (μM) range and below. For these reasons, we first explored all six compounds for toxicity at high concentrations (≥ 1mM). Depending on the impact on the bacterial growth, DNA glycation and spontaneous mutagenesis, the inhibitors were allocated to three two-member groups as follows: thiamine and pyridoxamine, aminoguanidine and pyridoxine, and acetyl salicylic acid and pyridoxal 5'-phosphate.

Thiamine and Pyridoxamine

We found that thiamine (B1) and pyridoxamine (PM) are well tolerated by bacteria at high concentrations. At the highest concentration tested (15 mM) thiamine proved to be the better inhibitor. It caused a tenfold decrease in the content of early glycation adducts in DNA and only about twofold decline in the spontaneous mutation rate. At the concentration of 10 mM pyridoxamine inhibited DNA glycation by ~30%, but had little effect on the spontaneous reversion rate to arginine prototrophy (Figure 3).

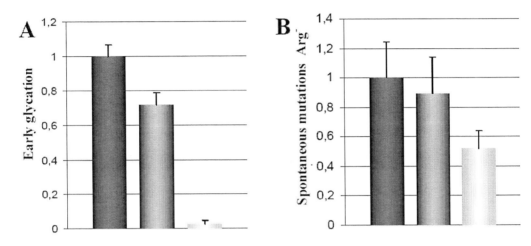

Figure 3. Impact of pyridoxamine and thiamine on DNA glycation and spontaneous mutagenesis in *E. coli* AB1157. Early glycation adducts in DNA (A) and rate of spontaneous reversions to Arg+ phenotype (B) of control *E. coli* cells (no inhibitors) (■) and of cells treated either with 10 mM PM (■) or 15 mM thiamine (B1)(□); EGAc = 1.5 ± 0.04 nmol/5 mg DNA (n=3); SRRc = 1.4 ± 0.3 Arg+ cells per 108 cells (n=5).

In view of the demonstrated efficacy, thiamine was further explored. Bacteria were continuously subcultured in LB medium supplemented with 15 mM thiamine. In the time course of four consecutive days bacteria were transferred to fresh thiamine supplemented media twice a day. At the end of each individual passage the optical density of the culture was measured, whereas the spontaneous mutation rate to Arg$^+$ phenotype was determined at the end of the whole culturing period. As seen in Figure 4, the prolonged growth of bacteria in nutrient broth with high thiamine content resulted in accelerated bacterial growth (Figure 4A) and additional decline in the spontaneous mutation rate (Figure 4B).

Experimental and clinical data reveal vitamin B1 and PM as promising candidates for intervention against the Maillard reaction *in vivo*. Early studies have shown that among various vitamin B1 and B6 derivatives, thiamine pyrophosphate and pyridoxamine potently inhibit the post-Amadori pathway of AGE formation *in vitro* and are more effective than the favorite compound aminoguanidine (Booth et al., 1996). This conclusion found further support in a mechanism-based approach *in vitro*, and the role of AGE breaker for thiamine has been proposed (Booth et al., 1997). Thiamine proved to be highly beneficial also *in vivo* and studies of P. Thornalley and co-workers provided first insights into its inhibitory mechanism.

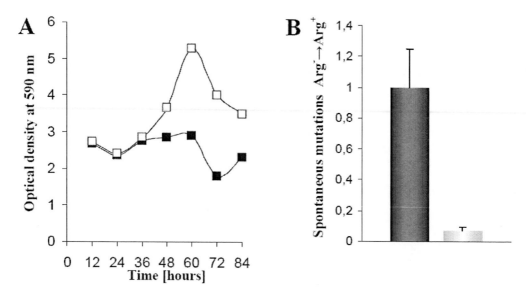

Figure 4. Effect of prolonged thiamine treatment on *E. coli* AB1157 growth and spontaneous reversion rate to arginine prototrophy. Optical density (OD_{590}) of stationary *E. coli* subcultures (A) and spontaneous reversion rate to Arg^+ phenotype (B) following four days of growth in LB medium (■) supplemented with 15 mM thiamine (□); SRRc = 1.4 ± 0.4 Arg^+ cells *per* 10^8 cells (n=5).

They observed normalization of the thriosephosphate pool in hyperglycemic cultures of human erythrocytes upon treatment with high dose of thiamine due to activation of the reductive pentosephosphate pathway (Thornalley et al., 2001). The thiamine derivative, benfothiamine, was found to correct increased AGE generation in endothelial cells, cultured in high glucose concentration (Pomero et al., 2001).

These *in vivo* effects of thiamine have been directly linked to its function of a cofactor for the enzyme transketolase, which shunts the thriosephosphate pool to the pentose pathway, thus preventing the accumulation of the reactive α-oxoaldehyde methylglyoxal (Thornalley et al., 2001). It appears likely that thiamine plays *in vivo* also a role in detoxification of glyoxal and methylglyoxal by stimulating the NAD(P)H dependent aldose/ketose reductases (Shangari et al., 2003). In consistency with *in vitro* studies, experiments with diabetic rats have demonstrated that high-dose thiamine and benfothiamine treatment inhibits the development of incipient nephropathy (Babaei-Jadidi et al., 2003), decreases the levels of AGEs in renal glomeruli, retina, peripheral nerve and plasma protein (Karachalias et al., 2003), and prevents experimental diabetic retinopathy (Hammes et al., 2003). Benfothiamine, a lipophylic thiamine derivative (S-benzoylthiamine-O-monophosphate) with better bioavailability upon oral administration, was tested in pilot studies for prevention of diabetic polyneuropathy, where it demonstrated encouraging results (Stracke et al., 1996; Simeonov et al., 1997; Sadekov et al., 1998; Winkler et al., 1999; Haupt et al., 2005).

At present, pyridoxamine is, perhaps, the most extensively studied glycation inhibitor. PM is formed *in vivo* in minute amounts by transamination and dephosphorylation of another B6 vitamer, pyridoxal 5'-phosphate (PLP). Booth and co-workers (1996) have developed an elegant method for avoiding interference of non-reacted free sugars, Schiff bases and their derivatives with the post-Amadori pathway. By using this mechanism-based approach, called "interrupted glycation", they have demonstrated that PM is a unique inhibitor of the post-Amadori AGE formation, which lowers both the rate of AGE appearance and the final levels of AGEs in model proteins (Booth et al., 1997). The authors suggested using the term "Amadorins" for such inhibitors (Khalifah et al., 1999). Experimental evidence indicates that PM inhibits the post-Amadori pathway by binding catalytic redox metal ions, thus blocking the oxidative degradation of Amadori intermediates to reactive carbonyl species (Voziyan et al., 2003; Voziyan and Hudson, 2005). PM has been shown also to scavenge the highly toxic dicarbonyls glyoxal, methilglyoxal, glycoaldehyde and 3-deoxyglucosone (Voziyan et al., 2002; Nagaraj et al., 2002; Amarnath et al., 2004; Chetyrkin et al., 2008a), and to inhibit the generation of reactive oxygen species (Jain and Lim, 2001; Kannan, and Jain, 2004; Chetyrkin et al., 2008b). Possibly, PM may interfere also with early stages of the Maillard reaction by interacting with glucose (Adrover et al., 2005). Preclinical studies revealed PM as a promising candidate for treatment of diabetic complications (Degenhardt et al., 2002; Stitt et al., 2002; Alderson et al., 2004; Waanders et al., 2008). The pharmaceutical formulation of PM, Pyridorin^{TM}, passed successfully Phase II clinical trails for treatment of diabetic nephropathy (McGill et al., 2004; Williams et al., 2007).

Thiamine and pyridoxamine proved to be the best inhibitors also in our study. The two vitamins were non-toxic at high concentrations, and lowered both the level of DNA glycation and the frequency of spontaneous mutations in *E. coli*. Therefore, one could expect that, besides other benefits, the clinical application of thiamine and PM at high dosages may result in better protection of the human genome against glycation and instability. Indeed, at 1 mM PM has been shown to protect DNA from damage, caused by high glucose concentration in cultured human umbilical vein endothelial cells (Shimoi et al., 2001). Interestingly, in this case PM prevented carbonyl (glyoxal) rather than oxidative damage of DNA. This latter result favors the idea that glycation deserves more attention as a possible source of DNA lesions. At high concentrations (>1 mM) the other four compounds, aminoguanidine, pyridoxine, acetyl salicylic acid and pyridoxal 5'-phosphate, significantly suppressed the *E. coli* growth. At the non-toxic concentration of 0.1 mM all four compounds inhibited DNA glycation, whereas lower concentrations were inefficient. The reason, the four inhibitors to fall within different groups, was their impact on DNA mutability.

Aminoguanidine and Pyridoxine

At the non-toxic concentration of 0.1 mM aminoguanidine (AG) and pyridoxine (PN) exhibited similar effects. They suppressed both DNA glycation and spontaneous mutagenesis in *E. coli* (Figure 5). Note that PN suppressed DNA mutability more efficiently at 1 mM rather than at 0.1 mM (indicated with an asterisk in Figure 5B). In addition, PN prove to be more toxic than AG. At 10 mM concentration PN did not only inhibit bacterial growth but displayed also a mutagenic effect (data not shown). It is worth mentioning that we always observed a good correlation between the rate of bacterial growth and the level of DNA glycation. High levels of DNA glycation definitely meant suppressed bacterial growth and increased DNA mutability. Therefore, the level of DNA glycation appears a sensitive "diagnostic" marker reflecting subtle changes in the bacterial physiology.

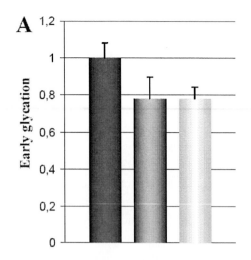 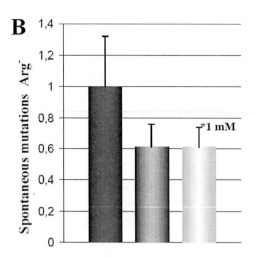

Figure 5. Impact of aminoguanidine and pyridoxine on DNA glycation and spontaneous mutagenesis in *E. coli* AB1157. Early glycation adducts in DNA (A) and rate of spontaneous reversions to Arg+ phenotype (B) of control *E. coli* cells (no inhibitors) (■) and of cells treated either with 0.1 mM AG (■) or 0.1 mM PN (□); EGAc = 1.7 ± 0.14 nmol/5 mg DNA (n=3); SRRc = 1.4 ± 0.4 Arg+ cells per 108 cells (n=5).

The inhibitory effect of AG and PN on DNA mutability at low concentrations, 0.1 mM and 1 mM respectively, was comparable to that of thiamine at the significantly higher concentration of 15 mM. In view of this fact, AG and PM could be considered more potent inhibitors than vitamin B1. However, the real inhibitory effect, achieved with all three inhibitors, independently of their potency, was only twofold decrease in the spontaneous mutation rate. Note also that during a single passage of bacteria in thiamine supplemented broth the level of early glycation adducts in DNA felt tenfold (Figure 3A), whereas the spontaneous reversion rate to Arg^+ dropped only two times (Figure 3B). These observations led us to conclude that glycation accounts for nearly half the rate of the spontaneous mutations in *E. coli*, even though glycation of DNA seems to be more robust. Probably, most glycation lesions in DNA are not fixed but rather mended by repair. Aminoguanidine is the "oldest" glycation inhibitor. Over many years it was the lead compound in the attempts to suppress the Maillard reaction *in vitro* as well as in animal models of diabetes. This exogenous chemical seemed to be promising because of its chemical structure specifying it as both hydrazine and guanidine. These two functionalities make AG capable of competing with other amines for binding glucose (as a hydrazine) or α,β-dycarbonyls (as a guanidine). In fact, it has been proven that AG binds efficiently α-oxoaldehydes like glyoxal, methylglyoxal and 3-deoxyglucoxone to form amino-triazine derivatives (Thornalley et al., 2000).

In early preclinical studies AG has been shown to ameliorate symptoms in diabetic rats (Brownlee et al., 1986; Soulis-Liparota et al., 1991; Edelstein and Brownlee, 1992; Osicka et al., 2000), although later studies revealed PM as a more potent inhibitor (Degenhardt et al., 2002). There is also experimental evidence that AG interacts with pyridoxal 5'-phosphate under both *in vitro* and *in vivo* conditions and thus may cause vitamin B6 deficiency (Okada and Ayabe, 1995; Taguchi et al., 1998). Despite this evidence, AG (Pimagedine) was introduced into clinical trials for prevention of diabetic nephropathy (Bolton et al., 2004), but withdrawn in the crucial Phase III because of neurotoxicity (Jakus and Rietbrock, 2004). On one side, the high carbonyl trapping activity of AG makes it a perfect glycation inhibitor *in vitro*. On the other, however, the high affinity of AG for life essential carbonyls like PLP renders it inapplicable *in vivo*. Thus, inhibition of glycation in living systems appears a serious challenge, which has to be faced by scientists working today in the Maillard field.

At high concentrations (> 1 mM) AG proved to be highly toxic also in our study. At concentrations higher than 1 mM, PN was found to be even more toxic than AG. At the concentration of 10 mM PN exhibited mutagenic effect in *E. coli* similar to that of the last group inhibitors, ASA and PLP. Early studies revealed PN as a less potent glycation inhibitor than other B6 derivatives (Boot et al., 1996, 1997), and little attention has been paid thereafter to this B6 vitamer. In addition, *in vivo* PN undergoes enzymatic conversation to PM and PLP, which are the true biologically active forms of vitamin B6. Therefore, we assume that the observed genotoxicity of PN at high concentrations may be linked to its partial transformation into PLP *in vivo* (see below).

Acetyl Salicylic Acid and Pyridoxal 5'-Phosphate

In contrast to AG and PN, both ASA and PLP inhibited DNA glycation at the non-toxic concentration of 0.1 mM but demonstrated an opposite effect on the mutation frequency to arginine prototrophy causing an increase in the reversion rate to this phenotype. Furthermore, the higher was the extent of DNA glycation inhibition, the pronounced was the rise in the mutation frequency (Figure 6).

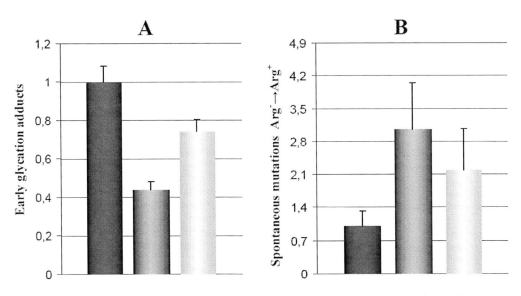

Figure 6. Impact of acetyl salicylic acid and pyridoxal 5'-phosphate on DNA glycation and spontaneous mutagenesis in *E. coli* AB1157. Early glycation adducts in DNA (A) and rate of spontaneous reversions to Arg+ phenotype (B) of control *E. coli* cells (no inhibitors) (■) and of cells treated either with 0.1 mM ASA (■) or 0.1 mM PLP (□). EGAc = 1.6 ± 0.13 nmol/5 mg DNA (n=3); SRRc = 1.4 ± 0.4 Arg+ cells per 10^8 cells (n=5).

Aspirin was found to be more toxic than PLP. At concentration 10 mM it suppressed bacterial growth completely, whereas in the presence of 10 mM PLP the growth of *E.coli* was only partially affected (data not shown). In our study, ASA was the second (following thiamine) potent glycation inhibitor. It inhibited DNA glycation by 60%, whereas the inhibition by thiamine reached 90%. The inhibitory activity of PLP was comparable to that of all other inhibitors, PM, PN and AG and did not exceed 20%.

What were the reasons for the increased DNA mutability, when bacteria were grown in media supplemented either with ASA or PLP? One could suggest that high ASA and PLP concentrations are stressful for bacterial cells and therefore trigger the mechanisms of adaptive mutagenesis (Foster, 2005; Foster, 2007). However, we could not rule out *a priori* the possibility that ASA and PLP exhibit direct mutagenic effect by interacting with DNA. Next experiments were designed to address this issue.

At High Concentrations Pyridoxal 5'-Phosphate and Aspirin May Interact with DNA

In order to explore the proposed interaction between ASA/PLP and DNA, plasmid pUC18 DNA was incubated with different concentrations of both glycation inhibitors. Agarose gel electrophoresis was performed to reveal a possible conversation of the covalently closed circular (ccc) plasmid DNA into a relaxed form. It is well known that chemical modifications affecting the DNA bases result in destabilization of the N-glycosidic bond. This promotes removal of the modified bases from the deoxyribose residues and creation of abasic lesions. In turn abasic DNA also becomes unstable and undergoes carbohydrate-phosphate backbone cleavage. Relaxation of the ccc form of plasmid DNA is a sensitive tool for establishing hydrolytic activity, because one nick *per* ccc plasmid molecule is sufficient to cause irreversible plasmid relaxation.

Pyridoxal 5'-Phosphate

Aqueous solutions of PLP (0.01 mM to 10 mM) adjusted to pH 7.5 with NaOH in order to avoid acidic hydrolysis of DNA were used and plasmid DNA was treated at 37°C for 12 h. We observed a substantial plasmid relaxation at high (10 mM) PLP concentrations only (Figure 7). PLP is a strong electrophilic compound, capable of forming Schiff bases with free NH_2 groups in proteins (Ganea et al., 1994; Ganea and Harding, 1996) and aminoguanidine (Taguchi et al., 1998). Three of the DNA bases (A, G and C) also contain free NH_2 groups and we supposed that the latter, though less nucleophilic than the NH_2 groups of amino acids and aminoguanidine, may interact with PLP in a similar manner. We further argued that, if PLP reacts with NH2 groups in DNA, this reaction should be inhibited in the presence of other, competing amines such as TRIS (trishydroxymethylaminomethane).

Following the above assumptions, we incubated pUC18 DNA in 10 mM PLP solutions adjusted to pH 7.5 either with TRIS or NaOH, and analyzed the change in the electrophoretic profile of the plasmid over eight consecutive days. In accordance with our hypothesis, after eight days of incubation in the NaOH titrated PLP solution, the plasmid DNA was almost fully relaxed (Figure 8A), whereas TRIS caused retardation of plasmid DNA relaxation (Figure 8B).

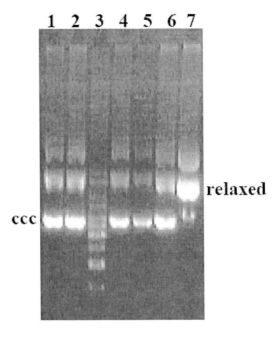

Figure 7. Effect of PLP on the electrophoretic profile of pUC18 plasmid DNA. Plasmid DNA (2 μg/lane) either stored at -20°C (1) or incubated at 0,2 mg/ml for 12 h at 37°C (2) in PLP solutions with pH 7.0 and concentrations 0.01 mM (4), 0.1 mM (5), 1 mM (6) and 10 mM (7); molecular standard (3).

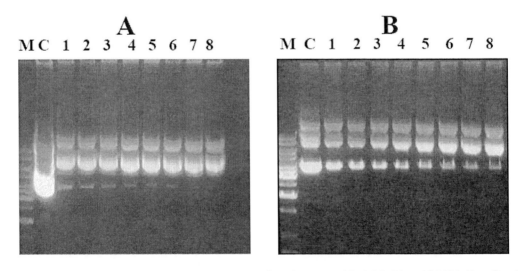

Figure 8. Effect of TRIS on plasmid pUC18 DNA relaxation caused by PLP. Plasmid DNA (2 μg/lane) (C) was incubated at 0.2 mg/ml for 8 days at 37°C in 10 mM PLP solutions with pH 7.5 either in the absence (A) or presence of 50 mM TRIS (B) for 1 (1), 2 (2), 3 (3), 4 (4), 5 (5), 6 (6), 7 (7) or 8 (8) consecutive days; M – molecular standard.

In searching for additional proof that PLP reacts with DNA by its aldehyde group, we incubated pUC18 DNA with B6 vitamers containing (pyridoxal, PL) or devoid of (PN and PM) an aldehyde group. Recall that at 10 mM concentration, PM, lacking a carbonyl moiety did not exhibit mutagenic effect (Figure 3B). Figure 9 demonstrates that, although less pronounced, PL also caused plasmid DNA relaxation. The stronger chemical reactivity of PLP as compared to that of PL has been documented in many studies (Christmann-Franck et al., 2007; Mizushina et al., 2003), and could be attributed to a catalytic effect of the phosphate group in PLP. Importantly, the plasmid relaxation activity of PL, like that of PLP, was suppressed by TRIS. The PLP trapping activity of TRIS has already been reported, and in this regard it is recommended TRIS to be avoided for buffering of enzyme solutions containing PLP as a cofactor (Michael et al., 1982).

We did not observe plasmid DNA relaxation in solutions prepared with the B6 vitamers PM and PN that lack an aldehyde group (Figure 10). Taken together, these results support our suggestion that PLP, through its aldehyde group, forms Schiff bases with free amino groups in DNA.

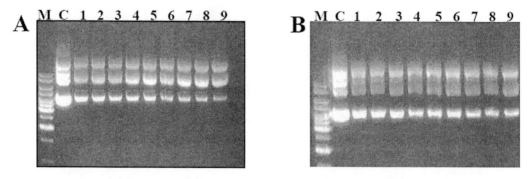

Figure. 9. Effect of TRIS on plasmid pUC18 DNA relaxation caused by PL. Plasmid DNA (2 µg/lane (C) was incubated at 0.2 mg/ml for 9 days at 37°C in 10 mM PL solutions with pH 7.5 either in the absence (A) or presence of 50 mM TRIS (B) for 1 (1), 2 (2), 3 (3), 4 (4), 5 (5), 6 (6), 7 (7), 8 (8) or 9, (9) consecutive days; M – molecular standard.

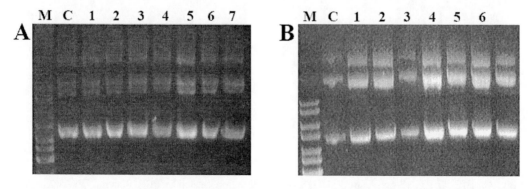

Figure 10. Effect of PN and PM on the electrophoretic profile of pUC18 plasmid DNA. Plasmid DNA (0.2 mg/ml) (C) was incubated at 37°C in 10 mM solutions (pH 7.0) of PN (A) or PM (B) for 1 (1), 2 (2), 3 (3), 4 (4), 5 (5), 6 (6) or 7 (7) consecutive days; M – molecular standard.

We further explored the mutagenic potential of PLP by transforming the PLP treated plasmid pUC18 DNA into *E. coli* cells. Plasmid DNA damage, caused by PLP, may result in decreased transformation efficacy and inactivation of the plasmid borne *lacZ'* gene. To evaluate the rate of mutations in the *lacZ'* gene, we used the α-complementation assay. Briefly, plasmid pUC18 carries the very 5'end of the *E. coli lacZ* gene, designated as *lacZ'*. The *lacZ* gene encodes the enzyme β-galactosidase, which catalyzes the breakdown of lactose to glucose and galactose. The 3'end of the same gene is encoded by the host genome of some strains (XL1-Blue, DH5α) with deleted chromosomal *lacZ* gene. Upon transformation of such strains with pUC18 both parts of the gene are brought together within the same cell, where they restore the β-galactosidase activity. Transformed bacteria are spread on a special solid broth, containing an inducer of the *lac* operon (IPTG) and a chromogenic substrate (X-gal). On such medium LacZ$^-$ colonies are colored in white, whereas LacZ$^+$ colonies turn blue. Mutations in the pUC18 borne *lacZ'* gene will result in increased number of white colonies.

Plasmid pUC18 DNA was incubated with PLP and transformed into the *E. coli* strain XL1-Blue. In this experiment, with the PLP treated plasmid, we observed about three times drop in the transformation efficiency and three times increase in the percentage of white colonies, as compared to the control plasmid (no PLP treatment) (Figure 11).

This result is consistent with the DNA hydrolytic activity of PLP observed *in vitro*, and additionally demonstrates the mutagenic potential of PLP. The successful clinical application of PM (Pyridorin^{TM}) was the reason for researchers to undertake investigations with the B6 vitamer PLP as an alternative for glycation inhibition *in vivo*. Our investigations, however, indicate that at high concentrations PLP might be mutagenic. Though being a natural compound, PLP appears a "bed" glycation inhibitor, because of its mode of action. Through its aldehydic functionality PLP competes with deleterious carbonyls for binding cellular amines including DNA. However, for DNA it does not matter whether it binds sugars or vitamins.

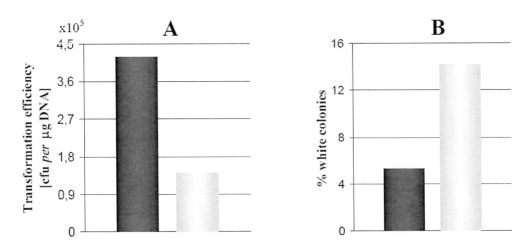

Figure 11. Effect of PLP on the transformation efficiency (A) and the rate of mutations in the lacZ' gene (B). *E coli* strain XL1-Blue was transformed with control plasmid pUC18 DNA (■) or with plasmid DNA treated with 10 mM PLP (pH 7.0) (□). PLP treatment was carried out at 37єC for 24 hours at plasmid DNA concentration of 0.2 mg/ml.

Chemical "vitamination" of DNA could be as deleterious as chemical glycation as both result in increased DNA mutability. Based on these results, we suggest that the clinical intervention with high dosages PLP may cause undesirable adverse events.

Acetyl Salicylic Acid

Acetyl salicylic acid (ASA) did not exhibit DNA hydrolytic activity within 24 hours of plasmid DNA incubation at 37°C at none of the concentrations tested (data not shown). Aspirin is an old medicine used for a long time in the experiments for inhibition of the Maillard reaction *in vitro* (Huby and Harding, 1988; Swamy and Abraham, 1989; Ajiboye and Harding, 1989) as well as *in vivo* (Blakytny and Harding, 1992; Cotlier, 1981). Early experiments have shown that in the presence of ASA proteins become irreversibly acetylated, whereas ASA binds weakly to DNA, most probably through hydrogen bonds (Pinckard et al., 1968). Later studies using Raman and infrared spectroscopy confirmed this observation demonstrating that ASA forms a complex with DNA (Neault et al., 1996). At a low ASA/DNA molar ration (1:40), ASA interacted mainly with the phosphate backbone and A:T base pairs. In this interaction adenine and thymine took part through the N-7 and O-2 atoms respectively, which are not normally involved in Watson-Crick hydrogen bonding. By stepwise increasing the ASA concentration with respect to DNA up to 2:1 molar ratio, in reaction with ASA were involved also the other DNA bases, guanine (through N-7) and cytosine. Alterations of the B-DNA structure towards A-DNA and helix destabilization were also observed. The authors suggested that ASA interacted with DNA through both the carboxylic anion CO^- and acetyl group ($COOCH_3$) donor atoms with those of the backbone phosphate groups and DNA bases donor sites (directly or indirectly *via* H_2O molecules) (Neault et al., 1996). The weak interaction of ASA with DNA could explain why we did not observe relaxation of the plasmid DNA. The probable interaction of ASA remained hidden in the plasmid relaxation assay used in our study.

We performed an additional experiment with ASA. In the above experiments the solutions of ASA were adjusted to pH 7.0 with NaOH, whereby the labile ester group in ASA is hydrolyzed to form the sodium salts of two acids, sodium acetate and sodium salicylate:

$$CH_3COO.C_6H_4.COOH + 2NaOH \rightarrow CH_3COONa + HO.C_6H_4.COONa + H_2O$$

In order to avoid deacetylation of ASA, we incubated plasmid pUC18 DNA in an acidic (not titrated) 10 mM ASA solution. The low pH of this solution (3.4) made acidic hydrolysis of plasmid DNA unavoidable. To evaluate the effect of the acidic DNA hydrolysis *per se*, we incubated the plasmid DNA also in 2 mM HCl solution having the same pH (3.4). Incubations were carried out at 37°C for either 4 or 30 hours and plasmid DNA was analyzed by agarose gel electrophoresis.

As shown in Figure 12, the hydrolysis of plasmid DNA in the ASA solution was accelerated. This result indicates that plasmid DNA decay is not simply due to acidic hydrolysis. Most probably, the structural alterations in DNA including helix distortions, as proposed by Neault et al. (1996), render DNA more susceptible to hydrolysis.

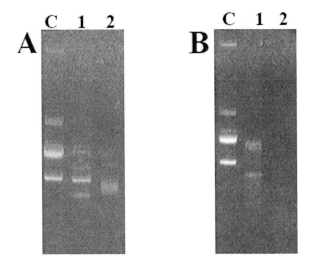

Figure 12. Effect of ASA and HCl at pH 3.4 on the electrophoretic profile of plasmid pUC18. Plasmid DNA (C) was incubated at 1 mg/ml for 4 h (A) or 30 h (B) at 37°C either in 2 mM HCl (pH 3, 4) (1) or 10 mM ASA (pH 3, 4) (2).

What is the exact mechanism of ASA-DNA interaction, how this interaction hinders DNA glycation but increases DNA mutability – these are important issues, which remain to be elucidated. It is worth mentioning that the therapeutic concentration of ASA in the human blood is considered to be about 1 mM. In our experiments, at this concentration in the bacterial broth, ASA caused a remarkable increase in the frequency of the mutations to Arg prototrophy (data not shown). Although these results, obtained with bacteria, could not be directly extrapolated to humans, they strengthen our previous conclusion that glycation inhibitors having the potential to interact with DNA and other biologically important amines deserve special concerns when applied at high concentrations *in vivo*.

E. coli Might Repair Glycated DNA by Using DNA Amadoriase(s)

We estimated that the chromosomal DNA isolated from stationary *E. coli* cells contains approximately one fructosamine residue *per* 10^4 to 10^5 nucleotides. This is a quite high concentration of chemical modifications, having in mind the fidelity of DNA replication, which approaches roughly 1 wrong base per 10^6 nucleotides. Data presented in this study as well as reported by others (Murata-Kamiya et al., 1998; Murata-Kamiya et al., 1999) point to the possible involvement of NER in repair of glycated DNA. However, one should wonder if *E. coli* does not possess special mechanisms to counteract involvement of DNA in the Maillard reaction at its early stages.

Amadoriases are a class of enzymes, discovered about 20 years ago (Horiuchi et al., 1989; Petersen et al., 1990; Szwergold et al., 1990; Horiuchi and Kurokawa), which catalyze the decomposition of Amadori products. These enzymes utilize different mechanisms to do so. For example, amadoriases can catalyze either hydrolytic or oxidative cleavage of aldoamines to composite amines and sugars (or their derivatives) or they can even act as kinases (Thornalley, 2003; Wu and Monnier, 2003; Monnier, 2005). In this latter case, phosphorylation of AP at the sugar moiety destabilizes the link between the sugar and the amine, which then spontaneously breaks down. This mode of action is typical for the human amadoriase, which represents actually a fructosamine-3-kinase (Delpierre et al., 2000a; Delpierre et al., 2000b; Szwergold et al., 2001). At the time, we started our investigations on DNA glycation in *E. coli*, Van Schaftingen and co-workers reported the identification of a new operon in *E. coli*, later called the fructoselysine (*frl*) operon (Wiame et al., 2002). The key gene in this operon, the *frlB* gene, encodes an amadoriase (FrlB), which catalyzes the hydrolysis of N^ε-fructoselysine 6-phosphate to lysine and glucose 6-phosphate. The FrlB amadoriase acts in concert with the FrlD kinase, which first phoshporylates the fructose moiety in N^ε-fructoselysine at the C-6 position. In a later publication, Van Schaftingen and co-workers reported that, in contrast to the *Bacillus subtilis* ortholog, the *E. coli* FrlB amadoriase is highly specific for AP linked to the ε-amino group of lysine but not to the α-amino groups of other amino acids (Wiame et al., 2004). Except for free amino acids, the FrlB amadoriase has not been tested for activity towards other amines, including DNA and nucleotides. In order to fill in this gap, we undertook the experiments, described below in this chapter. We cloned the *frlB* gene in a multicopy plasmid (pJP₁R₉) under a strong constitutive promoter (P1) and a consensus ribosome binding site (R9). We constructed also an *E. coli* knock out (KO) strain with a deleted *frlB* gene. This KO strain, isogenic to the *E. coli* strain BW28357 (Zhou et al., 2003), was designated BW37935. The *frlB* KO strain was transformed with the recombinant plasmid pJP1R9-*frlB* and the overexpression of the *frlB* amadoriase was confirmed by SDS-PAGE performed under reducing conditions. The frlB gene product is a polypeptide of approximately 39 kDa and the corresponding band is well visible on the gel in Figure 13. Later experiments allowed us to calculate that the recombinant amadoriase constitutes over 40% of the total bacterial protein.

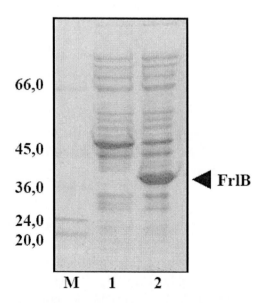

Figure 13. Overexpression of the FrlB amadoriase in *E. coli*. SDS-PAGE (12% gel) of total protein from *E. coli* strain BW37935 (1), transformed with the plasmid pJP1R9-frlB (2); M –molecular standard in kDa.

Van Schaftingen and co-workers used a procedure for measurement of the FrlB enzyme activity based on the reversal of the reaction at high, non-physiological concentrations of glucose 6-phosphate and lysine (Wiame et al., 2002). We used the same approach with some modification. In order to avoid working with radioactively labeled glucose 6-phosphate, we added this substrate at a high concentration (0.15 M) to the reaction mixture. This allowed us to monitor the reaction by applying the colorimetric NBT reduction assay for measurement of the reaction product, N^{ε}-fructoselysine 6-phosphate. Reaction mixtures were directly mixed with the NBT reagent and the absorption at 525 nm was read over time. Figure 14 shows the kinetic of the NBT reduction by reaction mixtures differing in the concentration of the *E. coli* total protein from cells overexpressing the FrlB amadoriase. For all protein concentrations tested, the NBT reduction followed a good linearity over 20 minutes (Figure 14A). The specificity of the reaction was confirmed by the lack of NBT reduction, when the enzyme assay was performed with total protein (200 µg/ml) from the *E. coli* KO strain BW37935. The rate of NBT reduction (ΔA_{525}/min) as dependent on the *E. coli* protein concentration also demonstrated an excellent linearity ($R^2=0,998$) (Figure 14B). In all subsequent analyses we worked with 100 µg/ml total *E. coli* protein in the reaction mixtures and calculated the rate of the NBT reduction (ΔA_{525}/min) as a measure for the enzyme activity.

The FrlB enzyme assay was performed by using either α-Boc-lysine or DNA as NH_2-substrates. We used salmon sperm DNA at a concentration of 25 mg/ml (~0.1 M nucleotides) which, for better solubility, was sonicated to an average size of about 2000 bp. The enzyme assay was carried out with total *E. coli* protein isolated either from BW37935/pJP$_1$R$_9$-*frlB* or *flrB* KO cells. Expectedly, when using α-Boc-lysine in the reaction, the total protein from the BW37935/pJP1R9-*frlB* cells demonstrated FrlB enzyme activity in contrast to the protein from the KO strain. However, we were quite surprised to find that the protein from both types of cells exhibited similar enzyme activity, when DNA was substituted for α-Boc-lysine in the reaction mixture.

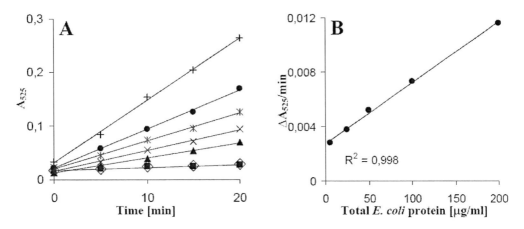

Figure 14. Enzyme activity of the FrlB amadoriase. (A) Kinetics of the NBT reduction by reaction mixture containing 0 (■), 5 (▲), 25 (x), 50 (*), 100 (●) or 200 (+) µg/ml total protein from *E.coli* BW37935/pJP1R9-frlB or 200 µg/ml total protein from *E. coli* BW37935 (◊). (B) Velocity of the NBT reduction as dependent on the concentration of the total protein from *E. coli* BW37935/pJP1R9-frlB.

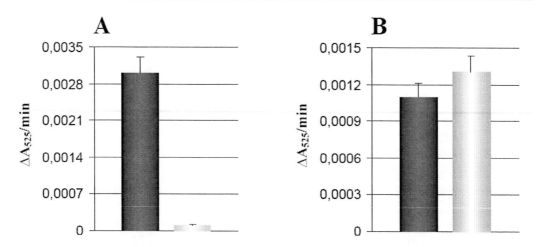

Figure 15. Activity of the FrlB amadoriase with α-Boc-lysine (A) and DNA (B) as NH2-substrates. Rate of NBT reduction by reaction mixtures containing total protein (100 μg/ml) from *E. coli* BW37935/pJP1R9-frlB (■) and *E. coli* BW37935 (□). Data are mean ± SD (n=3).

This result clearly demonstrates that the FrlB amadoriase does not recognize DNA as a substrate. Next conclusion that follows is that *E. coli* possesses another, different from the FrlB enzyme activity capable of catalyzing the attachment of glucose 6-phoshate to DNA. In the FrlB enzyme assay we used high, non-physiological concentration of glucose 6-phosphate. Therefore, we suppose that *in vivo* the observed activity most probably catalyses the removal of glucose 6-phosphate bound to DNA in the form of fructosamine 6-phosphate (Amadori product). We called this newly established enzyme activity a putative DNA amadoriase or deglycase. Following discovery of the amadoriases, the term "protein repair" has gain increasing popularity in the scientific literature. However, the existence of amadoriases, especially designed for repair of glycated DNA, has not been either proposed or demonstrated so far.

In the next experiment, we performed the enzyme assay with DNA instead of α-Boc-lysine as NH$_2$-substrate, and total protein form five *E. coli* strains, grown either in rich (LB) or minimal (M9G) medium. Figure 16 shows that all strains exhibited enzyme activity only when they were grown in LB medium. In a previous study we have found that DNA accumulates more glycation adducts, when bacteria are grown in rich rather than in minimal medium (Mironova et al., 2005). Therefore, the putative DNA amadoriase seems to be an inducible enzyme, which is switched on only when bacteria are exposed to enhanced risk of DNA damage by carbonyl compounds. In addition, the repair deficient strain 2410 demonstrated lower activity than all other strains and thus appears an interesting strain for further studies on the putative DNA amadoriase(s).

It is well known that DNA polymerases may incorporate into newly synthesized DNA chemically modified nucleotides. In fact, many methods for *in vitro* DNA labeling as well as the DNA sequencing procedure (Sanger et al., 1977) are based on this property of DNA polymerases. Inside cells, however, the incorporation of modified nucleotides into DNA is mutagenic and therefore should be precluded. Although most nucleotide modifications are removed at a DNA level, mechanisms also do exist for processing free chemically modified nucleotides.

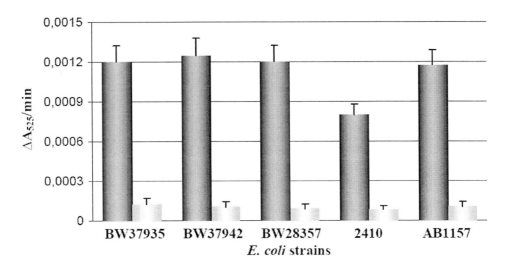

Figure 16. Putative DNA amadoriase activity of different E. coli strains. Velocity of NBT reduction by reaction mixtures containing total protein (100 µg/ml) from different E. coli strains cultured either in rich LB (■) or minimal M9G (■) medium. Data are mean ± SD (n=3).

For example, the *E. coli* MutT enzyme and its mammalian ortholog MTH1 hydrolyze 8-oxo-dGTP in the dNTPs pool to monophosphate thus preventing the incorporation of this oxidized nucleotide in DNA (Tsuzuki et al., 2007). That is why we asked the question whether the putative DNA amadoriase can utilize free nucleotides as substrates. In order to answer this question, we performed the amadoriase assay with the three NH_2 group containing nucleotides - dAMP, dCMP и dGMP.

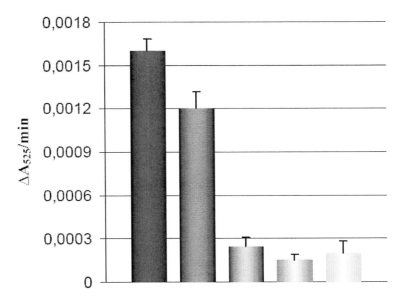

Figure 17. Substrate specificity of the putative DNA amadoriase. Velocity of NBT reduction by reaction mixtures containing total protein (100 µg/ml) from *E. coli* BW37935 and native DNA (■), denatured DNA (■), dGMP (■), dCMP (■) or dAMP (□) as NH_2-substrates. Data are mean ±SD (n=3).

In this experiment, we used also double strand DNA, either native or heat denatured, as NH_2-substrate. The putative DNA amadoriase exhibited highest activity towards native DNA and was almost inactive with dNMPs as substrates. Therefore, this putative enzyme seems to recognize Amadori products bound to a native double strand DNA rather than to nucleotides. Most base modifications in DNA are processed in this way by specific N-glycosidases as part of the BER machinery. It is believed that recognition of modified bases in native DNA relies on a base flipping mechanism, unmasking damaged bases and making them 'visible' for the cognate N-glycosidases. The putative DNA amadoriase may recognize AP modified DNA bases in a similar way and then catalyze the cleavage of the bond between the sugar and the amino group of the DNA bases. In other words, the enzyme may reverse the early step of the Maillard reaction, which ends up with the formation of relatively stable Amadori products. In this sense, the putative DNA amadoriase resembles the O^6-(methylguanine)-methyltransferase, which repairs DNA *in situ* by simply removing the methyl group from the affected base.

Conclusion

The present study reveals the carbonyl amine reaction, discovered a century ago by L. C. Maillard (1912), as an important endogenous source of DNA damage and mutability. The Maillard reaction affects nearly one *per* 10^5 to 10^4 nucleotides in the *E. coli* chromosome. This high magnitude of DNA damage is comparable to that imposed on DNA by base tautomerization. At the 9th International Symposium of the Maillard Reaction, P. Thornalley and co-workers reported that the concentration of a methylglyoxal-derived deoxyguanosine adduct in nuclear DNA of peripheral human mononuclear leukocytes is approximately five times higher than that of the oxidative DNA biomarker 8-oxo-dGTP (Felming et al. (2007). Therefore, it seems likely that the magnitude of DNA damage, caused by glycation, is even greater than that caused by oxidation. To date, more than 60 different oxidative DNA base lesions have been identified (Klungland and Bjelland, 2007). We suppose that the diversity of glycation lesions in DNA is even grater having in mind that three (A, G and C) of the four DNA bases are susceptible to glycation and that a cocktail of carbonyls floods the intracellular space.

In our study, by using glycation inhibitors, we were unable to suppress spontaneous mutation rate by more than twofold, even though early glycation could be inhibited up to tenfold. This fact tells us that the Maillard reaction accounts for nearly half of the spontaneous mutations in *E. coli* and that the majority of glycation adducts in DNA are not fixed but rather mended by repair. In contrast to oxidation, glycation introduces into DNA bulky chemical moieties that are most likely removed by the nucleotide excision repair. In addition to this common way for repair of chemically modified DNA, *E. coli* seems to have evolved special mechanisms to combat glycation. Based on results presented in this chapter, we postulate the existence in *E. coli* of a novel enzyme, DNA amadoriase (deglycase), designed to catalyze the removal of glucose 6-phosphate bound to DNA. Moreover, in view of the variety of cellular carbonyls that can bind DNA, there should be a family of DNA amadoriases responsible for removal of different Amadori products from DNA. Such enzyme family probably exists also in eukaryotes. Identification of DNA amadoriases and elucidation of their role for maintaining genome integrity are important issues to be addressed in the future.

Finally, this study provides evidence for the impact on DNA glycation and mutability of the most commonly used glycation inhibitors for suppression of the Maillard reaction under physiological conditions *in vivo*. Generally, inhibitors like thiamine and pyridoxamine that do not display affinity towards DNA are safety for clinical intervention. However, the use of compounds like acetyl salicylic acid and pyridoxal 5'-phosphate at high concentrations *in vivo* may be risky. Although these compounds inhibit DNA glycation, they do so by interacting with DNA, and are therefore potentially mutagenic.

Acknowledgements

This work is supported by Contracts TK-B-1602/06 and B-1501/05 from the Bulgarian Ministry of Education and Science. B. L. W. is supported by NIH GM62662 from the U. S. Public health Service.

References

Adrover, M; Vilanova, B; Munoz, F; Donoso, J. Inhibition of glycosylation processes: the reaction between pyridoxamine and glucose, *Chem Biodivers*, 2005, 2: 964-975.

Ahmed, MU; Thorpe, SR; Baynes, JW. Identification of Nε-carboxymethyl-lysine as a degradation product of fructoselysine in glycated protein. *J. Biol. Chem.*, 1986, 261: 4889-4894.

Ahmed, N; Thornalley, PJ.Quantitative screening of protein biomarkers of early glycation, advanced glycation, oxidation and nitrosation in cellular and extracellular proteins by tandem mass spectrometry multiple reaction monitoring. *Biochem. Soc. Trans*, 2003, 31(Pt 6): 1417-1422.

Ahmed, N; Thornalley PJ. Advanced glycation endproducts: what is their relevance to diabetic complications? *Diabetes Obes Metab.*, 2007, 9: 233-245.

Ajiboye, R; Harding, JJ. The non-enzymic glycosylation of bovine lens proteins by glucosamine and its inhibition by aspirin, ibuprofen and glutathione. *Exp. Eye Res*, 1989, 49: 31-41. Erratum in: *Exp. Eye Res*, 1989, 49: 704.

Alderson, NL; Chachich, ME; Frizzell, N; Canning, P; Metz, TO; Januszewski, AS; Youssef, NN; Stitt, AW; Baynes, JW; Thorpe, SR. Effect of antioxidants and ACE inhibition on chemical modification of proteins and progression of nephropathy in the streptozotocin diabetic rat. *Diabetologia*, 2004, 47: 1385-1395.

Amarnath, V; Amarnath, K; Amarnath, K; Davies, S; Roberts, LJ 2nd. Pyridoxamine: an extremely potent scavenger of 1,4-dicarbonyls. *Chem. Res. Toxicol*, 2004, 17: 410-415.

Anet, EFLJ. Degradation of carbohydrates. I. Isolation of 3-deoxyhexosones. *Australian J. Chem.*, 1960, 13: 396-403.

Babaei-Jadidi, R; Karachalias, N; Ahmed, N; Battah, S; Thornalley, PJ. (2003) Prevention of incipient diabetic nephropathy by high-dose thiamine and benfotiamine. *Diabetes*, 2003, 52: 2110-2120.

Barnes, DE; Lindahl T. Repair and genetic consequences of endogenous DNA base damage in mammalian cells. *Annu. Rev. Genet*, 2004, 38: 445-476.

Baynes, JW. From life to death – the straggle between chemistry and biology during aging: the Maillard reaction as an amplifier of genomic damage. *Biogerontology*, 2000, 1: 235-246.

Baynes, JW. The Maillard hypothesis on aging: time to focus on DNA. *Ann. NY Acad Sci.*, 2002, 959: 360-367.

Blakytny, R; Harding, JJ. Prevention of cataract in diabetic rats by aspirin, paracetamol (acetaminophen) and ibuprofen. *Exp. Eye Res*, 1992, 54: 509-518.

Bolton, WK; Daniel C. Cattran, DC; Williams, ME; Adler, SG; Appel, GB; Cartwright, K; Foiles, PG; Freedman, BI; Raskin, Ph; Ratner, RE; Spinowitz, BS; Whittier, FC; Jean-Paul Wuerth, J-P. Randomized trial of an inhibitor of formation of advanced glycation end products in diabetic nephropathy. *Am. J. Nephrol*, 2004, 24: 32-40.

Bonsignore, A; Leoncini, G; Siri, A; Ricci, D. Kinetic behaviour of glyceraldehyde 3-phosphate conversion into methylglyoxal. *Ital. J. Biochem.*, 1973, 22: 131-140.

Bookchin, RM; Gallop, PM. Structure of hemoglobin AIc: nature of the N-terminal beta chain blocking group. *Biochem. Biophys. Res. Commun.*, 1968, 32: 86-93.

Booth, AA; Khalifah, RG; Hudson, BG. Thiamine pyrophosphate and pyridoxamine inhibit the formation of antigenic advanced glycation end-products: comparison with aminoguanidine. *Biochem. Biophys Res. Commun.*, 1996, 220: 113-119.

Booth, AA; Khalifah, RG; Todd, P; Hudson, BG. In vitro kinetic studies of formation of antigenic advanced glycation end products (AGEs). Novel inhibition of post-Amadori glycation pathways. *J. Biol. Chem.*, 1997, 272: 5430-5437.

Brownlee, M; Vlassara, H; Kooney, A; Ulrich, P; Cerami, A. Aminoguanidine prevents diabetes-induced arterial wall protein cross-linking. *Science*, 1986, 232: 1629-1632.

Brownlee, M. Advanced protein glycosylation in diabetes and aging. *Annu. Rev. Med*, 1995, 146: 223-234.

Bucala, R; Model, P; Cerami, A. Modification of DNA by reducing sugars: a possible mechanism for nucleic acid aging and age-related dysfunction in gene expression. *Proc. Natl. Acad. Sci. USA*, 1984, 81: 105-109.

Bucala, R; Lee, AT; Rourke, L; Cerami, A. Transposition of an Alu-containing element induced by DNA-advanced glycosylation endproducts. *Proc. Natl. Acad. Sci. USA*, 1993, 90: 2666-2670.

Cerami, A. Aging of proteins and nucleic acids: what is the role of glucose? *TIBS*, 1986, 11: 311-314.

Chetyrkin, SV; Zhang, W; Hudson, BG; Serianni, AS; Voziyan, PA. Pyridoxamine protects proteins from functional damage by 3-deoxyglucosone: mechanism of action of pyridoxamine. *Biochemistry*, 2008a, 47: 997-1006.

Chetyrkin, SV; Mathis, ME; Ham, AJ; Hachey, DL; Hudson, BG; Voziyan, PA. Propagation of protein glycation damage involves modification of tryptophan residues via reactive oxygen species: inhibition by pyridoxamine. *Free Radic. Biol. Med.*, 2008b, 44: 1276-1285.

Christmann-Franck, S; Fermandjian, S; Mirambeau, G; Der Garabedian, PA. Molecular modeling studies on the interactions of human DNA topoisomerase IB with pyridoxal-compounds. *Biochimie*, 2007, 89: 468- 473.

Cotlier E. Senile cataracts: Evidence for acceleration by diabetes and deceleration by salicylate. *Can. J. Ophthalmol.*, 1981, 16: 113.

Datsenko, KA; Wanner, BL. One-step inactivation of chromosomal genes in Escherichia coli K-12 using PCR products. *Proc. Natl. Acad. Sci.*, 2000, 97: 6640-6645.

Davis, MD; Edmondson, DE; McCormick, DB. Reaction of pyridoxal 5'-phosphate with TRIS. *Chemical Monthly*, 1982, 113: 999-1004.

Degenhardt, TP; Alderson, NL; Arrington, DD; Beattie, RJ; Basgen, JM; Steffes, MW; Thorpe, SR; Baynes, JW. Pyridoxamine inhibits early renal disease and dyslipidemia in the streptozotocin-diabetic rat. *Kidney Int*, 2002, 61: 939-950.

Delpierre, G; Rider, MH; Collard, F; Stroobant, V; Vanstapel, F; Santos, H; Van Schaftingen, E. Identification, cloning, and heterologous expression of a mammalian fructosamine-3-kinase. *Diabetes*, 2000a, 49: 1627-1634.

Delpierre, G; Vanstapel, F; Stroobant, V; Van Schaftingen, E. Conversion of a synthetic fructosamine into its 3-phospho derivative in human erythrocytes. *Biochem. J.*, 2000b, 352 Pt 3: 835-839.

Edelstein, D; Brownlee, M. Aminoguanidine ameliorates albuminuria in diabetic hypertensive rats. *Diabetologia*, 1992, 35: 96-97.

Fang, JL; Vaca, CE; Valsta, LM; Mutanen, M. Determination of DNA adducts of malonaldehyde in humans: effects of dietary fatty acid composition. *Carcinogenesis*, 1996, 17: 1035-1040.

Fleming, T., Rabbani, N, Thoranalley, T. (2007) Quantitative screening of DNA biomarkers of glycation and oxidation. In: Abstracts of the 9th International Symposium on the Maillard Reaction, p. 99.

Foster PL. Stress responses and genetic variation in bacteria. *Mutat. Res.*, 2005, 569: 3-11.

Foster PL. Stress-induced mutagenesis in bacteria. *Crit. Rev. Biochem. Mol. Biol.*, 2007, 42: 373-397.

Fu, MX; Requena, JR; Jenkins, AJ; Lyons, TJ; Baynes, JW; Thorpe, SR. The advanced glycation end product, Nε-(carboxymethyl)lysine, is a product of both lipid peroxidation and glycoxidation reactions. *J. Biol. Chem.*, 1996, 271: 9982-9986.

Ganea, E; Rixon, KC; Harding; JJ. Binding of glucose, galactose and pyridoxal phosphate to lens crystallins. *Biochim. Biophys. Acta*, 1994, 1226: 286-290.

Ganea, E; Harding, JJ. Lens proteins changes induced by sugars and pyridoxal phosphate. *Ophthalmic Res.*, 1996, 28 Suppl 1: 65-68.

Goodman, MF. Error-prone repair DNA polymerases in prokaryotes and eukaryotes. *Annu Rev Biochem*, 2002, 71: 17-50.

Hammes, HP; Du, X; Edelstein, D; Taguchi, T; Matsumura, T; Ju, Q; Lin, J; Bierhaus, A; Nawroth, P; Hannak, D; Neumaier, M; Bergfeld, R; Giardino, I; Brownlee, M. Benfotiamine blocks three major pathways of hyperglycemic damage and prevents experimental diabetic retinopathy. *Nat. Med.*, 2003, 9: 294-299.

Hartmann, G; Honikel, KO; Knьsel, F; Nьesch, J. The specific inhibition of the DNA-directed RNA synthesis by rifamycin. *Biochim. Biophys. Acta*, 1967, 145: 843-844.

Haupt, E; Ledermann, H; Kopcke, W. Benfotiamine in the treatment of diabetic polyneuropathy -a three-week randomized, controlled pilot study (BEDIP study). *Int. J. Cli. Pharmaco Ther.*, 2005. 43: 71-77.

Hayashi, T; Namiki, M. Formation of two-carbon sugar fragments at an early stage of the browning reaction of sugar and amine. *Agric. Biol Chem.*, 1980, 44: 2575-2580.

Hodge, JE. Deydrated Foods: Chemistry of Browning reactions in model systems. *J. Agric. Food Chem.*, 1953, 1: 928-943.

Horiuchi, T; Kurokawa, T; Saito, N. Purification and properties of fructosyl-amino acid oxidase from CorynebacteriumCorynebacterium sp. 2-4-1. *Agric. Biol. Chem.*, 1989, 53: 103-110.

Horiuchi, T; Kurokawa, T. Purification and properties of fructosylamine oxidase from Aspergillus Aspergillus sp. 1005. *Agric. Biol. Chem.*, 1991, 55: 333-338.

Huby, R; Harding, JJ. Non-enzymic glycosylation (glycation) of lens proteins by galactose and protection by aspirin and reduced glutathione. *Exp. Eye Res.*, 1988, 47: 53-59.

Ivanov, I; Tam, J; Wishart, P; Jay, E. Chemical synthesis and expression of the human calcitonin gene. *Gene*, 1987, 59: 223-230.

Jain, SK; Lim, G. Pyridoxine and pyridoxamine inhibits superoxide radicals and prevents lipid peroxidation, protein glycosylation, and (Na+ + K+)-ATPase activity reduction in high glucose-treated human erythrocytes. *Free Radic Biol. Med.*, 2001, 30: 232-237.

Jakus, V; Rietbrock, N. Advanced glycation end-products and the progress of diabetic vascular complications. *Physiol Res.*, 2004, 53: 131-142.

Jin, DJ; Gross, CA. Mapping and sequencing of mutations in the Escherichia coli rpoB gene that lead to rifampicin resistance. *J. Mol. Biol.*, 1988, 202: 45-58.

Johnson, RN; Metcalf, PA; Baker, JR. Fructosamine: a new approach to the estimation of serum glycosylprotein. An index of diabetic control. *Clin. Chim. Acta*, 1982, 127: 87-95.

Kannan, K; Jain, SK. Effect of vitamin B6 on oxygen radicals, mitochondrial membrane potential, and lipid peroxidation in H2O2-treated U937 monocytes. *Free Radic. Biol. Med.*, 2004, 36: 423-428.

Karachalias, N; Babaei-Jadidi, R; Ahmed, N; Thornalley, PJ. Accumulation of fructosyl-lysine and advanced glycation end products in the kidney, retina and peripheral nerve of streptozotocin-induced diabetic rats. *Biochem. Soc. Trans*, 2003, 31(Pt 6): 1423-1425.

Kasai, H; Iwamoto-Tanaka, N; Fukada, S. DNA modifications by the mutagen glyoxal: adduction to G and C, deamination of C and GC and GA cross-linking. *Carcinogenesis*, 1998, 19: 1459-1465.

Kato, H. Studies on browning reactions between sugars and amino compounds. V. Isolation and characterisation of new carbonyl compounds, 3-deoxyglucosones formed from N-glycosides and their significance for browning reaction. *Bull Agric. Chem. Soc. Japan*, 1960, 24: 1-12.

Khalifah, RG; Todd, P; Booth, AA; Yang, SX; Mott, JD; Hudson, BG. Kinetics of nonenzymatic glycation of ribonuclease A leading to advanced glycation end products. Paradoxical inhibition by ribose leads to facile isolation of protein intermediate for rapid post-Amadori studies. *Biochemistry*, 1996, 35: 4645-4654.

Khalifah, RG; Baynes, JW; Hudson, BG. Amadorins: novel post-Amadori inhibitors of advanced glycation reactions. *Biochem. Biophys. Res. Commun*, 1999, 257: 251-258.

Khalifah, RG; Chen, Y; Wassenberg, JJ. Post-Amadori AGE inhibition as a therapeutic target for diabetic complications: a rational approach to second-generation Amadorin design. *Ann. N Y Acad Sci.*, 2005, 1043: 793-806.

Kuhn, R; Weygand, F. The Amadori rearrangement. *Ber,* 1937, 70B: 769-772.

Lee, AT; Cerami, A. Induction of gamma delta transposition in response to elevated glucose-6-phosphate levels. *Mutat. Res*, 1991, 249: 125-133.

Li, GM. Mechanisms and functions of DNA mismatch repair. *Cell Res.*, 2008, 18: 85-98.

Maillard LC, Action des acides amines sur les sucres: formation des melanoidines par voie methodique. *C R Acad. Sci.*, 1912, 154: 66-68.

McClure, WR; Cech, CL. On the mechanism of rifampicin inhibition of RNA synthesis. *J. Biol. Chem.*, 1978, 253: 8949-8956.

Michel, B; Boubakri, H; Baharoglu, Z; LeMasson, M; Lestini, R. Recombination proteins and rescue of arrested replication forks. *DNA Repair (Amst)*, 2007, 6: 967-980.

Mironova, R; Niwa, T; Handzhiyski, Y; Sredovska, A; Ivanov, I. Evidence for non-enzymatic glycosylation of *Escherichia coli* chromosomal DNA. *Mol. Microbiol.*, 2005, 55: 1801-1811.

Mizushina, Y; Xu, X; Matsubara, K; Murakami, C; Kuriyama, I; Oshige, M; Takemura, M; Kato, N; Yoshida, H; Sakaguchi, K. Pyridoxal 5'-phosphate is a selective inhibitor in vivo of DNA polymerase alpha and epsilon. *Biochem. Biophys. Res. Commun.*, 2003, 312: 1025-1032.

Monnier, VM; Cerami, A. Nonenzymatic browning in vivo: possible process for aging of long-lived proteins. *Science*, 1981, 211: 491-493.

Monnier, VM; Stevens, VJ; Cerami, A. Maillard reactions involving proteins and carbohydrates in vivo: relevance to diabetes mellitus and aging. *Prog. Food Nutr. Sci.*, 1981, 5: 315-327.

Monnier, VM. Intervention against the Maillard reaction in vivo. *Arch. Biochem. Biophys.*, 2003, 419: 1-15.

Monnier, VM. Bacterial enzymes that can deglycate glucose- and fructose-modified lysine. *Biochem. J.*, 2005, 392(Pt 2): e1-3.

Monnier VM; Sell, DR. Prevention and repair of protein damage by the Maillard reaction in vivo. *Rejuvenation Res.*, 2006, 9: 264-273.

Murata-Kamiya, N; Kamiya, H; Kaji, H; Kasai H. Nucleotide excision repair proteins may be involved in the fixation of glyoxal-induced mutagenesis in *Escherichia coli*. *Biochem Biophys. Res. Commun.*, 1998, 248: 412-417.

Murata-Kamiya, N; Kaji, H; Kasai, H. Deficient nucleotide excision repair increases base-pair substitutions but decreases TGGC frameshifts induced by methylglyoxal in *Escherichia coli*. *Mutat Res.*, 1999, 442: 19-28.

Nagaraj, RH; Sarkar, P; Mally, A; Biemel, KM; Lederer, MO; Padayatti, PS. Effect of pyridoxamine on chemical modification of proteins by carbonyls in diabetic rats: characterization of a major product from the reaction of pyridoxamine and methylglyoxal. *Arch. Biochem. Biophyl.*, 2002, 402: 110-119.

Nakayama, T; Hayase, F; Kato, H. Formation of Nε-(2-formyl-5-hydroxy-methyl-pyrrol-1-yl)-L-norleucine in the Maillard reaction between D-glucose and L-lysine. *Agric. Biol. Chem.*, 1980, 44: 1201-1202.

Nass, N; Bartling, B; Navarrete Santos, A; Scheubel, RJ; Burgermann, J; Silber, RE; Simm, A. Advanced glycation end products, diabetes and ageing. *Z Gerontol Geriatr.*, 2007, 40: 349-356.

Neault, J; Naoui, M; Manfait, M; Tajmir-Riahi, H. Aspirin-DNA interaction studied by FTIR and laser Raman difference spectroscopy. *FEBS Letters*, 1996, 382: 26-30.

Niwa, T; Sato, M; Katsuzaki, T; Tomoo, T; Miyazaki, T; Tatemichi, N; Takei, Y; Kondo, T. Amyloid β2-microglobulin is modified with Nε-(carboxymethyl)lysine in dialysis-related amyloidosis. *Kidney Int*, 1996, 50: 1303-1309.

Okada, M; Ayabe, Y. Effects of aminoguanidine and pyridoxal phosphate on glycation reaction of aspartate aminotransferase and serum albumin. *J. Nutr. Sci. Vitaminol (Tokyo)*, 1995, 41: 43-50.

Osicka, TM; Yu, Y; Panagiotopoulos, S; Clavant, SP; Kiriazis, Z; Pike, RN; Pratt, LM; Russo, LM, Kemp, BE, Comper, WD, Jerums, G. Prevention of albuminuria by aminoguanidine or ramipril in streptozotocin-induced diabetic rats is associated with the normalization of glomerular protein kinase C. *Diabetes*, 2000, 49: 87-93.

Petersen, A; Szwergold, BS; Kappler, F; Weingarten, M; Brown, TR. Identification of sorbitol 3-phosphate and fructose 3-phosphate in normal and diabetic human erythrocytes. *J. Biol. Chem.*, 1990, 265: 17424-17427.

Pfeifer, GP. Mutagenesis at methylated CpG sequences. *Curr. Top Microbiol. Immunol.*, 2006, 301:259-281.

Pinckard, N; Hawkins, D; Richard, F. *In vitro* acetylation of plasma proteins, enzymes and DNA by aspirin. *Nature*, 1968, 219: 68-69.

Pomero, F; Molinar Min, A; La Selva, M; Allione, A; Molinatti, G.M; Porta, M. Benfotiamine is similar to thiamine in correcting endothelial cell defects induced by high glucose. *Acta Diabetol*, 2001, 38: 135-138.

Pongor, S; Ulrich, PC; Benesath, FA; Cerami, A. Aging of protiens: isolation and identification of a fluorescent chromophore from the reaction of polypeptides with glucose. *Proc. Natl. Acad. Sci. USA*, 1984, 81: 2684-2688.

Pushkarsky, T; Rourke, L; Spiegel, LA; Seldin, MF; Bucala, R. Molecular characterization of a mouse genomic element mobilized by advanced glycation endproduct modified-DNA (AGE-DNA). *Mol. Med.*, 1997, 3: 740-749.

Rahbar, S. An abnormal haemoglobin in red cells of diabetes. *Clin. Chim. Acta*, 1968, 22: 296-298.

Rahbar, S. The discovery of glycated hemoglobin: a major event in the study of nonenzymatic chemistry in biological systems. *Ann. N Y Acad. Sci.*, 2005, 1043: 9-19.

Sadekov, RA; Danilov, AB; Vein, AM. Diabetic polyneuropathy treatment by milgamma-100 preparation. *Zh Nevrol Psikhiatr Im S S Korsakova*, 1998, 98: 30-32.

Sambrook, J., Fritsch, E.F., and Maniatis, T. (1989) Bacterial Media, Antibiotics and Bacterial Strains. *In Molecular Cloning: A Laboratory Manual*, 2nd Ed., Cold Spring Harbor Laboratory, Cold Spring Harbor, NY, pA11.

Sanger, F; Nicklen, S; Coulson, AR. DNA sequencing with chain-terminating inhibitors. *Proc Natl Acad Sci U S A*, 1977, 74: 5463-5467.

Seidel, W; Pischetsrieder, M. Immunochemical detection of N^2-[1-(1-carboxy)ethyl]guanosine, an advanced glycation end product formed by the reaction of DNA and reducing sugars or L-ascorbic acid *in vitro*. *Biochim. Biophys. Acta*, 1998, 1425: 478-484.

Sell, DR; Monnier, VM. Structure elucidation of a senescence crosslink from human extracellular atrix. Implication of pentoses in the aging process. *J. Biol. Chem.*, 1989, 264: 21597-21602.

Shangari, N; Bruce, WR; Poon, R; O'Brien, PJ. Toxicity of glyoxals - role of oxidative stress, etabolic detoxification and thiamine deficiency. *Biochem. Soc. Trans*, 2003, 31: 1390-1393.

Shimoi, K; Okitsu, A; Green, MH; Lowe, JE; Ohta, T; Kaji, K; Terato, H; Ide, H; Kinae, N. xidative DNA damage induced by high glucose and its suppression in human umbilical ein endothelial cells. *Mutat Res*, 2001, 480-481: 371-378.

Simeonov, S; Pavlova, M; Mitkov, M; Mincheva, L; Troev, D. Therapeutic efficacy of "Milgamma" in patients with painful diabetic neuropathy. *Folia Med (Plovdiv)*, 1997, 39: -10.

Soulis-Liparota, T; Cooper, M; Papazoglou, D; Clarke, B; Jerums, G. Retardation by minoguanidine of development of albuminuria, mesangial expansion, and tissue luorescence in streptozocin-induced diabetic rat. *Diabetes*, 1991, 10: 1328-1334.

Stitt, A; Gardiner, TA; Alderson, NL; Canning, P; Frizzell, N; Duffy, N; Boyle, C; Januszewski, S; Chachich, M; Baynes, JW; Thorpe, SR. The AGE inhibitor pyridoxamine inhibits evelopment of retinopathy in experimental diabetes. *Diabetes*, 2002, 51: 2826-2832.

Stracke, H; Lindemann, A; Federlin, K. A benfotiamine-vitamin B combination in treatment of iabetic polyneuropathy. *Exp. Clin. Endocrinol. Diabetes*, 1996, 104: 311-316.

Swamy; MS; Abraham, EC. Inhibition of lens crystallin glycation and high molecular weight ggregate formation by aspirin in vitro and in vivo. *Invest Ophthalmol Vis. Sci*, 1989, 30: 120-1126.

Szwergold, BS; Kappler, F; Brown, TR. Identification of fructose 3-phosphate in the lens of diabetic rats. *Science*, 1990, 247: 451-454.

Szwergold, BS; Howell, S; Beisswenger, PJ. Human fructosamine-3-kinase: purification, equencing, substrate specificity, and evidence of activity in vivo. *Diabetes*, 2001, 50: 2139-2147.

Taguchi, T; Sugiura, M; Hamada, Y; Miwa, I. In vivo formation of a Schiff base of aminoguanidine with pyridoxal phosphate. *Biochem. Pharmacol.*, 1998, 55: 1667-1671.

Thornalley, PJ; Wolff, SP; Crabbe, J; Stern, A. The autoxidation of glyceraldehyde and other simple monosaccharides under physiological conditions catalysed by buffer ions. *Biochim. Biophys. Acta*, 1984, 797: 276-287.

Thornalley, PJ; Langborg, A; Minhas, HS. Formation of glyoxal, methylglyoxal and 3-deoxyglucosone in the glycation of proteins by glucose. *Biochem. J.*, 1999, 344 Pt 1: 109-16.

Thornalley, PJ; Yurek-George, A; Argirov, OK. Kinetics and mechanism of the reaction of aminoguanidine with the alpha-oxoaldehydes glyoxal, methylglyoxal, and 3-deoxyglucosone under physiological conditions. *Biochem. Pharmacol.*, 2000, 60: 55-65.

Thornalley, PJ; Jahan, I; Ng, R. Suppression of the accumulation of triosephosphates and increased formation of methylglyoxal in human red blood cells during hyperglycaemia by thiamine in vitro. *J. Biochem.*, 2001, 129: 543-549.

Thornalley, PJ. The enzymatic defence against glycation in health, disease and therapeutics: a symposium to examine the concept. *Biochem. Soc. Trans*, 2003, 31(Pt 6): 1341-1342.

Thorpe, SR; Baynes, JW. Role of the Maillard reaction in diabetes mellitus and diseases of aging. *Drugs Aging*, 1996, 9: 69-77.

Thorpe, SR; Baynes, JW. Maillard reaction products in tissue proteins: new products and new perspectives. *Amino Acids*, 2003, 25: 275-281.

Tsuzuki, T; Nakatsu, Y; Nakabeppu, Y. Significance of error-avoiding mechanisms for oxidative DNA damage in carcinogenesis. *Cancer Sci*, 2007 98: 465-470.

Ulrich, P; Cerami, A. Protein glycation, diabetes, and aging. *Recent Prog Horm Res*, 2001, 56: 1-21.

Vaca, CE; Fang, JL; Conradi, M; Hou, SM. Development of a ^{32}P-postlabelling method for the analysis of 2'-deoxyguanosine-3'-monophosphate and DNA adducts of methylglyoxal. *Carcinogenesis*, 1994, 15: 1887-1894.

Voziyan, PA; Metz, TO; Baynes, JW; Hudson, BG. A post-Amadori inhibitor pyridoxamine also inhibits chemical modification of proteins by scavenging carbonyl intermediates of carbohydrate and lipid degradation. *J. Biol. Chem.*, 2002, 277: 3397-3403.

Voziyan, PA; Khalifah, RG; Thibaudeau, C; Yildiz, A; Jacob, J; Serianni, AS; Hudson, BG. Modification of proteins in vitro by physiological levels of glucose: pyridoxamine inhibits conversion of Amadori intermediate to advanced glycation end-products through binding of redox metal ions. *J. Biol. Chem.*, 2003, 278: 46616-46624.

Voziyan, PA; Hudson, BG. Pyridoxamine as a multifunctional pharmaceutical: targeting pathogenic glycation and oxidative damage. *Cell Mol. Life Sci.*, 2005, 62: 1671-1681.

Waanders, F; van den Berg, E; Nagai, R; van Veen, I; Navis, G; van Goor, H. Renoprotective effects of the AGE-inhibitor pyridoxamine in experimental chronic allograft nephropathy in rats. *Nephrol Dial Transplant*, 2008, 23: 518-524.

Whittle, CA; Johnston, CD. Moving forward in determining the causes of mutations: the features of plants that make them suitable for assessing the impact of environmental factors and cell age. *J Exp Bot*, 2006, 57: 1847-1855.

Wiame, E; Delpierre, G; Van Schaftingen, C; Van Schaftingen, E. Identification of a pathway for the utilization of the Amadori product fructoselysine in Escherichia coli. *J. Biol Chem*, 2002, 277: 42523-42529.

Wiame, E; Duquenne, A; Delpierre, G; Van Schaftingen, E. Identification of enzymes acting on alpha-glycated amino acids in Bacillus subtilis. *FEBS Lett*, 2004, 577: 469-472.

Williams, ME; Bolton, WK; Khalifah, RG; Degenhardt, TP; Schotzinger, RJ; McGill, JB. Effects of pyridoxamine in combined phase 2 studies of patients with type 1 and type 2 diabetes and overt nephropathy. *Am J Nephrol*, 2007, 27: 605-614.

Winkler, G; Pal, B; Nagybeganyi, E; Ory, I; Porochnavec, M; Kempler, P. Effectiveness of different benfotiamine dosage regimens in the treatment of painful diabetic neuropathy. *Arzneimittelforschung*, 1999, 49: 220-224.

Wolff, SP; Dean, RT. Glucose autoxidation and protein modification. The potential role of 'autoxidative glycosylation' in diabetes. *Biochem J.*, 1987, 245: 243-250.

Wu, X; Monnier, VM. Enzymatic deglycation of proteins. *Arch. Biochem. Biophys.*, 2003, 419: 16-24.

Zhou, L; Lei, XH; Bochner, BR; Wanner, BL. Phenotype microarray analysis of Escherichia coli K-12 mutants with deletions of all two-component systems. *J. Bacteriol.*, 2003, 185: 4956-49572.

Reviewed by
Ashkan Golshani,
Department of Biology,
Carleton University, Ottawa, Canada

In: Bacterial DNA, DNA Polimerase and DNA Helicases ISBN 978-1-60741-094-2
Editors: W. D. Knutsen and S. S. Bruns © 2009 Nova Science Publishers, Inc.

Chapter III

Nucleoid Architecture and Dynamics in Bacteria

Ryosuke L. Ohniwa[*1,2], *Kazuya Morikawa*[2], *Toshiko Ohta*[2], *Chieko Wada*[1,3] *and Kunio Takeyasu*[1]

[1]Laboratory of Plasma Membrane and Nuclear Signaling,
Kyoto University Graduate School of Biostudies, Yoshida-Konoe,
Sakyo-ku, Kyoto, Japan
[2]Institute of Basic Medical Sciences,
Graduate School of Comprehensive Human studies,
University of Tsukuba, 1-1-1 Tennodai, Tsukuba, Japan
[3]Yoshida Biological Laboratory, Yamashina, Kyoto, Japan

Abstract

Both prokaryotes and eukaryotes store their genomic DNA in an environment, which balances the physical properties of double-stranded DNA with the mechanical effects of DNA-protein interactions. The origin of genome packing remains a mystery; it seems to go back to the very beginning of life itself. Since then, the principle of higher-order genome construction and architecture has been maintained and shared among three domains of living things. The genomes are stored as flexible higher-order structures that can shuttle between relatively active and inactive states. The major differences reside in the structural protein components of the architecture. Bacteria and Eukaria utilize HU and histones as the most fundamental structural proteins, respectively, forming the nucleoid in bacterial cell and the nucleosomes in the nucleus. On the other hand Archaea employ either HU or histones depending on their phylogenetic origins. Nevertheless, the hierarchies of nucleoid fibers look extremely similar among different species in different domains. Focusing upon the mechanism of bacterial genome folding, recent structural investigations have revealed a step-wise folding of the genome DNA; from 10 nm to 30

---

[*] Correspond to: R.L. Ohniwa, Institute of Basic Medical Sciences, Graduate School of Comprehensive Human Sciences, University of Tsukuba, Tsukuba, 305-8575 Japan Phone and Fax: +81-29-853-3928; e-mail: ohniwa@md.tsukuba.ac.jp

nm fibers, and to 80 nm and further condensed beaded structures, depending upon the growth conditions and differential contribution of nascent single-stranded RNA. The nucleoid condensation is mainly brought about by another nucleoid protein, Dps, which occurs widely in the bacterial domain, but not in Archaea or Eukarya. The regulatory mechanisms of the *dps* gene are greatly different among bacterial species. In γ-Proteobacteria such as *Escherichia coli*, the *dps* gene is governed by two regulatory systems, IHF-σs and OxyR, and is induced towards the stationary phase and under oxidative stress. The DNA topology control by *E. coli* Fis also participates in the regulation of the nucleoid condensation. In contrast, in Firmicutes such as *Staphylococcus aureus*, the gene is mainly controlled by oxidative stress-response system, PerR, and the Dps expression is directly coupled with the nucleoid condensation. Here the key features of the nucleoid dynamics will be discussed from the biochemical and structural biological points of views.

1. Introduction

Most bacterial genomes are circular, and, with varying sizes (*Mycoplasma genitalium*; 0.2 mm, *Staphylococcus aureus*; 1 mm, *Escherichia coli*; 1.6 mm, *Bradyrhizobium japonicum*; 3 mm, per a genome respectively) are packed into cells of a few μm as a form of "nucleoid" [1, 2]. On the other hand, the ~m long entire eukaryotic genomes are spitted into several linear "chromosomes", and stored in the nucleus with a diameter of several μm. The search for the proteins involved in the genome architectures has revealed that no structural protein is shared by bacteria and eukaryotes. Namely the major components are HU in bacteria and histones in eukaryotes. The third domain of life, archaea, provides interesting information that may bridge between the gaps in the composition and the hierarchy of genome structures, because some archaea possess histones and others HU (Table 3).

Eukaryotic mitochondria and chloroplast genomes contain none of these proteins; neither HU nor histone has been identified in these intracellular organelles, while their genetic machineries are shared with their ancestral bacterial factors. Irrespective to such complete difference of the nucleoid components and the genetic systems, the hierarchy of the genome structures are generally common among living organisms and organelles [3].

In this chapter, we center our focus on the structural hierarchy in bacterial nucleoids, and discuss the dynamic properties of nucleoids especially in comparison with the eukaryotic chromosomes.

2. Components of Nucleoid

2.1. Proteins in the Nucleoid

The isolated *E. coli* nucleoids have a set of DNA-binding proteins including the RNA polymerase subunits and about 300 species of transcription factors [4]. Among them, HU (heat-unstable nucleoid protein), IHF (integration host factor protein), H-NS (histone-like nucleoid structuring protein), Fis (factor for inversion stimulation), Hfq (host factor for phage

RNA Qb replication), StpA (suppressor of T4 *td* mutant phenotype A, H-NS homolog) and Dps (DNA-binding protein from starved cells) are the major small nucleoid components [5-8]. These proteins termed as nucleoid proteins are involved in the organization and condensation of bacterial chromosome, and are often referred to as histone-like proteins based on the superficial phenotypic similarities to the eukaryotic histone proteins (basic protein, high intracellular abundance, DNA binding properties, and low molecular weight). However they have no sequence similarity to the eukaryotic histones. The molecular features of these proteins are summarized in Table 1.

It is important to note that the abundance of different kinds of nucleoid proteins in *E. coli* is a reflection of the characteristics of γ-Proteobacteria. Nevertheless, as discussed in Section 2, the hierarchies of nucleoid fibers look extremely similar among different species.

Bacteria also possess SMC (Structural Maintenance of Chromosome) proteins such as MukB in *E. coli* [9] and BsSMC in *Bacillus subtilis*. These SMC proteins have little sequence similarity to the eukaryotic SMCs [10], but share the common structural characteristics (the existence of DNA binding domains and ATP binding domains that are connected by a long coiled-coil and a flexible center hinge). The enzymatic and DNA-binding mechanisms of these proteins are similar to the eukaryotic condensins and their ATPase activities appear to play a key role in DNA binding and compaction [11, 12]. The lack of SMC caused a irregular cell division and nucleoid partitioning [13, 14]. Topoisomerases play critical roles in the construction, maintenance, and re-construction of well-organized higher-order structures of the genome [15, 16].

Interestingly, Archaea employs either HU or histones depending on their phylogenetic origins. Investigation on the nucleoid architectures in Archaea species will provide an insight into the biochemical and physical basis of the genome folding mechanism in general.

Table 1. Molecular features of nucleoid proteins

Proteins	MW	multimer	Nucleotide binding ability		Genetic regulation		Number of multimer per 10 kbp in *E. coli*			
	kDa		DNA	RNA	Transcription	Translation	LA	L	ES	LS
Hu	9,5	dimer	yes	yes	yes	yes	38	59	32	16
IHF	11	dimer	yes	yes	yes	?	8	11	59	32
H-NS	15,4	dimer	yes	yes	yes	yes	19	22	16	11
StpA	15,3	dimer	yes	yes	?	yes	23	27	14	11
Fis	22	dimer	yes	?	yes	?	8	67	un	un
Hfq	11,2	hexamer	yes	yes	?	yes	12	15	8	8
Dps	19	dodecamer	yes	?	?	?	1	1	21	32

2.2. Nucleoid Proteins Bending a DNA, Bundling DNA Fibers and Controling DNA Topology

Many nucleoid proteins bind DNA in a sequence-independent manner at high concentrations, and bend DNA and/or bridge DNA strands. HU binds and bends DNA, and stabilizes the "bending" in a similar manner to high mobility group (HMG) proteins in eukaryotes [17-19]. H-NS "zips up" the double–stranded DNA to form a bridge between adjacent strands[20]. Fis prefers to bind to A/T-rich regions with a highly degenerated dyad symmetry, bends DNA, and stabilize DNA loops at high concentration [21-23]. The bended and bundled DNA-protein complex can be identified by AFM (Figure 1D, E, expanding Figure).

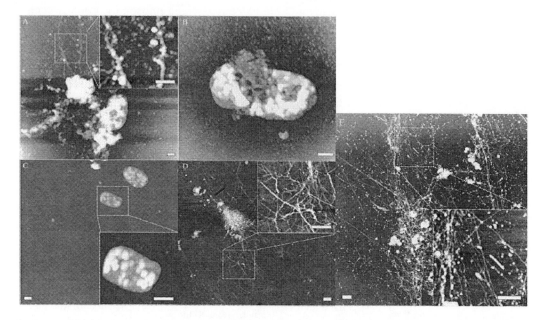

Figure 1. AFM images of the E. co li nucleoids. The lysed E. coli cells were observed by AFM: (A, B) in log phase, (C) in stationary phase, (D) after treatment with RNase A, and (E) in log phase after rifampicin treatment in the cultural medium. 30 nm- and 80 nm-fibers (A) and loop structures (B) are identified in the log phase. In the stationary phase, fibrous nucleoid is condensed (C) due to the expression of Dps (see text). 10 nm fibers are exposed by the disruption of single stranded RNA (D) or inhibition of transpcription (E). Scale bars represent 500 nm.

The DNA binding of the nucleoid proteins is known to control DNA topology. The binding of HU forces the negatively supercoiled DNA to the relaxed form, whereas H-NS increases the apparent supercoiling [24, 25]. *In vitro*, Fis changes the overall shape of supercoiled DNA in a equence-independent manner [26, 27], and prevents the topological changes catalyzed by DNA gyrase and Topoisomerase I [28]. SMC and H-NS can form topological barriers because of their DNA bridging function. It has been revealed that the distributions of the nucleoid proteins over the entire genome are uneven and restricted to certain chromosomal loci [29-31]. Such uneven localization may be related to the structural boundaries on the nucleoid (Figure 2) [32]. DNA gyrase has also been reported to contribute to such boundary formation [33].

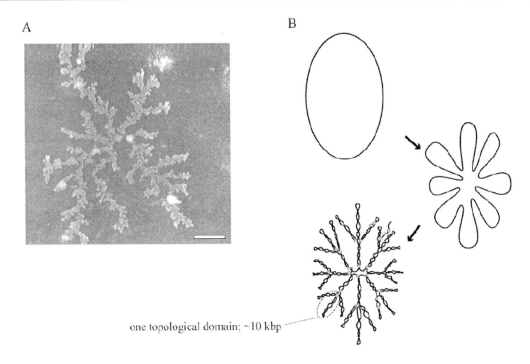

Figure 2. The rosette-like structure of the prepared nucleoid. (A) AFM image of the nucleoid as a rosette-like structure. (B) Folding model of the rosette-like structure. The topological domain of ~10kbp DNA is assumed.

2.3. Nucleoid Proteins as Regulators for Transcription, Translation and Replication

Most nucleoid proteins have been reported to function as regulators of translation, transcription or replication at low concentrations. In the DNA replication initiation, HU, IHF and FIS bind to the E. coli chromosomal origin (oriC region) and interact with initiator protein, DnaA, to facilitate the strand opening at oriC required for the DNA replication [34-40]. The genes of nucleoid proteins are regulated by the nucleoid proteins themselves; ihfA/B, hns and fis are regulated by IHF, H-NS and Fis, respectively (Table 2). Fis is also a transcriptional regulator for hns, hupA and hupB, and fis itself is regulated by IHF. Fis preferably binds to the 15 bp consensus sequence, and works as a transcription regulator for various genes including hupA (HU-α), hupB (HU-β), hns (H-NS), and the topA, gyrA and gyrB genes [41, 42]. Other transcriptional regulators like CRP, CspA and GadX also participate in the regulation of these genes. HU does not have specific binding sites on the E. coli genome, but can change the DNA topology, which changes the global activity of transcription [43]. H-NS is known to work with Hha protein [44-46], which binds preferably to AT-rich regions and acts as a repressor for horizontally acquired genes [47]. IHF, a homologue of HU, binds to the specific DNA binding sites, and regulates various genes.

Table 2. Transcriptional network of the nucleoid proteins in *E.coli*

	Regulatory proteins[*1]			Regulon[*2]									
				Information Strage & Processing			Cellular Process & Singnaling			Metabolism	Poorly Characterized	RNA gene		
	activator	repressor	dual	up-regulation	down-regulation	dual[*3]	up-regulation	down-regulation	dual[*3]			up-regulation	down-regulation	dual[*3]
HU (*hupA*, *hupB*)	Fis, CRP (*hupA*), CRP (*hupB*),	Fis (*hupB*),	-	(0)	*seqA-pgm* (1)	(0)	(0)	(0)	(0)	(7)	(0)	*micF* (1)	(0)	(0)
IHF (*ihfA*, *ihfB*)	-	IHF (*ihfA*), IHF (*ihfB*)	-	*fimB, rpoH, sra, dusB-fis, hipB-hipA, rtcB-rtcA hemA-prfA-prmC, pspA-pspB-pspC-pspD-pspE, tdcA-tdcB-tdcC-tdcD-tdcE-tdcF-tdcG*, (13)	*ihfA, ihfB, uspB, ompR-envZ, flhD-flhC* (6)	(0)	*uspA, ibpB, glnH-glnP-glnQ, pspA-pspB-pspC-pspD-pspE, sufA-sufB-sufC-sufD-sufS-sufE, fimA-fimI-fimC-fimD-fimF-fimG-fimH* (15)	*ompC, osmE, ompR-envZ, flhD-flhC, caiT-caiA-caiB-caiC-caiD-caiE* (7)	*ompF* (1)	*dps*, etc (115)	(42)	(0)	*micF* (1)	(0)

Regulatory proteins[*1]		Regulon[*2]							
		Information Storage & Processing			Cellular Process & Signaling		Metabolism	Poorly Characterized	RNA gene	
H-NS (hns)	Fis, CspA, GadX (hns)	H-NS (hns)	-	cspD, smtA-mukF-mukE-mukB, flhD-flhC, fliA-fliZ-fliY, srlA-srlE-srlB-srlD-gutM-srlR-gutQ, relA-chpR-chpA (9)	hns, stpA, caiF, rcsA, fimB, fimE, gadX, bglG-bglF-bglB (8)	(0)	fliC, fliA-fliZ-fliY, hisJ-hisQ-hisM-hisP, srlA-srlE-srlB-srlD-gutM-srlR-gutQ, relA-chpR-chpA, fimA-fimI-fimC-fimD-fimF-fimG-fimH (17)	appY, bolA, degP, hchA, osmC, gspA-gspB, gspC-gspD-gspE-gspF-gspG-gspH-gspI-gspJ-gspK-gspL-gspM-gspO (18)	(0)	(50)	(16)	(0)	micF, rnpB, rRNA (15), tRNA (8)	(0)
Fis (fis)	IHF (fis)	Fis (fis)	CRP (fis)	hupA, hns, queA, sra, trmA, marR-marA-marB, metY-yhbc-nusA-infB-rbfA-truB-rpsO-pnp (13)	hupB, gyrA, gyrB, crp, dusB-fis, xylF-xylG-xylH-xylR, mtlA-mtlD-mtlR, bglG-bglF-bglB (9)	topA (1)	chpR-chpA-mazG, (2)	osmE, mltA-mltD-mltR (2)	(0)	(55)	(8)	rnpB, rRNA (20), tRNA (54)	tRNA (1)	(0)

*1; transcription factors regulating the genes of nucleoid protein.
*2; genes regulated by the nucleoid proteins.
*3; the way of regulation (up or down) is dependent on environmental conditions.
Grey colored genes; genes not in its functional category.
Genes connected '-' are operons.
The numbers in ( ) represent the total number of genes categorized into each functional category.

Toward the stationary phase, the expression level of IHF becomes higher and this in turn activates the dps gene expression [48, 49]. This regulation is one of the key mechanisms for inducing Dps-induced nucleoid condensation in E. coli [50] (see Section 3).

In addition to 'Metabolic Genes', many of the genes categorized in 'Information Storage and Processing' and 'Cellular Processing and Signaling' are under the regulation of IHF, H-NS and Fis. They regulate both single genes and operons. Single genes include: dps (Dps), sra (30S ribosomal subunit protein S22), cspD (DNA replication inhibitor), osmE (osmotically inducible lipoprotein OsmE), fimB (type 1 fimbriae regulatory protein FimB) etc. Examples of operons are; mukF-mukE-mukB, fimA-fimI-fimC-fimD-fimF-fimG-fimH (fimbrial protein), gspA-gspB and gspC-gspD-gspE-gspF-gspG-gspH-gspI-gspJ-gspK-gspL-gspM-gspO (general secretion pathway proteins (relative to type II secretion proteins)). In addition, H-NS and Fis regulate the RNA genes such as the rRNA/tRNA genes, non-coding RNA genes and rnpB encoding RNA for RNase P, and HU, IHF and H-NS control the micF (an antisense regulator of the translation of OmpF porin) expression.

Some of the nucleoid proteins bind to RNA. HU and H-NS recognize the 5'-UTR secondary structure of mRNA (such as rpoS mRNA) and non-coding small RNA (such as DsrA RNA), and regulate translational activities [51-55]. Hfq acts as an RNA chaperone to bridge mRNA and non-coding small RNA, in translation control [56-60]. This implies that Hfq may be tethering non-coding small RNA, mRNA, and proteins to form nucleoid complexes. Indeed, StpA, a homologue of H-NS, works as an RNA chaperon [61].

3. Nucleoid Architecture

The first observation of the compacted nucleoid in the cell was accomplished when DNA-specific Feulgen stain procedure was introduced into cytochemistry in 1930s [62]. In the 1970s and 1980s, the isolated nucleoids, prepared under relatively high salt conditions, were actively studied to elucidate some details of their structure. Electron Microscopy (EM) observations of isolated nucleoid revealed that the circular fibrous genome in bacteria, as a whole, is bundled to form a rosette-like structure with interwound loops emanating radially from the core portion (Figure 2) [63-66]. This observation is consistent with the later findings that the introduction of a single-strand DNA break into the E. coli genome could only relax the restricted portion; thus revealing the existence of independent supercoiling domains, which has recently been determined as c.a.10 kbp [67-69].

3.1. Nucleoid Fibers

Recent developments in nano-scale imaging techniques have facilitated the understanding of the detailed structure of the nucleoid [3, 70] (Figure 1). When *E. coli* cells are harvested in the log phase, and lysed under physiological salt conditions, the fibers with the widths of 30 nm and 80 nm (not naked DNA) are released (Figure 1A). Loop structures composed of the 80 nm fibers are also detectable inside the cell (Figure 1B). These fibers are further condensed beaded structures depending upon the growth conditions (Figure 1C).

Interestingly, the 30 and 80nm fiber structures are also commonly found in *S. aureus* and *Clostridium perfringens* [71] This implies the existence of step-wise folding mechanisms of DNA, in achieving the higher-order architectures, regardless of the protein components (Figure 3).

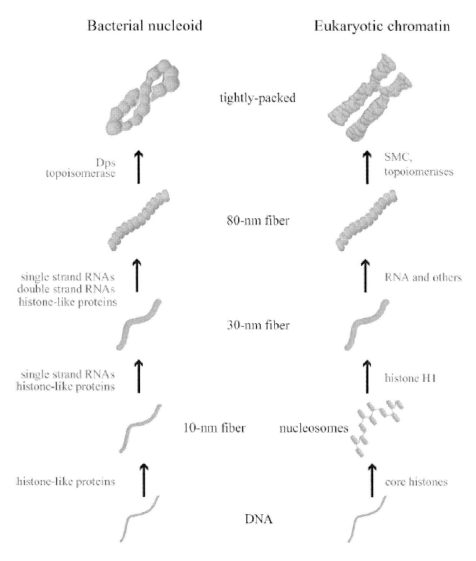

Figure 3. Hierarchy of genome architecture. In the bacterial nucleoid, the 10-nm fiber is composed of DNA and nucleoid proteins, and further folds up into the 30-nm fiber with the help of RNA molecules. The 30-nm fiber folds into a 80-nm fiber. The expression of Dps induces a highly condensed structure. In eukaryotes, a basic unit of chromatin fibers is a nucleosome which is composed of core histones, each wrapped by 150 bp of DNA [149-155]. Such nucleosomes have been identified as a 'beads-on-a-string' structure, which is formed by DNA and nucleosomes with the width of 2 and 11 nm, respectively [146, 156-158]. This basic chromatin fiber is folded into 30-, 80--nm fibers and mitotic condensed chromosomes, with the help of a series of proteins such as histone H1, TopoII and Condensines [159-165].

In addition, commonly the higher-order structures are sensitive to the RNase A-treatment [3]. An enzymatic digestion with RNase A, but not with RNase H, produce 10-nm fibers from the 30-nm fibers (Figure 1D), suggesting that the thinnest fiber without RNA possesses a diameter of 10 nm. Also in an Archaea species, a similar hierarchy of genome DNA architectures can be observed under AFM (unpublished data), supporting the notion of the step-wise folding of the genome DNA; from 10 nm to 30 nm fibers, and to 80 nm.

3.2. Organization of the Fibers inside the Cell

The *in situ* nucleoid structure has also been studied. Based on the EM observation that the nucleoid is shrunken by a rifampicin treatment that inhibits transcription elongation (by blocking the RNA exit channel of RNA polymerase), it is generally accepted that RNA and proteins must participate in anchoring the nucleoid to the cell membrane [72-75]; via co-transcription and translation of membrane- and periplasmic-proteins [72, 76]. But it should be noted that the EM sample preparation includes dehydration and embedding that can aggregate the nucleoid structure, and the rifampicin treatment makes the nucleoid more sensitive to it. The rifampicin-treated nucleoid structure can be observed as dispersed when using DAPI (4',6-diamino-2-phenylindole) staining [77, 78], indicating that the in situ nucleoid structure is disturbed by the rifampicin treatment (also discussed in section 3).

Regarding the native states of the nucleoid *in situ*, TEM observation detected randomly oriented, whirled and bundled fiber structures inside the cell [79-81]. Although it is suspected that these observed structures were artifacts because of the fixation process of the specimen, a mild preparation with CFS (the combination of cryo-immobilization and freeze substitution) exhibits a granular and fibrous nucleoid in the cell [82]. Recently, GEMOVIS (cryoelectron microscopy of vitreous sections), a method which enables us to observe unfixed, unstained, fully hydrated biological material close to their native state, was applied to the nucleoids observation [83-88], and thus successfully allowed the observation of locally ordered DNA filaments. This technique may provide us with valuable information in the near future.

In addition to EM, recent developments in cellular biology has also contributed to elucidating the *in situ* nucleoid structure. The applications of FISH (fluorescent *in situ* hydridization) and FROS (fluorescent DNA-binding proteins to their binding sites), to label the individual chromosome loci in the cell, have demonstrated that the origin and terminus of replication are consistently found at defined locations in newborn cells, with the terminus oriented towards one of the cell poles and the origin positioned at the opposite side or in the middle of the cell [89-96]. The analysis of *Caulobacter crescentus* to label the 112 different chromosome positions showed that the chromosomal locus in the cell is well defined and each position is linearly correlated with its position on the genome map relative to the origin of replication [97]. The two different loci were frequently stained as a single dot even though the loci are sufficiently separated on the chromosome [98-101]. These observations indicate that the nucleoid is properly arranged in the cell.

In contrast to the abundance of eukaryotic histones [102], the number of each nucleoid protein is too small to cover the entire genomic region by itself [5] (Table 1). It is an interesting question how the nucleoid fibers, specifically the thinnest 10 nm fibers, are constructed. It may be that distinct protein components can build up physically similar fibrous units.

4. Nucleoid Dynamics

The nucleoid structures are not static or rigid, but dynamically change in response to environmental conditions [3, 70, 103-106]. Active transcription is essential to maintain the structures higher than 10 nm fibers. The nucleoid fibers are further folded up into a condensed state towards "stationary phase" in *E. coli* or in response to oxidative stress in S. aureus (Figure 1C and 5).

4.1. Nucleoid Architecture Coupled with Transcription

The nucleoid proteins constitute the nucleoid architecture and regulate transcription and translation. On the other hand, the activities of transcription and translation themselves seems to contribute to the nucleoid architecture. Evidence for this includes: (i) transcriptional inhibition by rifampicin treatment disperses nucleoid fibers, and (ii) translational inhibition by chloramphenicol causes nucleoid instability, and induces a nucleoid compaction that leads to a ring-like structure in the cell [107, 108].

Rifampicin treatment of bacterial cells completely breaks down 80-nm fibers into thinner fibers (Figure 1E), and the 80 nm fiber is sensitive to the RNase treatment [104] as discussed in Section 2; i.e., the observed 80 nm fibers contain RNA molecules. We have recently found that a chloramphenicol treatment also completely breaks down the 80 nm fibers (unpublished results). Therefore, it is reasonable to assume that the 80 nm fiber is a heterogeneous complex of RNA, DNA, RNA polymerases, ribosomes, and nucleoid proteins and that the 80-nm fiber is the place where transcription and translation occur simultaneously.

The 80 nm fibers constitute roughly about 30% of the total nucleoid structure in the log phase [104]. This is consistent with the fact that 25 % of the ORF is actively transcribed in the log phase [109]. Recently, it has been proposed that active transcription complexes are assembled to form transcription loci *in situ* [110-112]. The GFP-fused RNA polymerase is expressed as several dots "transcription loci" [110, 113], and these dots disperse after rifampicin treatment.

ChIP-chip (chromatin immunoprecipitation and high-density microarrays) assays have determined the distributions of RNA polymerase and three nucleoid proteins (H-HS, Fis, and IHF) on the whole chromosome of *E. coli*. The positions of RNA polymerase on the genomic DNA are unevenly distributed as is the case for other nucleoid proteins, but overlapped with those of H-NS and Fis [31, 114, 115]. Interestingly, H-NS binding regions are well correlated with the transcriptionally silencing genes, predominantly AT-rich regions, [29, 47], and Fis associates with transcriptionally active genomic regions [30]. It has been estimated that there is one RNA polymerase per 85 bp on the rRNA operon, whereas in the other regions there is one molecule per 10-20 kbp [116]. Thus, the rRNA operons may be the candidates for the transcription loci [117].

4.2. Nucleoid Dynamics in Response to Environmental Stress

In order to protect the genetic information, the nucleoid undergoes structural re-organization, depending on the growth states and environmental conditions (reviewed in [118]). In spore-forming bacteria such as *B. subtilis*, the genomic DNA, within the spores, is tightly condensed into a ring-like structure with dipicolinic acid and acid soluble proteins; whereas the ordinary nucleoid proteins are displaced. The DNA-RecA repairosome lattice may provide an example of such a bacterial chromatin re-organization, and Dps-dependent nucleoid condensation may also be another case (Figure 3). In this type of structural conversion, the nucleoid sustains its trascriptional activities; in clear contrast to the complete genome silencing in the spores.

Dps is a protein with a molecular weight of 19 kDa and is a member of the Fe-binding protein family that forms dodecameric complexes [119, 120]. Dps can bind DNA and protect it against oxidative stress [121], nuclease cleavage, UV damage, thermal shock [122] and acidic environments [123]. Dps is one of a very few nucleoid proteins that are conserved in many bacterial species (Table 3) [71], e.g., *S. aureus* has a *dps* homologue termed *mrgA* (<u>m</u>etal ion <u>r</u>esponsive <u>g</u>ene) [124]. However, Dps is induced in a distinct manner depending on the bacterial species. Figure 4 summarizes the difference of Dps induction between *S. aureus* and *E. coli*. This distinct expression profile simply correlates with the condensation states in the stationary phase.

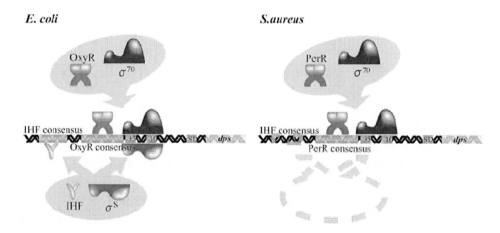

Figure 4. Regulatory mechanisms of dps genes in E. coli and S. aureus. In E. coli, the dps gene expression towards stationary phases is regulated by IHF-binding to the −50 region as well as □s-binding to the −10/-35 region, whereas, in the H2O2 stress, OxyR and □70 stimulate the dps gene expression. In S. aureus, only the PerR/(□70) pathway exists due to the lack of a gene for the IHF/□s pathway.

First, the *E. coli* Dps is the most abundant nucleoid protein in the stationary phase when the nucleoid is tightly condensed [50, 125], whereas *Staphylococcal* MrgA is not induced towards the stationary phase and the nucleoid sustains the fibrous structures [106]. This is due to the lack of the transcription factor (IHF) for the Dps expression in the stationary phase in *S. auresus* but not in *E. coli* (Table 1).

Table 3. Distribution of the nucleoid proteins, histones, topoisomerase and smc proteins

			Histone	Nucleoid Proteins							Topoisomerases								SMC			SMC homologue			RecN	MukB	Rad50/Sbc
				Hu/IHF	H-NS/StpA	Fis	Dps	Hfq	Alba	Sso7	IA I(B)/III	IA I(E)	Revgyr	II	IB V	IIA gyrase/IV	IIB VI		SMC	SMC1	SMC2	SMC3	SMC4	SMC5	SMC6			
Eukaryotes	Animals	Mammals	*H.sapiens*	+	-	-	-	-	-	-	-	+	+	+	+	-	+	-	-	+	+	+	+	+	+	-	-	+
			M.musculus	+	-	-	-	-	-	-	-	+	+	+	+	-	+	-	-	+	+	+	+	+	+	-	-	+
		Insect	*D.melanogaster*	+	-	-	-	-	-	-	-	+	+	+	+	-	+	-	-	+	+	+	+	+	+	-	-	+
		Nematode	*C.elegans*	+	-	-	-	-	-	-	-	+	+	+	+	-	+	-	-	+	+	+	+	+	+	-	-	+
	Plants	Dicotyledon	*A.thaliana*	+	-	-	-	-	-	-	-	+	+	+	+	-	+	-	-	+	+	+	+	+	+	-	-	+
			O.sativa	+	-	-	-	-	-	-	-	+	+	+	+	-	+	-	-	+	+	+	+	+	+	-	-	+
	Protists	Red algae	*C.merolae*	+	-	-	-	-	-	-	-	+	+	+	+	-	+	-	-	+	+	+	+	+	+	-	-	-
		Protozoan	*P.falciparum*	+	-	-	-	-	-	-	-	+	+	+	+	-	+	-	-	+	+	+	+	+	+	-	-	+
		Cellular slime mold	*D.discoideum*	+	-	-	-	-	-	-	-	+	+	+	+	-	+	-	-	+	+	+	+	+	+	-	-	+
	Fungi	Budding yeast	*S.cerevisiae*	+	-	-	-	-	-	-	-	+	+	+	+	-	+	-	-	+	+	+	+	+	+	-	-	+
		Fission yeast	*S.pombe*	+	-	-	-	-	-	-	-	+	+	+	+	-	+	-	-	+	+	+	+	+	+	-	-	+
		Microsporidia	*E.cuniculi*	+	-	-	-	-	-	-	-	+	+	-	+	-	+	-	-	+	+	+	+	+	+	-	-	-
Bacteria	Proteobacteria gamma		*E.coli*	-	+	+	+	+	+	-	-	+	-	-	+	-	+	+	-	-	+	+	+	-	-	-	+	+	+
			Y.pestis	-	+	+	+	+	+	-	-	+	-	-	+	-	+	+	-	-	+	+	+	-	-	-	+	+	+
			Buchnera	-	+	-	-	-	-	-	-	+	-	-	+	-	+	+	-	-	+	+	+	-	-	-	+	+	+
			X.fastidiosa	-	+	-	-	+	+	-	-	+	-	-	+	-	+	+	-	-	+	+	+	-	-	-	+	+	+
			V.cholerae	-	+	-	+	+	+	-	-	+	-	-	+	-	+	+	-	-	+	+	+	-	-	-	+	+	+
			P.multocida	-	+	+	+	+	+	-	-	+	-	-	+	-	+	+	-	-	+	+	+	-	-	-	+	+	+
	epsilon		*H.pylori*	-	+	-	+	+	+	-	-	+	-	-	+	-	+	+	-	-	+	+	+	-	-	-	-	-	+
	alpha		*R.prowazekii*	-	+	-	-	-	+	-	-	+	-	-	+	-	+	+	-	-	+	+	+	-	-	-	-	-	+
	Firmicutes	Bacillales	*B.subtilis*	-	+	-	-	-	+	-	-	+	-	-	+	-	+	+	-	-	+	+	+	-	-	-	+	-	+
			S.aureus	-	+	-	-	-	+	-	-	+	-	-	+	-	+	+	-	-	+	+	+	-	-	-	+	-	+
		Lactobacillales	*S.pyogenes*	-	+	-	-	-	+	-	-	+	-	-	+	-	+	+	-	-	+	+	+	-	-	-	+	-	+
		Clostridia	*C.acetobutylicum*	-	+	-	-	-	+	-	-	+	-	-	+	-	+	+	-	-	+	+	+	-	-	-	+	-	+
		Mollicutes	*M.genitalium*	-	+	-	-	-	-	-	-	+	-	-	+	-	+	+	-	-	+	+	+	-	-	-	-	-	-
	Actinobacteria		*M.tuberculosis*	-	+	-	-	-	+	-	-	+	-	-	+	-	+	+	-	-	+	+	+	-	-	-	+	-	+
			C.glutamicum	-	+	-	-	-	+	-	-	+	-	-	+	-	+	+	-	-	+	+	+	-	-	-	+	-	+
	Chlamydia		*C.muridarum*	-	+	-	-	-	-	-	-	+	-	-	+	-	+	+	-	-	+	+	+	-	-	-	+	-	+
	Spirohete		*B.burgdorferi*	-	+	-	-	-	-	-	-	+	-	-	+	-	+	+	-	-	+	+	+	-	-	-	-	-	+
	Cyanobacteria		*T.elongatus*	-	+	-	+	-	+	-	-	+	-	-	+	-	+	+	-	-	+	+	+	-	-	-	+	-	+
	Green sulfur bacteria		*D.radiodurans*	-	+	-	+	-	-	-	-	+	-	-	+	-	+	+	-	-	+	+	+	-	-	-	+	-	+
	Hyperthermophilic bacteria		*T.maritima*	-	+	-	+	-	+	-	-	+	-	+	+	-	+	+	-	-	+	+	+	-	-	-	-	-	+
Archaea	Euryarchaeota		*Methanococcus jannaschii*	+	-	-	-	-	-	+	-	+	-	+	-	+	-	-	+	+	-	-	-	-	-	-	-	-	+
			Methanosarcina acetivorans	+	-	-	-	-	-	+	-	+	-	+	-	+	-	-	+	+	-	-	-	-	-	-	-	-	+
			Methanosarcina mazei	+	-	-	-	-	-	+	-	+	-	+	-	+	-	-	+	+	-	-	-	-	-	-	-	-	+
			Methanobacterium thermoautotrophicum	+	-	-	-	-	-	+	-	+	-	+	-	+	-	-	+	+	-	-	-	-	-	-	-	-	+
			Methanopyrus kandleri	+	-	-	-	-	-	+	-	+	-	+	-	+	-	-	+	+	-	-	-	-	-	-	-	-	+
			Archaeoglobus fulgidus	+	-	-	-	-	-	+	-	+	-	+	-	+	-	-	+	+	-	-	-	-	-	-	-	-	+
			Halobacterium sp. NRC-1	+	-	-	-	-	-	+	-	+	-	+	-	+	-	-	+	+	-	-	-	-	-	-	-	-	+
			Thermoplasma acidophilum	+	-	-	-	-	-	+	-	+	-	+	-	+	-	-	+	+	-	-	-	-	-	-	-	-	+
			Pyrococcus horikoshii	+	-	-	-	-	-	+	-	+	-	+	-	+	-	-	+	+	-	-	-	-	-	-	-	-	+
			Thermococcus kodakaraensis	+	-	-	-	-	-	+	-	+	-	+	-	+	-	-	+	+	-	-	-	-	-	-	-	-	+
	Crenarchaeota		*Aeropyrum pernix*	-	-	-	-	-	+	+	-	+	-	+	-	+	-	-	+	+	-	-	-	-	-	-	-	-	+
			Sulfolobus solfataricus	-	-	-	-	-	+	+	-	+	-	+	-	+	-	-	+	+	-	-	-	-	-	-	-	-	+
			Pyrobaculum calidifontis	-	-	-	-	-	+	+	-	+	-	+	-	+	-	-	+	+	-	-	-	-	-	-	-	-	+
	Nanoarchaeota		*Nanoarchaeum equitans*	+	-	-	-	-	-	-	-	+	-	-	-	-	-	-	-	-	-	-	-	-	-	-	-	-	+

Indeed, the nucleoid of *S. aureus* undergoes the condensation procesess by molecular genetic manipulations (*mrgA* overexpression or *perR* knockout) and interestingly cell proliferation is still allowed, indicating that the MrgA-dependent nucleoid condensation can permit the gene expression and genome duplication [106].

Second, both *E. coli dps* and *S. aureus mrgA* are specifically induced by oxidative stresses (Figure 4). The oxidative stress then causes nucleoid condensation through the induction of MrgA (Figure 5).

Interestingly, however, it cannot induce the nucleoid condensation in the log phase of *E. coli*, even though the *dps*-gene expression is up-regulated. This is simply due to the presence of Fis (Figure 6), which is γ-Proteobacteria specific and the most abundant nucleoid protein in the log phase of *E. coli* (Table 1 and 3). Fis interrupts the changes of DNA topology required for the Dps-induced nucleoid condensation by repressing the expressions of DNA gyrase and Topoiosmerase I [105].

In addition, the binding of Fis to DNA itself prevents the topological changes [28]. It is interesting that Fis is not detectable in the stationary phase of *E. coli* and that there is no Fis homologue in *S. aureus*.

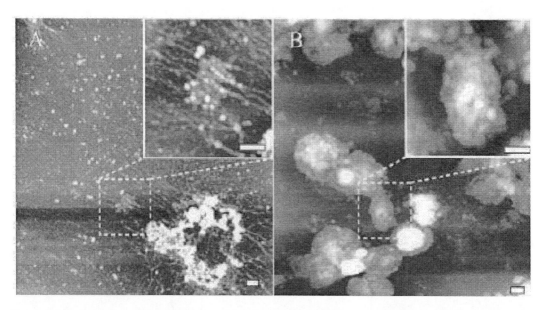

Figure 5. Nucleoid dynamics of S. aureus. AFM images of S. aureus nucleoid (A) in the stationary phase and (B) that of after the treatment with an oxidative stress element (phenanthrenquinone). Scale bars represent 500 nm.

When bacteria are confronted with stresses such as starvation (*E. coli*) or oxidative stress (*S. aureus*), nucleoid condensation seems to play a protective role. Dps-dependent nucleoid compaction would be an effective way to achieve the flexible change of the nucleoid structure towards a protective form. At the same time, this type of condensation is beneficial in that bacteria can still respond to the environmental stresses by expressing the genetic information embedded in the compacted nucleoid [126].

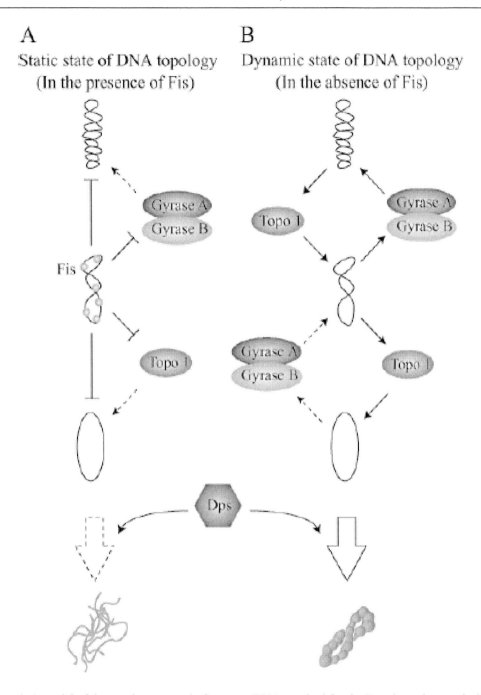

Figure 6. A model of the topology control of genome DNA required for the Dps-dependent nucleoid condensation. (A) Static state. In the presence of Fis, DNA topology is maintained statically by the repression of Topo I and DNA gyrase expressions, and also by the physical blocking against the topological changes of DNA topoisomers. Therefore, the expression of Dps does not lead to a nucleoid condensation. (B) Dynamic state. In the absence of Fis, Topo I and DNA gyrase are induced and may easily change the DNA topology. This dynamic status facilitates the nucleoid condensation by Dps (Topo I and DNA gyrase coexist).

5. Specificity and Flexibility of Nucleoid Proteins for Nuclear Architectures

Only HU, Dps, Topo I and Topo IIA (Gyrase) are conserved in many bacterial species, and most nucleoid proteins including Fis and H-NS are not shared among all the bacteria (Table 3). Nevertheless, the hierarchies of nucleoid fibers look extremely similar among bacterial species. The thinnest fiber attained with DNA and nucleoid proteins exhibits a diameter of 10 nm, without the formation of eukaryote-type nucleosomes, and turns into a 30-nm fiber with RNA and other proteins.

The mechanism for the 10-nm fiber formation has not been elucidated, but it will be an interesting issue in the future investigations. In eukaryotes in contrast, a 10-nm beads-on-a-string fiber is formed with core histones, and converted into a 30-nm fiber. The core histones are essential for the eukaryotic chromatin structure [127-131], and, so far, no eukaryotic species which lacks the core histones has been discovered (Table 3). Topo IA, Topo IIA and at least one of the SMC homologues are also essential and present in all eukaryotes [132, 133].

Surprisingly, none of the nucleoid proteins are essential in *E. coli*; even HU, the most popular nucleoid component protein, seems not to be essential for survival, although it is believed to play a critical role for the nucleoid formation. A set of proteins belonging to distinct families may co-operate toward the construction of the common nucleoid architecture, in which a lack of particular protein would be compensated by others.

This notion is supported by the facts that: (i) a deletion of a single kind of nucleoid protein usually does not lead to a lethal phenotype, but (ii) a combination mutant of deletions of different protein families causes lethality. In *E. coli*, for example, a deletion of one of 4 genes encoding HU homologues (Hu-1, Hu-2, IHF-A and IHF-B) is not lethal, although the growth speed and the morphology of the cells are severely affected [134]. However, double deletions in combination with a non-lethal H-NS mutation exhibit a lethality [135]. HU/MukB double mutant is also lethal [136], although the MukB deletion alone is not. Such flexibility in the nucloid protein components in sustaining the robust nucleoid architecture might have allowed the diversification of the nucleoid proteins. Some of the species-specific nucleoid proteins may be advantageous to survival in certain situation.

It has been thought that, in eukatyotes, the access of the transcription factors to their binding sites is restricted by the nucleosomes, to which 'nucleosome-remodeling factors' attack to achieve the RNA production (reviewed in [137]). In contrast, the active transcription is required to sustain the higher order structures of the bacterial nucleoid. In such nucleoid status, transcription factors can freely access to their binding sites [138].

In general, bacteria can express genes more quickly than eukaryotes in response to stimuli. The expression of bacterial UV responsive genes takes only a few minutes [139, 140], whereas the eukaryote response takes over ten minutes [141-143].

These nucleoid characteristics may allow bacterial cells to quickly respond to environmental changes [139-141, 144, 145].

Conclusion

Both prokaryotes and eukaryotes store their genomic DNA in an environment, which balances the physical properties of double-stranded DNA with the mechanical effects of DNA-protein interactions [146, 147]. There are critical differences between eukaryotic chromatin, bacterial and archaeal nucleoids in terms of the protein compositions. It is of great interest that eukaryotic mitochondria and chloroplast contain neither HU nor histone, but their nuceoid architectures are similar to those of bacteria and eukaryotes [3].

It has been thought that mitochondria, chloroplast and eukaryotic nucleoid/chromatin underwent symbiotic events where their ancestral bacterial and/or achaeal genetic information had been re-organized into the current state [148]. The further comparative investigations on the architecture and dynamics of genomes in three domains of life will provide a line of useful information about the principles of genome folding mechanisms that have been underlying throughout the evolutionary processes emanating three domains of living organisms.

References

[1] Poplawski A, Bernander R: Nucleoid structure and distribution in thermophilic Archaea. *J. Bacteriol* 1997, 179(24):7625-7630.

[2] Robinow C, Kellenberger E: The bacterial nucleoid revisited. *Microbiol. Rev.* 1994, 58(2):211-232.

[3] Ohniwa RL, Morikawa K, Kim J, Kobori T, Hizume K, Matsumi R, Atomi H, Imanaka T, Ohta T, Wada C et al: Atomic force microscopy dissects the hierarchy of genome architectures in eukaryote, prokaryote, and chloroplast. *Microsc. Microanal.* 2007, 13(1):3-12.

[4] Kundu TK, Kusano S, Ishihama A: Promoter selectivity of Escherichia coli RNA polymerase sigmaF holoenzyme involved in transcription of flagellar and chemotaxis genes. *J. Bacteriol.* 1997, 179(13):4264-4269.

[5] Azam TA, Ishihama A: Twelve species of the nucleoid-associated protein from Escherichia coli. Sequence recognition specificity and DNA binding affinity. *J. Biol. Chem.* 1999, 274(46):33105-33113.

[6] Drlica K, Rouviere-Yaniv J: Histonelike proteins of bacteria. *Microbiol Rev* 1987, 51(3):301-319.

[7] Rouviere-Yaniv J, Yaniv M, Germond JE: E. coli DNA binding protein HU forms nucleosomelike structure with circular double-stranded DNA. *Cell* 1979, 17(2):265-274.

[8] Talukder AA, Iwata A, Nishimura A, Ueda S, Ishihama A: Growth phase-dependent variation in protein composition of the Escherichia coli nucleoid. *J. Bacteriol.* 1999, 181(20):6361-6370.

[9] Hiraga S: Dynamic localization of bacterial and plasmid chromosomes. *Annu. Rev. Genet.* 2000, 34:21-59.

[10] Hirano M, Hirano T: ATP-dependent aggregation of single-stranded DNA by a bacterial SMC homodimer. *Embo J.* 1998, 17(23):7139-7148.
[11] Hirano M, Hirano T: Positive and negative regulation of SMC-DNA interactions by ATP and accessory proteins. *Embo J.* 2004, 23(13):2664-2673.
[12] Strick TR, Kawaguchi T, Hirano T: Real-time detection of single-molecule DNA compaction by condensin I. *Curr. Biol.* 2004, 14(10):874-880.
[13] Britton RA, Lin DC, Grossman AD: Characterization of a prokaryotic SMC protein involved in chromosome partitioning. *Genes Dev.* 1998, 12(9):1254-1259.
[14] Niki H, Jaffe A, Imamura R, Ogura T, Hiraga S: The new gene mukB codes for a 177 kd protein with coiled-coil domains involved in chromosome partitioning of E. coli. *Embo J.* 1991, 10(1):183-193.
[15] Hayat MA, Mancarella DA: Nucleoid proteins. *Micron.* 1995, 26(5):461-480.
[16] Tabuchi H, Handa H, Hirose S: Underwinding of DNA on binding of yeast TFIID to the TATA element. *Biochem. Biophys. Res. Commun.* 1993, 192(3):1432-1438.
[17] Paull TT, Haykinson MJ, Johnson RC: The nonspecific DNA-binding and -bending proteins HMG1 and HMG2 promote the assembly of complex nucleoprotein structures. *Genes Dev.* 1993, 7(8):1521-1534.
[18] Swinger KK, Rice PA: IHF and HU: flexible architects of bent DNA. *Curr. Opin. Struct Biol.* 2004, 14(1):28-35.
[19] Swinger KK, Lemberg KM, Zhang Y, Rice PA: Flexible DNA bending in HU-DNA cocrystal structures. *Embo J.* 2003, 22(14):3749-3760.
[20] Dame RT, Noom MC, Wuite GJ: Bacterial chromatin organization by H-NS protein unravelled using dual DNA manipulation. *Nature* 2006, 444(7117):387-390.
[21] Hubner P, Arber W: Mutational analysis of a prokaryotic recombinational enhancer element with two functions. *Embo J.* 1989, 8(2):577-585.
[22] Skoko D, Yan J, Johnson RC, Marko JF: Low-force DNA condensation and discontinuous high-force decondensation reveal a loop-stabilizing function of the protein Fis. *Phys Rev. Lett* 2005, 95(20):208101.
[23] Skoko D, Yoo D, Bai H, Schnurr B, Yan J, McLeod SM, Marko JF, Johnson RC: Mechanism of chromosome compaction and looping by the Escherichia coli nucleoid protein Fis. *J. Mol. Biol.* 2006, 364(4):777-798.
[24] Dame RT, Goosen N: HU: promoting or counteracting DNA compaction? *FEBS Lett* 2002, 529(2-3):151-156.
[25] van Noort J, Verbrugge S, Goosen N, Dekker C, Dame RT: Dual architectural roles of HU: formation of flexible hinges and rigid filaments. *Proc. Natl. Acad. Sci. USA* 2004, 101(18):6969-6974.
[26] Hardy CD, Cozzarelli NR: A genetic selection for supercoiling mutants of Escherichia coli reveals proteins implicated in chromosome structure. *Mol. Microbiol* 2005, 57(6):1636-1652.
[27] Schneider R, Lurz R, Luder G, Tolksdorf C, Travers A, Muskhelishvili G: An architectural role of the Escherichia coli chromatin protein FIS in organising DNA. *Nucleic Acids Res.* 2001, 29(24):5107-5114.

[28] Schneider R, Travers A, Muskhelishvili G: FIS modulates growth phase-dependent topological transitions of DNA in Escherichia coli. *Mol. Microbiol* 1997, 26(3):519-530.

[29] Oshima T, Ishikawa S, Kurokawa K, Aiba H, Ogasawara N: Escherichia coli histone-like protein H-NS preferentially binds to horizontally acquired DNA in association with RNA polymerase. *DNA Res.* 2006, 13(4):141-153.

[30] Grainger DC, Hurd D, Goldberg MD, Busby SJ: Association of nucleoid proteins with coding and non-coding segments of the Escherichia coli genome. *Nucleic Acids Res* 2006, 34(16):4642-4652.

[31] Grainger DC, Hurd D, Harrison M, Holdstock J, Busby SJ: Studies of the distribution of Escherichia coli cAMP-receptor protein and RNA polymerase along the E. coli chromosome. *Proc. Natl. Acad. Sci. USA* 2005, 102(49):17693-17698.

[32] Luijsterburg MS, Noom MC, Wuite GJ, Dame RT: The architectural role of nucleoid-associated proteins in the organization of bacterial chromatin: a molecular perspective. *J. Struct. Biol.* 2006, 156(2):262-272.

[33] Hsu YH, Chung MW, Li TK: Distribution of gyrase and topoisomerase IV on bacterial nucleoid: implications for nucleoid organization. *Nucleic Acids Res.* 2006, 34(10):3128-3138.

[34] Gille H, Egan JB, Roth A, Messer W: The FIS protein binds and bends the origin of chromosomal DNA replication, oriC, of Escherichia coli. *Nucleic Acids Res.* 1991, 19(15):4167-4172.

[35] Chodavarapu S, Felczak MM, Yaniv JR, Kaguni JM: Escherichia coli DnaA interacts with HU in initiation at the E. coli replication origin. *Mol. Microbiol.* 2008, 67(4):781-792.

[36] Chodavarapu S, Gomez R, Vicente M, Kaguni JM: Escherichia coli Dps interacts with DnaA protein to impede initiation: a model of adaptive mutation. *Mol. Microbiol.* 2008, 67(6):1331-1346.

[37] Hwang DS, Kornberg A: Opening of the replication origin of Escherichia coli by DnaA protein with protein HU or IHF. *J. Biol. Chem.* 1992, 267(32):23083-23086.

[38] Polaczek P: Bending of the origin of replication of E. coli by binding of IHF at a specific site. *New Biol.* 1990, 2(3):265-271.

[39] Roth A, Urmoneit B, Messer W: Functions of histone-like proteins in the initiation of DNA replication at oriC of Escherichia coli. *Biochimie* 1994, 76(10-11):917-923.

[40] Filutowicz M, Ross W, Wild J, Gourse RL: Involvement of Fis protein in replication of the Escherichia coli chromosome. *J Bacteriol* 1992, 174(2):398-407.

[41] Schneider R, Travers A, Kutateladze T, Muskhelishvili G: A DNA architectural protein couples cellular physiology and DNA topology in Escherichia coli. *Mol. Microbiol.* 1999, 34(5):953-964.

[42] Weinstein-Fischer D, Elgrably-Weiss M, Altuvia S: Escherichia coli response to hydrogen peroxide: a role for DNA supercoiling, topoisomerase I and Fis. *Mol. Microbiol.* 2000, 35(6):1413-1420.

[43] Kar S, Edgar R, Adhya S: Nucleoid remodeling by an altered HU protein: reorganization of the transcription program. *Proc. Natl. Acad. Sci. USA* 2005, 102(45):16397-16402.

[44] Nieto JM, Mourino M, Balsalobre C, Madrid C, Prenafeta A, Munoa FJ, Juarez A: Construction of a double hha hns mutant of Escherichia coli: effect on DNA supercoiling and alpha-haemolysin production. *FEMS Microbiol Lett* 1997, 155(1):39-44.

[45] Nieto JM, Madrid C, Miquelay E, Parra JL, Rodriguez S, Juarez A: Evidence for direct protein-protein interaction between members of the enterobacterial Hha/YmoA and H-NS families of proteins. *J. Bacteriol* 2002, 184(3):629-635.

[46] Olekhnovich IN, Kadner RJ: Role of nucleoid-associated proteins Hha and H-NS in expression of Salmonella enterica activators HilD, HilC, and RtsA required for cell invasion. *J. Bacteriol.* 2007, 189(19):6882-6890.

[47] Navarre WW, Porwollik S, Wang Y, McClelland M, Rosen H, Libby SJ, Fang FC: Selective silencing of foreign DNA with low GC content by the H-NS protein in Salmonella. *Science* 2006, 313(5784):236-238.

[48] Altuvia S, Almiron M, Huisman G, Kolter R, Storz G: The dps promoter is activated by OxyR during growth and by IHF and sigma S in stationary phase. *Mol. Microbiol.* 1994, 13(2):265-272.

[49] Lomovskaya OL, Kidwell JP, Matin A: Characterization of the sigma 38-dependent expression of a core Escherichia coli starvation gene, pexB. *J. Bacteriol.* 1994, 176(13):3928-3935.

[50] Kim J, Yoshimura SH, Hizume K, Ohniwa RL, Ishihama A, Takeyasu K: Fundamental structural units of the Escherichia coli nucleoid revealed by atomic force microscopy. *Nucleic Acids Res.* 2004, 32(6):1982-1992.

[51] Balandina A, Claret L, Hengge-Aronis R, Rouviere-Yaniv J: The Escherichia coli histone-like protein HU regulates rpoS translation. *Mol. Microbiol.* 2001, 39(4):1069-1079.

[52] Balandina A, Kamashev D, Rouviere-Yaniv J: The bacterial histone-like protein HU specifically recognizes similar structures in all nucleic acids. DNA, RNA, and their hybrids. *J. Biol. Chem.* 2002, 277(31):27622-27628.

[53] Brescia CC, Kaw MK, Sledjeski DD: The DNA binding protein H-NS binds to and alters the stability of RNA in vitro and in vivo. *J. Mol. Biol.* 2004, 339(3):505-514.

[54] Nakamura K, Yahagi S, Yamazaki T, Yamane K: Bacillus subtilis histone-like protein, HBsu, is an integral component of a SRP-like particle that can bind the Alu domain of small cytoplasmic RNA. *J. Biol. Chem.* 1999, 274(19):13569-13576.

[55] Yamashino T, Ueguchi C, Mizuno T: Quantitative control of the stationary phase-specific sigma factor, sigma S, in Escherichia coli: involvement of the nucleoid protein H-NS. *Embo J.* 1995, 14(3):594-602.

[56] Arluison V, Hohng S, Roy R, Pellegrini O, Regnier P, Ha T: Spectroscopic observation of RNA chaperone activities of Hfq in post-transcriptional regulation by a small non-coding RNA. *Nucleic Acids Res.* 2007, 35(3):999-1006.

[57] Brescia CC, Mikulecky PJ, Feig AL, Sledjeski DD: Identification of the Hfq-binding site on DsrA RNA: Hfq binds without altering DsrA secondary structure. *Rna* 2003, 9(1):33-43.

[58] Moll I, Afonyushkin T, Vytvytska O, Kaberdin VR, Blasi U: Coincident Hfq binding and RNase E cleavage sites on mRNA and small regulatory RNAs. *Rna* 2003, 9(11):1308-1314.

[59] Morita T, Mochizuki Y, Aiba H: Translational repression is sufficient for gene silencing by bacterial small noncoding RNAs in the absence of mRNA destruction. *Proc. Natl. Acad. Sci. USA* 2006, 103(13):4858-4863.

[60] Zhang A, Wassarman KM, Rosenow C, Tjaden BC, Storz G, Gottesman S: Global analysis of small RNA and mRNA targets of Hfq. *Mol. Microbiol.* 2003, 50(4):1111-1124.

[61] Mayer O, Waldsich C, Grossberger R, Schroeder R: Folding of the td pre-RNA with the help of the RNA chaperone StpA. *Biochem. Soc. Trans* 2002, 30(Pt 6):1175-1180.

[62] Piekarski G: Zytologische Untersuchungen an Paratyphus- und Colibacterien. *Arch Mikrobiol.* 1937, 8.:428-429.

[63] Sloof P, Maagdelijn A, Boswinkel E: Folding of prokaryotic DNA. Isolation and characterization of nucleoids from Bacillus licheniformis. *J. Mol. Biol.* 1983, 163(2):277-297.

[64] Kavenoff R, Ryder OA: Electron microscopy of membrane-associated folded chromosomes of Escherichia coli. *Chromosoma* 1976, 55(1):13-25.

[65] Kavenoff R, Bowen BC: Electron microscopy of membrane-free folded chromosomes from Escherichia coli. *Chromosoma* 1976, 59(2):89-101.

[66] Delius H, Worcel A: Letter: Electron microscopic visualization of the folded chromosome of Escherichia coli. *J. Mol. Biol.* 1974, 82(1):107-109.

[67] Postow L, Hardy CD, Arsuaga J, Cozzarelli NR: Topological domain structure of the Escherichia coli chromosome. *Genes Dev* 2004, 18(14):1766-1779.

[68] Sinden RR, Pettijohn DE: Chromosomes in living Escherichia coli cells are segregated into domains of supercoiling. *Proc. Natl. Acad. Sci. USA* 1981, 78(1):224-228.

[69] Worcel A, Burgi E: On the structure of the folded chromosome of Escherichia coli. *J. Mol. Biol.* 1972, 71(2):127-147.

[70] Kim J, Yoshimura SH, Ohniwa RL, Wada A, Ishihama A, Takeyasu K: The Characterization of Fundamental Structure of Bacterial Chromosome by the Plasmid Structural Analysis with Atomic Force Microscopy. *44TH ANNUAL MEETING of The American Society for Cell Biology* 2004:poster 2503.

[71] Takeyasu K, Kim J, Ohniwa RL, Kobori T, Inose Y, Morikawa K, Ohta T, Ishihama A, Yoshimura SH: Genome architecture studied by nanoscale imaging: analyses among bacterial phyla and their implication to eukaryotic genome folding. *Cytogenet Genome Res* 2004, 107(1-2):38-48.

[72] Dworsky P, Schaechter M: Effect of rifampin on the structure and membrane attachment of the nucleoid of Escherichia coli. *J. Bacteriol.* 1973, 116(3):1364-1374.

[73] Kleppe K, Ovrebo S, Lossius I: The bacterial nucleoid. *J. Gen. Microbiol.* 1979, 112(1):1-13.

[74] Vos-Scheperkeuter GH, Witholt B: Co-translational insertion of envelope proteins: theoretical consideration and implications. *Ann. Microbiol. (Paris)* 1982, 133A(1):129-138.

[75] Zimmerman SB, Murphy LD: Macromolecular crowding and the mandatory condensation of DNA in bacteria. *FEBS Lett.* 1996, 390(3):245-248.

[76] Lynch AS, Wang JC: Anchoring of DNA to the bacterial cytoplasmic membrane through cotranscriptional synthesis of polypeptides encoding membrane proteins or proteins for export: a mechanism of plasmid hypernegative supercoiling in mutants deficient in DNA topoisomerase I. *J. Bacteriol.* 1993, 175(6):1645-1655.

[77] Kruse T, Blagoev B, Lobner-Olesen A, Wachi M, Sasaki K, Iwai N, Mann M, Gerdes K: Actin homolog MreB and RNA polymerase interact and are both required for chromosome segregation in Escherichia coli. *Genes Dev.* 2006, 20(1):113-124.

[78] Sun Q, Margolin W: Effects of perturbing nucleoid structure on nucleoid occlusion-mediated toporegulation of FtsZ ring assembly. *J. Bacteriol.* 2004, 186(12):3951-3959.

[79] Schreil WH: Studies On The Fixation Of Artificial And Bacterial Dna Plasms For The Electron Microscopy Of Thin Sections. *J. Cell Biol.* 1964, 22:1-20.

[80] Ryter A, Kellenberger E, Birchandersen A, Maaloe O: Etude au microscope électronique de plasmas contenant de l'acide désoxyribonucléique. I. Les nucléoides des bactéries en croissance active. *Z. Naturforsch B* 1958, 13:597-605.

[81] Giesbrecht P, Piekarski G: Zur Organisation des Zellkerns von Bacillus megaterium. *Arch Microbiol* 1958, 31:68-81.

[82] Kellenberger E, Arnold-Schulz-Gahmen B: Chromatins of low-protein content: special features of their compaction and condensation. *FEMS Microbiol Lett.* 1992, 79(1-3):361-370.

[83] Matias VR, Beveridge TJ: Native cell wall organization shown by cryo-electron microscopy confirms the existence of a periplasmic space in Staphylococcus aureus. *J. Bacteriol.* 2006, 188(3):1011-1021.

[84] Eltsov M, Dubochet J: Study of the Deinococcus radiodurans nucleoid by cryoelectron microscopy of vitreous sections: Supplementary comments. *J. Bacteriol.* 2006, 188(17):6053-6058; discussion 6059.

[85] Matias VR, Beveridge TJ: Cryo-electron microscopy reveals native polymeric cell wall structure in Bacillus subtilis 168 and the existence of a periplasmic space. *Mol. Microbiol.* 2005, 56(1):240-251.

[86] Eltsov M, Dubochet J: Fine structure of the Deinococcus radiodurans nucleoid revealed by cryoelectron microscopy of vitreous sections. *J. Bacteriol.* 2005, 187(23):8047-8054.

[87] Al-Amoudi A, Chang JJ, Leforestier A, McDowall A, Salamin LM, Norlen LP, Richter K, Blanc NS, Studer D, Dubochet J: Cryo-electron microscopy of vitreous sections. *Embo J.* 2004, 23(18):3583-3588.

[88] Matias A, de la Riva J, Martinez E, Torrez M, Dujardin JP: Domiciliation process of Rhodnius stali (Hemiptera: Reduviidae) in Alto Beni, La Paz, Bolivia. *Trop. Med. Int. Health* 2003, 8(3):264-268.

[89] Fogel MA, Waldor MK: Distinct segregation dynamics of the two Vibrio cholerae chromosomes. *Mol. Microbiol.* 2005, 55(1):125-136.

[90] Gordon GS, Sitnikov D, Webb CD, Teleman A, Straight A, Losick R, Murray AW, Wright A: Chromosome and low copy plasmid segregation in E. coli: visual evidence for distinct mechanisms. *Cell* 1997, 90(6):1113-1121.

[91] Jensen RB, Shapiro L: The Caulobacter crescentus smc gene is required for cell cycle progression and chromosome segregation. *Proc. Natl. Acad. Sci. USA* 1999, 96(19):10661-10666.

[92] Kahng LS, Shapiro L: Polar localization of replicon origins in the multipartite genomes of Agrobacterium tumefaciens and Sinorhizobium meliloti. *J. Bacteriol.* 2003, 185(11):3384-3391.

[93] Li Y, Sergueev K, Austin S: The segregation of the Escherichia coli origin and terminus of replication. *Mol. Microbiol.* 2002, 46(4):985-996.

[94] Niki H, Hiraga S: Polar localization of the replication origin and terminus in Escherichia coli nucleoids during chromosome partitioning. *Genes Dev.* 1998, 12(7):1036-1045.

[95] Wang X, Possoz C, Sherratt DJ: Dancing around the divisome: asymmetric chromosome segregation in Escherichia coli. *Genes Dev.* 2005, 19(19):2367-2377.

[96] Webb CD, Teleman A, Gordon S, Straight A, Belmont A, Lin DC, Grossman AD, Wright A, Losick R: Bipolar localization of the replication origin regions of chromosomes in vegetative and sporulating cells of B. subtilis. *Cell* 1997, 88(5):667-674.

[97] Viollier PH, Thanbichler M, McGrath PT, West L, Meewan M, McAdams HH, Shapiro L: Rapid and sequential movement of individual chromosomal loci to specific subcellular locations during bacterial DNA replication. *Proc. Natl. Acad. Sci. USA* 2004, 101(25):9257-9262.

[98] Lau IF, Filipe SR, Soballe B, Okstad OA, Barre FX, Sherratt DJ: Spatial and temporal organization of replicating Escherichia coli chromosomes. *Mol. Microbiol.* 2003, 49(3):731-743.

[99] Michaelis C, Ciosk R, Nasmyth K: Cohesins: chromosomal proteins that prevent premature separation of sister chromatids. *Cell* 1997, 91(1):35-45.

[100] Robinett CC, Straight A, Li G, Willhelm C, Sudlow G, Murray A, Belmont AS: In vivo localization of DNA sequences and visualization of large-scale chromatin organization using lac operator/repressor recognition. *J. Cell Biol.* 1996, 135(6 Pt 2):1685-1700.

[101] Straight AF, Belmont AS, Robinett CC, Murray AW: GFP tagging of budding yeast chromosomes reveals that protein-protein interactions can mediate sister chromatid cohesion. *Curr. Biol.* 1996, 6(12):1599-1608.

[102] Bernstein BE, Liu CL, Humphrey EL, Perlstein EO, Schreiber SL: Global nucleosome occupancy in yeast. *Genome Biol.* 2004, 5(9):R62.

[103] Morikawa K, Ohniwa RL, Kim J, Takeshita SL, Maruyama A, Inose Y, Takeyasu K, Ohta T: Biochemical, molecular genetic, and structural analyses of the staphylococcal nucleoid. *Microsc. Microanal* 2007, 13(1):30-35.

[104] Ohniwa RL, Morikawa K, Takeshita SL, Kim J, Ohta T, Wada C, Takeyasu K: Transcription coupled nucleoid archtiecture in bacteria. *Genes Cells* 2007, 12(10):In press.

[105] Ohniwa RL, Morikawa K, Kim J, Ohta T, Ishihama A, Wada C, Takeyasu K: Dynamic state of DNA topology is essential for genome condensation in bacteria. *Embo J.* 2006, 25(23):5591-5602.

[106] Morikawa K, Ohniwa RL, Kim J, Maruyama A, Ohta T, Takeyasu K: Bacterial nucleoid dynamics: oxidative stress response in Staphylococcus aureus. *Genes Cells* 2006, 11(4):409-423.

[107] Morgan C, Rosenkranz HS, Carr HS, Rose HM: Electron microscopy of chloramphenicol-treated Escherichia coli. *J. Bacteriol.* 1967, 93(6):1987-2002.

[108] Zimmerman SB: Shape and compaction of Escherichia coli nucleoids. *J. Struct Biol.* 2006, 156(2):255-261.

[109] Richmond CS, Glasner JD, Mau R, Jin H, Blattner FR: Genome-wide expression profiling in Escherichia coli K-12. *Nucleic Acids Res* 1999, 27(19):3821-3835.

[110] Jin DJ, Cabrera JE: Coupling the distribution of RNA polymerase to global gene regulation and the dynamic structure of the bacterial nucleoid in Escherichia coli. *J. Struct. Biol.* 2006, 156(2):284-291.

[111] Marenduzzo D, Faro-Trindade I, Cook PR: What are the molecular ties that maintain genomic loops? *Trends Genet.* 2007, 23(3):126-133.

[112] Deng S, Stein RA, Higgins NP: Organization of supercoil domains and their reorganization by transcription. *Mol. Microbiol.* 2005, 57(6):1511-1521.

[113] Cabrera JE, Jin DJ: The distribution of RNA polymerase in Escherichia coli is dynamic and sensitive to environmental cues. *Mol. Microbiol.* 2003, 50(5):1493-1505.

[114] Bon M, McGowan SJ, Cook PR: Many expressed genes in bacteria and yeast are transcribed only once per cell cycle. *Faseb J.* 2006, 20(10):1721-1723.

[115] Herring CD, Raffaelle M, Allen TE, Kanin EI, Landick R, Ansari AZ, Palsson BO: Immobilization of Escherichia coli RNA polymerase and location of binding sites by use of chromatin immunoprecipitation and microarrays. *J. Bacteriol.* 2005, 187(17):6166-6174.

[116] French SL, Miller OL, Jr.: Transcription mapping of the Escherichia coli chromosome by electron microscopy. *J. Bacteriol.* 1989, 171(8):4207-4216.

[117] Bremer H, Dennis P: Modulation of chemical composition and other parameters of the cell by growth rate: American Society for Microbiology, Washington, DC; 1996.

[118] Frenkiel-Krispin D, Minsky A: Nucleoid organization and the maintenance of DNA integrity in E. coli, B. subtilis and D. radiodurans. *J. Struct Biol* 2006, 156(2):311-319.

[119] Almiron M, Link AJ, Furlong D, Kolter R: A novel DNA-binding protein with regulatory and protective roles in starved Escherichia coli. *Genes Dev.* 1992, 6(12B):2646-2654.

[120] Grant RA, Filman DJ, Finkel SE, Kolter R, Hogle JM: The crystal structure of Dps, a ferritin homolog that binds and protects DNA. *Nat. Struct. Biol.* 1998, 5(4):294-303.

[121] Martinez A, Kolter R: Protection of DNA during oxidative stress by the nonspecific DNA-binding protein Dps. *J. Bacteriol.* 1997, 179(16):5188-5194.

[122] Nair S, Finkel SE: Dps protects cells against multiple stresses during stationary phase. *J. Bacteriol.* 2004, 186(13):4192-4198.

[123] Choi SH, Baumler DJ, Kaspar CW: Contribution of dps to acid stress tolerance and oxidative stress tolerance in Escherichia coli O157:H7. *Appl. Environ. Microbiol.* 2000, 66(9):3911-3916.

[124] Chen L, James LP, Helmann JD: Metalloregulation in Bacillus subtilis: isolation and characterization of two genes differentially repressed by metal ions. *J. Bacteriol.* 1993, 175(17):5428-5437.

[125] Wolf SG, Frenkiel D, Arad T, Finkel SE, Kolter R, Minsky A: DNA protection by stress-induced biocrystallization. *Nature* 1999, 400(6739):83-85.

[126] Sangurdekar DP, Srienc F, Khodursky AB: A classification based framework for quantitative description of large-scale microarray data. *Genome Biol.* 2006, 7(4):R32.

[127] Rykowski MC, Wallis JW, Choe J, Grunstein M: Histone H2B subtypes are dispensable during the yeast cell cycle. *Cell* 1981, 25(2):477-487.

[128] Kolodrubetz D, Rykowski MC, Grunstein M: Histone H2A subtypes associate interchangeably in vivo with histone H2B subtypes. *Proc. Natl. Acad. Sci. USA* 1982, 79(24):7814-7818.

[129] Smith MM, Stirling VB: Histone H3 and H4 gene deletions in Saccharomyces cerevisiae. *J. Cell Biol.* 1988, 106(3):557-566.

[130] Stoler S, Keith KC, Curnick KE, Fitzgerald-Hayes M: A mutation in CSE4, an essential gene encoding a novel chromatin-associated protein in yeast, causes chromosome nondisjunction and cell cycle arrest at mitosis. *Genes Dev.* 1995, 9(5):573-586.

[131] Smith MM, Santisteban MS: Genetic dissection of histone function. *Methods* 1998, 15(4):269-281.

[132] Hirano T: At the heart of the chromosome: SMC proteins in action. *Nat. Rev. Mol. Cell Biol* 2006, 7(5):311-322.

[133] Yanagida M: Basic mechanism of eukaryotic chromosome segregation. *Philos Trans R Soc. Lond B Biol. Sci.* 2005, 360(1455):609-621.

[134] Kano Y, Imamoto F: Requirement of integration host factor (IHF) for growth of Escherichia coli deficient in HU protein. *Gene* 1990, 89(1):133-137.

[135] Yasuzawa K, Hayashi N, Goshima N, Kohno K, Imamoto F, Kano Y: Histone-like proteins are required for cell growth and constraint of supercoils in DNA. *Gene* 1992, 122(1):9-15.

[136] Graumann PL: SMC proteins in bacteria: condensation motors for chromosome segregation? *Biochimie* 2001, 83(1):53-59.

[137] Li B, Carey M, Workman JL: The role of chromatin during transcription. *Cell* 2007, 128(4):707-719.

[138] Wade JT, Reppas NB, Church GM, Struhl K: Genomic analysis of LexA binding reveals the permissive nature of the Escherichia coli genome and identifies unconventional target sites. *Genes Dev.* 2005, 19(21):2619-2630.

[139] Courcelle J, Khodursky A, Peter B, Brown PO, Hanawalt PC: Comparative gene expression profiles following UV exposure in wild-type and SOS-deficient Escherichia coli. *Genetics* 2001, 158(1):41-64.

[140] Michan C, Manchado M, Dorado G, Pueyo C: In vivo transcription of the Escherichia coli oxyR regulon as a function of growth phase and in response to oxidative stress. *J. Bacteriol.* 1999, 181(9):2759-2764.

[141] Gasch AP, Spellman PT, Kao CM, Carmel-Harel O, Eisen MB, Storz G, Botstein D, Brown PO: Genomic expression programs in the response of yeast cells to environmental changes. *Mol. Biol. Cell* 2000, 11(12):4241-4257.

[142] Garcia R, Bermejo C, Grau C, Perez R, Rodriguez-Pena JM, Francois J, Nombela C, Arroyo J: The global transcriptional response to transient cell wall damage in Saccharomyces cerevisiae and its regulation by the cell integrity signaling pathway. *J. Biol. Chem.* 2004, 279(15):15183-15195.

[143] Mendes-Ferreira A, del Olmo M, Garcia-Martinez J, Jimenez-Marti E, Mendes-Faia A, Perez-Ortin JE, Leao C: Transcriptional response of Saccharomyces cerevisiae to different nitrogen concentrations during alcoholic fermentation. *Appl. Environ. Microbiol.* 2007, 73(9):3049-3060.

[144] Murray JI, Whitfield ML, Trinklein ND, Myers RM, Brown PO, Botstein D: Diverse and specific gene expression responses to stresses in cultured human cells. *Mol Biol Cell* 2004, 15(5):2361-2374.

[145] Vandenbroucke K, Robbens S, Vandepoele K, Inze D, Van de Peer Y, Van Breusegem F: Hydrogen peroxide-induced gene expression across kingdoms: a comparative analysis. *Mol. Biol. Evol.* 2008, 25(3):507-516.

[146] Hizume K, Yoshimura SH, Takeyasu K: Atomic force microscopy demonstrates a critical role of DNA superhelicity in nucleosome dynamics. *Cell Biochem. Biophys.* 2004, 40(3):249-262.

[147] Kornberg RD, Lorch Y: Chromatin-modifying and -remodeling complexes. *Curr. Opin. Genet Dev.* 1999, 9(2):148-151.

[148] Horiike T, Hamada K, Kanaya S, Shinozawa T: Origin of eukaryotic cell nuclei by symbiosis of Archaea in Bacteria is revealed by homology-hit analysis. *Nat. Cell Biol.* 2001, 3(2):210-214.

[149] Gross DS, Garrard WT: Nuclease hypersensitive sites in chromatin. *Annu. Rev. Biochem.* 1988, 57:159-197.

[150] Kornberg RD: Chromatin structure: a repeating unit of histones and DNA. *Science* 1974, 184(139):868-871.

[151] Simpson RT: Nucleosome positioning in vivo and in vitro. *Bioessays* 1986, 4(4):172-176.

[152] Svaren J, Chalkley R: The structure and assembly of active chromatin. *Trends Genet* 1990, 6(2):52-56.

[153] Travers AA, Klug A: The bending of DNA in nucleosomes and its wider implications. *Philos Trans R Soc. Lond B Biol. Sci.* 1987, 317(1187):537-561.

[154] Wolffe AP: Packaging principle: how DNA methylation and histone acetylation control the transcriptional activity of chromatin. *J. Exp. Zool.* 1998, 282(1-2):239-244.

[155] Zhang L, Gralla JD: In situ nucleoprotein structure involving origin-proximal SV40 DNA control elements. *Nucleic Acids Res* 1990, 18(7):1797-1803.

[156] Karymov MA, Tomschik M, Leuba SH, Caiafa P, Zlatanova J: DNA methylation-dependent chromatin fiber compaction in vivo and in vitro: requirement for linker histone. *Faseb J.* 2001, 15(14):2631-2641.

[157] Sato MH, Ura K, Hohmura KI, Tokumasu F, Yoshimura SH, Hanaoka F, Takeyasu K: Atomic force microscopy sees nucleosome positioning and histone H1-induced compaction in reconstituted chromatin. *FEBS Lett.* 1999, 452(3):267-271.

[158] Zlatanova J, Leuba SH, van Holde K: Chromatin fiber structure: morphology, molecular determinants, structural transitions. *Biophys J.* 1998, 74(5):2554-2566.

[159] Belmont AS, Braunfeld MB, Sedat JW, Agard DA: Large-scale chromatin structural domains within mitotic and interphase chromosomes in vivo and in vitro. *Chromosoma* 1989, 98(2):129-143.

[160] Hirano T, Kobayashi R, Hirano M: Condensins, chromosome condensation protein complexes containing XCAP-C, XCAP-E and a Xenopus homolog of the Drosophila Barren protein. *Cell* 1997, 89(4):511-521.

[161] Hizume K, Yoshimura SH, Takeyasu K: Linker histone H1 per se can induce three-dimensional folding of chromatin fiber. *Biochemistry* 2005, 44(39):12978-12989.

[162] Kimura K, Hirano T: ATP-dependent positive supercoiling of DNA by 13S condensin: a biochemical implication for chromosome condensation. *Cell* 1997, 90(4):625-634.

[163] Maeshima K, Laemmli UK: A two-step scaffolding model for mitotic chromosome assembly. *Dev. Cell* 2003, 4(4):467-480.

[164] Wang JC: Cellular roles of DNA topoisomerases: a molecular perspective. *Nat Rev Mol Cell Biol.* 2002, 3(6):430-440.

[165] Yoshimura SH, Kim J, Takeyasu K: On-substrate lysis treatment combined with scanning probe microscopy revealed chromosome structures in eukaryotes and prokaryotes. *J. Electron. Microsc. (Tokyo)* 2003, 52(4):415-423.

[166] Karp PD, Keseler IM, Shearer A, Latendresse M, Krummenacker M, Paley SM, Paulsen I, Collado-Vides J, Gama-Castro S, Peralta-Gil M et al: Multidimensional annotation of the Escherichia coli K-12 genome. *Nucleic Acids Res.* 2007, 35(22):7577-7590.

[167] Tatusov RL, Fedorova ND, Jackson JD, Jacobs AR, Kiryutin B, Koonin EV, Krylov DM, Mazumder R, Mekhedov SL, Nikolskaya AN et al: The COG database: an updated version includes eukaryotes. *BMC Bioinformatics* 2003, 4:41.

Chapter IV

Genome Reduction in *Bacillus subtilis* and Enhanced Productivities of Recombinant Proteins

Yasushi Kageyama[1], *Takuya Morimoto*[1,2], *Katsutoshi Ara*[1], *Katsuya Ozaki*[1] *and Naotake Ogasawara*[2]

[1]Biological Science Laboratories, Kao Corporation,
2606 Akabane, Ichikai, Haga, Tochigi 321-3497, Japan
[2]School of Information Science,
Nara Institute of Science and Technology, 8916-5 Takayama,
Ikoma, Nara 630-0101, Japan

Abstract

Bacillus subtilis, which is one of the most important host microorganisms for large-scale industrial production of useful proteins, has a genome of 4.2 Mb with approximately 4106 protein-coding genes. Some of these genes are expected to be unnecessary for industrial production of proteins under controlled conditions and may be wasteful with regard to energy consumption. We attempted to reduce the genome size of *B. subtilis* by deleting unnecessary regions of the genome to allow the construction of simplified host cells as a platform for the further development of novel genetic systems with increased productivity.

First, we generated the strain MGB469 with deletion of all prophage (SPβ and PBSX) and prophage-like (pro1-7 and skin) sequences, with the exception of pro7, as well as two large operons that produce secondary metabolites (*pks* and *pps*). These cells showed normal growth, but no beneficial effects were observed with regard to recombinant protein production from plasmids carrying the corresponding genes. Second, we constructed several multiple-deletion mutants containing additional deletions in the MGB469 genome, resulting in total genome size reductions of 0.78 to 0.99 Mb. In most of the multiple-deletion series, extensive deletion mutants showed no beneficial improvements in traits as host strains. The strain MG1M with a total genome size reduction of 0.99 Mb showed unstable phenotypes with regard to growth rate, cell morphology, and recombinant protein productivity after successive culture, making it inappropriate for further studies. In addition, strain MGB943 derived from another lineage with genome reduction of 0.94 Mb showed reduced recombinant cellulase productivity. Finally, we generated another multiple-deletion series including the mutant MGB874 with a total genome deletion of 0.87 Mb. In comparison to wild-type cells, the metabolic network of the mutant strain was reorganized after entry into the transition state due to the synergistic effects of multiple deletions. Moreover, the levels of production of extracellular cellulase and protease from transformed plasmids carrying the corresponding genes were markedly increased.

Our results demonstrated the effectiveness of a synthetic genomic approach with reduction of genome size to generate novel and useful bacteria for industrial uses.

Introduction

The new field of synthetic genomics is expected to contribute to future sustainability through design and creation of novel microorganisms for industrial use [1–3]. To facilitate the design of various novel genetic systems for specific purposes, it will be useful to construct simplified cells with predictable behavior by genome reduction in microorganisms by introduction of rationally designed deletions.

Microorganisms generally harbor a number of genes that are activated only under specific sets of conditions or in response to particular environmental conditions. Under conditions such as those experienced in an industrial production system where sufficient nutrients and a good growth environment are provided, some of these genes are expected to be unnecessary and may be wasteful in terms of energy consumption. In addition, genomes often contain horizontally acquired sequences and genes for secondary metabolite production. Genome reduction by stepwise deletion has been reported in *Escherichia coli* and *Bacillus subtilis*.

Posfai et al. reported that an *E. coli* strain, MDS42, with deletion of about 15% of the genome (0.71 Mb) showed normal cell growth and protein expression comparable to the parental strain [4]. Moreover, genome reduction also resulted in unexpected beneficial properties, including high electroporation efficiency and accurate propagation of plasmids that were unstable in other strains [4]. Another group in Japan reported that *E. coli* strain MGF-01, with deletion of 22% of the genome (1.03 Mb), also exhibited unanticipated phenotypes, i.e., improved growth and elevated threonine production as compared with the wild-type strain [5].

B. subtilis, a spore-forming Gram-positive bacterium, has superior ability to secrete various enzymes as well as a high level of safety with GRAS (Generally Regarded As Safe) status. These properties have been widely exploited for the production of various useful enzymes in industrial applications. Moreover, *B. subtilis* is one of the best-characterized model microorganisms by biochemical, genetic, and molecular biological studies. In 1997, the complete genome sequence of *B. subtilis* was published by an international consortium including Japanese and European laboratories [6]. The *B. subtilis* genome database (SubtiList, http://genolist.pasteur.fr/SubtiList/) lists a total of 4106 annotated protein-coding genes on the 4.2 Mb genome. Based on the results of a study involving systematic inactivation of *B. subtilis* genes, Kobayashi et al. reported that 271 of the genes were essential for growth in LB medium [7].

Westers et al. reported a *B. subtilis* Δ6 mutant strain with 7.7% genome reduction (0.53 Mb) produced by deleting two prophage (SPβ and PBSX), three prophage-like sequences (pro1, pro6, skin), and the *pks* operon. However, phenotypic characterization of the Δ6 cells revealed no unique properties, including secretion of AmyQ protein, relative to wild-type 168 cells [8].

To assess the effects of each gene on recombinant protein productivity, we evaluated recombinant cellulase productivities of single-gene deletion/disruption strains with plasmids harboring the corresponding gene. By deleting unnecessary regions based on the above findings, we have reduced the genome size of *B. subtilis* for construction of simplified host cells to provide a platform for further development of novel genetic systems with increased productivity.

First, we generated a strain, MGB469, in which all prophage and prophage-like sequences, with the exception of pro7, as well as *pks* and *pps* operons, were deleted. These cells showed normal cell growth, but no beneficial effects were observed, including exogenous protein production from plasmids harboring the corresponding genes [9].

Second, we constructed several multiple-deletion mutants containing additional deletions in the MGB469 genome with total genome size reductions of 0.78 to 0.99 Mb. However, in most of the multiple-deletion series, extensive deletion mutants showed traits that made them inappropriate as host strains for industrial applications. For example, strain MG1M with a 0.99 Mb reduction showed unstable phenotypes with regard to growth rate, cell morphology, and recombinant protein production after successive culture, and it was therefore inappropriate for further studies [9]. In addition, strain MGB943 derived from another lineage with a genome reduction of 0.94 Mb showed reduced recombinant cellulase production.

Finally, we generated another multiple-deletion mutant MGB874 with a total genome deletion of 0.87 Mb. In comparison to wild-type cells, the metabolic network of the mutant strain was reorganized after entry into the transition state due to the synergistic effects of multiple deletions [10]. Moreover, the levels of production of extracellular cellulase and protease from transformed plasmids harboring the corresponding genes were markedly enhanced in these cells [10]. Our results indicated the effectiveness of a synthetic genomic approach with genome size reduction to generate novel and useful bacteria for industrial use.

Here, we describe the genome reduction of the *B. subtilis* 168 strain, enhanced production of useful protein using a host strain with a reduced genome, and the properties of the strain as determined by transcriptome analysis.

Effects of Single-Gene Deletion/Disruption on Productivity of Recombinant Proteins

The *B. subtilis* genome has been reported to possess approximately 4106 protein-coding genes [6], for approximately 60% of which the functions of the gene products have been annotated based on results of previous studies and homology with known proteins. In addition, a previous study involving the systematic disruption of individual genes of *B. subtilis* demonstrated that only 271 genes were essential for growth in LB medium at 37°C [7]. Thus, most of the *B. subtilis* genome is predicted to be comprised of unnecessary genes, and cells with deletion of these unnecessary genes will still be able to grow.

We performed large-scale analysis of recombinant cellulase production by 2792 single-gene deleted/disrupted mutants. We introduced pHYS237, a multicopy plasmid pHY300PLK (*ca.* 50 copies per cell) harboring a gene encoding thermostable alkaline cellulase Egl237 [11] under the control of its own constitutive and SigA-dependent promoter. Cellulase activities were measured after 75 h in culture in modified 2xL-Mal medium [9], a model medium for industrial protein production. We identified 116 single-gene deleted/disrupted mutants with increased production and 96 with decreased production of recombinant cellulase (Figure 1).

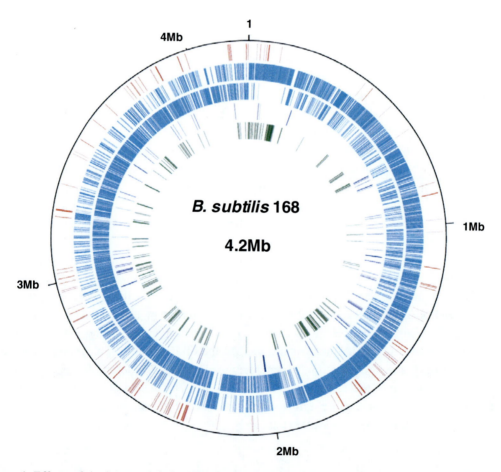

Figure 1. Effects of single-gene deletion/disruption on production of recombinant proteins. (A) Outer concentric ring: genome coordinate (bases) of the *B. subtilis* 168 genome. Ring 2 (red): positions of single-gene deletion/disruption leading to increased production of recombinant cellulase (>110% of the wild-type strain). Rings 3 and 4 (light blue): protein coding regions in clockwise (Ring 3) and counterclockwise (Ring 4) orientations. Ring 5 (dark blue): positions of single-gene deletion/disruption leading to decreased production of recombinant cellulase (<80% of the wild-type strain). Ring 6 (green): essential genes [7].

Effects of Single-Region Deletions on Cell Growth

To construct a series of multiple-deletion mutants, we initially identified contiguous genome sequences longer than 10 kb that did not encode rRNAs, tRNAs, essential proteins, all known and possible proteins involved in primary metabolism to maintain growth in synthetic medium such as Spizizen's minimal medium (SMM) [12], and the proteins related to DNA metabolism to avoid genome instability. Thus, we selected a total of 74 regions, including prophage, prophage-like, and secondary metabolite-producing sequences, and performed individual replacement experiments with the tetracycline resistance gene [13] by selection on LB plates with tetracycline. We obtained deletion mutants for 63 regions, totaling up to 2 Mb (Figure 2). Nine of the deletion mutants did not grow in poor synthetic medium, while others showed impaired growth, even in liquid LB medium (Figure 3).

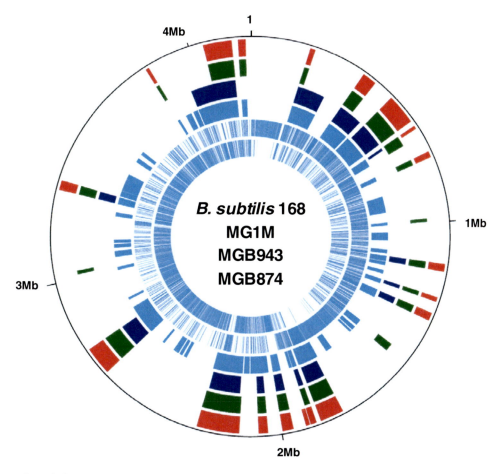

Figure 2. Deletion regions. Outer concentric ring: genome coordinate (bases) of the *B. subtilis* 168 genome. Ring 2 (red): Deletion regions in MGB874. Ring 3 (green): Deletion regions in MGB943. Ring 4 (dark blue): Deletion regions in MG1M. Ring 4 (light blue): Single deletion regions. Rings 5 and 6 (light blue): protein coding regions in clockwise (Ring 5) and counterclockwise (Ring 6) orientations.

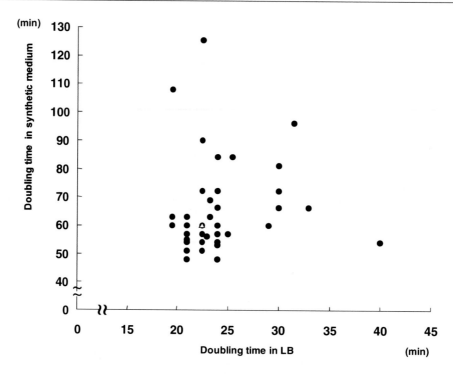

Figure 3. Doubling time (min) of single-region deletion mutants. The mutants (closed circles) and wild-type 168 (open triangles) were cultured in LB medium or in synthetic medium containing Spizizen's minimal salts [12], 0.5% glucose, 0.01% potassium glutamate, and trace elements [12] at 37°C. The deletion mutants which did not grow in the synthetic medium were not shown. Figure 3 was reproduced from data in [10].

Step-by-Step Genome Reduction in *B. subtilis*

To construct reduced genome *B. subtilis* strains, we performed step-by-step multiple deletion of some of the regions where single deletions did not affect cell growth. We used the *upp* (encoding uracil phosphoribosyltransferase) cassette and 5-fluorouracil (5-FU) selection [14] to remove the drug resistance markers used to introduce primary deletions in a step-by-step manner (Figure 4) [10].

First, we generated the strain MGB469 with deletions of prophage and prophage-like regions as well as two large operons involved in secondary metabolite production, affecting neither cell growth nor recombinant protein production [9]. Using MGB469 as the starting strain, we then constructed multiple-deletion series by additional deletion of various regions of the MGB469 genome resulting in final total genome reductions of 0.78 to 0.99 Mb [9]. However, many multiple-deletion series showed traits that made them inappropriate as host strains for industrial applications. After several attempts to generate an improved host strain with a reduced genome, we generated a multiple-deletion series including a mutant derivative of MGB874 with a total genome deletion of 0.87 Mb that showed enhanced recombinant protein production [10].

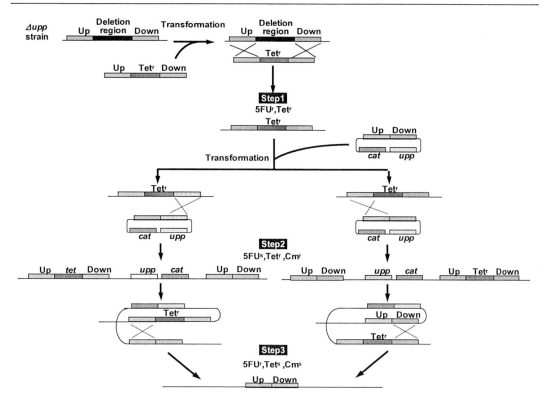

Figure 4. Sequential introduction of large-scale deletions into the *B. subtilis* genome [10]. A derivative of *B. subtilis* 168, 168 Δupp, in which the *upp* gene encoding uracil phosphoribosyltransferase was inactivated by replacement with the erythromycin resistance gene was used as the starting strain for generation of the deletion mutant series. The entire length of the tetracycline (Tet) resistance gene cassette (Tet^r) with its upstream regulatory region was amplified from the pBEST307 plasmid [13]. Regions of at least 500 bp of flanking sequence on both sides of the region to be deleted were amplified by PCR and joined upstream and downstream of the Tet^r cassette by ligation using overlapping sequences in primers. The *B. subtilis* 168 Δupp strain and its derivatives were transformed with the resultant fragment to obtain a strain in which the target sequence was replaced with the Tet resistance gene (Step 1). Next, to obtain markerless mutants, fragments upstream and downstream of the target sequence were amplified, ligated, and cloned into the pBRcat/upp plasmid harboring the *upp* and chloramphenicol (Cm) resistance (*cat*) genes. The resultant plasmid was integrated into the genome of the primary transformant with selection for tetracycline and chloramphenicol resistance (Step 2). The resultant strain became sensitive to 5-FU due to introduction of the functional *upp* gene, and mutants without the plasmid sequence were selected on LB plates containing 10 mM 5FU (Step 3). The crosses indicate the sites of recombination.

a) MGB469

First, we generated the strain MGB469 with deletion of all prophage (SPβ and PBSX) and prophage-like sequences (pro1-7 and skin), with the exception of pro7 as well as two large operons that produce secondary metabolites (*pks* and *pps*) (Figure 2, 5) [9]. These regions spanned approx. 469 kb, representing about 11% of the *B. subtilis* genome. MGB469 cells showed normal growth and did not form spores (data not shown) due to the deletion of genes essential for spore formation, including *spoIVCB* and *spoIIIC* encoding the N- and C-terminal regions of sporulation-specific sigma factor-K, respectively.

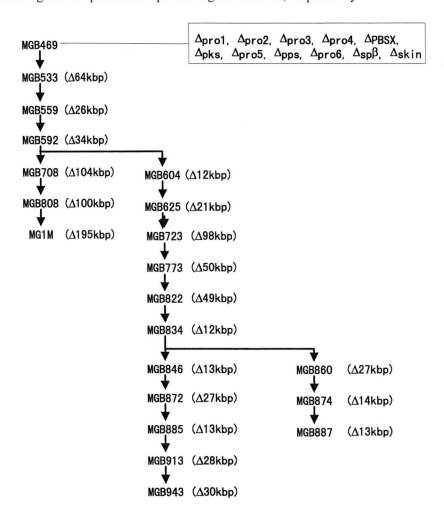

Figure 5. Reduction of *B. subtilis* genome size [9, 10]. The length of each deletion is shown.

The production of recombinant proteins from plasmids (pHYS237 for cellulase from alkaliphilic *Bacillus* sp. strain KSM-S237 [11] or pHP237-K16 for protease [15] from *Bacillus clausii* KSM-K16 [16]) were assayed in the strain MGB469. The productivities were similar to those of wild-type 168 [Figure 6]. The results indicated that the deleted region of approximately 469 kb is dispensable.

b) MG1M and MGB943

We constructed several multiple-deletion series of mutants in the MGB469 strain with additional deletions, resulting in total reductions in genome size of 0.78 to 0.99 Mb. However, in many of the multiple-deletion series, extensive deletion mutants showed traits that made them inappropriate as host strains for industrial applications. The strain MG1M with a genome reduction of 0.99 Mb showed complete inhibition of growth in poor synthetic medium (Figure 2, 5) [9]. In the enzyme production medium (modified 2×L-Mal medium), the strain MG1M showed a slight reduction in growth but no marked morphological changes. This strain produced cellulase and protease at levels comparable to those obtained with *B. subtilis* strain 168 (Figure 6) [9]. To our knowledge, MG1M is the first *B. subtilis* strain with large regions representing about 25% of the genome. However, MG1M displayed unstable phenotypes with regard to growth rate, cell morphology, and recombinant protein production after successive culture, making it inappropriate for further studies. Another strain, MGB943, from another lineage with genome reduction of 0.94 Mb (Figure 2, 5) showed decreased production of recombinant cellulase (Figure 6).

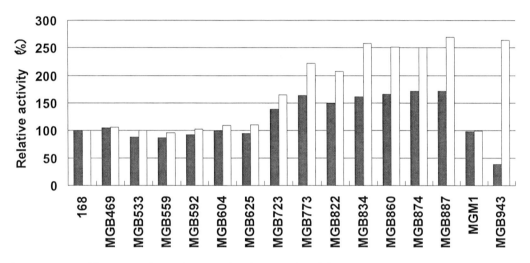

Figure 6. Recombinant protein production of multiple-region deletion mutants [9, 10]. Relative activities of cellulase Egl237 [11] (black bars) and M-protease [15] (white bars) in growth medium of the multiple-deletion series of mutants are shown. The activity of each enzyme after 75 h cultivation of wild-type 168 in modified 2xL-Mal medium, was set as 100%.

c) MG874 and Further Deletion

Finally, we generated another multiple-deletion mutant, MGB874, by sequential deletion of the 11 non-essential gene clusters in the MGB469 genome. MGB874 had deletion of 874 kb (20.7%) of the sequence including 865 genes from the original *B. subtilis* 168 genome (Figure 2, 5) [10]. Although MGB874 cells showed reduced growth rate (30% in LB and 50% in SMM) as compared with the wild-type 168 strain, both cell morphology and chromosome distribution were normal (data not shown). The deduced growth rate suggested that some of the non-annotated and deleted genes contribute to the metabolic capacity of *B. subtilis* cells under normal growth conditions. It is also possible that this phenotype was due to unexpected synergistic effects of the deletion of annotated genes.

To assess the exogenous protein secretion production by MGB874 cells, we introduced the plasmids pHYS237 for cellulase production [11] or pHP237-K16 for protease production [15] into the MGB469 to MGB874 multiple-deletion strains and determined the levels of protease and cellulase activity after 75 h in culture in modified 2xL-Mal, a model medium for industrial protein production (Figure 6). Contrary to our expectations, the levels of production of both enzymes increased in proportion to extent of genome deletion, with maximum levels in the MGB874 strain. The cellulase and protease activities in the culture medium of MGB874 cells were about 1.7- and 2.5-fold higher than those associated with wild-type cells, respectively (Figure 6) [10].

We are in the process of introducing further rational deletions using MGB874 as the starting strain. For example, the strain MGB887 has a total genome reduction of 887 kb and shows levels of recombinant enzyme production similar to those of MGB874 (Figures 5, 6). Using transcriptome data, gene function information, and comparative genomics approaches, we are generating simple, predictable strains the genomes of which contain genes with defined functions as a new platform for the development of bacterial strains for industrial applications.

30-L Jar Fermentation of MGB874 and Wild-Type 168

To evaluate the properties of the MGB874 strain in recombinant cellulase production culture, MGB874 and wild-type 168 cells were cultured in 15 L of modified 2xL-Mal medium in a 30-L jar fermenter [10]. Both MGB874 and wild-type strains showed similar increases in cell mass (Figure 7A). Cellulase production was arrested in wild-type 168 cells from 28 h in culture (Figure 7B), while the cellulase level continued to increase for MGB874 throughout the culture period to about double that obtained with wild-type 168. The period of protease production was similarly increased in the mutant strain as compared to the wild-type control (data not shown). Furthermore, maltose consumption in the culture medium was enhanced in MGB874 cells (Figure 7C) [10], indicating that the efficiency of carbon source utilization was also improved with genome reduction. Carbon dioxide concentration in exhausted gas was higher in MGB874 than in wild-type 168 after 32 h (Figure 7D), suggesting that the MGB874 cells maintained metabolic activity.

MGB874 and Wild-Type 168 Transcriptome Analyses

We investigated the molecular events underlying these phenomena by comparing the transcriptome profiles of MGB874 and wild-type strains during growth in modified 2xL-Mal medium [10]. Total RNA was extracted from *B. subtilis* cells (OD_{600}=10) at the time points of 1, 7.5, 13, 26, 40, and 60 h [18]. We used custom Affymetrix tiling chips containing a total of 55430 25-mer probes for the coding strand of protein coding regions at intervals of 25–30 bp and 72218 probes for both strands of intergenic regions at intervals of 2–3 bp [19]. The expression levels of each gene in strains MGB874 and 168 were determined as described previously [20, 21].

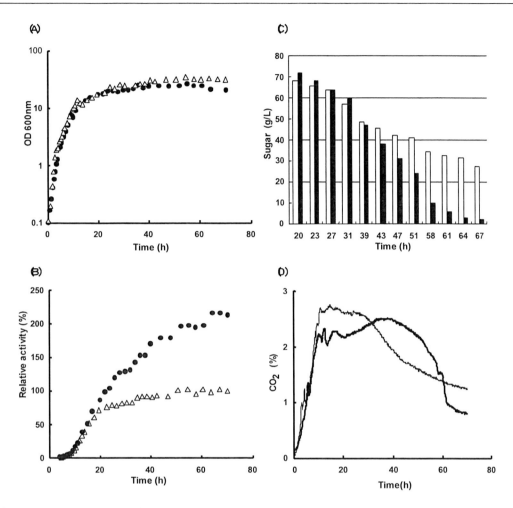

Figure 7. Recombinant cellulase productions using MGB874 and wild-type 168 in jar fermenters. (A) Growth of wild-type 168 (open triangles) and MGB874 (closed circles). (B) Relative activities of recombinant cellulase Egl237 [11] in the growth medium of wild-type 168 (open triangles) and in that of MGB874 (closed circles). The final level of cellulose activity in wild-type 168 was set as 100%. (C) Total sugar content in the supernatants of wild-type 168 (open triangles) and MGB874 (closed circles) cultures. (D) Carbon dioxide concentration in the exhaust gas from the culture of wild-type 168 (open triangles) and MGB874 (closed circles). Figure 7A to C were reproduced from data in [10].

Some transcriptome profiles indicating differences between MGB874 and wild-type 168 are shown in Figure 8. Comparison of the transcriptome profiles indicated earlier development of genetic competence, delayed entry to sporulation, and maintenance of metabolic activity in MGB874 as compared to the wild-type control.

a) Earlier Development of Competence

Genetic competence in *B. subtilis*, i.e., the ability to incorporate exogenous DNA from outside the cell, is regulated by a number of regulatory factors including ComA, ComS, and ComK, and by extracellular peptide signaling of Rap-Phr systems (Figure 9) [22].

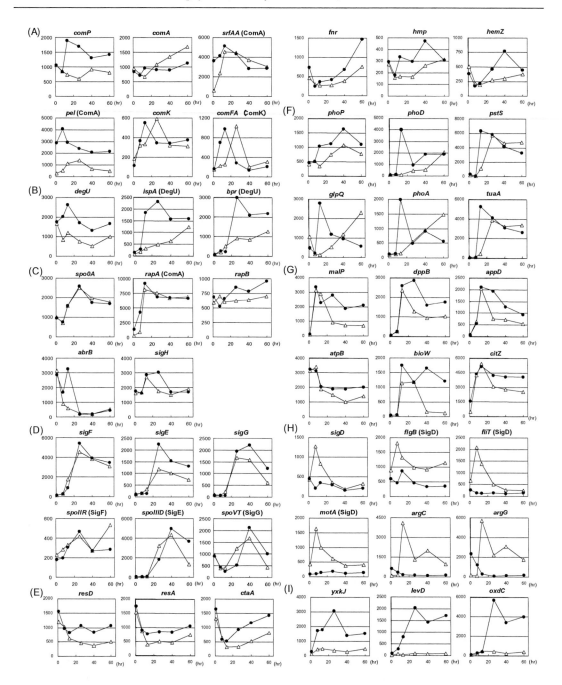

Figure 8. Identification of genes with significantly disrupted expression in MGB874 cells, as compared with wild-type cells. The average signal intensities of probes in each coding sequence in wild-type 168 (open triangles) and MGB874 (closed circles) cells grown in modified 2xL-Mal medium [9] for 1, 7.5, 13, 26, 40, and 60 h are indicated. Regulators of expression are specified in parentheses. (A) ComA and ComK-dependent genes. (B) DegU-dependent genes. (C) Transition state regulators. (D) Sporulation genes. (E) Respiration-related genes. (F) Phosphate metabolism-related genes. (G) Genes with markedly elongated expression in MGB874 cells. (H) Genes with markedly suppressed expression in MGB874 cells. (I) Genes that are specifically induced in MGB874 cells. Reproduced from data in [10].

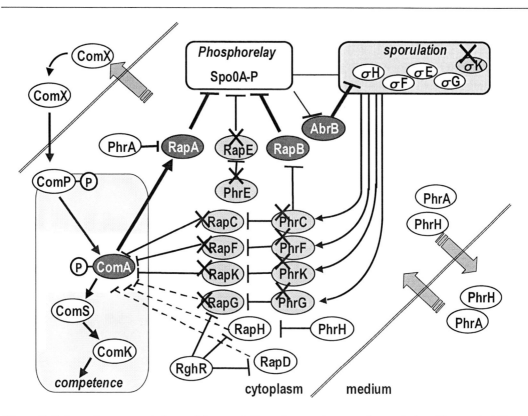

Figure 9. Schematic representations of the regulation of competence and sporulation. The symbol X indicates the proteins or peptides, of which genes were deleted in MGB874.

On activation by interaction with the extracellular signaling peptide ComX, the phosphate from the membrane-bound receptor histidine kinase ComP is transferred to ComA. Rap family proteins, including RapC, RapF, RapK, RapG, RapH, and RapD, have been reported to regulate the activation (phosphorylation) timing of ComA [22]. The phosphorylated form of ComA (ComA-P) activates transcription of *comS* in the *srf* operon. The competence regulator ComS indirectly stimulates self-activation of another competence gene, *comK*, the gene product of which, ComK, activates transformation genes and operons.

The genes regulated by ComA-P [23], including *comS* in the *srf* operon, *pel*, and *rapA*, were activated earlier in MGB874 than in wild-type strain 168 (Figure 8A, C), and thus ComA-P accumulation occurred earlier in MGB874. Four of the six *rap* genes, *rapC*, *rapF*, *rapK*, and *rapG*, were deleted in MGB874 (Figure 9), which may have been responsible for the earlier accumulation of ComA-P in this mutant strain. Moreover, activation of *comK* and the transformation genes and operons positively regulated by ComK, including *comC*, *comEA*, *comFA*, *comGA*, and *nucA*, occurred earlier in MGB874 than in the wild-type strain 168 (Figure 8A). These observations strongly suggest the earlier development of genetic competence in MGB874.

The DegU regulator, which participates in competence activation in the non-phosphorylated form [24] and protein production in the highly phosphorylated form (DegU-P) [25], binds to the promoter of *comK* and increases the affinity of ComK for the *comK* promoter [24, 26]. In MGB874, DegU was induced earlier and its expression level remained high throughout the entire culture period (Figure 8B). The alteration of *degU* expression may also contribute to the earlier development of competence in MGB874 cells. In addition, deletion of the *rapG* gene in MGB874 may contribute to the activity of DegU by preventing the binding of RapG to DegU [22].

However, degradative protein production, possibly dependent on DegU-P, was similar in both MGB874 and wild-type strains, with the exception of the major intracellular serine protease (IspA) and bacillopeptidase F (Bpr), which were strongly induced in MGB874 cells in the late growth phase (Figure 8B) [24].

b) Delayed Entry to Sporulation

In *B. subtilis*, a transcription factor Spo0A is activated by phosphorylation via a multicomponent phosphorelay. A phosphorylated intermediate response regulator of the phosphorelay, Spo0F-P, is dephosphorylated by aspartyl phosphate phosphatases, RapA and RapB (Figure 9). The phosphorylated form of Spo0A, i.e., Spo0A-P, suppresses AbrB, a repressor of transition state genes in the exponential growth phase. The alternative sigma factor SigH is repressed by AbrB and activated by Spo0A-P. Spo0A-P triggers a regulatory program for spore formation involving the activation of SigH, and sporulation-specific sigma factors, SigF and SigG in prespores, and SigE and SigK in mother cells [27].

In MGB874, an aspartyl phosphate phosphatase gene *rapA* was activated early and another aspartyl phosphate phosphatase gene *rapB* was activated after the transition state in MGB874 (Figure 8C). In *B. subtilis*, *rapA* is activated by ComA-P, and RapB is inhibited by a Phr peptide PhrC, the gene of which, *phrC*, is deleted in MGB874 (Figure 9) [28]. The expression of *rapA* and *rapB* may have contributed to delay of the phosphorelay through the dephosphorylation of Spo0F-P in MGB874.

The inhibition of AbrB expression was delayed and SigH expression was disrupted in MGB874 (Figure 8C). In addition, activation of sporulation-specific sigma factors was delayed in MGB874 cells, as indicated by the delay in expression of genes under their control, with the exception of those regulated by SigK, which was deleted in MGB874 (Figure 8D) [10]. These findings suggest that the transition state is extended in MGB874 cells as compared with the wild-type strain.

c) Maintained Activities of Metabolism

The rates of recombinant cellulase production, maltose consumption, and carbon dioxide emission were maintained for longer in jar-fermenter culture of MGB874 than in that of wild-type 168 [Fig.7]. We also observed the higher levels of expression of respiration-related genes under the control of the ResD-ResE two-component signal transduction system [29] in MGB874 cells, including *resABCDE* encoding cytochrome c biosynthetic genes and ResD-ResE two-component system, *fnr* (transcriptional regulator of anaerobic genes), *hmp* (flavohemoglobin), *hemZ* (coproporphyrinogen III oxidase), and *ctaA* (cytochrome caa3 oxidase) (Fig.8E). In addition, expression of the operons *appDFABC* and *dppBCDE* encoding

ABC transporters for oligopeptide and dipeptide, respectively, *malP* encoding a maltose-specific phosphotransferase (PTS) enzyme IIB component, and *bioWFADB* encoding enzymes involved in biotin biosynthesis were activated throughout the culture period (Figure 8E). Thus, it appears that metabolic activities are maintained in the extended transition state of MGB874 cells, resulting in increased enzyme production.

d) Earlier Expression of Phosphate Metabolism-Related Genes

In addition, the marked alterations in gene expression observed in MGB874 cells may be related to increased protein production in this strain. We observed the earlier and stronger induction of genes related to phosphate metabolism under the control of the PhoP-PhoR two-component system [30] in MGB874 cells, including *glpQ* encoding glycerophosphoryl diesterphosphodiesterase, *phoD* (phosphodiesterase), *phoA* (alkaline phosphatase A), *pstSCAB* (phosphate transporter), and *tuaABCDEFG* (teichuronic acid biosynthesis) (Figure 8F). The PhoP-PhoR system is positively regulated by ResD (Figure 8E) and a transition state regulator, AbrB (Figure 8C) [31]. In addition, the expressions of *resABCDE* and *abrB* are negatively regulated by a response regulator Spo0A-P [31]. Therefore, the delay of increase of Spo0A-P, the expression of *resD*, and the prolonged expression of *abrB* in MGB874 may contribute the earlier and stronger induction of the phosphate metabolism-related genes.

e) Expression of Other Genes

B. subtilis SigD, which is activated in the transition state, is responsible for the transcription of genes involved in flagellar assembly, motility, chemotaxis, and autolysis [32, 33]. MGB874 cells showed significantly reduced transcription of SigD and genes under its control (Figure 8H). The *argGH* and *argCJBD* operons required for arginine biosynthesis were induced in the transition state in wild-type cells, but not in the mutant MGB874 cells (Figure 8H). A number of genes were strongly induced in MGB874 cells, including *yxkJ* encoding L-malate and citrate transporter, *levDEFG* encoding fructose-specific PTS enzyme, and *oxdC* encoding cytosolic oxalate decarboxylase (Figure 8I) [10].

The observations indicated that considerable reprogramming of the transcriptional regulatory network has occurred in MGB874 cells, probably due to the synergistic effects of multiple deletions, although further studies are required to determine the molecular basis for these changes and their relationships to increased protein production.

Conclusion

The results of the present study indicated that genome reduction actually contributes to the generation of bacterial cells suitable for practical industrial applications. It is not yet clear whether the phenomena described here were due to global synergistic effects of large-scale genome reduction or were mainly due to the deletion of several regulators. Further systematic analysis of the changes in the transcriptional regulatory network in MGB874 cells in relation to protein production should facilitate the generation of improved *B. subtili*s cells as hosts for industrial protein production.

We are currently engaged in further rational deletion studies based on transcriptome data, gene function information, and comparative genomics approaches, with a view toward generating simple, predictable cells containing genes with defined functions as a new platform for the development of bacterial strains for industrial applications.

Acknowledgments

This study was performed in collaboration with Keiji Endo, Masatoshi Tohata, Kazuhisa Sawada, Kenji Manabe, Shengao Liu, Tadahiro Ozawa, Ryosuke Kadoya, Takeko Kodama, and Hiroshi Kakeshita. We are grateful to Shu Ishikawa and Taku Oshima for help in transcriptome analysis, and to Hiroshi Kodama and Kazuhiro Saito for cultivation in 30-L jar fermenters. We thank Kouji Nakamura, Shigehiko Kanaya, Fujio Kawamura, Yasutaro Fujita, and Junichi Sekiguchi for valuable discussions. We also thank Yoshiharu Kimura and Yoshinori Takema for valuable advice.

Funding

This work was performed as part of the subproject "Development of a Technology for the Creation of a Host Cell" included within the industrial technology project "Development of Generic Technology for Production Process Starting Productive Function" of the Ministry of Economy, Trade and Industry, entrusted by the New Energy and Industrial Technology Development Organization (NEDO), Japan.

References

[1] Ball, P. (2007). Synthetic biology: designs for life, *Nature*, 448, 32–33.
[2] Drubin, D. A., Way, J. C. and Silver, P. A. (2007). Designing biological systems, *Genes Dev.*, 21, 242–254.
[3] Forster, A. C. and Church, G. M. (2007). Synthetic biology projects *in vitro*, *Genome Res.*, 17, 1–6.
[4] Posfai, G., Plunkett, G. III, Feher, T., Frisch, D., Keil, G. M., Umenhoffer, K., Kolisnychenko, V., Stahl, B., Sharma, S. S., de Arruda, M., et al. (2006). Emergent properties of reduced-genome *Escherichia coli*, *Science*, 312, 1044–1046.
[5] Mizoguchi, H., Mori, H. and Fujio, T. (2007). *Escherichia coli* minimum genome factory, Biotechnol. *Appl.Biochem.*, 46, 157–167.
[6] Kunst, F., Ogasawara, N., Moszer, I., Albertini, A. M., Alloni, G., Azevedo, V., Bertero, M. G., Bessieres, P., Bolotin, A., Borchert, S., et al. (1997). The complete

genome sequence of the gram-positive bacterium *Bacillus subtilis*, *Nature*, 390, 249–256.

[7] Kobayashi, K., Ehrlich, S. D., Albertini, A., Amati, G., Andersen, K. K., Arnaud, M., Asai, K., Ashikaga, S., Aymerich, S., Bessieres, P., et al. (2003). Essential *Bacillus subtilis* genes, *Proc. Natl. Acad. Sci. USA*, 100, 4678–4683.

[8] Westers, H., Dorenbos, R., van Dijl, J. M., Kabel, J., Flanagan, T., Devine, K. M., Jude, F., Seror, S. J., Beekman, A. C., Darmon, E., et al. (2003). Genome engineering reveals large dispensable regions in *Bacillus subtilis*, *Mol. Biol. Evol.*, 20, 2076–2090.

[9] Ara, K., Ozaki, K., Nakamura, K., Yamane, K., Sekiguchi, J. and Ogasawara, N. (2007). *Bacillus* minimum genome factory: effective utilization of microbial genome information, *Biotechnol. Appl. Biochem.*, 46, 169–178.

[10] Morimoto T, Kadoya R, Endo K, Tohata M, Sawada K, Liu S, Ozawa T, Kodama T, Kakeshita H, Kageyama Y, Manabe K, Kanaya S, Ara K, Ozaki K, Ogasawara N. (2008). Enhanced recombinant protein productivity by genome reduction in *Bacillus subtilis*. *DNA Res.* 15:73–81.

[11] Hakamada, Y., Hatada, Y., Koike, K., Yoshimatsu, T., Kawai, S., Kobayashi, T. and Ito, S. (2000). Deduced amino acid sequence and possible catalytic residues of a thermostable, alkaline cellulase from an alkaliphilic *Bacillus* strain, *Biosci. Biotech. Biochem.*, 64, 2281–2289.

[12] Harwood, C. R., and Cutting S. M. (Eds.). (1990). Molecular biological methods for *Bacillus*. Chichester, England: John Wiley and Sons.

[13] Itaya, M. (1992). Construction of a novel tetracycline resistance gene cassette useful as a marker on the *Bacillus subtilis* chromosome, *Biosci. Biotech. Biochem.*, 56, 685–686.

[14] Fabret, C., Ehrlich, S. D. and Noirot, P. A. (2002). A new mutation delivery system for genome-scale approaches in *Bacillus subtilis*, *Mol. Microbiol.*, 46, 25–36.

[15] Kobayashi, T., Hakamada, Y., Adachi, S., Hitomi, J., Yoshimatsu, T., Koike, K., Kawai, S. and Ito, S. (1995). Purification and properties of an alkaline protease from alkalophilic *Bacillus* sp. KSM-K16., *Appl. Microbiol. Biotechnol.*, 43, 473–481.

[16] Kageyama Y, Takaki Y, Shimamura S, Nishi S, Nogi Y, Uchimura K, Kobayashi T, Hitomi J, Ozaki K, Kawai S, Ito S, Horikoshi K. (2007). Intragenomic diversity of the V1 regions of 16S rRNA genes in high-alkaline protease-producing *Bacillus clausii* spp., *Extremophiles.*, 11, 597-603.

[17] Hodge, J. E. and Hofreiter, B. T. (1962). Analysis and preparation of sugars, In R. L. Whistler, and M. L. Wolfrom (Eds), *Methods in Carbohydrate Chemistry* (pp. 388–389). New York, San Francisco, London: Academic Press.

[18] Igo, M. M. and Losick, R. (1986). Regulation of a promoter that is utilized by minor forms of RNA polymerase holoenzyme in *Bacillus subtilis*, *J. Mol. Biol.*, 191, 615–624.

[19] Ishikawa, S., Ogura, Y., Yoshimura, M., Okumura, H., Cho, E., Kawai, Y., Kurokawa, K., Oshima, T. and Ogasawara, N. (2007). Distribution of stable DnaA binding sites on the *Bacillus subtilis* genome detected using a modified ChIP-chip method, *DNA Res.*, 14, 155–168.

[20] Quackenbush, J. (2002). Microarray data normalization and transformation, *Nature Genet*, 32, Suppl., 496–501.

[21] Hirai, M. Y., Klein, M., Fujikawa, Y., Yano, M., Goodenowe, D. B., Yamazaki, Y., Kanaya, S., Nakamura, Y., Kitayama, M., Suzuki, H., et al. (2005). Elucidation of gene-to-gene and metabolite-to-gene networks in *Arabidopsis* by integration of metabolomics and transcriptomics, *J. Biol. Chem.*, 280, 25590–25595.

[22] Ogura, M. (2007). Competence development: Extracellular regulation and the emergence of bistability, In Y. Fujita (Eds), *Global regulatory networks in Bacillus subtilis* (pp. 171–188). Kerala, India: Transworld Research Network.

[23] Ogura, M., Yamaguchi, H., Yoshida, K., Fujita, Y. and Tanaka, T. (2001). DNA microarray analysis of *Bacillus subtilis* DegU, ComA and PhoP regulons: an approach to comprehensive analysis of *B. subtilis* two-component regulatory systems, *Nucleic Acids Res.*, 29, 3804–3813.

[24] Hamoen, L. W., Van Werkhoven, A. F., Venema, G. and Dubnau, D. (2000). The pleiotropic response regulator DegU functions as a priming protein in competence development in *Bacillus subtilis*, *Proc. Natl. Acad. Sci. USA*, 97, 9246–9251.

[25] Olmos, J., de Anda, R., Ferrari, E., Bolivar, F. and Valle, F. (1997). Effects of the sinR and degU32 (Hy) mutations on the regulation of the aprE gene in *Bacillus subtilis*, *Mol. Gen. Genet.*, 253, 562–567.

[26] Shimane, K. and Ogura, M. (2004). Mutational analysis of the helix-turn-helix region of *Bacillus subtilis* response regulator DegU, and identification of *cis*-acting sequences for DegU in the *aprE* and *comK* promoters. *J Biochem.*, 136:387-397.

[27] Piggot, P. J. and Hilbert, D. W. (2004). Sporulation of *Bacillus subtilis*, *Curr. Opin. Microbiol.*, 7, 579–586.

[28] Auchtung, J. M., Lee C. A , and Grossman A.D. (2006). Modulation of the ComA-dependent quorum response in *Bacillus subtilis* by multiple Rap proteins and Phr peptides. *J. Bacteriol.* 188, 5273-5285.

[29] Nakano, M. M. and Zuber, P. (2002). Anaerobiosis. In A. L. Sonenshein, J.A. Hoch, R. Losick (Eds.), *Bacillus subtilis and its closest relatives: from genes to cells* (pp. 393-404). Washinton, DC: ASM Press.

[30] Antelmann, H., Scharf, C. and Hecker, M. (2000). Phosphate starvation-inducible proteins of *Bacillus subtilis*: Proteomics and transcriptional analysis, *J. Bacteriol.*, 182, 4478–4490.

[31] Hulett, F. M. (2002). The *pho* regulon. In A. L. Sonenshein, J.A. Hoch, R. Losick (Eds.), *Bacillus subtilis and its closest relatives: from genes to cells* (pp. 193-201). Washinton, DC: ASM Press.

[32] Mirel, D. B., Estacio, W. F., Mathieu, M., Olmsted, E., Ramirez, J. and Marquez-Magana, L. M. (2000). Environmental regulation of *Bacillus subtilis* sigma(D)-dependent gene expression, *J. Bacteriol.,* 182,3055–3062.

[33] Serizawa, M., Yamamoto, H., Yamaguchi, H., Fujita, Y., Kobayashi, K., Ogasawara, N. and Sekiguchi, J. (2004). Systematic analysis of SigD-regulated genes in *Bacillus subtilis* by DNA microarray and northern blotting analyses, *Gene*, 329, 125–136.

In: Bacterial DNA, DNA Polymerase and DNA Helicases
Editor: Walter D. Knudsen and Sam S. Bruns

ISBN 978-1-60741-094-2
© 2009 Nova Science Publishers, Inc.

Chapter V

Molecular Detection and Characterization of Food and Waterborne Viruses

Alain Houde[1], Kirsten Mattison[2], Pierre Ward[1] and Daniel Plante[3]

[1]Agriculture and Agri-Food Canada, Food Research and Development Centre, St-Hyacinthe, Quebec, Canada

[2]Bureau of Microbial Hazards, Health Products and Food Branch, Food Directorate, Health Canada, Ottawa, Ontario, Canada

[3]Health Canada, Quebec Region, Longueuil, Quebec, Canada

Abstract

Enterically transmitted viruses such as norovirus (NoV), rotavirus (RV), hepatitis A virus (HAV) and hepatitis E virus (HEV) are currently recognized as a major cause of food and waterborne illness in humans worldwide. Most of these viruses cannot be cultured *in vitro*, are stable in the environment for long periods of time, are generally infectious at low doses (only a few viral particles are needed to trigger illness) and would be usually present in low numbers in contaminated food and water samples. There are also increasing concerns about the possible zoonotic transmission or the emergence of new recombinant strains from animal enteric viruses closely related to human pathogenic strains. Until recently, the viral contamination of food and water remained undetected due to a lack of appropriate detection methods. Electronic microscopy (EM) and ELISA assays commonly used on stool suspensions were rather insensitive for food and water applications. Due to their increased sensitivity, conventional and now real-time quantitative reverse transcription PCR (qRT-PCR) assays were developed and widely used for the detection and identification of the nucleic acid of these challenging viruses. Along with microarrays, these new approaches have also allowed significant advances in the molecular typing of these organisms. Rapid, sensitive and reliable molecular detection and typing methods will have a significant impact in the mitigation of outbreaks and could contribute in the prevention of food/waterborne transmission. This review will provide recent trends and advances in the field of DNA polymerase-based detection and characterization methods of these 4 main food and waterborne viruses. It will focus on the use of different conventional and qRT-PCR methods and the microarray technology.

Introduction

Enterically transmitted viruses are now recognized as a major cause of food and waterborne illness in humans. In the United States, foodborne infections were estimated annually at 76 million, causing 325 000 hospitalizations and 5 000 deaths (Mead et al., 1999). In the United Kingdom, foodborne gastroenteritis causes an estimated 1.7 million illnesses resulting in 21 997 hospitalizations and 687 deaths each year (Adak et al., 2005) while in Australia the number of gastroenteritis cases, hospitalizations, and deaths due to foodborne pathogens in year 2000 was estimated at 5.4 million, 15 000 and 76, respectively (Hall et al., 2005). In these surveillance studies, the causative etiological agent remained unidentified in 81.5% of the cases in US, 48.7% in UK and 72.2% in Australia. From the remaining 18.5%, 51.3% and 27.8% of foodborne infections where the pathogenic agents have been identified, 67.2% were associated with viruses, 30.2% with bacteria and 2.6% with parasites in US (Mead et al., 1999), 9.9% were associated with viruses, 89.6% with bacteria and 0.5% with parasites in UK (Adak et al., 2005), and finally 31.7% were associated with viruses, 64.2% with bacteria and 4.4% with parasites in Australia (Hall et al., 2005). However, viruses are suspected of being responsible of a high proportion of the unexplained food-related illnesses due to a lack of effective, standardized and recognized detection methods.

Over the last decade, many documented outbreaks and reports around the world have clearly shown the transmission of pathogenic viruses to humans through food and water consumption. Among European countries conducting outbreak surveillance, the proportions of viral gastroenteritis outbreaks associated with food or water transmission were estimated at 24% in Finland, 17% in the Netherlands, 14% in Slovenia, 7% in Spain, and 7% in England and Wales (Lopman et al., 2003). The cost of these events is always important to the community. A Canadian report estimated that 11 million of foodborne gastroenteritis cases per year are occurring in Canada and that acute gastroenteritis represents a significant disease burden in the Canadian population with estimated associated treatment annual costs of > Can$ 3.7 billion (Majowicz et al., 2006). A foodborne outbreak of hepatitis A virus resulted in 43 cases and exposed up to 5 000 persons in Colorado. In this latter event, the total costs assessed in this outbreak from a societal perspective were US$809 706 (Dalton et al., 1996).

The main food and waterborne viral agents identified so far are mostly responsible of gastroenteritis and enterically transmitted hepatitis and include norovirus (NoV), hepatitis A virus (HAV), rotavirus (RV), hepatitis E virus (HEV), sapovirus (SaV), astrovirus (HAstV), enteric adenovirus (types 40 and 41), enterovirus and aichivirus (Koopmans and Duizer, 2004; Carter, 2005; Bosh et al., 2008). These enteric and hepatic viruses are highly resistant to acidic environment, disinfectant, heat, pressure and temperature (Koopmans et al., 2002). They are strict intracellular parasites i.e., need specific human living cells for their replication, and therefore they cannot grow in food and water. Only a few viral infectious particles (estimated to be around 10-100 infectious viral particles for NoV and HAV) are necessary to trigger illness. Once their replicative cycle has been established in their specific target cells, viral particles are shed in very large quantities in the stools of infected individuals (up to 10^{11} particles per gram of stool for RV and HAV) (Sair et al., 2002; Koopmans and Duizer, 2004; Carter, 2005; Pinto and Saiz, 2007).

Historically, enterically transmitted viruses were diagnosed in clinical samples by scanning stool suspensions under an electron microscope (EM) (Atmar and Estes, 2001). However, diagnosis by electron microscopy requires at least 10^6 viral particles/g of stool (Duizer et al., 2004). Different routine enzyme-linked immunosorbent assays (ELISA) were also developed but the limits of detection of these methods were around 10^4-10^5 viral particles/g of sample. Virus detection in food and water samples had always been more challenging than their detection in clinical samples because the levels of virus present in food and water samples are orders of magnitude lower and the detection system should be able to detect a potentially infectious dose (10-100 particles) in a size portion for consumption. Since these viruses can either not be cultivated *in vitro* or cultivated only with difficulty, most detection methods developed have focussed on viral nucleic acid amplification technologies. In order to achieve a suitable analytical sensitivity for the diagnostic of viral contamination in food and water, a comprehensive approach to enhance the detection of enterically transmitted viruses should combine an efficient viral concentration and extraction procedure with a sensitive molecular detection method.

This chapter will focus on the four main food and waterborne viruses and provides a detailed description of the different molecular detection and characterization techniques available.

Norovirus (NoV)

Overview

The family *Caliciviridae* consists of four currently recognized genera. The vesiviruses and lagoviruses are animal pathogens, while the noroviruses and sapoviruses infect both humans and animals (Fauquet et al., 2005). NoV are the leading cause of infectious gastroenteritis in humans worldwide (Duizer et al., 2007). They may be transmitted through food (Friedman et al., 2005), water (Kim et al., 2005), aerosols (Marks et al., 2003), fomites (Isakbaeva et al., 2005) or person-to-person contact (Tsugawa et al., 2006). The NoV cause illness in all age groups, which results in the rapid spread of infection and the development of large outbreaks when food or water sources are contaminated (Atmar and Estes, 2006). In addition, there is extensive strain variation within the NoV genus and newly emerging strains may contribute to periodic increases in disease incidence (Siebenga et al., 2007; Lindesmith et al., 2008). These variants may arise through errors in RNA replication by the viral polymerase or through recombination between different strains (Bull et al., 2007; Siebenga et al., 2007). There are animal strains of NoV closely related to the human strains and the potential for zoonotic transmission is not well understood (Mattison et al., 2007; Wang et al., 2005; Wang et al., 2007).

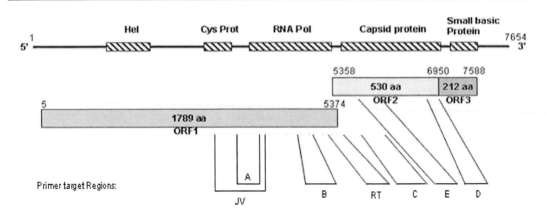

Figure 1. Genomic organization of norovirus and localization of target regions. Numbers refer to the corresponding nucleotide position of Norwalk/68 virus (GI) (GenBank accession number M87661).

The molecular detection and characterization of NoV is therefore of paramount importance in understanding and controlling the spread of disease. Significant progress has been made in recent years in the development of sensitive and specific detection tools and strain typing procedures. The main drawback to these techniques is that they are not yet sufficiently standardized to enable rapid worldwide communication regarding strain types. This limitation is recognized and current initiatives are expected to harmonize nomenclature in the near future.

The NoV are small virions, 27 nm in diameter, consisting mainly of a single major capsid protein with a few copies of the minor capsid protein per particle (Glass et al., 2000; Jiang et al., 1992). These enclose a genome of approximately 7.6 kb that codes for three open reading frames (ORF) (Figure 1) (Jiang et al., 1993). ORF1 is translated as a polyprotein and subsequently cleaved by the viral protease to yield: N-terminal protein, ATPase/helicase, VPg, 3C-like protease, RNA-dependent-RNA-polymerase (RdRp) (Seah et al., 1999). ORF2 codes for the 58 kDa major capsid protein and ORF3 is a minor structural protein. Two transcripts are generated during NoV infection, the full-length mRNA and a subgenomic mRNA initiated in the region of overlap between ORF1 and ORF2 (Katayama et al., 2006). The 5' end of this subgenomic mRNA contains a highly conserved sequence motif that is shared by all NoV genotypes. As such, it is a hotspot for recombination and this provides a significant mechanism for strain variation (Bull et al., 2005).

Norovirus Classification

There are currently two up-to-date classification schemes for the general identification of NoV genotypes (Kageyama et al., 2004; Zheng et al., 2006). These are derived from different types of analysis, yet they have a great deal in common. A third system, focused on the more specialized typing of NoV recombinants will not be described here (Bull et al., 2007). The two NoV genotyping standards use sequence data from the capsid region. They separate the NoV into genogroups, numbered with roman numerals as in GI. They then classify clusters within each genogroup by adding an Arabic numeral, as in GI.1. These two systems of assigning NoV nomenclature agree on the type strains for clusters GI.1 through GI.7 and GII.1 through GII.10 and GII.12 and use *Fields Virology*, 4th edition (Green et al., 2001) as a common reference. A summary of the reference strains for each system and a consensus classification approach are presented in Table 1.

Table 1. Classification of norovirus genotypes

Reference Strain	Accession no.	Kageyama classification	Zheng classification	Consensus classification
Norwalk	M87661	GI.1	GI.1	GI.1
Southampton	L07418	GI.2	GI.2	GI.2
Desert Shield	U04469	GI.3	GI.3	GI.3
Chiba	AB042808	GI.4	GI.4	GI.4
Musgrove	AJ277614	GI.5	GI.5	GI.5
Hesse	AF093797	GI.6	GI.6	GI.6
WUG1	AB081723	GI.8	GI.6	GI.6
Winchester	AJ277609	GI.7	GI.7	GI.7
Boxer	AF538679	GI.10	GI.8	GI.8
SzUG1	AB039774	GI.9	n/a	GI.9
KU8	AB067547	GI.11	n/a	GI.11
KU19	AB058525	GI.12	n/a	GI.12
T35	AB112132	GI.13	n/a	GI.13
T25	AB112100	GI.14	n/a	GI.14
Hawaii	U07611	GII.1	GII.1	GII.1
Melksham	X81879	GII.2	GII.2	GII.2
Toronto	U02030	GII.3	GII.3	GII.3
Bristol	X76716	GII.4	GII.4	GII.4
Hillingdon	AJ277607	GII.5	GII.5	GII.5
Seacroft	AJ277620	GII.6	GII.6	GII.6
Leeds	AJ277608	GII.7	GII.7	GII.7
Amsterdam	AF195848	GII.8	GII.8	GII.8
Virginia Beach	AY038599	GII.9	GII.9	GII.9
Erfurt	AF427118	GII.10	GII.10	GII.10
SW918	AB074893	n/a	GII.11	GII.11
Wortley	AJ277618	GII.12	GII.12	GII.12
Fayetteville	AY113106	GII.14	GII.13	GII.13
M7	AY130761	GII.13	GII.14	GII.14
J23	AY130762	n/a	GII.15	GII.15
Tiffin	AY502010	GII.15	GII.16	GII.16
CSE1	AY502009	GII.11	GII.17	GII.17
SW101	AY823305	n/a	n/a	GII.18
SW170	AY823306	n/a	n/a	GII.19
T5	AB112260	GII.16	n/a	GII.20
Jena	AJ011099	n/a	GIII.1	GIII.1
CH126	AF320625	n/a	GIII.2	GIII.2
Alphatron	AF195847	GII.17	GIV.1	GIV.1
MNV-1	AY228235	n/a	GV.1	GV.1

The Kageyama classification system relies on approximately 300 bp (depending on the genogroup) of nucleotide sequence from the 5' end of the capsid, known as the N/S region (Kageyama et al., 2004). It separates the human NoV sequences into 2 genogroups, GI and GII. Fourteen GI clusters are assigned based on pairwise distances of less than 12.1%. Seventeen GII clusters are assigned using a pairwise distance of less than 11.7%. These include only human NoV sequences.

The Zheng classification system uses the entire amino acid sequence of the capsid protein to assign genogroup and cluster information (Zheng et al., 2006). Genogroups have uncorrected pairwise distances of 4.9 – 61.4% and clusters have uncorrected pairwise distances of 14.3 – 43.8%. This scheme separates the human and animal NoV into five genogroups, GI – GV. GI, GII and GIV viruses infect humans and swine, GIII viruses infect cattle and GV viruses infect mice. This classification system identifies eight GI clusters, seventeen GII clusters, two GIII clusters and one cluster each within GIV and GV.

In addition to the viruses characterized by these two comprehensive publications (Kageyama et al., 2004; Zheng et al., 2006), two novel swine genotypes have also been identified and tentatively assigned genotypes GII.18 and GII.19 (Wang et al., 2005).

A consensus classification system can be defined based on the Zheng system, complemented with the additional strains that have been defined by Kageyama and Wang. A consensus classification is presented in the right hand column of Table 1. It is important to note that the cluster assignments after GI.7 and GII.10 are tentative until a definitive classification system has been recognized by the scientific community. Although such a presentation does not account for the variation in recombinant NoV types, these can be classified based on their capsid sequences, which are of previously identified types (Bull et al., 2007).

Norovirus Detection and Typing

Early detection methods for NoV used electron microscopy and serology to detect and characterize types (Kapikian et al., 1972). A modern version of this type of detection is the use of an ELISA for NoV detection (Milne et al., 2007). These methods suffer from a lack of sensitivity, requiring an estimated 10^5 particles per gram of stool for detection (Duizer et al., 2007). Antibodies also cannot distinguish between all of the standard genotypes that have been determined based on sequence analysis (Ando et al., 2000; Hansman et al., 2006). The more sensitive detection methods target and amplify a region of the NoV genome. This has the added advantage that products can be sequenced to provide genotype information.

Multiple platforms have been presented and evaluated for the amplification of NoV RNA (Table 2). They include reverse-transcription-loop-mediated isothermal amplification (RT-LAMP) (Fukuda et al., 2006; Yoda et al., 2007; Iturriza-Gomara et al., 2008; Fukuda et al., 2008), nucleic acid sequence-based amplification (NASBA) (Greene et al., 2003; Moore et al., 2004; Houde et al., 2006; Patterson et al., 2006; Rutjes et al., 2006) and reverse-transcription polymerase chain reaction (RT-PCR) (Moe at al., 1994; Ando et al., 1995; Green et al., 1995a, 1995b; Vinje and Koopmans, 1996; Noel et al., 1997; Kojima et al., 2002; Vennema et al., 2002; Anderson et al., 2003; Vinje et al., 2004).

Table 2. Molecular system available for the amplification of NoV RNA

Targeted region	Molecular amplification system	Reference
RdRp	Conventional RT-PCR	Moe et al., 1994
RdRp	Conventional RT-PCR	Ando et al., 1995 Vennema et al., 2002
RdRp	Conventional RT-PCR	Green et al., 1995a; 1995b
Protease/RdRp	Conventional RT-PCR	Vinje and Koopmans, 1996
Capsid	Conventional RT-PCR	Noel et al., 1997
Capsid	Conventional RT-PCR	Kojima et al., 2002
Not available	Light cycler	Miller et al., 2002
RdRp	Conventional RT-PCR	Anderson et al., 2003
ORF1/ORF2 junction	TaqMan	Kageyama et al., 2003 Jothikumar et al., 2005b Trujillo et al., 2006
RdRp	SYBR Green	Gunson et al., 2003
RdRp	NASBA	Greene et al., 2003 Moore et al., 2004
Capsid	Conventional RT-PCR	Vinje et al., 2004
RdRp	TaqMan	Hohne and Schreier, 2004
RdRp	SYBR Green	Laverick et al., 2004
RdRp	SYBR Green	Richards et al., 2004
RdRp	Real-time NASBA	Rutjes et al., 2006
RdRp	Real-time NASBA	Patterson et al., 2006
Capsid	Real-time RT-LAMP	Fukuda et al., 2006
RdRp and ORF1/ORF2 junction	NASBA	Houde et al., 2006
ORF1/ORF2 junction	SYBR Green	Menton et al., 2007
ORF1/ORF2 junction	Eclipse	Hymas et al., 2007
Not available	RT-LAMP	Yoda et al., 2007
Not available	RT-LAMP	Iturriza-Gomara et al., 2008
Capsid	RT-LAMP/NASBA	Fukuda et al., 2008
RdRp	LUX RT-PCR	Nordgren et al., 2008

The majority of laboratories performing molecular detection use RT-PCR in a conventional or real-time format. There are a number of protocols presented in the literature and some genomic regions are targeted more frequently (Table 2).

Conventional RT-PCR (cRT-PCR) was the first major advance for the sensitive and specific detection of NoV in fecal, alimentary and environmental samples. RT-PCR detection methods have targeted the 3' end of ORF1, where the coding sequence for the viral RNA polymerase is located. This sequence is the most conserved between the various NoV types. Widely cited systems targeting this region are: the JV12/JV13 primer pair in its original (Vinje and Koopmans, 1996) or updated (Vennema et al., 2002) format; the SR33/SR46/SR48/SR50/SR52 (Region A) primer set (Ando et al., 1995) and the Mon431/Mon432/Mon433/Mon434 (Region B) primer set (Anderson et al., 2003) (Figure 1). Although there are publications that use sequence analysis in these regions to group NoV, this does not accurately discriminate between all of the known viral types. Figure 2 demonstrates the phylogenetic analysis of reference strains using the available nucleotide sequences from Region JV (Figure 2A), Region A (Figure 2B) and Region B (Figure 2C).

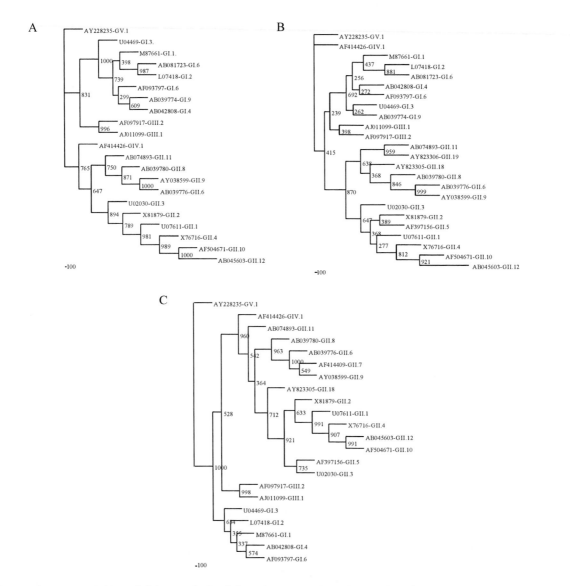

Figure 2. Neighbour-joining analysis of the reference strains using nucleotides from (A) Region JV, (B) Region A and (C) Region B. Bootstrap scores for branches are shown from 1000 replicates. Reference strains (GenBank accession numbers) used were: GI.1 (M87661), GI.2 (L07418), GI.3 (U04469), GI.4 (AB042808), GI.6 (AF093797), GI.6 (AB081723), GI.9 (AB039774), GII.1 (U07611), GII.2 (X81879), GII.3 (U02030), GII.4 (X76716), GII.5 (AF397156), GII.6 (AB039776), GII.7 AF414409), GII.8 (AB039780), GII.9 (AY038599), GII.10 (AF504671), GII.11 (AB074893), GII.12 (AB045603), GII.18 (AY823305), GII.19 (AY823306), GIII.1 (AJ011099), GIII.2 (AF097917), GIV.1 (AF414426), GV.1 (AY228235).

Unfortunately, a complete comparison cannot be made because some sequences are not available in the public data bank. For the strains shown in Figure 2A, the JV region discriminates between all standards. In Figure 2B we can see from the low bootstrap scores that GI.4 and GI.6, GI.3 and GI.9, GIII.1 and GIII.2, GII.2 and GII.5 are not well distinguished using Region A sequence. A similar situation exists with the region B analysis, where GI.2 and GI.1 and GI.4 and GI.6, GII.7 and GII.9 are not easily differentiated. This is a consequence of the sequence conservation in these regions between strains. This results in good targets for molecular detection but poor discrimination in typing reactions.

Many testing laboratories are currently using or moving toward the use of real-time RT-PCR (qRT-PCR) technology for the detection of NoV. The published qRT-PCR assays almost exclusively target the RdRp region at the overlap between ORF1 and ORF2 (Figure 1). This is the site of initiation for the synthesis of the sub-genomic RNA and the most conserved region in the NoV genome (Kageyama et al., 2003; Lambden et al., 1995). The original published protocol targeting 89 bp (GI) and 98 bp (GII) amplicons in the RT region with confirmation by TaqMan probes (Kageyama et al., 2003) has been adapted and improved in a number of ways. Primers and probes have been modified slightly to increase sensitivity or ubiquity (Jothikumar et al., 2005b). Primer and probe sets have been added to detect additional human or animal caliciviruses (Trujillo et al., 2006; Wolf et al., 2007). The reactions have been modified slightly to be used in a multiplex format (Antonishyn et al., 2006; Mohamed et al., 2006; Pang et al., 2005; Rolfe et al., 2007). Alternative oligonucleotide or fluorescent chemistries have been explored (Hoehne and Schreier, 2006; Hymas et al., 2007; Nordgren et al., 2008).

The system has been expanded to include primers and probes for the detection of other viruses (Logan et al., 2007). Although SYBR Green protocols are also available in the literature (Gunson et al., 2003; Laverick et al., 2004; Richards et al., 2004; Menton et al., 2007), there is a general consensus among international reference laboratories that the probe-based qRT-PCR amplification is the most reliable and robust in use today for enterically transmitted virus detection.

As indicated above, the accurate genetic typing of NoV depends upon the amplification of less conserved regions. The three most common targets are all located in the capsid coding sequence. This region has sequence heterogeneity that may be relevant for the spread of disease since this major capsid protein is the primary immune target (Lindesmith et al., 2008; Xerry et al., 2008). They are the SKF/SKR (Region C) primers (Kojima et al., 2002), the CapABCD (Region D) primers (Vinje et al., 2004) and the Mon381/383 (Region E) primers (Noel et al., 1997) (Figure 1). The Region E system was only designed to amplify GII NoV and so a meaningful comparison cannot be made between this and the Region C and D systems.

Figure 3 shows Region C and Region D classification of the reference strains available. Although the Region D sequence is less conserved overall, sequence from either region is suitable for discriminating between the NoV clusters. For differentiation of the GII.4 noroviruses into strains corresponding to epidemic years, as exemplified by Siebenga et al. (2007) and Lindesmith et al. (2008), only the Region D sequence provides the required degree of sequence variation.

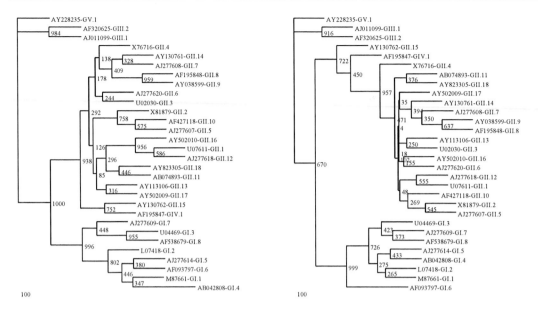

Figure 3. Neighbour-joining analysis of the reference strains using nucleotides from (A) Region C, (B) Region D. Bootstrap scores for branches are shown from 1000 replicates. Reference strains (GenBank accession numbers) used were: GI.1 (M87661), GI.2 (L07418),GI.3 (U04469), GI.4 (AB042808), GI.5 (AJ277614), GI.6 (AF093797), GI.6 (AB081723), GI.7 (AJ277609), GI.8 (AF538679), GI.9 (AB039774), GII.1 (U07611), GII.2 (X81879), GII.3 (U02030), GII.4 (X76716), GII.5 (AJ277607), GII.6 (AJ277620), GII.7 (AJ277608), GII.8 (AF195848), GII.9 (AY038599), GII.10 (AF427118), GII.11 (AB074893), GII.12 (AJ277618), GII.13 (AY113106), GII.14 (AY130761), GII.15 (AY130762), GII.16 (AY502010), GII.17 (AY502009), GII.18 (AY823305), GII.19 (AY823306), GIII.1 (AJ011099), GIII.2 (AF320625), GIV.1 (AF195847), GV.1 (AY228235).

Rotavirus (RV)

Overview

RV are the most important etiological agent of acute gastroenteritis in young infants and children worldwide. As is often the case with gastrointestinal diseases, RV are transmitted through the fecal-oral route. It is estimated that RV accounts for approximately 25% of all diarrhoea-related deaths worldwide (Bass et al., 2007). In the developed countries the mortality rate is low but even in the United States, RV still account for approximately 500 000 outpatient visits per year (Bass et al., 2007). In a country-wide UK study, RV, with a prevalence rate of 31% among patients exhibiting gastroenteritis symptoms were second only to NoV in terms of frequency (Amar et al., 2007). RV are mostly found in children < 5 yr of age but are also frequently found in asymptomatic individuals.

RV are members of the family *Reoviridae*, characterized by non-enveloped isocahedral capsids and segmented double-stranded RNA (dsRNA) genomes. The RV genome consists of 11 fragments ranging in size from 667 to 3 302 base pairs, with a total genome size of 18 522 base pairs. Each fragment starts with a 5' guanidine residue and has conserved non-coding sequences. All fragments also end with conserved non-coding sequences and a 3' cytidine residue. There is no poly-(A) tail as opposed to a number of RNA viruses. Each RNA segment contains only 1 ORF with the exception of segment 11 which contains two more ORF: one in-frame ORF and one out-of-frame ORF. These 11 segments encode 6 structural proteins (VP1-4, VP6 and VP7 (VP5 and VP8 are generated through the proteolytic cleavage of VP4), and 5 non-structural proteins (NSP1-5) (Estes and Cohen, 1989).

The 75 nm capsid is comprised of a central core which contains the genetic material, an inner layer and an outer layer. The core is mostly composed of the VP2 protein (encoded by segment 2) with a smaller proportion of VP3 (segment 3). The inner capsid contains mostly the structural protein VP6 (segment 6) which can trigger a strong immunological response. The outer layer is composed of two structural proteins, the more abundant glycoprotein VP7 (segment 9, or 7 or 8 depending on strain) and the protease-sensitive VP4 (segment 4) (which is cleaved to generate VP5 and VP8).

Rotavirus Classification

On the basis of VP6, seven serogroups have been described (A-G). Serogroup A is responsible for the vast majority of human infections and is associated with severe gastroenteritis in children; serogroup B has mostly been linked to severe epidemics among adults in China but a recent study described its relatively frequent occurrence in children under 3 yr of age in India (Barman et al., 2006); serogroup C is mostly porcine with some cases in infants and children. The other serogroups are limited to animals (Estes and Cohen, 1989).

Within serogroup A, RV are further classified into G and P types based on VP7 and VP4, respectively, as both proteins are capable of eliciting a neutralizing antibody response (Gray et al., 2008). At least 15 G types and 26 P types have been recognized to date (Araujo et al., 2007) but a small proportion of those accounts for the vast majority of human infections. Whereas genotypes and serotypes of the G antigen usually match, this is not the case for P antigens. For this reason, serogroup A rotaviruses are classified by indicating their G serotype number and their P genotype in brackets. The most common types currently circulating in the human population worldwide are G1-G4 and G9, P[4], P[6] and P[8] (Fisher and Gentsch, 2004). Among these, the most common combinations are G1P[8], G2P[4], G3P[8], G4P[8] and G9P[8] (Iturriza-Gomara et al., 2000b; Simmonds et al., 2008).

Analysis of circulating RV G- and P-types has lead to the observation that the prevalence of the various strains varies geographically and over time (Samajdar et al., 2008). Indeed, RV are known to evolve rapidly. Due to their segmented nature, reassortment of the genomic fragments constitutes a strong evolutionary mechanism (Gouvea and Brantly, 1995). Co-infection of a single cell with two different strains may lead to the reassortment of genes from different viruses (Palombo, 2002). Through RV typing, strong evidence has also been gathered that zoonotic transmission occurs. Porcine and bovine (Steyer et al., 2008) or canine and feline (Tsugawa and Hoshino, 2008) RV strains have recently been detected in humans. It is believed that zoonotic transmission is likely more common in regions where close contact between humans and livestock is frequent (Iturriza-Gomara et al., 2000b). See also Cook et al. (2004) for a recent review. Hence, strains circulating throughout the world are in constant evolution and require continuous monitoring. For this reason rapid and effective typing methods, molecular or otherwise, must be developed, constantly revised and updated (Iturriza-Gomara et al., 2004).

Rotavirus Detection and Typing

Detection of RV in human stool samples has traditionally relied on electron microscopy, extraction of total RNA and separation of the viral RNA fragments by polyacrylamide gel electrophosis (Herring et al., 1982), or detection of group A VP6 with various enzyme immuno-assays, of which many are commercial. These techniques often lack sensitivity or are cumbersome and lengthy. Although in the clinical settings enzymatic detection of viral antigens is still the most widely used reference method, various implementations of the RT-PCR technology are rapidly gaining ground (Gray et al., 2008). RT-PCR has been shown to be more sensitive than ELISA assays, especially for the detection of asymptomatic RV infections (Gladstone et al., 2008; Gutiérrez-Aguirre et al., 2008; Stockman et al., 2008).

The first RT-PCR systems were generally gel-based. Many studies use an extension of group A ELISA assays as they target the VP6 gene sequence; most are based on the work of Iturriza-Gomara et al. (2002). They can be conventional gel-based RT-PCR (Amar et al., 2007; Stockman et al., 2008), qRT-PCR with SYBR Green (Gladstone et al., 2008) or they may use TaqMan probes (Gutiérrez-Aguirre et al., 2008) (Table 4). They tend to be used in clinical settings. Many studies involving RT-PCR detection are also based on various modifications of the primer sets developed by Gouvea et al. (1990) for the typing of the VP7 antigen. They are either cRT-PCR with one round of amplification (Brassard et al., 2005), semi-nested (Coelho et al., 2003; Costantini et al., 2007) or nested (Le Guyader et al., 1994; Baggi and Peduzzi, 2000; van Zyl et al., 2006). Some RT-PCR detection systems were also developed around the primers sets of Gentsch et al. (1992) initially developed for the typing of the VP4 antigen. They are either conventional (Borchardt et al., 2003) or qRT-PCR with SYBR Green detection chemistry (Min et al., 2006). Another often cited RT-PCR system is the one from Pang et al. (2004). The system is based on the NSP3 gene segment, a non structural protein that binds the consensus sequence at the 3' end of the viral mRNA and also interacts with the host cell translation machinery, thus facilitating translation (Jayaram et al,. 2004). This system uses qRT-PCR detection with TaqMan probes (Butot et al., 2007; Butot et al., 2008; Zeng et al., 2008).

Finally, a few studies also used nucleic acid sequence based amplification (NASBA), an isothermal RNA amplification technology, for the detection of RV (Jean et al., 2002a; Jean et al., 2002b). Immuno-PCR, a combination of immunoassay detection and PCR, has also been reported (Adler et al., 2005; see also Barletta (2006) for a recent review of the technology). One study also described the use of microarrays based on the VP7 gene sequence (Chizikov et al., 2002), however, these methods are not widely used.

Molecular typing of the G antigen as it is done by RT-PCR is widely based on the work of Gouvea et al. (1990) and common prototype strains are presented in Table 3.

Table 3. G- and P-types of common prototype rotavirus strains

Reference strain	G serogroup	Accession no.	P genotype	Accession no.	P genotype[a]
Wa	G1	K02033	P[8]	L34161	1
KU	G1	D16343	P[8]	AB222784	1
M37	G1	M37344	P[6]	L20877	3
K8	G1	D16344	P[9]	D90960	4
DS-1	G2	EF672581	P[4]	AB118025	2
S2	G2	DQ498990	P[4]	n/a	2
1076	G2	n/a	P[6]	M88480	3
P	G3	EF672602	P[8]	EF672598	1
SA11-H96 (simian)	G3	NC_011503	P[2]	NC_011510	n/a
RRV (simian)	G3	M21650	P[3]	AY033150	n/a
YO	G3	D86284	P[8]	AB008279	1
AU-1	G3	D86271	P[9]	D10970	4
Hochi	G4	AB012078	P[8]	AB008295	1
ST3	G4	EF672616	P[6]	L33895	3
VA70	G4	M37369	P[8]	AJ540229	1
UK (bovine)	G6	n/a	P[5]	M22306	n/a
69M	G8	EF672560	P[10]	EF672556	5
WI61	G9	EF672623	P[8]	EF672619	1
F45	G9	AB180970	P[8]	U30716	1

[a] As initially defined by Gentsch et al. (1992).
n/a: not available.

Table 4. Molecular system available for the amplification of RV RNA

Targeted region	Sample type	Molecular amplification system	Reference
VP6			
	Clinical	Conventional RT-PCR	Iturriza-Gomara et al., 2002
	Clinical	Immuno-PCR	Adler et al., 2005
	Clinical	Conventional RT-PCR	Amar et al., 2007
	Clinical	SYBR Green	Gladstone et al., 2008
	Clinical	TaqMan	Gutierrez-Aguirre et al., 2008
	Clinical	Conventional RT-PCR	Stockman et al., 2008
NSP3			
	Clinical	TaqMan	Pang et al., 2004
	Berries and vegetables	TaqMan	Butot et al., 2007
	Berries and herbs	TaqMan	Butot et al., 2008
	Clinical	TaqMan	Zeng et al., 2008
VP7			
	Shellfish; sediments	Conventional nested RT-PCR	LeGuyader et al., 1994

Table 4. (Continued)

Targeted region	Sample type	Molecular amplification system	Reference
	Clinical; water	Conventional nested RT-PCR	Baggi and Peduzzi, 2000
	Purified virus	Microarray	Chizikov et al., 2002
	Spring water	Conventional RT-PCR	Brassard et al., 2005
	Purified virus	NASBA	Jean et al., 2002a
	Sewage treatment effluents	NASBA	Jean et al., 2002b
	Artificially seeded oysters	Conventional semi-nested RT-PCR	Coelho et al., 2003
	Water, raw vegetables	Conventional nested RT-PCR	Van Zyl et al., 2006
	Manure; post-treatment water	Conventional semi-nested RT-PCR	Costantini et al., 2007
VP4			
	Ground water	Conventional RT-PCR	Borchardt et al., 2003
	Clinical	SYBR Green	Min et al., 2006

The scheme involves the cRT-PCR amplification of a 1062 bp fragment of gene segment 9 (or 7 or 8, depending on strain) encoding VP7 with primers Beg9 / End9 which fall into highly conserved regions at both ends of the gene. The first round amplification product is then re-amplified with a mixture of six serotype-specific forward primers and one conserved reverse primer (Figure 4).

Each serotype-specific forward primer falls into one of 6 highly variable regions which are nonetheless well conserved within each serotype. The amplification products are separated by agarose gel electrophoresis and identified based on product size. Because of the constant evolution of the RV serotypes and genotypes, untypable strains are frequently encountered.

To account for emerging G types and the accumulation of point mutations in the corresponding gene, the typing system of Gouvea et al. (1990) has been improved and expanded over the years (Iturriza-Gomara et al., 2001; Das, 2004; Iturriza-Gomara et al., 2004; Banerjee et al., 2007).

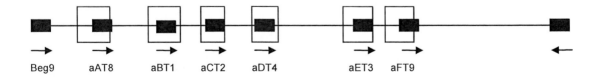

Figure 4. RT-PCR G-typing system. The bar represents rotavirus gene 9 encoding VP7. Black boxes represent PCR primers, open boxes represent regions of high variability associated with the most common human G-type (adapted from Gouvea et al., 1990).

Typing of the VP4 antigen by cRT-PCR is done in a manner very similar to VP7. The most widely cited system was developed by Gentsch et al. (1992) and like the G-typing system, it has also been modified and expanded over time (Iturriza-Gomara et al., 2000a; Iturriza-Gomara et al., 2004; Simmonds et al., 2008).

Very recently, a new molecular typing system based on all 11 gene segments was presented (Matthijnssens et al., 2008a). It is based on sequencing and phylogenetic analysis of all fragments and takes the form of Gx-P[x]-Ix-Rx-Cx-Mx-Ax-Nx-Tx-Ex-Hx which correspond to the VP7-VP4-VP6-VP1-VP2-VP3-NSP1-NSP2-NSP3-NSP4-NSP5/6 genes. The authors have identified nucleotide identity cut off percentages for all fragments to allow the classification of sequences into defined groups. This system relies on new gene sequences being sent to a newly formed Rotavirus Classification Working Group (RCWG) who will analyze the sequences and assign each fragment to a known or new group as required (Matthijnssens et al., 2008b). This system has the potential to be very discriminating and to allow a very fine description of each strain's genotype. However, sequencing all gene fragments is very tedious and this may limit its application. A microarray platform may be more suitable for current applications of this RV molecular genotyping and characterization system.

Hepatitis A Virus (HAV)

Overview

The *Picornaviridae* family consists of eight currently recognized genera: *Aphthovirus*, *Cardiovirus*, *Enterovirus* (which is now including *Rhinovirus*), *Erbovirus*, *Hepatovirus*, *Kobuvirus*, *Parechovirus* and *Teschovirus* (Stanway et al., 2005). Hepatitis A virus (HAV) is a small (27 to 32 nm in diameter) non-enveloped single-stranded RNA virus belonging to the genus *Hepatovirus* (Stanway et al., 2005). HAV is responsible for around half of the hepatitis infections diagnosed worldwide (Sanchez et al., 2007). Fever, nausea, fatigue, anorexia, abdominal cramps, dark urine and jaundice represent the main symptoms associated with HAV infections. In rare complications, it may induce a fulminant hepatitis which could result in permanent liver failure and/or death (for recent reviews, see Nainan et al., 2006; Pinto and Saiz, 2007; Sanchez et al., 2007). High efficacy inactivated HAV vaccines are available since the early 1990s and provide long-term immunity against hepatitis A infection (Bell and Feinstone, 2004).

HAV is mainly transmitted through the fecal-oral route, by person-to-person contact or by the consumption of contaminated food (Koopmans et al., 2002; Fiore, 2004) and water (Bloch et al., 1990; De Serres et al., 1999; Tallon et al., 2008). Shellfish (Shieh et al., 2007; Chironna et al., 2002, 2003; Desenclos et al., 1991; Halliday et al., 1991), berries (Calder et al., 2003; Hutin et al., 1999; Niu et al., 1992; Ramsay and Upton, 1989; Reid and Robinson, 1987) and produce (Dentinger et al., 2001; Rosenblum et al., 1990; Wheeler et al., 2005) have been implicated in HAV outbreaks.

A small icosahedral capsid is protecting a positive-sense single-stranded polyadenylated RNA genome of approximately 7.5 kb containing two untranslated region (5' UTR and

3' UTR). The 5' end is uncapped but covalently linked to a small viral protein (VPg) (Weitz et al., 1986). The rest of the genome is constituted of a single ORF (ORF1 and ORF2 are equivalent) with three distinct regions (P1, P2, and P3) translated into a large polyprotein (2226 to 2228 amino acids) (Baroudy et al., 1985; Cohen et al., 1987) that is cleaved by an encoded viral protease (protease 3C) (Schultheiss et al., 1994) in capsid (VP1, VP2, VP3, VP4a and VP4b) and nonstructural proteins (VPg, putative protease 2A, cysteine protease 3C and RNA-dependent-RNA-polymerase) (Figure 5).

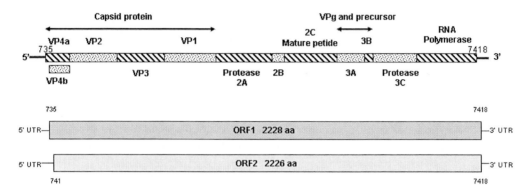

Figure 5. Genomic organization of Hepatitis A virus. Numbers refer to the corresponding nucleotide position of Hepatitis A virus reference strain (GenBank accession number NC001489).

Recently, HAV recombinant strains were isolated from infected humans (Costa-Mattioli et al., 2003). The recombination event between two different genotypes was found to occur within the VP1 capsid protein. It is believed that the capsid recombination may play a significant role in shaping the genetic diversity of HAV which may have important implications for its evolution, biology and control (Cristina and Costa-Mattioli, 2007). Despite its high degree of genetic heterogeneity at the nucleotide level, HAV possesses only a single serotype (Nainan et al., 2006).

Hepatitis A virus Classification

HAV was first subdivided into seven different genotypes including four of human origin (I, II, III and VII) and three of simian origin (IV, V and VI) (Robertson et al., 1992), based on the sequence information of the VP1/P2A junction. Recently, genotype VII was reclassified as a subgenotype of genotype II based on the sequence information of the complete capsid protein VP1 (Costa-Mattioli et al., 2003) and the complete genome (Lu et al., 2004). Therefore, HAV is now subdivided into six genotypes, three isolated from humans (I, II, III) and three from a simian origin (IV, V, VI) (Cristina and Costa-Mattioli, 2007). Most of the human circulating strains belong to genotypes I and III. These genotypes have been subdivided in subgenotypes IA, IB, IIIA and IIIB (Costa-Mattioli et al., 2003). Genotype IA tends to be responsible for the majority of HAV infections worldwide (Costa-Mattioli et al., 2003; Robertson et al., 1992). The prototype strains of each genotype and subgenotype are presented in Table 5. Most RdRp are lacking a 3'-5' exonucleolytic proof reading-repair activity which affect their copying fidelity and generate genomic variants known as quasispecies (Domingo and Gomez, 2007). This defect of the RdRp is contributing to the genetic diversity and evolution of RNA viruses.

Table 5. Classification of HAV genotype/subgenotype

Prototype Strain	Accession no.	Genotype/subgenotype
GBM/1976	X75215	IA
LA/1975	K02990	IA
HM175/1976	M14707	IB
CF53/Berne/1979	AY644676	IIA
SLF88/1988	AY644670	IIB
PA21	M34084	IIIA
NOR-21/1997	AJ299464	IIIA
HA-JNG06-90F/1990	AB258387	IIIB
HAJ85-1F/1985	AB279735	IIIB
CY145	M59286	IV
AGM-27/1985	D00924	V
JM55	L07731	VI

Hepatitis A Virus Detection and Typing

HAV circulating strains grow poorly in cell culture and rarely produce cytopathic effects (Nainan et al., 2006). The presence of HAV can be demonstrated in clinical, environmental and food samples by using immunoassays for viral antigen detection or molecular detection systems of the viral nucleic acid. However, amplification of HAV RNA by RT-PCR was proven more sensitive than immunoassays and is currently the most commonly used method for the detection of HAV in clinical, food and environmental samples (Nainan et al., 2006).

Over the last two decades, several sets of primers and probes targeting different regions of the genome (Table 6) have been reported and used for the detection of HAV in cRT-PCR assays (Jansen et al., 1990; Robertson et al., 1991, 1992; Cromeans et al., 1997; Hutin et al., 1999; Pina et al., 2001; de Paula et al., 2002; Guévremont et al., 2006; Nainan et al., 2006), in qRT-PCR assays using either a SYBR Green chemistry (Brooks et al., 2005; Casas et al., 2007) or fluorescent probes (Costa-Mattioli et al., 2002b; Silberstein et al., 2003; Hewitt and Greening, 2004; Abd El Galil et al., 2005; Jothikumar et al., 2005a; Costafreda et al., 2006; Houde et al., 2007; Villar et al., 2006; Amado et al., 2008) and finally isothermal amplifications such as NASBA (Jean et al., 2001) and RT-LAMP (Yoneyama et al., 2007). In addition, two LightCycler FRET probes detection kits were also commercialized by Roche Diagnostics and Artus GmbH and tested (Heitmann et al., 2005; Sanchez et al., 2006).

Nucleic acid sequence of different genomic regions had been used to genotype circulating HAV isolates (Nainan et al., 2006). However, amplification and sequencing of a 168 bp fragment corresponding to the VP1/P2A region formed the basis of the genotyping classification during many years (Robertson et al., 1992). Recently, Costa-Mattioli et al.

(2002a) used the entire VP1 sequence including its P2A junction (a total fragment of 900 bp) and performed phylogenetic analysis (Figure 6). This comprehensive study proposed a new molecular typing system for HAV isolates that was further confirmed with another study based on the entire genome analysis (Lu et al., 2004).

Table 6. Molecular system available for the amplification of HAV RNA

Targeted region	Molecular amplification system	Reference
VP3	Conventional RT-PCR	Jansen et al., 1990
VP1	Conventional nested RT-PCR	Robertson et al., 1991
VP1/2A	Conventional nested RT-PCR	Robertson et al., 1992
VP3/VP1	Conventional RT-PCR	Cromeans et al., 1997
VP1/2B	Conventional nested RT-PCR	Hutin et al., 1999
5' UTR	Conventional nested RT-PCR	Pina et al., 2001
2B	Conventional nested RT-PCR	Pina et al., 2001
Not available	FRET probes	LightCycler® HAV (Roche Diagnostics, 2001) Heitmann et al., 2005 Sanchez et al., 2006
VP2	NASBA	Jean et al., 2001
VP1/2A	Conventional nested RT-PCR	de Paula et al., 2002
Entire VP1	Conventional nested RT-PCR	Costa-Mattioli et al., 2002a
5' UTR	TaqMan	Costa-Mattioli et al., 2002b
5'UTR/VP4	TaqMan	Silberstein et al., 2003
Entire genome	Conventional RT-PCR	Lu et al., 2004
5' UTR	TaqMan	Hewitt and Greening, 2004
VP1	FRET probes	RealArt HAV LC (artus Gmbh, 2004) Sanchez et al., 2006
5' UTR	Molecular Beacon	Abd El Galil et al., 2005
5' UTR	TaqMan	Jothikumar et al., 2005a
VP3/VP1	SYBR Green	Brooks et al., 2005
5' UTR	TaqMan and Molecular Beacon	Costafreda et al., 2006
5' UTR	TaqMan	Houde et al., 2007
2A	Conventional and TaqMan	Guévremont et al., 2006; Houde et al., 2007
RdRp	Conventional and TaqMan	Guévremont et al., 2006; Houde et al., 2007
5' UTR	TaqMan	Villar et al., 2006
VP3/2B	Conventional nested RT-PCR	Nainan et al., 2006
5' UTR	SYBR Green	Casas et al., 2007
5' UTR	RT-LAMP	Yoneyama et al., 2007
Entire genome	Conventional RT-PCR	Garcia-Aguirre and Cristina, 2008
5' UTR	TaqMan	Amado et al., 2008

These two typing systems are presently the current recognized standards for the identification of HAV genotypes and subgenotypes and for the molecular characterization of field isolates.

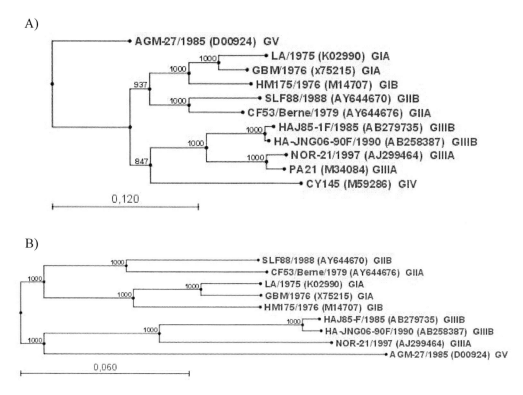

Figure 6. Neighbour-joining analysis of the reference strains using nucleotides from (A) VP1 and (B) complete genome. Bootstrap scores for branches are shown from 1000 replicates. Prototype strains (GenBank accession numbers) used were: IA (X75215 and K02990), IB (M14707), IIA (AY644676), IIB (AY644670), IIIA (M34084 and AJ299464), IIIB (AB258387 and AB279735), IV (M59286), V (D00924) and VI (L07731).

Hepatitis E Virus (HEV)

Overview

The *Hepeviridae* is a new officially approved family with one recognized genus *Hepevirus* (ICTV, 2008). HEV was classified originally in the family *Caliciviridae* but was reclassified recently as the sole member of the genus *Hepevirus* (Emerson et al., 2005a; ICTV, 2008). HEV is a non-enveloped icosahedral virus of approximately 27-34 nm in diameter discovered in 1983 by immune electron microscopy (Balayan et al., 1983).

HEV, previously known as enterically transmitted non-A, non-B hepatitis virus, is a major public health problem in many developing countries (Shrestha et al., 2007). There appear to be a direct correlation between poor sanitation and a high incidence of hepatitis E illness (Goens and Perdue, 2004). HEV is principally transmitted by the faecal-oral route or contaminated water. The possibility of HEV transmission through food (shellfish, pig liver, wild deer and boar meat) and from person-to-person (blood transfusion and perinatal transmission) were also reported in different studies (Aggarwal and Krawczynski, 2000; Jothikumar et al., 2000; Smith, 2001; Matsuda et al., 2003; Tei et al., 2003; Yazaki et al., 2003; Goens and Perdue, 2004; Takahashi et al., 2004; Li et al., 2005; Colson et al., 2007; Feagins et al., 2007; Pinto and Saiz, 2007; Vasickova et al., 2007; Matsubayashi et al., 2008; Mushahwar, 2008). There were differences in thermal stability among HEV strains and temperature that would be required to inactivate virus embedded in meat or viscera from infected animals is expected to be higher than 60°C (Emerson et al., 2005b; Tanaka et al., 2007). HEV may occur in three different forms: large epidemics, smaller outbreaks or sporadic infections (Okamoto, 2007). Incubation period ranges from 2 to 10 weeks and the symptoms of infection are fever, nausea and vomiting, fatigue and anorexia, abdominal pain, jaundice, dark urine and elevated liver enzymes. Most of the time, it is difficult to differentiate between HEV and HAV infections based only on symptom analysis (Aggarwal and Krawczynski, 2000; Goens and Perdue, 2004; Smith, 2001; Mushahwar, 2008). The case-fatality rate is usually low (0.2 – 3%) (Smith, 2001; Pinto and Saiz, 2007; Mushahwar, 2008) but higher than with HAV infections (0.2%) (Goens and Perdue, 2004). The severity of HEV infection can be associated with host factors of the infected patients such as pregnancy or aging (Inoue et al., 2006a). Premature birth and mortality rate among pregnant women who acquired HEV infection is as high as 25% in the third trimester of pregnancy (Aggarwal and Krawczynski, 2000; Smith, 2001; Goens and Perdue, 2004; Myint and Gibbons, 2008).

It seems that genotype 4 strains tended to be more often associated with acute or fulminant hepatitis than HEV of other genotypes (Inoue et al., 2006a; Mizuo et al., 2005; Okamoto, 2007). No effective treatment exists for hepatitis E. However, a vaccine using a baculovirus-expressed recombinant HEV 56 kDa capsid protein was recently studied in a high-risk population in Nepal. In a phase 2 randomized, double-blind, placebo-control trials the vaccine has shown an efficacy of 95.5% (Mushahwar, 2008; Myint and Gibbons, 2008; Shrestha et al., 2007).

The virion capsid seems to be composed of a single structural protein (Aggarwal and Krawczynski, 2000; Emerson and Purcell, 2003) and encloses a single-stranded RNA genome of approximately 7.2 kb. Viral RNA contains a short (27 to 35 nucleotides) 5' UTR followed by three partially overlapping ORF, a 3' UTR (65 to 74 nucleotides) that is terminated by a poly-(A) tract (Figure 7).

ORF1 (5073-5124 nucleotides long) encodes non-structural proteins, ORF2 (1977-1980 nucleotides) encodes the capsid protein and ORF3 (366-369 nucleotides) a small cytoskeleton-associated phosphoprotein (Tam et al., 1991; Aggarwal and Krawczynski, 2000; Smith, 2001; Emerson and Purcell, 2003; Lu et al., 2006; Pinto and Saiz, 2007; Vasickova et al., 2007).

The first complete genomic sequence was determined in 1991 (Tam et al., 1991) and the first animal strain of HEV was identified and characterized in 1997 from a pig in the United States (Meng et al., 1997). The exact mode of replication and expression has not been recognised yet (Vasickova et al., 2007). A cell-culture system with the PLC/PRF/5 cell line was recently developed (Tanaka et al., 2007) and would be useful for the study of virological characteristics of HEV.

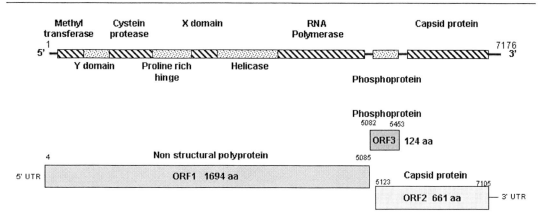

Figure 7. Genomic organization of Hepatitis E virus. Numbers refer to the corresponding nucleotide position of Hepatitis E virus reference strain (GenBank accession number NC001434).

Hepatitis E Virus Classification

HEV is classified into four genotypes based on the sequence of the complete genome (Emerson et al., 2005a; Lu et al., 2006; Zhai et al., 2006) and a single serotype is described with extensive cross-reactivity among circulating human snd swine strains (Emerson et al., 2005a). There is no consensus on how many subtypes each genotype contains. Recently five, two, ten and seven subtypes were proposed for HEV genotype 1, 2, 3 and 4, respectively (Lu et al., 2006). These 24 subtypes (1a, 1b, 1c, 1d, 1e, 2a, 2b, 3a, 3b, 3c, 3d, 3e, 3f, 3g, 3h, 3i, 3j, 4a, 4b, 4c, 4d, 4e, 4f and 4g) are based upon nucleotides differences from five phylogenies 5' ORF1, 3' ORF1, 5' ORF2, 3' ORF2 and complete genome. Common prototype strains or isolates for each subtype are presented in Table 7. Genotypes 1 and 2, mostly associated with epidemics and outbreaks of hepatitis E, are usually due to fecal contamination of water supplies.

These two genotypes are more conserved and comprised until quite recently only human isolates (Lu et al., 2006; Caron et al., 2006). Identification of genotype 1 HEV virus in a swine sample from Cambodia was reported in 2006 (Caron et al., 2006). Genotypes 3 and 4 include human and animal (swine, wild boar, deer and mongoose) isolates (Takahashi et al., 2004; Nishizawa et al., 2005; Lu et al., 2006; Nakamura et al., 2006; Okamoto, 2007; de Deus et al., 2008) and show a high diversity most likely due to their zoonotic origin. Genotype 1 was found in tropical and several subtropical countries in Asia and Africa; genotype 2 in Mexico and Nigeria; genotype 3 was isolated in Asia, Europe, Oceania, North and South America, and finally genotype 4 has been found exclusively in Asia (Lu et al., 2006: Okamoto, 2007; Vasickova et al., 2007).

Avian HEV has been isolated from chickens in USA (Haqshenas et al., 2001) and shared approximately only 50% nucleotide sequence identity with human and swine HEV (Huang et al., 2004; Vasickova et al., 2007; Myint and Gibbons, 2008). Avian HEV is recognized as a new species in the *Hepeviridae* family. In the ICTV 8th Report (2008), it was designated as a "tentative species" in the genus *Hepevirus*.

Table 7. Classification of HEV genotype/subtype

Strain or isolate	Accession no.	Country	Host	Classification (genotype/subtype) [a]
Bur-82 (B1) [b]	M73218	Burna	Human	1a
Sar-55 (P1) [b]	M80581	Pakistan	Human	1b
Hev037 (I1)	X98292	India	Human	1c
Morocco	AY230202	Morocco	Human	1d
T3	AY204877	Chad (N'Djamena)	Human	1e
CAM-3F15	DQ145799	Cambodia	Swine	*1
Mexican (M1) [b]	M74506	Mexico	Human	2a
Nig7	AF173231	Nigeria	Human	2b
HEV-US1 [b]	AF060668	USA	Human	3a
Meng (swUS1) [b]	AF082843	USA	Swine	3a
JRA1	AP003430	Japan	Human	3b
swJ570	AB073912	Japan	Pig	3b
JBOAR-1Hyo04	AB189070	Japan	Boar	3b
JDEER-Hyo03L	AB189071	Japan	Deer	3b
MaR7307574 (HuMar)	EU116336	France	Human	3c
NLSW20	AF336001/AF336290	Netherlands	Swine	3c
TW3SW	AF296167	Taiwan	Swine	3d
HEV-Sendai	AB093535	Japan	Human	3e
swJ8-2	AB094227	Japan	Swine	3e
VH1 (Sp1)	AF195064/AF195061/AF491000	Spain	Human	3f
NLSW15	AF336000/AF332620	Netherlands	Swine	3f
Osh 205	AF455784	Kyrgyzstan	Swine	3g
It1	AF110387/AF110390	Italy	Human	3h
New Zealand (swNZ)	AF215661/AF200704	New Zealand	Swine	3h
Au1	AF279122/AF279123	Australia	Human	3i
swAr	AY258006	Argentina	Swine	3i
Arkell	AY115488	Canada	Swine	3j
JMNG-Oki02C	AB236320	Japan	Mongoose	*3
TW8E-2	AF117275	Taiwan	Human	4a
TW11SW	AF302068	Taiwan	Swine	4a
Ch181	AJ344188	China	Human	4b
swSJ14	AJ428856	China	Swine	4b
JAK-Sai	AB074915	Japan	Human	4c
swJ13-1	AB097811	Japan	Swine	4c
T1 [b]	AJ272108	China	Human	4d
swCH25	AY594199	China	Swine	4d
IND-SW16	AF505859	India	Swine	4e
HE-JA2	AB082547/AB082558	Japan	Human	4f
CCC220	AB108537	China	Human	4g
wbOK128	AB184831	Japan	Boar	*4
Avian HEV	AY535004	USA	Avian	n/a

[a]: classification suggested by Lu et al., 2006.
[b]: reference strain.
*: strains not included in the study of Lu et al., 2006.

Hepatitis E Virus Detection and Typing

The most commonly used procedures in the diagnosis of HEV infections are enzyme immunoassays for the detection of anti-HEV IgG and IgM in serum samples (Pinto and Saiz, 2007). However, these tests often have limited sensitivity (Wang et al., 2001; Innis et al., 2002; Orru et al., 2004), cannot be used for genotype determination and therefore provide limited information for epidemiological studies. The detection of RNA in serum, bile or faecal samples by RT-PCR enables both early diagnosis and genotype determination. In the last sixteen years many cRT-PCR (Jameel et al., 1992; Zanetti et al., 1999; Wang et al., 2000; Nishizawa et al., 2003; Phan et al., 2005; Chobe et al., 2006; Xia et al., 2008) or nested cRT-PCR (Huang et al., 1995; Meng et al., 1997; Gouvea et al., 1997; Chatterjee et al., 1997; Meng et al., 1998; Aggarwal et al., 1999; Erker et al., 1999; Hsieh et al., 1999; Takahashi et al., 2001; Yoo et al., 2001; Kabrane-Lazizi et al., 2001; Huang et al., 2002; Takahashi et al., 2002; Mizuo et al., 2002; Choi et al., 2003; Ahn et al., 2005; Li et al., 2005; Zhai et al., 2006; Inoue et al., 2006a, 2006b; de Deus et al., 2007) and qRT-PCR assays using SYBR Green (Orrù et al., 2004), TaqMan probes (Gardner et al., 2003; Mansuy et al., 2004; Enouf et al., 2006; Ahn et al., 2006; Jothikumar et al., 2006; Zhao et al., 2007; Colson et al., 2007; Gyarmati et al., 2007; Matsubayashi et al., 2008) or Primer-Probe Energy Transfer (Gyarmati et al., 2007), were developed for the detection of HEV RNA (Table 8).

Table 8. Molecular system available for the amplification of HEV RNA

Targeted region	Molecular amplification system	Reference
ORF1 (RNA polymerase)	Conventional RT-PCR	Jameel et al., 1992
ORF1 (RNA polymerase)	Conventional nested RT-PCR	Huang et al., 1995
ORF2 (capsid)	Conventional nested RT-PCR	Meng et al., 1997
ORF1 (RNA polymerase)	Conventional nested RT-PCR	Gouvea et al., 1997
ORF2 (capsid)	Conventional nested RT-PCR	Gouvea et al., 1997
ORF1 (methyltransferase, hypervariable region)	Conventional nested RT-PCR	Chatterjee et al., 1997
ORF1/ORF2/ORF3	Conventional nested RT-PCR	Chatterjee et al., 1997
ORF2 (capsid)	Conventional nested RT-PCR	Meng et al., 1998
ORF1 (RNA polymerase)	Conventional nested RT-PCR	Aggarwal et al., 1999
ORF1 (methyltransferase)	Conventional RT-PCR	Zanetti et al., 1999
ORF1 (methyltransferase)	Conventional nested RT-PCR	Erker et al., 1999
ORF2 (capsid)	Conventional nested RT-PCR	Erker et al., 1999
ORF2/ORF3	Conventional nested RT-PCR	Erker et al., 1999
ORF2 (capsid)	Conventional nested RT-PCR	Hsieh et al., 1999
Entire genome	Conventional RT-PCR	Wang et al., 2000
ORF1 (methyltransferase)	Conventional nested RT-PCR	Takahashi et al., 2001
ORF2 (capsid)	Conventional nested RT-PCR	Yoo et al., 2001
ORF1 (hypervariable region)	Conventional nested RT-PCR	Kabrane-Lazizi et al., 2001
ORF2	Conventional nested RT-PCR	Kabrane-Lazizi et al., 2001
ORF1/ORF2/ORF3	Conventional nested RT-PCR	Kabrane-Lazizi et al., 2001
ORF2 (capsid)	Conventional nested RT-PCR	Huang et al., 2002
ORF1/ORF2/ORF3	Conventional nested RT-PCR	Takahashi et al., 2002
ORF1	Conventional nested RT-PCR	Mizuo et al., 2002
ORF2 (capsid)	Conventional nested RT-PCR	Mizuo et al., 2002

Table 8. (Continued)

Targeted region	Molecular amplification system	Reference
Entire genome	Conventional RT-PCR	Nishizawa et al., 2003
ORF2 (capsid)	Conventional nested RT-PCR	Choi et al., 2003
ORF2/ORF3	TaqMan	Gardner et al., 2003
ORF2/ORF3	SYBR Green	Orru et al., 2004
ORF2 (capsid)	TaqMan	Mansuy et al., 2004
ORF2 (capsid)	Conventional RT-PCR	Phan et al., 2005
ORF2 (capsid)	Conventional nested RT-PCR	Ahn et al., 2005
ORF2 (capsid)	Conventional nested RT-PCR	Li et al., 2005
Entire genome	Conventional RT-PCR	Chobe et al., 2006
ORF1 (methyltransferase, hypervariable region, RdRp)	Conventional nested RT-PCR	Zhai et al., 2006
ORF2 (capsid)	Conventional nested RT-PCR	Zhai et al., 2006
ORF1 (helicase)	Conventional nested RT-PCR	Inoue et al., 2006a
ORF2/ORF3	Conventional nested RT-PCR	Inoue et al., 2006b
ORF2/ORF3	TaqMan	Enouf et al., 2006
ORF2 (capsid)	TaqMan	Ahn et al., 2006
ORF2/ORF3	TaqMan	Jothikumar et al., 2006
ORF2/ORF3	TaqMan	Zhao et al., 2007
ORF2 (capsid)	TaqMan	Colson et al., 2007
ORF2 (capsid)	TaqMan and Primer-Probe Energy Transfert (PriProET)	Gyarmati et al., 2007
ORF2 (capsid)	Conventional semi-nested RT-PCR	de Deus et al., 2007
ORF2 (capsid)	TaqMan	Matsubayashi et al., 2008
Entire genome	Conventional RT-PCR	Xia et al., 2008

A)

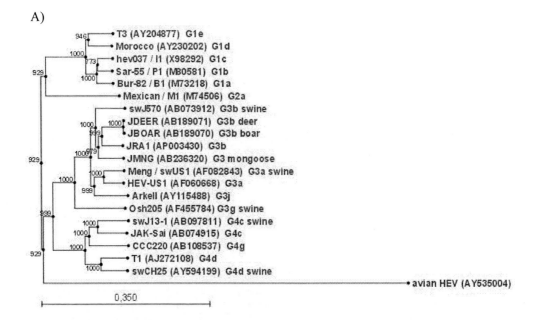

Molecular Detection and Characterization of Food and Waterborne Viruses   161

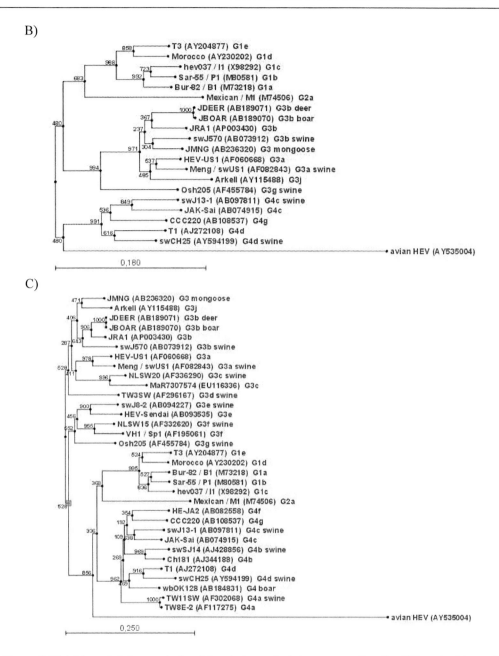

Figure 8. Neighbour-joining analysis of the strains or isolates representing the different subtypes for HEV genotype 1, 2, 3 and 4 using nucleotides from (A) complete genome (B) 3' end of ORF1 and (C) 5' end of ORF2. Bootstrap scores for branches are shown from 1000 replicates. Strains or isolates (GenBank accession numbers) used were: Bur-82 / B1 (M73218), Sar-55 / P1 (M80581), hev037 / I1 (X98292), Morocco (AY 230202), T3 (AY204877), Mexican / M1 (M74506), HEV-US1 (AF060668), Meng / swUS1 (AF082843), JRA1 (AP003430), swJ570 (AB073912), JBOAR-1Hyo04 (AB189070), JDEER-Hyo3/L (AB189071), MaR7307574 (EU116336), NLSW20 (AF336290), TW3SW (AF296167), HEV-Sendai (AB093535), swJ8-2 (AB094227), VH1 / Sp1 (AF195061), NLSW15 (AF332620), Osh 205 (AF455784), Arkell (AY115488), JMNG-Oki02C (AB236320), TW8E-2 (AF117275), TW11SW (AF302068), Ch181 (AJ344188), swSJ14 (AJ428856), JAK-Sai (AB074915), swJ13-1 (AB097811), T1 (AJ272108), swCH25 (AY594199), He-Ja2 (AB082558), CCC220 (AB108537), wbOK128 (AB184831), Avian HEV (AY535004).

These different RT-PCR targeted ORF2 or ORF3 genes and some of them were designed for the detection of the four genotypes (Huang et al., 2002; Enouf et al., 2006; Inoue et al., 2006b; Jothikumar et al., 2006; Zhao et al., 2006; de Deus et al., 2007; Gyarmati et al., 2007; Matsubayashi et al., 2008). The sensitivity of these detection assays can be affected by the quality of extracted RNA, RNase contamination or RT-PCR inhibitors in environmental and clinical samples, especially in faecal material (Escobar-Herrera et al., 2006; Rutjes et al., 2007; Scipioni et al., 2008). Statistical analysis showed that the genomic region corresponding to nt 4254 to 4560 (sequence M73218 in GenBank) located in the HEV RdRp domain (3' end of ORF1 region) could substitute the full length genome for HEV genotyping (Zhai et al., 2006) (Figure 8A and 8B). The 5' end of the ORF2 region (corresponding to nt 5994 to 6294) can also give a good reproduction of the complete genome (Lu et al., 2006) (Figure 8C) and the number of available sequences in GenBank for this region is larger than the other genome sections. Immune electron microscopy is sometimes used for the detection of viral particles in faecal material but was shown less sensitive than RT-PCR (Goens and Perdue, 2004).

Use of RT-PCR for the Detection and Characterization

Efficient, fast and reliable molecular tools are the cornerstone for the detection and the mitigation of food and waterborne viruses. Over the last 20 years, a variety of DNA-polymerase based (RT-PCR and RT-LAMP) and RNA-polymerase based (NASBA) amplification assays targeting the viral nucleic acids were developed (Tables 2, 4, 6, 8). Most of them were shown to possess the ability of amplifying their respective viral nucleic acid in clinical samples. However, the viral loads in clinical samples, such as faecal material, are usually very high compared to those found in environmental, food or water samples. Therefore, a comprehensive detection system should also consider other important aspects such as sample preparation and processing (extraction and concentration of the viral particles and their nucleic acid), specificity and analytical sensitivity of the molecular amplification system, and the possible presence of inhibitors to amplification reactions.

Conventional (direct or nested) RT-PCR assays were and are still widely used for the detection and identification of enterically transmitted viruses. However, post-amplification steps, such as gel electrophoresis combined to probe hybridization or sequencing, are necessary to confirm the amplification products and to prevent the misinterpretation of false positive results due to non-specific amplification. Because of their high analytical sensitivity, alternative isothermal amplification methods (NASBA and LAMP) have been designed and proposed for the routine detection of enterically transmitted viruses. The detection of amplified products is achieved using an electrochemiluminescent or fluorescent probes or dot blot probe hybridization. It is important to mention that even after a positive detection signal obtained with these isothermal amplification technologies, cRT-PCR amplification will still be necessary for further strain genotyping, characterization and comparison in support to epidemiological investigations, outbreak mitigation and surveillance programs (Houde et al.,

2006). PCR and RT-PCR remain as the actual gold standard for food and waterborne virus detection.

It was reported that one of the major applications of real-time PCR technology is in the field of virology for the detection of viruses and viral loads (Mackay, 2004; Mackay et al., 2002; Espy et al., 2006). Real-time RT-PCR offers many advantages over the other two techniques for the rapid and specific detection of viral particles: 1) a high analytical sensitivity; 2) the possibility of simultaneously amplifying and detecting the targeted nucleic acids in a single step procedure; 3) a wide dynamic range (> 10^7-fold) allowing a straightforward comparison between samples containing a large range of viral RNA concentrations; 4) the possibility of generating quantitative data by comparison to a standard curve; and, 5) the reduced risk of carry-over contamination through the use of a closed system (Bustin et al., 2005; Espy et al., 2006; Mackay, 2004; Mackay et al., 2002; Valasek and Repa, 2005). SYBR Green assays, which use a non-specific double stranded DNA intercalating dye, raised concerns about possible diagnostic errors from clinical and field samples due to potential non-specific amplifications. Real-time assays integrating target specific fluorescent probes (TaqMan, molecular beacon or FRET probes) were shown highly specific, thereby eliminating the need of post-amplification confirmatory steps, and also offer the possibility of multiplexing amplification reactions (Bustin et al., 2005; Espy et al., 2006; Mackay, 2004; Mackay et al., 2002; Valasek and Repa, 2005).

qRT-PCR technology can also generate quantitative data by comparing the data obtained to a standard curve of known concentrations of the targeted region. Defined concentrations of the purified viral genome (in the case of cultivable viruses such as laboratory-adapted variant of HAV or prototypes strains of RV), or of ssRNA molecules produced by *in vitro* transcription of the cloned cDNA of the target amplicon or of the cloned cDNA amplicon itself could be used to contruct the standard curves. However, following an extensive evaluation of these 3 schemes, Costafreda et al. (2006) have recommended the use of cloned amplicons to generate the external standard curves because of a better reproducibility and accuracy for quantification.

Although most of the TaqMan systems proposed for the detection of their respective food and waterborne viruses were shown able to detect approximately 1 to 10 genomic equivalent copy(ies) per reaction, a thorough parallel comparison and evaluation on the same purified RNA templates highlighted significant differences (1 to several logs) in their respective analytical sensitivities and limits of detection (Houde et al., 2006; Houde et al., 2007; Ward et al., 2009). These differences may be due to genomic RNA conformation, hairpin structures and nucleotide composition of the region which might influence the availability of the targeted region to the reverse transcriptase. This raises an important concern on the interpretation and comparison of quantitative data obtained using different TaqMan primer and probe systems. These studies also stressed the importance of performing optimization and validation asessment before selecting a detection system.

Another essential element for the significance of food and waterborne virus detection results obtained by molecular amplification methods is the integration of a control to monitor the efficiency of the different critical steps. The quality of either qualitative or quantitative detection results obtained by conventional or qRT-PCR will directly be affected by the recovery rate of viral particles from the sample, the nucleic acid extraction, the reverse transcription of RNA into DNA, the DNA amplification reaction and the presence of co-extracted inhibitors that may intefere with the polymerases. This could be achieved by the addition of a known concentration of a similar RNA virus in the sample before starting the extraction procedure. Costafreda et al. (2006) used the Mengo virus to monitor the whole performance of their HAV detection assay. In Canada, the Health Canada's Technical Group in Virology is recommending the use of the Feline calicivirus (FCV) as a sample processing control in all detection methods and to provide a common quality control for testing laboratories. FCV had successfully been included as sample process control to evaluate the performance of qRT-PCR assays for the detection of HEV (Ward et al., 2009) and F-coliphages of animal origin (Jones et al., 2009) in swine faecal samples under a multiplexed format and for the detection of HAV in water and strawberry samples (Mattison et al., 2009).

Without the use of an adequate control, qualitative and quantitative data should always be analyzed and interpreted carefully. On the qualitative aspect (presence/absence), the sample process control will be extremely helpful to identify complications arising during viral RNA extraction or the presence of RT-PCR inhibitors that may result in false negative results.

On the quantitative aspect, the integration of a sample process control will enable the correction of genome copies in accordance with the efficiency of the entire process thus providing a better accuracy in the quantification of viral particle loads. However, molecular techniques are detecting nucleic acids of all viral particles present and cannot discriminate between infectious and non-infectious particles. This represents the main limitation of these methods. A lack of correlation between infectivity and genome copies was also clearly demonstrated for HAV, FCV, poliovirus, adenovirus and, recently, murine norovirus (Gasilloud et al., 2003; Choi and Jiang, 2005; Hewitt and Greening, 2006; Baert et al., 2008).

While the detection methods are designed on the most conserved viral genomic regions for a broad reactivity, molecular typing systems should target less conserved regions to provide a better discrimination for genotyping isolates. Historically, amplicons obtained from cRT-PCR detection systems were sequenced and widely used for isolate identification and molecular characterization. Due to the accumulation of available sequence data and epidemiological information as well as improvements and refinements in molecular methods, the development of new and better typing approaches became not only essential but also achievable.

A comprehensive strategy in the study of food and waterborne viruses should now include two different amplification systems: one for the detection and a second for their genotyping. Despite the lack of standard methods, the actual trends in molecular genotyping are moving toward the phylogenetic analysis of viral capsid gene and full-length genome sequences (Kageyama et al., 2004; Vinje et al., 2004; Zheng et al., 2006; Matthijnssens et al., 2008a, 2008b; Simmonds et al., 2008; Costa-Mattioli et al., 2002a; Lu et al., 2004; Lu et al., 2006; Zhai et al., 2006). The development of a single assay combining at the same time the detection of enterically transmitted viruses and their genotyping would represent the next major improvement.

Microarray Technology

DNA microarray technology encompasses methods that allow for the simultaneous hybridization and detection of multiple probe/target pairs. The power of the technology lies in its capacity to provide both confirmation and molecular typing of RT-PCR amplicons in a single step. Multiple applications of microarray methodology have been described for food and waterborne viruses, the details of which are provided below. In general, the probes are fixed to a solid support in a format that allows for many probes to be analysed in a single hybridization step. The test amplicon (target) is hybridized to the array under conditions of defined stringency. Depending on probe design and stringency used, the signal from a hybridization event can confirm the presence of viral sequence and can provide genetic typing information without the requirement for sequence analysis.

The simplest form of microarray is the reverse line blot hybridization (Menton et al., 2007; Vinje and Koopmans, 2000). In these examples, 18-25 probes corresponding to various norovirus genotypes were immobilized on a membrane. Labelled RT-PCR amplification products are then hybridized to the membrane and hybridization patterns confirm the norovirus sequence and assign it to a genotype. A miniaturized and more complete system uses hundreds of probes printed on a glass slide to simultaneously confirm and genotype norovirus sequences (Pagotto et al., 2008). These systems all use the principle of direct probe hybridization for the detection of sequence homology. They have many advantages over other methods of molecular typing. The reactions conserve both target amplicon and reagents compared to performing a series of individual hybridizations. The stringency of hybridization and the methodology for interpretation of results can be predetermined, which allows the molecular typing procedure to be standardized and automated. They have the disadvantage that they may fail to identify emerging sequences that are not included in the assay design. This risk is reduced as the number of oligonucleotide probes in the array increases.

Other systems have been developed using more complex enzymatic amplification or hybridization principles (Brinkman and Fout, 2009; Jaaskelainen and Maunula, 2006). In the first example, primers are fixed to the solid support and transcription and primer extension reactions are performed on the microarray surface to amplify the signal (Jaaskelainen and Maunula, 2006). The main shortcoming of this system is the same as for the reverse line blot; the number of probes is extremely limited and identification is therefore strain-dependent (Jaaskelainen and Maunula, 2006). This is less of a concern where highly conserved sequences exist, but for the genetically variable noroviruses there are strains that will not be detected. An additional hurdle to the implementation of this system is the complexity of the detection procedure. The results obtained from any microarray experiment are highly dependent on the hybridization conditions. Adding additional steps that could increase lab-to-lab variability may reduce the usefulness of the procedure. In another recent protocol, labelled oligonucleotides act as a bridge to link viral amplicons to generic probes on a standardized microarray (Brinkman and Fout, 2009). This also increases the potential for variability in the system since it adds a second order to the kinetics of hybridization. The strategy is attractive because it allows for multiple analyses using the same microarray platform. This reduces the up-front cost inherent in microarray development and may allow the technology to be tested in labs without array production capabilities.

As for all molecular methods, the significance of the results from a microarray experiment depends largely on the quality of the system design (appropriate primer and probe sets) as well as on the quality of the input sample (purified, intact nucleic acid). A robust microarray platform is relatively easy to evaluate, it will give consistent and accurate typing results when tested in multiple laboratories. This should be evaluated against standard sequencing procedures until data is available to demonstrate that the microarray procedure provides accurate data for multiple isolates of the type strains. Such a validation protocol is particularly important for virus families with a high degree of sequence variation, such as the noroviruses. The quality of input sample can be more difficult to address and must be determined for each individual test. Controls should be added as early in the analysis as possible, to give an indication of the effects of each purification or preparatory step. They should be amplified, labelled and detected using the same system as the test samples whenever possible. Since a great degree of time and effort goes in to the development of microarray platforms in particular, they must always be designed to incorporate appropriate sample processing controls.

Conclusion

Food and waterborne viruses are recognized as an important concern for public health. Most of the circulating wild type strains of NoV, RV, HAV and HEV are not or difficult to cultivate *in vitro* and possess a low infectious dose. Because of their low analytical sensitivity, molecular DNA-polymerase based amplification techniques such as RT-PCR were the cornerstone for the detection and molecular typing methods of these challenging viruses in food and water samples. Isothermal amplification techniques (NASBA and LAMP) have been explored but most of the reference laboratories are now moving toward qRT-PCR technologies based on labelled fluorescent probes because of their high sensitivity, high specificity and their potential to generate quantitative data. However, the nucleic acid preparation from food and water samples and the integration of a sample process control as a quality control to monitor the efficiencies of the different steps from extraction to enzymatic amplification are two other critical elements that must be considered in detection systems. Harmonization and standardization of molecular detection and typing procedures for food and waterborne viruses also represent two other important burdens for outbreak mitigation, epidemiological studies and surveillance programs. The detection of food and waterborne viruses through a qRT-PCR technology targeting the most conserved genomic region and the phylogenetic sequence analysis of the capsid gene or the complete genome through cRT-PCR and amplicon sequencing for the genotyping represent the actual international trends. DNA microarray technology is a powerful tool and may represent the next major advance in the field by combining in a single assay the detection of multiple enterically transmitted viruses and their full genome typing.

References

Abd El Galil, K.H., El Sokkary, M.A., Kheira, S.M., Salazar, A.M., Yates, M.V., Chen, W. and Mulchandani, A. (2004). Combined immunomagnetic separation-molecular beacon-reverse transcription PCR assay for detection of hepatitis A virus from environmental samples. *Appl. Environ. Microbiol., 70*, 4371–4374.

Adak, G.K., Meakins, S.M., Yip, H., Lopman, B.A. and O'Brien, S.J. (2005). Disease risks from foods, England and Whales, 1996-2000. *Emerg. Infect. Dis., 11*, 365-372.

Adler, M., Schulz, S., Fischer, R. and Niemeyer, C.M. (2005). Detection of rotavirus from stool samples using a standardized immuno-PCR ("*Imperacer*") method with end-point and real-time detection. *Biochem. Biophys. Res. Comm., 333*, 1289-1294.

Aggarwal, R. and Krawczynski, K. (2000). Hepatitis E: An overview and recent advances in clinical and laboratory research. *J. Gastroenterol. Hepatol., 15*, 9-20.

Aggarwal, R., McCaustland, K.A., Dilawari, J.B., Sinha, S.D. and Robertson, B.H. (1999). Genetic variability of hepatitis E virus within and between three epidemics in India. *Virus Res., 59*, 35-48.

Amado, L.A., Villar, L.M., de Paula, V.S. and Gaspar, A.M.C. (2008). Comparison between serum and saliva for the detection of hepatitis A virus RNA. *J. Virol. Methods, 148*, 74-80.

Amar, C.F.L., East, C.L., Gray, J., Iturriza-Gomara, M., Maclure, E.A. and McLauchlin, J. (2007). Detection by PCR of eight groups of enteric pathogens in 4,627 faecal samples: re-examination of the English case-control infectious intestinal disease study (1993-1996). *Eur. J. Clin. Microbiol. Infect. Dis., 26*, 311-323.

Anderson, A.D., Heryford, A.G., Sarisky, J.P., Higgins, C., Monroe, S.S., Beard, R.S., Newport, C.M., Cashdollar, J.L., Fout, G.S., Robbins, D.E., Seys, S.A., Musgrave, K.J., Medus, C., Vinje, J., Bresee, J.S., Mainzer, H.M. and Glass, R.I. (2003). A waterborne outbreak of Norwalk-like virus among snowmobilers-Wyoming, 2001. *J. Infect. Dis., 187*, 303-306.

Ando, T., Monroe, S.S., Gentsch, J.R., Jin, Q., Lewis, D.C. and Glass, R.I. (1995). Detection and differentiation of antigenically distinct small round-structured viruses (Norwalk-like viruses) by reverse transcription-PCR and southern hybridization. *J. Clin. Microbiol., 33*, 64-71.

Ando, T., Noel, J.S. and Fankhauser, R.L. (2000). Genetic classification of "Norwalk-like viruses. *J. Infect. Dis., 181 Suppl. 2*, S336-S348.

Ahn, J.M, Kang, S.G., Lee, D.Y., Shin, S.J. and Yoo, H.S. (2005). Identification of novel human hepatitis E virus (HEV) isolates and determination of the seroprevalence of HEV in Korea. *J. Clin. Microbiol., 43*, 3042-3048.

Ahn, J.M., Rayamajhi, N., Kang, S.G. and Yoo, HS. (2006). Comparaison of real-time reverse transcriptase-polymerase chain reaction and nested or commercial reverse transcriptase-polymerase chain reaction for the detection of hepatitis E virus particle in human serum. *Diagn. Microbiol. Infec. Dis., 56*, 269-274.

Antonishyn, N.A., Crozier, N.A., McDonald, R.R., Levett, P.N. and Horsman, G.B. (2006). Rapid detection of Norovirus based on an automated extraction protocol and a real-time multiplexed single-step RT-PCR. *J. Clin. Virol., 37*, 156-161.

Araujo, I.T., Santos Assis, R.M., Fialho, A.M., Mascarenhas, J.D.P, Heinemann, M.B. and Leite, J.P.G. (2007). Brazilian P[8],G1, P[8],G5, P[8],G9, and P[4],G2 rotavirus strains: nucleotide sequence and phylogenetic analysis. *J. Med. Virol., 79*, 995-1001.

Atmar, R.L. and Estes, M.K. (2001). Diagnosis of noncultivatable gastroenteritis viruses, the human caliciviruses. *Clin. Microbiol. Rev., 14*, 15-37.

Atmar, R.L. and Estes, M.K. (2006). The epidemiologic and clinical importance of norovirus infection. *Gastroenterol. Clin. North Am., 35*, 275-290.

Baert, L., Wobus, C.E., Van Coillie, E., Thackray, L.B., Debevere, J. and Uyttendaele, M. (2008). Detection of murine norovirus 1 by using plaque assay, transfection assay, and real-time reverse transcription-PCR before and after heat exposure. *Appl. Environ. Microbiol., 74*, 543-546.

Baggi, F. and Peduzzi, R. (2000). Genotyping of rotaviruses in environmental water and stool samples in Southern Switzerland by necleotide sequence analysis of 189 base pairs at the 5' end of the VP7 gene. *J. Clin. Microbiol., 38*, 3681-3685.

Balayan, M.S., Andjaparidze, A.G. and Savinskaya, S.S. (1983). Evidence for a virus in non-A, non-B hepatitis transmitted via the fecal-oral route. *Intervirology, 20*, 23-31.

Banerjee, I., Ramani, S., Primrose, B., Iturriza-Gomara, M., Gray, J.J., Brown, D.W. and Kang, G. (2007). Modification of rotavirus multiplex RT-PCR for the detection of G12 strains based on characteriziation of emerging G12 rotavirus strains from south India. *J. Med. Virol., 79*, 1413-1421.

Barletta, J. (2006). Applications of real-time immuno-polymerase chain reaction (rt-IPCR) for the rapid diagnoses of viral antigens and pathologic proteins. *Mol. Aspects Med., 27*, 224-253.

Barman, P., Ghosh, S., Samajdar, S., Mitra, U., Dutta, P., Bhattacharya, S.K., Krishnan, T., Kobayashi, N. and Naik, T.N. (2006). RT-PCR based diagnosis revealed importance of human group B rotavirus infection in childhood diarrhoea. *J. Clin. Virol., 36*, 222-227.

Baroudy, B.M., Ticehurst, J.R., Miele, T.A., Maizel, J.V., Purcell, R.H. and Feinstone, S.M. (1985). Sequence analysis of hepatitis A virus cDNA coding for capsid proteins and RNA polymerase. *Proc. Natl. Acad. Sci. USA, 82,* 2143-2147.

Bass, E.S., Pappano, D.A. and Humiston, S.G. (2007). Rotavirus. *Pediatr. Rev., 28*, 183-191.

Bell, B.P. and Feinstone, S.M. (2004). Hepatitis A vaccine. In Vaccine, Plotkin S.A., Orenstein, W.A. and Offit, P.A., editors. Philadelphia, PA: Saunders; 269-297.

Bloch, A.B., Stramer, S.L., Smith, J.D., Margolis, H.S., Fields, H.A., McKinley, T.W., Gerba, C.P., Maynard, J.E. and Sikes, R.K. (1990). Recovery of hepatitis A virus from a water supply responsible for a common source outbreak of hepatitis A. *Am. J. Public Health, 80*, 428–430.

Borchardt, M.A., Bertz, P.D., Spencer, S.K. and Battigelli, D.A. (2003). Incidence of enteric viruses in groundwater from household wells in Wisconsin. *Appl. Environ. Microbiol., 69*, 1172-1180.

Bosch, A., Guix, S., Sano, D. and Pinto, R.M. (2008). New tools for the study and direct surveillance of viral pathogens in water. *Curr. Opin. Biotechnol., 19*, 295-301.

Brassard, J., Seyer, K., Houde, A., Simard, C. and Trottier, Y.-L. (2005). Concentration and detection of hepatitis A virus and rotavirus in sping water samples by reverse transcription-PCR. *J. Virol. Met., 123*, 163-169.

Brinkman, N.E. and Fout, G.S. (2009). Development and evaluation of a generic tag array to detect and genotype noroviruses in water. *J. Virol. Methods, 156*, 8-18.

Brooks, H.A., Gersberg, R.M. and Dhar, A.K. (2005). Detection and quantification of hepatitis A virus in seawater via real-time RT-PCR. *J. Virol. Methods, 127*, 109-118.

Bull, R.A., Hansman, G.S., Clancy, L.E., Tanaka, M.M., Rawlinson, W.D. and White, P.A. (2005). Norovirus recombination in ORF1/ORF2 overlap. *Emerg. Infect. Dis., 11*, 1079-1085.

Bull, R.A., Tanaka, M.M. and White, P.A. (2007). Norovirus recombination. *J. Gen. Virol., 88*, 3347-3359.

Bustin, S.A., Benes, V., Nolan, T. and Pfaffi, M.W. (2005). Quantitative real-time RT-PCR – a perspective. *J. Mol. Endocrinol., 34*, 597-601.

Butot, S., Putallaz, T. and Sanchez, G. (2007). Procedure for rapid concentration and detection of enteric viruses from berries and vegetables. *Appl. Environ. Microbiol., 73*, 186-192.

Butot, S., Putallaz, T. and Sanchez, G. (2008). Effects of sanitation, freezing and frozen storage on enteric viruses in berries and herbs. *Int. J. Food Microbiol., 126*, 30-35.

Calder, L., Simmons, G., Thornley, C., Taylor, P., Pritchard, K., Greening, G. and Bishop, J. (2003). An outbreak of hepatitis A associated with consumption of raw blueberries. *Epidemiol. Infect., 131*, 745–751.

Caron, M., Enouf, V., Than, S.C., Dellamonica, L., Buisson, Y. and Nicand, E. (2006). Identification of genotype 1 hepatitis E virus in samples from swine in Cambodia. *J. Clin. Microbiol., 44*, 3440-3442.

Carter, M.J. (2005). Enterically infecting viruses: pathogenicity, transmission and significance for food and waterborne infections. *J. Appl. Microbiol., 98*, 1354-1380.

Casas, N., Amarita, F. and de Maranon, I.M. (2007). Evaluation of an extracting method for the detection of hepatitis A virus in shellfish by SYBR-Green real-time RT-PCR. *Int. J. Food Microbiol., 120*, 179-185.

Chatterjee, R., Tsarev, S., Pillot, J., Coursaget, P., Emerson, S.J. and Purcell, R.H. (1997). African strains of hepatitis E virus that are distinct from Asian strains. *J. Med. Virol., 53*, 139-144.

Chironna, M., Germinario, C., De Medici, D., Fiore, A., Di Pasquale, S., Quarto, M. and Barbuti, S. (2002). Detection of hepatitis A virus in mussels from different sources marketed in Apulia region (South Italy). *Int. J. Food Microbiol., 75*, 11–18.

Chironna, A., Grottola, A., Lanave, C., Villa, E., Barbuti, S. and Quarto M. (2003). Genetic analysis of HAV strains recovered from patients with acute hepatitis from Southern Italy. *J. Med. Virol., 70*, 343–349.

Chizikov, V., Wagner, M., Ivshina, A., Hoshino, Y., Kapikian, A.Z. and Chumakov, K. (2002). Detection and genotyping of human group A rotaviruses by oligonucleotide microarray hybridization. *J. Clin. Microbiol., 40*, 2398-2407.

Chobe, L.P., Lole, K.S. and Arankalle, V.A. (2006). Full genome sequence and analysis of Indian swine hepatitis E virus isolate of genotype 4. *Vet. Microbiol., 114*, 240-251.

Choi, I.S., Kwon, H.J., Shin, N.R. and Yoo, H.S. (2003). Identification of swine hepatitis E virus (HEV) and prevalence of anti-HEV antibodies in swine and human populations in Korea. *J. Clin. Microbiol., 41*, 3602-3608.

Choi, S. and Jiang, S.C. (2005). Real-time PCR quantification of human adenoviruses in urban rivers indicates genome prevalence but low infectivity. *Appl. Environ. Microbiol., 71*, 7426–7433.

Coelho, C., Vinatea, C.E.B., Heinert, A.P., Simões, C.M.O. and Barardi, C.R.M. (2003). Comparison between specific and multiplex reverse transcription-polymerase chain reaction for detection of hepatitis A virus, poliovirus and rotavirus in experimentally seeded oysters. *Mem. Inst. Oswaldo Cruz, Rio de Janeiro, 98*, 465-468.

Cohen, J.I., Ticehurst, J.R., Purcell, R.H., Buckler-White, A. and Baroudy, B.M. (1987). Complete nucleotide sequence of wild-type hepatitis A virus: comparison with different strains of hepatitis A virus and other picornaviruses. *J. Virol., 61*, 50-59.

Colson, P., Coze, C., Gallian, P., Henry, M., De Micco, P. and Tamalet, C. (2007). Transfusion-associated hepatitis E, France. *Emerg. Infect. Dis., 13*, 648-649.

Cook, N., Bridger, J., Kendall, K., Iturriza-Gomara, M., El-Attar, L. and Gray, J. (2004). The zoonotic potential of rotavirus. *J. Infect., 48*, 289-302.

Costafreda, M.I., Bosch, A. and Pinto, R.M. (2006). Development, evaluation and standardization of a real-time TaqMan reverse transcription-PCR assay for quantification of hepatitis A virus in clinical and shellfish samples. *Appl. Environ. Microbiol., 72*, 3846–3855.

Costa-Mattioli, M., Cristina, J., Romero, J., Perez-Bercoff, R., Casane, D., Colina, R., Garcia, L., Vega, I., Glikman, G., Romanowski, V., Castello, A., Nicand, E., Gassin, M., Ferré, V. and Billaudel, S. (2002a). Molecular evolution of hepatitis A virus: a new classification based on the complete VP1 protein. *J. Virol., 76*, 9516–9525.

Costa-Mattioli, M., Monpoeho, S., Nicand, E., Aleman, M.H., Billaudel, S. and Ferre, V. (2002b). Quantification and duration of viraemia during hepatitis A infection as determined by real-time RT-PCR. *J. Viral Hep., 9*, 101–106.

Costa-Mattioli, M., Di Napoli, A., Ferre, V., Billaudel, S., Perez-Bercoff, R. and Cristina, J. (2003). Genetic variability of hepatitis A virus. *J. Gen. Virol., 84*, 3191–3201.

Costantini, V.P., Azevedo, A.C., Li, X., Williams, M.C., Michel, F.C. Jr. and Saif, L.J. (2007). Effects of different animal waste treatment technologies on detection and viability of porcine enteric viruses. *Appl. Environ. Microbiol., 73*, 5284-5291.

Cristina, J. and Costa-Mattioli M. (2007). Genetic variability and molecular evolution of Hepatitis A virus. *Virus Res., 127*, 151-157.

Cromeans, T.L., Nainan, O.V. and Margolis, H.S. (1997). Detection of hepatitis A virus RNA in oyster meat. *Appl. Environ. Microbiol., 63*, 2460–2463.

Dalton, C.B., Haddix, A., Hoffman, R.E. and Mast, E.E. (1996). The cost of a foodborne outbreak of hepatitis A in Denver Colorado. *Arch. Intern. Med., 156*, 1013–1016.

de Deus, N. Peralta, B., Pina, S., Allepuz, A., Mateu, E., Vidal, D., Ruiz-Fons, F., Martin, M., Gortazar, G. and Segalés, J. (2008). Epidemiological study of hepatitis E virus infection in European wild boars (*Sus scrofa*) in Spain. *Vet. Microbiol., 129*, 163-170.

de Deus, N., Seminati, C., Pina, S., Mateu, E., Martin, M. and Segalés, J. (2007). Detection of hepatitis E virus in liver, mesenteric lymph node, serum, bile and faeces of naturally infected pigs affected by different pathological conditions. *Vet. Microbiol., 119*, 105-114.

Dentinger, C.M., Bower, W.A., Nainan, O.V., Cotter, S.M., Myers, G., Dubusky, L.M., Fowler, S., Salehi, E.D. and Bell, B.P. (2001). An outbreak of hepatitis A associated with green onions. *J. Infect. Dis., 183*, 1273–1276.

de Paula, V. S., Baptista, M.L., Lampe, E., Niel, C. and Gaspar, A.M.C. (2002). Characterization of hepatitis A virus isolates from sub-genotypes IA and IB in Rio de Janeiro, Brazil. *J. Med. Virol., 66*, 22–27.

Desenclos, J. C., Klontz, K.C., Wilder, M.H., Nainan, O.V., Margolis, H.S. and Gunn, R.A. (1991). A multistate outbreak of hepatitis A caused by the consumption of raw oysters. *Am. J. Public Health, 81*, 1268–1272.

De Serres, G., Cromeans, T.L., Levesque, B., Brassard, N., Barthe, C., Dionne, M., Prud'homme, H., Paradis, D., Shapiro, C.N., Nainan, O.V. and Margolis, H.S. (1999). Molecular confirmation of hepatitis A virus from well water: epidemiology and public health implications. *J. Infect. Dis., 179*, 37–43.

Domingo, E. and Gomez, J. (2007). Quasispecies and its impact on viral hepatitis. *Virus Res., 127*, 131-150.

Duizer, E., Pielaat, A., Vennema, H., Kroneman, A. and Koopmans, M. (2007). Probabilities in norovirus outbreak diagnosis. *J. Clin. Virol., 40*, 38-42.

Duizer, E., Schwab, K.J., Neill, F.H., Atmar, R.L., Koopmans, M.P.G. and Estes, M.K. (2004). Laboratory efforts to cultivate noroviruses. *J. Gen. Virol., 85*, 79-97.

Emerson S.U., Anderson D., Arankalle A., Meng X.J., Purdy M, Schauder G.G., and Tsarev S.A.. Hepevirus. (2005a). In Fauquet C.M., Mayo M.A., Maniloff J., Desselberger U., Ball L.A., eds. Virus taxonomy, Eighth Report of the International Committee on Taxonomy of Viruses, London: Elsevier/Academic Press. 853-857.

Emerson, S.U., Arankalle, V.A. and Purcell, R.H. (2005b). Thermal stability of hepatitis E virus. *J. Infec. Dis., 192*, 930-933.

Emerson, S.U. and Purcell, R.H. (2003). Hepatitis E virus. *Rev. Med. Virol., 13*, 145-154.

Enouf, V., Reis, G.D., Guthmann, J.P., Guerin, P.J., Caron, M., Marechal, V. and Nicand, E. (2006). Validation of single real-time TaqMan PCR assay for the detection and quantification of four major genotypes of hepatitis E virus in clinical specimens. *J. Med. Virol., 78*, 1076-1082.

Erker, J.C., Desai, S.M. and Mushahwar I.K. 1999. Rapid detection of hepatitis E virus RNA by reverse transcription-polymerase chain reaction using universal oligonucleotide primers. *J. Virol. Methods, 81*, 109-113.

Escobar-Herrera, J., Cancio, C., Guzman, G.I., Villegas-Sepulveda, N., Estrada-Garcia, T., Garcia-Lozano, H., Gomez-Santiago, F. and Gutierrez-Escolano, A.L. (2006). Construction of an internal RT-PCR standard control for the detection of human caliciviruses in stool. *J. Virol. Methods, 137*, 334-338.

Espy, M.J., Uhl, J.R., Sloan, L.M., Buckwalter, S.P., Jones, M.F., Vetter, E.A., Yao, J.D.C., Wengenack, N.L., Rosenblatt, J.E., Cockerill III, F.R. and Smith, T.F. (2006). Real-time PCR in clinical microbiology: applications for routine laboratory testing. *Clin. Microbiol. Rev., 19*, 165-256.

Estes, M.K. and Cohen, J. (1989). Rotavirus gene structure and function. *Microbiol. Rev., 53*, 410-449.

Fauquet, C.M., Mayo, M.A., Maniloff, J., Desselberger, U. and Ball, L.A. (2005). Virus Taxonomy: Eigth Report of the International Committee on the Taxonomy of Viruses. Elsevier Inc., San Diego, CA.

Feagins, A.R., Opriessnig, T., Guenette, D.K., Halbur, P.G. and Meng, X.J. (2007). Detection and characterization of infectious hepatitis E virus from commercial pig livers sold in local grocery stores in the USA. *J. Gen. Virol., 88*, 912-917.

Fiore, A. E. (2004). Hepatitis A transmitted by food. *Clin. Infect. Dis., 38*, 705–715.

Fischer, T.K. and Gentsch, J.R. (2004). Rotavirus typing methods and algorithms. *Rev. Med. Virol., 14*, 71-82.

Friedman, D.S., Heisey-Grove, D., Argyros, F., Berl, E., Nsubuga, J., Stiles, T., Fontana, J., Beard, R.S., Monroe, S., McGrath, M.E., Sutherby, H., Dicker, R.C., DeMaria, A. and Matyas, B.T. (2005). An outbreak of norovirus gastroenteritis associated with wedding cakes. *Epidemiol. Infect., 133*, 1057-1063.

Fukuda, S., Sasaki, Y. and Sen, M. (2008). **Rapid and sensitive detection of norovirus genomes in oysters by a two-step isothermal amplification assay system combining nucleic acid sequence-based amplification and reverse transcription-loop-mediated isothermal amplification assays.** *Appl. Environ. Microbiol., 74*, 3912-3914.

Fukuda, S., Takao, S., Kuwayama, M., Shimazu, Y. and Miyazaki, K. (2006). Rapid detection of norovirus from fecal specimens by real-time reverse transcription-loop-mediated isothermal amplification assay. *J. Clin. Microbiol., 44*, 1376-1381.

Garcia-Aguirre, L. and Cristina, J. (2008). Analysis of the full-lenght genome of hepatits A virus isolated in South America: heterogeneity and evolutionary constraints. *Arch. Virol., 153*, 1473-1478.

Gardner, S.N. Kuczmarski, T.A., Vitalis, E.A. and Slezak, T.R. (2003). Limitations of TaqMan PCR for detecting divergent viral pathogens illustrated by hepatitis A, B, C and E viruses and human immunodeficiency virus. *J. Clin. Microbiol., 41*, 2417-2427.

Gassilloud, B., Schwartzbrod, L. and Gantzer, C. (2003). Presence of viral genomes in mineral water: a sufficient condition to assume infectious risk? *Appl. Environ. Microbiol., 69*, 3965–3969.

Gentsch, J.R., Glass, R.I., Woods, P., Gouvea, V., Gorziglia, M., Flores, J., Das, B.K. and Bhan, M.K. (1992). **Identification of group A rotavirus gene 4 types by polymerase chain reaction.** *J. Clin. Microbiol., 30*, 1365-1373.

Gladstone, B.P., Iturriza-Gomara, M., Ramani, S., Monica, B., Banerjee, I., Brown, D.W., Gray, J.J., Muliyil, J. and Kang, G. (2008). Polymerase chain reaction in the detection of an 'outbreak' of asymptomatic viral infections in a community birth cohort in south India. *Epidemiol. Infect., 136*, 399-405.

Glass, P.J., White, L.J., Ball, J.M., Leparc-Goffart, I., Hardy, M.E. and Estes, M.K. (2000). Norwalk virus open reading frame 3 encodes a minor structural protein. *J. Virol., 74*, 6581-6591.

Goens, S.D. and Perdue, M.L. (2004). Hepatitis E virus in humans and animals. *Anim. Health Res. Rev., 5*, 145-156.

Gouvea, V. and Brantly, M. (1995). Is rotavirus a polulation of reassortants? *Trends Microbiol., 3*, 159-162.

Gouvea, V., Cohen, S.J., Santos, N., Myint, K.S.A., Hoke, C. and Innis, B.L. (1997). Identification of hepatitis E virus in clinical specimens: amplification of hydroxyapatite-purified virus RNA and restriction endonuclease analysis. *J. Virol. Methods, 69*, 53-61.

Gouvea, V., Glass, R.I., Woods, P., Taniguchi, K., Clark, H.F., Forrester, B. and Fang, Z. (1990). Polymerase chain reaction amplification and typing of rotavirus nucleic acid from stool specimens. *J. Clin. Microbiol., 28*, 276-282.

Gray, J., Vesikari, T., Van Damme, P., Giaquinto, C., Mrukowicz, J., Guarino, A., Dagan, R., Szajewska, H. and Usonis, V. (2008). Rotavirus. *J. Pediatr. Gastroent. Nut., 46*, S24-S31.

Green, J., Gallimore, C.I., Norcott, J.P., Lewis, D. and Brown, D.W.G. (1995a). Broadly reactive reverse transcriptase polymerase chain reaction for the diagnosis of SRSV-associated gastroenteritis. *J. Med. Virol., 47*, 392-398.

Green, K.Y., Chanock, R.M., Kapikan, A.Z. and Howley, P.M. (2001). Human caliciviruses. In: D.M. Knipe (Ed), Fields Virology, Vol. 1, Lippincott Williams and Wilkins, Philadelphia, Pa., pp. 841-874.

Green, S.M., Lambden, P.R., Deng, Y., Lowes, J.A., Lineham, S., Bushell, J., Rogers, J., Caul, E.O., Ashley, C.R. and Clarke, I.N. (1995b). Polymerase chain reaction detection of small round-structured viruses from two related hospital outbreaks of gastroenteritis using inosine-containing primers. *J. Med. Virol., 45*, 197-202.

Guevremont, E., Brassard, J., Houde, A., Simard, C. and Trottier, Y.-L. (2006). Development of an extraction and concentration procedure and comparison of RT-PCR primer systems for the detection of hepatitis A virus and norovirus GII in green onions. *J. Virol. Methods, 134*, 130–135.

Gunson, R.N., Miller, J. and Carman, W.F. (2003). Comparison of real-time PCR and EIA for the detection of outbreaks of acute gastroenteritis caused by norovirus. *Commun. Dis. Public Health, 6*, 297–299.

Guttierrez-Aguirre, I., Steyer, A., Boben, J., Gruden, K., Poljšak-Prijatelj, M. and Ravnikar, M. (2008). Sensitive detection of multiple rotavirus genotypes with a single reverse transcription-real-time quantitative PCR assay. *J. Clin. Microbiol., 46*, 2547-2554.

Gyarmati, P., Mohammed, N., Norder, H., Blomberg, J., Belak, S. and Widén, F. (2007). Universal detection of hepatitis E virus by two real-time PCR assays: TaqMan and Primer-Probe Energy transfer. *J. Virol. Methods, 146*, 226-235.

Hall, G., Kirk, M.D., Becker, N., Gregory, J.E., Unicomb, L., Millard, G., Stafford, F., Lalor, K. and the Oz FoodNet Working Group. (2005). Estimating foodborne gastroenteritis, Australia. *Emerg. Infect. Dis., 11*, 1257-1264.

Halliday, M.L., Kang, L.Y., Zhou, T.K., Hu, M.D., Pan, Q.C., Fu, T.Y., Huang, Y.S. and Hu, S.L. (1991). An epidemic of hepatitis A attributable to the ingestion of raw clams in Shanghai, China. *J. Infect. Dis., 164*, 852–859.

Hansman, G.S., Natori, K., Shirato-Horikoshi, H., Ogawa, S., Oka, T., Katayama, K., Tanaka, T., Miyoshi, T., Sakae, K., Kobayashi, S., Shinohara, M., Uchida, K., Sakurai, N., Shinozaki, K., Okada, M., Seto, Y., Kamata, K., Nagata, N., Tanaka, K., Miyamura, T. and Takeda, N. (2006). Genetic and antigenic diversity among noroviruses. *J. Gen. Virol., 87*, 909-919.

Haqshenas, G., Shivaprasad, H.L., Woolcock, P.R., Read, D.H. and Meng, X.J. (2001). Genetic identification and characterization of a novel virus related to human hepatitis E virus from chickens with hepatitis-splenomegaly syndrome in the United States. *J. Gen. Virol., 82*, 2449-2462.

Heitmann, A., Laue, T., Schottstedt, V., Dotzauer, A. and Pichl, L. (2005). Occurrence of hepatitis A virus genotype III in Germany requires the adaptation of commercially available diagnostic test systems. *Transfusion, 45*, 1097-1105.

Herring, A.J., Inglis, N.F., Ojeh, C.K., Snodgrass, D.R. and Menzies, J.D. (1982). Rapid diagnosis of rotavirus infection by direct detection of viral nucleic acid in silver-stained polyacrylamide gels. *J. Clin. Microbiol., 16*, 473-477.

Hewitt, J. and Greening, G.E. (2004). Survival and persistence of norovirus, hepatitis A virus, and feline calicivirus in marinated mussels. *J. Food Prot., 67*, 1743–1750.

Hewitt, J. and Greening, G.E. (2006). Effect of heat treatment on hepatitis A virus and norovirus in New Zealand greenshell mussels (*Perna canaliculus*) by quantitative real-time reverse transcription PCR and cell culture. *J. Food Prot., 69*, 2217–2223.

Hohne, M. and Schreier, E. (2004). Detection and characterization of norovirus outbreaks in Germany: application of a one-tube RT-PCR using fluorigenic real-time detection system. *J. Med. Virol., 72*, 312-319.

Hohne, M. and Schreier, E. (2006). Detection of Norovirus genogroup I and II by multiplex real-time RT- PCR using a 3'-minor groove binder-DNA probe. *BMC Infect. Dis., 6*, 69.

Houde, A., Guevremont, E., Poitras, E., Leblanc, D., Ward, P., Simard, C. and Trottier, Y.-L. (2007). Comparative evaluation of new TaqMan assays for the detection of hepatitis A virus. *J. Virol. Methods, 140*, 80–89.

Houde, A., Leblanc, D., Poitras, E., Ward, P., Brassard, J., Simard, C. and Trottier, Y.-L. (2006). Comparative evaluation of RT-PCR, nucleic acid sequence based amplification (NASBA) and real-time RT-PCR for detection of noroviruses in faecal material. *J. Virol. Methods, 135*, 163–172.

Hsieh, S.Y., Meng, X.J., Wu, Y.H., Liu, S.H., Tam, A.W., Lin, D.Y and Liaw, Y.F. (1999). Identity of a novel swine hepatitis E virus in Taiwan forming a monophyletic group with Taiwan isolates of human hepatitis E virus. *J. Clin. Microbiol., 37*, 3828-3834.

Huang, F.F., Haqshenas, G., Guenette, D.K., Halbur, P.G., Schommer, S.K., Pierson, F.W., Toth, T.E. and Meng, X.J. (2002). Detection by reverse transcription-PCR and genetic characterization of field isolates of swine hepatitis E virus from pigs in different geographic regions of the United States. *J. Clin. Microbiol., 40*, 1326-1332.

Huang, F.F., Sun, Z.F., Emerson, S.U., Purcell, R.H., Shivaprasad, H.L., Pierson, F.W., Toth, T.E. and Meng, X. J. (2004). Determination and analysis of the complete genomic sequence of avian hepatitis E virus (avian HEV) and attempts to infect rhesus monkeys with avian HEV. *J. Gen. Virol., 85*, 1609-1618.

Huang, R., Nakazono, N., Ishii, K., Kawamata, O., Kawaguchi, R. and Tsukada, Y. (1995). II. Existing variations on the gene structure of hepatitis E virus strains from some regions of China. *J. Med. Virol., 47*, 303-308.

Hutin, Y.J., Pool, V., Cramer, E.H., Nainan, O.V., Weth, J., Williams, I.T., Goldstein, ST., Gensheimer, K.F., Bell, B.P., Shapiro, C.N., Alter, M.J. and Margolis, H.S. (1999). A multistate, foodborne outbreak of hepatitis A. *N. Engl. J. Med., 340*, 595–602.

Hymas, W., Atkinson, A., Stevenson, J. and Hillyard, D. (2007). Use of modified oligonucleotides to compensate for sequence polymorphisms in the real-time detection of norovirus. *J. Virol. Methods, 142*, 10-14.

Innis, B.L., Seriwatana, J., Robinson, R.A., Shrestha, M.P., Yarbough, P.O., Longer, C.F., Scott, R.M., Vaughn, D.W. and Myint, K.S. (2002). Quantitation of immunoglobulin to hepatitis E virus by enzyme immunoassay. *Clin. Diagn. Lab. Immunol., 9*, 639-648.

Inoue., J., Nishizawa, T., Takahashi, M., Aikawa, T., Mizuo, H., Suzuki, K., Shimosegawa, T. and Okamoto, H. (2006a). Analysis of the full-length genome of genotype 4 hepatitis E virus isolates from patients with fulminant or acute self-limited hepatitis E. *J. Med. Virol., 78*, 476-484.

Inoue, J., Takahashi, M., Yazaki, Y., Tsuda, F. and Okamoto, H. (2006b). Development and validation of an improved RT-PCR assay with nested universal primers for detection of hepatitis E virus strains with significant sequence divergence. *J. Virol. Methods, 137*, 325-333.

International Committee on Taxonomy of Viruses (ICTV). 2008. http //www.ncbi.nlm.nih.gov/ICTVdb/index

Isakbaeva, E.T., Widdowson, M.A., Beard, R.S., Bulens, S.N., Mullins, J., Monroe, S.S., Bresee, J., Sassano, P., Cramer, E.H. and Glass, R.I. (2005). Norovirus transmission on cruise ship. *Emerg. Infect. Dis., 11*, 154-158.

Iturriza-Gomara, M., Cubitt, D., Desselberger, U. and Gray, J. (2001). Amino-acid substitution within the VP7 protein of G2 rotavirus strains associated with failure to serotype. *J. Clin. Microbiol., 39*, 3796-3798.

Iturriza-Gomara, M., Green, J., Brown, D.W.G., Desselberger, U. and Gray, J. (2000a). Diversity within the VP4 gene of rotavirus P[8] strains: implications for reverse transcription-PCR genotyping. *J. Clin. Microbiol., 38*, 898-901.

Iturriza-Gomara, M., Green, J., Brown, D.W.G., Ramsay, M., Desselberger, U. and Gray, J.J. (2000b). Molecular epidemiology of human group A rotavirus infections in the United Kingdom between 1995 and 1998. *J. Clin. Microbiol., 38*, 4394-4401.

Iturriza-Gomara M., Kang, G. and Gray, J. (2004). Rotavirus geotyping: keeping up with an evolving population of human rotaviruses. *J. Clin. Virol., 31*, 259-265.

Iturriza-Gomara, M., Wong, C., Blome, S., Desselberger, U. and Gray, J. (2002). Molecular characterization of VP6 genes of human rotavirus isolates: correlation of genogroups with subgroups and evidence of independent segratation. *J. Virol., 76*, 6596-6601.

Iturriza-Gomara, M., Xerry, J., Gallimore, C.I., Dockery, C. and Gray, J. (2008). Evaluation of the Loopamp® (loop-mediated isothermal amplification) kit for detecting Norovirus RNA in faecal samples. *J. Clin. Virol., 42*, 389-393.

Jaaskelainen, A.J. and Maunula, L. (2006). Applicability of microarray technique for the detection of noro- and astroviruses. *J. Virol. Methods, 136*, 210-216.

Jameel, S., Durgapal, H., Habibullah, C.M., Khuroo, M.S. and Panda, S.K. (1992). Enteric non-A, non-B hepatitis: Epidemics, animal transmission, and hepatitis E virus detection by the polymerase chain reaction. *J. Med. Virol., 37*, 263-270.

Jansen, R.W., Siegl, G. and Lemon, S.M. (1990). Molecular epidemiology of human hepatitis A virus defined by an antigen-capture polymerase chain reaction method. *Proc. Natl. Acad. Sci. USA 87*, 2867–2871.

Jayaram, H., Estes, M.K. and Prasad, B.V.V. (2004). Ermerging themes in rotavirus cell entry, genome organization, transcription and replication. *Vir. Res., 101*, 67-81.

Jean, J., Blais, B., Darveau, A. and Fliss, I. (2001). Detection of hepatitis A virus by the nucleic acid sequence-based amplification technique and comparison with reverse transcription-PCR. *Appl. Environ. Microbiol., 67 (12)*, 5593–5600.

Jean, J., Blais, B., Darveau, A. and Fliss, I. (2002a). Simultaneous detection and identification of hepatitis A virus and rotavirus by multiplex nucleic acid sequence-based amplification (NASBA) and microtiter plate hybridization system. *J. Virol. Methods, 105*, 123-132.

Jean, J., Blais, B., Darveau, A. and Fliss, I. (2002b). Rapid detection of human rotavirus using colorimetric nucleic acid sequence-based amplification (NASBA)-enzyme-linked immunosorbent assay in sewage treatment effluent. *FEMS Microbiol. Lett., 210*, 143-147.

Jiang, X., Wang, M., Graham, D.Y. and Estes, M.K. (1992). Expression, self-assembly, and antigenicity of the Norwalk virus capsid protein. *J. Virol., 66*, 6527-6532.

Jiang, X., Wang, M., Wang, K. and Estes, M.K. (1993). Sequence and genomic organization of Norwalk virus. *Virology, 195*, 51-61.

Jones, T.H., Houde, A., Poitras, E., Ward, P. and Johns, M.W. (2009). Development and evaluation of a multiplexed real-time TaqMan RT-PCR assay with a sample process control for detection of F-specific RNA Coliphage genogroups I and IV. *Food Environ. Virol., 1*, 57-65.

Jothikumar, N., Cromeans, T.L., Robertson, B.H., Meng, X.J. and Hill, V.R. (2006). A broadly reactive one-step real-time RT-PCR assay for rapid and sensitive detection of hepatitis E virus. *J. Virol. Methods., 131*, 65-71.

Jothikumar, N., Cromeans, T.L., Sobsey, M.D. and Robertson, B.H. (2005a). Development and evaluation of a broadly reactive TaqMan assay for rapid detection of hepatitis A virus. *Appl. Environ. Microbiol., 71*, 3359–3363.

Jothikumar, N., Lowther, J.A., Henshilwood, K., Lees, D.N., Hill, V.R. and Vinje, J. (2005b). Rapid and sensitive detection of noroviruses by using TaqMan-based one-step reverse transcription-PCR assays and application to naturally contaminated shellfish samples. *Appl. Environ. Microbiol., 71*, 1870-1875.

Jothikumar, N., Paulmurugan, R., Padmanabhan, P., Balathiripura Sundari, R., Kamatchiammal, S. and Subba Rao, K. (2000). Duplex RT-PCR for simultaneous detection of hepatitis A and hepatitis E virus isolated from drinking water samples. *J. Environ. Monit., 2*, 587-590.

Kabrane-Lazizi, Y., Zhang, M., Purcell, R.H., Miller, K.D., Davey, R.T. and Emerson, S.U. (2001). Acute hepatitis caused by a novel strain of hepatitis E virus most closely related to United States strains. *J. Gen. Virol., 82*, 1687-1693.

Kageyama, T., Kojima, S., Shinohara, M., Uchida, K., Fukushi, S., Hoshino, F.B., Takeda, N. and Katayama, K. (2003). Broadly reactive and highly sensitive assay for Norwalk-like viruses based on real-time quantitative reverse transcription-PCR. *J. Clin. Microbiol., 41*, 1548-1557.

Kageyama, T., Shinohara, M., Uchida, K., Fukushi, S., Hoshino, F.B., Kojima, S., Takai, R., Oka, T., Takeda, N. and Katayama, K. (2004). Coexistence of multiple genotypes, including newly identified genotypes, in outbreaks of gastroenteritis due to Norovirus in Japan. *J. Clin. Microbiol., 42*, 2988-2995.

Kapikian, A.Z., Wyatt, R.G., Dolin, R., Thornhill, T.S., Kalica, A.R. and Chanock, R.M. (1972). Visualization by immune electron microscopy of a 27-nm particle associated with acute infectious nonbacterial gastroenteritis. *J. Virol., 10*, 1075-1081.

Katayama, K., Hansman, G.S., Oka, T., Ogawa, S. and Takeda, N. (2006). Investigation of norovirus replication in a human cell line. *Arch. Virol., 151*, 1291-1308.

Kim, S.H., Cheon, D.S., Kim, J.H., Lee, D.H., Jheong, W.H., Heo, Y.J., Chung, H.M., Jee, Y. and Lee, J.S. (2005). Outbreaks of gastroenteritis that occurred during school excursions in Korea were associated with several waterborne strains of norovirus. *J. Clin. Microbiol., 43*, 4836-4839.

Kojima, S., Kageyama, T., Fukushi, S., Hoshino, F.B., Shinohara, M., Uchida, K., Natori, K., Takeda, N. and Katayama, K. (2002). Genogroup-specific PCR primers for detection of Norwalk-like viruses. *J. Virol. Methods, 100*, 107-114.

Koopmans, M. and Duizer, E. (2004). Foodborne viruses: an emerging problem. *Int. J. Food Microbiol., 90*, 23-41.

Koopmans, M.P.G., von Bonsdorff, C.H., Vinje, J., De Medici, D. and Monroe, S.S. (2002). Foodborne enteric viruses. *FEMS Microbiol. Rev., 26*, 187-205.

Lambden, P.R., Liu, B. and Clarke, I.N. (1995). A conserved sequence motif at the 5' terminus of the Southampton virus genome is characteristic of the Caliciviridae. *Virus Genes, 10*, 149-152.

Laverick, M.A., Wyn-Jones, A.P. and Carter, M.J. (2004). Quantitative RT-PCR for the enumeration of noroviruses (Norwalk-like viruses) in water and sewage. *Lett. Appl. Microbiol., 39*, 127-136.

Le Guyader, F., Dubois, E., Menard, D. and Pommepuy, M. (1994). Detection of hepatitis A virus, rotavirus, and enterovirus in naturally contaminated shellfish and sediment by reverse transcription-seminested PCR. *Appl. Environ. Microbiol., 60*, 3665-3671.

Li, T.C., Chijiwa, K., Sera, N., Ishibashi., T., Etoh, Y., Shinohara, Y., Kurata, Y., Ishida, M., Sakamoto, S., Takeda, N. and Miyamura, T. (2005). Hepatitis E virus transmission from wild boar meat. *Emerg. Infect. Dis., 11*, 1958-1960.

Lindesmith, L.C., Donaldson, E.F., Lobue, A.D., Cannon, J.L., Zheng, D.P., Vinje, J. and Baric, R.S. (2008). Mechanisms of GII.4 norovirus persistence in human populations. *PLoS Med., 5*, e31.

Logan, C., O'Leary, J.J. and O'Sullivan, N. (2007). Real-time reverse transcription PCR detection of norovirus, sapovirus and astrovirus as causative agents of acute viral gastroenteritis. *J. Virol. Methods, 146*, 36-44.

Lopman, B.A., Reacher, M.H., van Duijnhoven, Y., Hanon, F.X., Brown, D. and Koopmans, M. (2003). Viral gastroenteritis outbreaks in Europe, 1995–2000. *Emerg. Infect. Dis., 9*, 90–96.

Lu, L., Ching, K.Z., Salete de Paula, V., Nakano, T., Siegl, G., Weitz, M. and Robertson, B.H. (2004). Characterization of the complete genomic sequence of genotype II hepatitis A virus (CF53/Berne isolate). *J. Gen. Virol., 85,* 2943–2952.

Lu, L., Li, C and Hagedorn, C.H. (2006). Phylogenetic analysis of global hepatitis E virus sequences: genetic diversity, subtypes and zoonosis. *Review Med. Virol., 16*, 5-36.

Mackay, I.M. (2004). Real-time PCR in the microbiology laboratory. *Clin. Microbiol. Infect., 10*, 190-212.

Mackay, I.M., Arden, K.E. and Nitsche, A. (2002). Real time PCR in virology. *Nucleic Acids Res., 30*, 1292–1305.

Majowicz, S.E., McNab, W.P., Sockett, P., Henson, S., Doré, K., Edge, V.L., Buffett, M.C., Fazil, A., Read, S., McEwen, S., Stacey, D. and Wilson, J.B. (2006). Burden and cost of gastroenteritis in a canadian community. *J. Food Protect., 69*, 651-659.

Mansuy, J.M., Peron, J.M., Abravanel, F., Poirson, H., Dubois, M., Miedouge M., Vischi, F., Alric, L., Vinel, J.P. and Izopet, J. (2004). Hepatitis E in the south west of France in individuals who have never visited a endemic area. *J. Med. Virol., 74*, 419-424.

Marks, P.J., Vipond, I.B., Regan, F.M., Wedgwood, K., Fey, R.E. and Caul, E.O. (2003). A school outbreak of Norwalk-like virus: evidence for airborne transmission. *Epidemiol. Infect., 131*, 727-736.

Matsubayashi, K., Kang, J-H., Sakata, H., Takahashi, K., Shindo, M., Kato, M., Sato, S., Kato, T., Nishimori, H., Tsuji, K., Maguchi, H., Yoshida, J-I, Maekubo, H., Mishiro, S. and Ikeda, H. (2008). A case of transfusion-transmitted hepatitis E caused by blood from a donor infected with hepatitis E virus via zoonotic food-borne route. *Transfusion, 48*, 1368-1375.

Matsuda, H., Okada, K., Takahashi, K. and Mishiro, S. (2003). Severe hepatitis E virus infection after ingestion of uncooked liver from a wild boar. *J. Infect. Dis., 188*, 944.

Matthijnssens, J., Ciarlet, M., Heiman, E., Arijs, I., Delbeke, T., McDonald, S.M., Palombo, E.A., Iturriza-Gomara, M., Maes, P., Patton, J.T., Rahman, M. and Van Ranst, M. (2008a). Full genome-based classification of rotaviruses reveals a common origin between human wa-like and porcine rotavirus strains and human DS-1-like and bovine rotavirus strains. *J. Virol., 82*, 3204-3219.

Matthijnssens, J., Ciarlet, M., Rahman, M., Attoui, H., Bányai, K., Estes, M.K., Gentsch, J.R., Iturriza-Gomara, M., Kirkwood, C.D., Martella, V., Mertens, P.P.C., Nakagomi, O., Patton, J.T., Ruggeri, F.M., Saif, M.S., Steyer, A., Taniguchi, K., Desselberger, U. and Van Ranst, M. (2008b). Recommendations for the classification of group A rotaviruses using all 11 genomic RNA segments. *Arch. Virol., 153*, 1621-1629.

Mattison, K., Brassard, J., Gagné, M.J., Ward, P., Houde, A., Lessard, L., Simard, C., Shukla, A., Pagotto, F., Jones, T.H. and Trottier, Y.L. (2009). The Feline calicivirus as a sample process control for the detection of food and waterborne RNA viruses. *Int. J. Food Microbiol., 132*, 73-77.

Mattison, K., Shukla, A., Cook, A., Pollari, F., Friendship, R., Kelton, D., Bidawid, S. and Farber, J.M. (2007). Human noroviruses in swine and cattle. *Emerg. Infect. Dis., 13*, 1184-1188.

Mead, P.S., Slutsker, L., Dietz, V., McCaig, L.F., Bresee, J.S., Shapiro, C., Griffin, P.M. and Tauxe, R.V. (1999). Food-related illness and death in the United States. *Emerg. Infect. Dis., 5*, 607–625.

Meng, X.J., Halbur, P.G., Shapiro, M.S., Govindarajan, S., Bruna, J. D., Mushahwar, I.K., Purcell, R.H. and Emerson S.U. (1998). Genetic and experimental evidence for cross-species infection by swine hepatitis E virus. *J. Virol., 72*, 9714-9721.

Meng, X.J., Purcell, R.H., Halbur, P.G., Lehman, J.R., Webb, D.M., Tsareva, T.S., Haynes, J.S., Thacker, B.J. and Emerson S.U. (1997). A novel virus in swine is closely related to the human hepatitis E virus. *Proc. Natl. Acad. Sci. USA, 94*, 9860-9865.

Menton, J.F., Kearney, K. and Morgan, J.G. (2007). Development of a real-time RT-PCR and Reverse Line probe Hybridisation assay for the routine detection and genotyping of Noroviruses in Ireland. *Virol. J., 4*, 86.

Miller, I., Gunson, R. and Carman, W.F. (2002). Norwalk like virus by light cycler PCR. *J. Clin. Virol., 25*, 231-232.

Milne, S.A., Gallacher, S., Cash, P., Lees, D.N., Henshilwood, K. and Porter, A.J. (2007). A sensitive and reliable reverse transcriptase PCR-enzyme-linked immunosorbent assay for the detection of human pathogenic viruses in bivalve molluscs. *J. Food Prot., 70*, 1475-1482.

Min, B.S., Noh, Y.J., Shin, J.H., Baek, S.Y., Min, K.I., Ryu, S.R., Kim, B.G., Park, M.K., Choi, S.E., Yang, E.H., Park, S.N., Hur, S.J. and Ahn, B.Y. (2006). Assessment of the quantitative real-time polymerase chain reaction using a cDNA standard for human group A rotavirus. *J. Virol. Methods, 137*, 280-286.

Mizuo, H., Suzuki, K., Takikawa, Y., Sugai, Y., Tokita, H., Akahane, Y., Itoh, K., Gotanda, Y., Takahashi, M., Nishizawa, T. and Okamoto, H. (2002). Polyphyletic strains of hepatitis E virus are responsible for sporadic cases of acute hepatitis in Japan. *J. Clin. Microbiol., 40*, 3209-3218.

Mizuo, H., Yazaki, Y., Sugawara, K., Tsuda, F., Takahashi, M., Nishizawa, T. and Okamoto, H. (2005). Possible risk factors for the transmission of hepatitis E virus and for the severe form of hepatitis E acquired locally in Hokkaido, Japan. *J. Med. Virol., 76*, 341-349.

Moe, C.L., Gentsch, J., Ando, T., Grohmann, G., Monroe, S.S., Jiang, X., Wang, J., Estes, M.K., Seto, Y., Humphrey, C., Stine, S. and Glass, R.I. (1994). Application of PCR to detect Norwalk virus in fecal specimens from outbreaks of gastroenteritis. *J. Clin. Microbiol., 32*, 642-/648.

Mohamed, N., Belak, S., Hedlund, K.O. and Blomberg, J. (2006). Experience from the development of a diagnostic single tube real-time PCR for human caliciviruses, Norovirus genogroups I and II. *J. Virol. Methods, 132*, 69-76.

Moore, C., Clarke, C.M., Gallimore, C.I., Corden, S.A., Gray, J.J. and Westmoreland, D. (2004). Evaluation of a bradly reactive nucleic acid sequence based amplification (NASBA) assay for the detection of Noroviruses in faecal material. *J. Clin. Microbiol., 29*, 290-296.

Mushahwar, I.K. (2008). Hepatitis E virus: Molecular virology, clinical features, diagnosis, transmission, epidemiology, and prevention. *J. Med. Virol., 80*, 646-658.

Myint, K.S.A. and Gibbons, R.V. (2008). Hepatitis E: a neglected threat. *Trans. Roy. Soc. Trop. Med. Hyg., 102*, 211-212.

Nainan, O.V., Xia, G., Vaughan, G. and Margolis, H.S. (2006). Diagnosis of hepatitis A virus infection: a molecular approach. *Clin. Microbiol. Rev., 19*, 63–79.

Nakamura, M., Takahashi, E., Taira, K., Taira, M., Ohno, A., Sakugawa, H., Arai, M. and Mishiro, S. (2006). Hepatitis E virus infection in wild mongoose of Okinawa, Japan: Demonstration of anti-HEV antibodies and a full-genome nucleotide sequence. *Hepatol. Res., 34*, 137-140.

Nishizawa, T., Takahashi, M., Endo, K., Fujiwara, S., Sakuma, N., Kawazuma, F., Sakamoto, H., Sato, Y., Bando, M. and Okamoto, H. (2005). Analysis of the full-length genome of hepatitis E virus isolates obtained from wild boars in Japan. *J. Gen. Virol., 86*, 3321-3326.

Nishizawa, T., Takahashi, M., Mizuo, H., Miyajima, H., Gotanda, Y. and Okamoto, H. (2003). Characterization of Japanese swine and human hepatitis E virus isolates of genotype IV with 99% identity over the entire genome. *J. Gen. Virol., 84*, 1245-1251.

Niu, M. T., Polish, L.B., Robertson, B.H., Khanna, B.K., Woodruff, B.A., Shapiro, C.N., Miller, M.A., Smith, J.D., Gedrose, J.K., Alter, M.J. and Margolis, H.S. (1992). Multistate outbreak of hepatitis A associated with frozen strawberries. *J. Infect. Dis., 166*, 518–524.

Noel, J.S., Ando, T., Leite, J.P., Green, K.Y., Dingle, K.E., Estes, M.K., Seto, Y., Monroe, S.S. and Glass, R.I. (1997). Correlation of patient immune responses with genetically characterized small round-structured viruses involved in outbreaks of nonbacterial acute gastroenteritis in the United States, 1990 to 1995. *J. Med. Virol., 53*, 372-383.

Nordgren, J., Bucardo, F., Dienus, O., Svensson, L. and Lindgren, P.E. (2008). Novel light-upon-extension real-time PCR assays for detection and quantification of genogroup I and II noroviruses in clinical specimens. *J. Clin. Microbiol., 46*, 164-170.

Okamoto, H. (2007). Genetic variability and evolution of hepatitis E virus. *Virus Res., 127*, 216-228.

Orru, G., Masia, G., Orrù G., Romano, L., Piras, V. and Coppola, R.C. (2004). Detection and quantitation of hepatitis E virus in human faeces by real-time quantitative PCR. *J. Virol. Methods, 118*, 77-82.

Pagotto, F., Mattison, K., Corneau, N. and Bidawid, S. (2008). Development of a DNA microarray for the simultaneous detection and genotyping of noroviruses. *J. Food Prot., 71*, 1434-1441.

Palombo, E. (2002). Genetic analysis of group A rotaviruses: evidence for interspecies transmission of rotavirus genes. *Virus Genes, 24*, 11-20.

Pang, X.L., Lee, B., Boroumand, N., Leblanc, B., Preiksaitis, J.K. and Yu Ip, C.C. (2004). Increased detection of rotavirus using a real time reverse transcription-polymerase chain reaction (RT-PCR) assay in stool specimens from children with diarrhea. *J. Med. Virol., 72*, 495-501.

Pang, X.L., Preiksaitis, J.K. and Lee, B. (2005). Multiplex real time RT-PCR for the detection and quantitation of norovirus genogroups I and II in patients with acute gastroenteritis. *J. Clin. Virol., 33*, 168-171.

Patterson, S.S., Smith, M.W., Casper, E.T., Huffman, D., Stark, L., Fries, D. and Paul, J.H. (2006). A nucleic acid sequence-based amplification assay for real-time detection of norovirus genogroup II. *J. Appl. Microbiol., 101*, 956-963.

Phan, T.G., Nguyen, T.A., Yan, H., Okitsu, S. and Ushijima, H. (2005). A novel RT-multiplex PCR for enteroviruses, hepatitis A and E viruses and influenza A virus among infants and children with diarrhea in Vietnam. *Arch. Virol., 150*, 1175-1185.

Pina, S., Buti, M., Jardi, R., Clemente-Casares, P., Jofre, J. and Girones, R. (2001). Genetic analysis of hepatitis A virus strains recovered from the environment and from patients with acute hepatitis. *J. Gen. Virol.*, 82, 2955-2963.

Pinto, R.M. and Saiz, J.C. (2007). Enteric hepatitis viruses. *Perspectives in Med. Virol., 17,* 39-67.

Ramsay, C.N. and Upton, P.A. (1989). Hepatitis A and frozen raspberries. *Lancet, 333,* 43–44.

Reid, T. M. and Robinson H.G. (1987). Frozen raspberries and hepatitis A. *Epidemiol. Infect., 98,* 109–112.

Richards, G.P., Watson, M.A. and Kingsley, D.H. (2004). A SYBR green, real-time RT-PCR method to detect and quantitate Norwalk virus in stools. *J. Virol. Methods, 116,* 63-70.

Robertson, B.H., Jansen, R.W., Khanna, B., Totsuka, A., Nainan, O.V., Siegl, G., Widell, A., Margolis, H.S., Isomura, S., Ito, K., Ishizu, T., Moritsugu, Y. and Lemon, S. M. (1992). Genetic relatedness of hepatitis A virus strains recovered from different geographical regions. *J. Gen. Virol., 73,* 1365–1377.

Robertson, B.H., Khanna, B., Nainan, O.V. and Margolis, H.S. (1991). Epidemiologic patterns of wild-type hepatitis A virus determined by genetic variation. *J. Infect. Dis., 163,* 286–292.

Rolfe, K.J., Parmar, S., Mururi, D., Wreghitt, T.G., Jalal, H., Zhang, H. and Curran, M.D. (2007). An internally controlled, one-step, real-time RT-PCR assay for norovirus detection and genogrouping. *J. Clin. Virol., 39,* 318-321.

Rosenblum, L.S., Mirkin, I.R., Allen, D.T., Safford, S. and Hadler, S.C. (1990). A multifocal outbreak of hepatitis A traced to commercially distributed lettuce. *Am. J. Public Health, 80,* 1075–1079.

Rutjes, S.A., Lodder, W.J., Bouwknegt, M. and de Roda Husman, A.M. (2007). Increased hepatitis E virus prevalence on Dutch pig farms from 33 to 55% by using appropriate internal quality controls for RT-PCR. *J. Virol. Methods, 143,* 112-116.

Rutjes, S.A., van den Berg, H.H., Lodder, W.J. and de Roda Husman, A.M. (2006). Real-time detection of noroviruses in surface water by use of a broadly reactive nucleic acid sequence-based amplification assay. *Appl. Environ. Microbiol., 72,* 5349-5358.

Sair, A.L., D'Souza, D.H. and Jaykus, L.A. (2002). Human enteric viruses as causes of foodborne disease. *Comp. Rev. Food Sci. Food Saf., 1,* 73–89.

Samajdar, S., Ghosh, S., Chawla-Sarkar, M., Mitra,U., Dutta, P., Kobayashi, N. and Naik, T.N. (2008). Increase in prevalence of human group A rotavirus G9 strains as an important VP7 genotype among children in eastern India. *J. Clin. Virol., 43,* 334-339.

Sanchez, G., Bosch, A. and Pinto, R.M. (2007). Hepatitis A virus detection in food: current and future prospects. *Letters Appl. Microbiol., 45,* 1-5.

Sanchez, G., Populaire, S., Butot, S., Putallaz, T. and Joosten, H. (2006). Detection and differentiation of human hepatitis A strains by commercial quantitative real-time RT-PCR tests. *J. Virol. Methods, 132,* 160–165.

Schultheiss, T., Kusov, Y.Y. and Gauss-Muller, V. (1994). Proteinase 3C of hepatitis A virus (HAV) cleaves the HAV polyprotein P2-P3 at all sites including VP1/2A and 2A/2B. *Virology, 198,* 275-281.

Scipioni, A., Bourgot, I., Mauroy, A., Ziant, D., Saegerman, C., Daube, G. and Thiry, E. (2008). Detection and quantification of human and bovine noroviruses by a TaqMan RT-PCR assay with a control for inhibition. *Mol Cell Probes. Mol. Cell. Probes, 22,* 215-222.

Seah, E.L., Marshall, J.A. and Wright, P.J. (1999). Open reading frame 1 of the Norwalk-like virus Camberwell: completion of sequence and expression in mammalian cells. *J. Virol., 73*, 10531-10535.

Shieh, Y.C., Khudyakov, Y.E., Xia, G., Ganova-Raeva, L.M., Khambaty, F.M., Woods, J.W., Veazey, J.E., Motes, M.L., Glatzer, M.B., Bialek, S.R. and Fiore, A.E. (2007). Molecular confirmation of oyster as the vector for hepatitis A in 2005 multistate outbreak. *J. Food Protection, 70*, 145-150.

Shrestha, M.P., Scott, R.M., Joshi, D.M., Mammen, M.P., Thapa, G.B., Thapa, N., Myint, K.S.A., Fourneau, M., Kuschner, R.A. Shrestha, S.K., David, M.P., Seriwatana, J., Vaughn, D.W., Safary, A., Endy, T.P. and Innis, B.L. (2007). Safety and efficacy of a recombinant hepatitis E vaccine. *N. Engl. J. Med., 356*, 895-903.

Siebenga, J.J., Vennema, H., Renckens, B., de Bruin, E., van der Veer, B., Siezen, R.J. and Koopmans, M. (2007). Epochal evolution of GGII.4 norovirus capsid proteins from 1995 to 2006. *J. Virol., 81*, 9932-9941.

Silberstein, E., Xing, L., van de Beek, W., Lu, J., Cheng, H. and Kaplan, G.G. (2003). Alteration of hepatitis A virus (HAV) particles by a soluble form of HAV cellular receptor 1 containing the immunoglobulin- and mucin-like regions. *J. Virol., 77*, 8765–8774.

Simmonds, M.K., Armah, G., Asmah, R., Banerjee, I., Damanka, S., Esona, M., Gentsch, J.R., Gray, J.J., Kirkwood, C., Page, N. and Iturriza-Gomara (2008). New oligonucleotide primers for P-typing of ratavirus strains: strategies for typing previously untypeable strains. *J. Clin. Virol., 42*, 368-373.

Smith, J.L. (2001). A review of hepatitis E virus. *J. Food Prot., 64*, 572-586.

Sonoda, H., Abe, M., Sugimoto, T., Sato, Y., Bando, M., Fukui, E., Mizuo, H., Takahashi, M., Nishizawa, T. and Okamoto, H. (2004). Prevalence of hepatitis E virus (HEV) infection in wild boars and deer and genetic identification of a genotype 3 HEV from a boar in Japan. *J. Clin. Microbiol., 42*, 5371-5374.

Stanway, G., Brown, F., Christian, P., Hovi, T., Hyypiä, T., King, A.M.Q., Knowles, N.J., Lemon, S.M., Minor, P.D., Pallansch, M.A., Palmenberg, A.C. and Skern, T. (2005). Family *Picornaviridae*. In: "Virus Taxonomy. Eighth Report of the International Committee on Taxonomy of Viruses". Eds. Fauquet, C.M., Mayo, M.A., Maniloff, J., Desselberger, U. and Ball, L.A. Elsevier/Academic Press, London. p. 757-778.

Steyer, A., Poljšak-Prijatelj, M., Barlič-Maganja, D. and Marin, J. (2008). Human, porcine and bovine rotaviruses in Slovenia: evidence of interspecies transmission and genome reassortment. *J. Gen. Virol., 89*, 1690-1698.

Stockman, L.J., Staat, M.A., Holloway, M., Bernstein, D.I., Kerin, T., Hull, J., Yee, E., Gentsch, J. and Parashar, U.D. (2008). Optimum diagnostic assay and clinical specimen for routine rotavirus surveillance. *J. Clin. Microbiol., 46*, 1842-1843.

Takahashi, K., Iwata, K., Watanabe, N., Hatahara, T., Ohta, Y., Baba, K., and Mishiro, S. (2001). Full-genome nucleotide sequence of a hepatitis E virus strain that may be indigenous to Japan. *Virology, 287*, 9-12.

Takahashi, K., Kitajima, N., Abe, N. and Mishiro, S. (2004). Complete or near-complete nucleotide sequences of hepatitis E virus genome recovered from a wild boar, a deer, and four patients who ate the deer. *Virology, 330*, 501-505.

Takahashi, M., Nishizawa, T., Yoshikawa, A., Sato, S., Isoda, N., Ido, K., Sugano, K. and Okamoto, H. (2002). Identification of two distinct genotypes of hepatitis E virus in a Japanese patient with acute hepatitis who had not travelled abroad. *J. Gen. Virol., 83*, 1931-1940.

Tallon, L.A., Love, D.C., Moore, Z.S. and Sobsey, M.D. (2008). Recovery and sequence analysis of Hepatitis A virus from springwater implicated in an outbreak of acute viral hepatitis. *Appl. Environ. Microbiol., 74*, 6158-6160.

Tam, A.W., Smith, M.M., Guerra, M.E., Huang, C-C., Bradley, D.W., Fry, K.E. and Reyes, G.R. (1991). Hepatitis E virus (HEV): Molecular cloning and sequencing of the full-length viral genome. *Virology, 185*, 120-131.

Tanaka, T., Takahashi, M., Kusano, E. and Okamoto, H. (2007). Development and evaluation of an efficient cell-culture system for hepatitis E virus. *J. Gen. Virol., 88*, 903-911.

Tei, S., Kitajima, N., Takahashi, K. and Mishiro, S. (2003). Zoonotic transmission of hepatitis E virus from deer to human beings. *Lancet, 362*, 371-373.

Trujillo, A.A., McCaustland, K.A., Zheng, D.P., Hadley, L.A., Vaughn, G., Adams, S.M., Ando, T., Glass, R.I. and Monroe, S.S. (2006). Use of TaqMan real-time reverse transcription-PCR for rapid detection, quantification, and typing of norovirus. *J. Clin. Microbiol., 44*, 1405-1412.

Tsugawa, T. and Hoshino, Y. (2008). Whole genome sequence and phylogenetic analyses reveal human rotavirus G3P[3] strains Ro1845 and HCR3A are examples of direct virion transmission of canaine/feline rotaviruses to humans. *Virology, 380*, 344-353..

Tsugawa, T., Numata-Kinoshita, K., Honma, S., Nakata, S., Tatsumi, M., Sakai, Y., Natori, K., Takeda, N., Kobayashi, S. and Tsutsumi, H. (2006). Virological, serological, and clinical features of an outbreak of acute gastroenteritis due to recombinant genogroup II norovirus in an infant home. *J. Clin. Microbiol., 44*, 177-182.

Valasek, M.A. and Repa, J.J. (2005). The power of real-time PCR. *Adv. Physiol. Educ., 29*, 151-159.

van Zyl, W.B., Page, N.A., Grabow, W.O.K., Steele, A.D. and Taylor, M.B. (2006). Molecular epidemiology of group a rotaviruses in water sources and selected raw vegetables in southern Africa. *Appl. Environ. Microbiol., 72*, 4554-4560.

Vasickova, P., Psikal, I., Kralik, P., Widen, F., Hubalek, Z. and Pavlik, I. (2007). Hepatitis E virus: a review. *Veterinari Medicina, 9*, 365-384.

Vennema, H., de Bruin, E. and Koopmans, M. (2002). Rational optimization of generic primers used for Norwalk-like virus detection by reverse transcriptase polymerase chain reaction. *J. Clin. Virol., 25*, 233-235.

Villar, L.M., De Paula, V.S., Diniz-Mendes, L., Lampe, E. and Gaspar, A.M. (2006). Evaluation of methods used to concentrate and detect hepatitis A virus in water samples. *J. Virol. Methods, 137*, 169–176.

Vinje, J., Hamidjaja, R.A. and Sobsey, M.D. (2004). Development and application of a capsid VP1 (region D) based reverse transcription PCR assay for genotyping of genogroup I and II noroviruses. *J. Virol. Methods., 116*, 109-117.

Vinje, J. and Koopmans, M.P. (1996). Molecular detection and epidemiology of small round-structured viruses in outbreaks of gastroenteritis in the Netherlands. *J. Infect. Dis., 174*, 610-615.

Vinje, J. and Koopmans, M.P. (2000). Simultaneous detection and genotyping of "Norwalk-like viruses" by oligonucleotide array in a reverse line blot hybridization format. *J. Clin. Microbiol., 38*, 2595-2601.

Wang, Q.H., Costantini, V. and Saif, L.J. (2007). Porcine enteric caliciviruses: genetic and antigenic relatedness to human caliciviruses, diagnosis and epidemiology. *Vaccine, 25*, 5453-5466.

Wang, Q.H., Han, M.G., Cheetham, S., Souza, M., Funk, J.A. and Saif, L.J. (2005). Porcine noroviruses related to human noroviruses. *Emerg. Infect. Dis., 11*, 1874-1881.

Wang, Y., Zhang, H., Li, Z., Gu, W., Lan, H., Hao, W., Ling, R., Li, H. and Harrison, T.J. (2001). Detection of sporadic cases of hepatitis E virus (HEV) infection in China using immunoassays based on recombinant open reading frame 2 and 3 polypeptides from HEV genotype 4. *J. Clin. Microbiol., 39*, 4370-4379.

Wang, Y., Zhang, H., Ling, R., Li, H. and Harrison, T.J. (2000). The complete sequence of hepatitis E virus genotype 4 reveals an alternative strategy for translation of open reading frames 2 and 3. *J. Gen. Virol., 81*, 1675-1686.

Ward, P., Poitras, E., Leblanc, D., Letellier, A., Brassard, J., Plante, D. and Houde, A. (2009). Comparative analysis of different TaqMan real-time RT-PCR assays for the detection of swine Hepatitis E virus and integration of Feline calicivirus as internal control. *J. Appl. Microbiol., 106*, 1360-1369.

Weitz, M., Baroudy, B.M., Maloy, W.L., Ticehurst, J.R. and Purcell, R.H. (1986). Detection of a genome-linked protein (VPg) of hepatitis A virus and its comparison with other picornaviral VPgs. *J. Virol., 60*, 124-130.

Wheeler, C., Vogt, T., Armstrong, G.L., Vaughan, G., Weltman, A., Nainan, O.V., Dato, V., Xia, G., Waller, K., Amon, J., Lee, T.M., Highbaugh-Battle, A., Hembree, C., Evenson, S., Ruta, M.A., Williams, I.T., Fiore, A. and Bell, B.P. (2005). An outbreak of hepatitis A associated with green onions. *N. Engl. J. Med., 353*, 890–897.

Wolf, S., Williamson, W.M., Hewitt, J., Rivera-Aban, M., Lin, S., Ball, A., Scholes, P. and Greening, G.E. (2007). Sensitive multiplex real-time reverse transcription-PCR assay for the detection of human and animal noroviruses in clinical and environmental samples. *Appl. Environ. Microbiol., 73*, 5464-5470.

Xerry, J., Gallimore, C.I., Iturriza-Gomara, M., Allen, D.J. and Gray, J.J. (2008). Transmission events within outbreaks of gastroenteritis determined through analysis of nucleotide sequences of the P2 domain of genogroup II noroviruses. *J. Clin. Microbiol., 46*, 947-953.

Xia, H., Liu, L., Linde, A.M., Belak, S., Norder, H. and Widen, F. (2008). Molecular characterization and phylogenetic analysis of the complete genome of hepatitis E virus from European swine. *Virus Gene, 37*, 39-48.

Yazaki, Y., Mizuo, H., Takahashi, M., Nishizawa, T., Sasaki, N., Gotanda, Y. and Okamoto, H. (2003). Sporadic acute or fulminant hepatitis E in Hokkaido, Japan, may be food-borne, as suggested the presence of hepatitis E virus in pig liver as food. *J. Gen. Virol., 84*, 2351-2357.

Yoda, T., Suzuki, Y., Yamazaki, K., Sakon, N., Kanki, M., Aoyama, I. and Tsukamoto, T. (2007). Evaluation and application of reverse transcription loop-mediated isothermal amplification for detection of noroviruses. *J. Med. Virol., 79*, 326-334.

Yoneyama, T., Kiyohara, T., Shimasaki, N., Kobayashi, G., Ota, Y., Notomi, T., Totsuka, A. and Wakita, T. (2007). Rapid and real-time detection of hepatitis A virus by reverse transcription loop-mediated isothermal amplification assay. *J. Virol. Methods, 145*, 162-168.

Yoo, D., Willson, P., Pei, Y., Hayes, M.A., Deckert, A., Dewey, C.E., Friendship, R.M., Yoon, Y., Gottschalk, M., Yason, C. and Giulivi, A. (2001). Prevalence of hepatitis E virus antibodies in Canadian swine herds and identification of a novel variant of swine hepatitis E virus. *Clin. Diagn. Lab. Immunol., 8*, 1213-1219.

Zanetti, A.R., Schlauder, G.G., Romano, L., Tanzi, E., Fabris, P., Dawson, G.J. and Mushahwar, I.K. 1999. Identification of a novel variant of hepatitis E virus in Italy. *J. Med. Virol., 57*, 356-360.

Zeng, S.Q., Halkosalo, A., Salminen, M., Szakal, E.D., Puustinen, L. and Vesikari, T. (2008). One-step quantitative RT-PCR for the detection of rotaviruses in acute gastroenteritis. *J. Virol. Methods, 153*, 238-240.

Zhai, L., Dai, X and Meng J. (2006). Hepatitis E virus genotyping based on full-length genome and partial genomic regions. *Virus Res., 120*, 57-69.

Zhao, C., Li, Z., Yan, B., Harrison, T.J., Guo, X., Zhang, F., Yin, J., Yan, Y. and Wang, Y. (2007). Comparaison of real-time fluorescent RT-PCR and conventional RT-PCR for the detection of hepatitis E virus genotypes prevalent in China. *J. Med. Virol., 79*, 1966-1973.

Zheng, D.P., Ando, T., Fankhauser, R.L., Beard, R.S., Glass, R.I. and Monroe, S.S. (2006). Norovirus classification and proposed strain nomenclature. *Virology, 346*, 312-323.

In: Bacterial DNA, DNA Polymerase and DNA Helicases
Editor: Walter D. Knudsen and Sam S. Bruns
ISBN 978-1-60741-094-2
© 2009 Nova Science Publishers, Inc.

Chapter VI

hTERT in Cancer Chemotherapy: A Novel Target of Histone Deacetylase Inhibitors

Jun Murakami[1]*, Jun-ichi Asaumi[*1], Hidetsugu Tsujigiwa[2], Masao Yamada[2], Susumu Kokeguchi[3], Hitoshi Nagatsuka[4], Tatsuo Yamamoto[5] and You-Jin Lee[6]

[1]Department of Oral and Maxillofacial Radiology,
[2] Department of Virology,
[3] Department of Oral Microbiology,
[4] Department of Oral Pathology,
[5] Department of Preventive Dentistry,
Graduate Schools of Medicine, Dentistry and Pharmaceutical Sciences,
Okayama University, Okayama, Japan
[6]Jeonnam Biotechnology Research Center, Korea

Abstract

Chromatin structure plays an important role in the regulation of gene transcription. Chromatin structure can be modified by various post-translational modifications, including histone acetylation, phosphorylation, methylation and ribosylation. Among those modifications, histone acetylation/deacetylation is the most important mechanism for regulating transcription and is regulated by a group of enzymes known as histone acetyltransferases/histone deacetylases (HDACs).

Recently, HDAC inhibitors have been shown to be a novel and promising new class of anti-cancer agent that can regulate the transcription of genes by disrupting the balance of acetylation/deacetylation in particular regions of chromatin. A number of HDAC inhibitors are currently in phase I and II clinical trials against a variety of cancers. Although some promising candidates have been identified (e.g., p21^{WAF1} and c-Myc), the

---

* E-mail: asaumi@md.okayama-u.ac.jp and jun-m@md.okayama-u.ac.jp, Telephone number: +81-86-235-6621, Fax number: +81-86-235-6709

precise molecular targets remain uncertain. In this article, we focus on one of the DNA polymerases, telomerase, as a new candidate molecular target for HDAC inhibitors. Telomerase is composed primarily of the catalytic subunit (hTERT) and the RNA template (hTERC), and its activity correlates with levels of hTERT mRNA. hTERT expression is apparently governed by complicated regulatory pathways. Based on recent studies, the hTERT gene is likely to be targeted by histone acetylation/deacetylation.

1. Telomere and Telomerase

Telomere and Telomerase

Fifty years have passed since the first DNA polymerase (pol) was discovered. In the last few years the number of known pols, including terminal transferase and telomerase, has increased [1]. For DNA replication, maintenance of the chromosomal ends, the telomeres, is important [2].

Telomeres are specialized structures at chromosome ends that are composed of tandem repeats of the sequence TTAGGG [3]. Telomeres are essential elements that protect linear eukaryotic chromosomal ends from degradation, end-to-end fusions, rearrangements, and chromosome loss [4]. However, DNA polymerase fails to completely replicate DNA termini, and telomeres undergo progressive shortening with cell division [5]. Reduction of telomere length with cell division induces replicative senescence.

Synthesis and maintenance of telomeric repeats are mediated by a specialized ribonucleoprotein complex known as telomerase [3]. Telomerase is an RNA-dependent DNA polymerase that directs the de novo synthesis of telomeric repeats at chromosome ends [6]. Numerous studies have demonstrated that telomerase is not active in most normal somatic tissues [7]. In most normal human somatic cells, the lack of telomere-maintaining mechanisms leads to the progressive loss of telomeres during DNA replication and limits the life span of these cells [8-10].

Telomerase is highly active in the vast majority of human cancers (>90% of cancers), allowing them to stably maintain their telomeres and gain immortality [8, 11, 12]. Telomerase activation is thought to be a critical step in cellular immortalization and carcinogenesis. Furthermore, abrogation of the telomerase activity in cancer cells has been reported to induce cell growth arrest and apoptosis [13].

Human Telomerase Reverse Transcriptase (hTERT)

Recently, three major subunits comprising the human telomerase complex have been identified: the RNA template for the addition of new telomeric repeats (hTERC) [14]; a telomerase-associated protein (TEP1) [15, 16]; and a catalytic subunit known as human telomerase reverse transcriptase (hTERT) [17, 18]. The most important component responsible for the enzymatic activity of telomerase is hTERT [17, 18]. Many studies have found that hTERT is preferentially expressed in cancer cells and immortalized cell lines, but not expressed in normal tissues [17-19]. We previously examined the relative levels of hTERT mRNA among seven oral cancer cell lines by RT-PCR (Figure 1). The great majority of oral cancer cell lines contained detectable amounts of hTERT transcript; HSC4 and SAS cells showed especially high levels of mRNA. Hep2 showed no evidence of hTERT message under the specific experimental conditions employed.

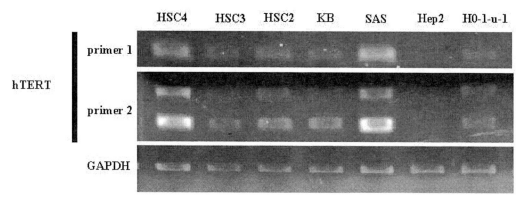

Figure 1. Comparison of hTERT expression among oral cancer cell lines. Extracted RNA was reverse-transcribed to cDNA and the relative levels of hTERT mRNA among seven oral cancer cell lines (HSC4, HSC3, HSC2, KB, SAS, Hep2 and HO-1-u-1) were determined by RT-PCR. Following incubation of the cells with agents under each experimental condition, extraction of total cellular RNA was carried out using Trizol reagent (Invitrogen Corp., Carlsbad, CA) according to the manufacturer's instructions. 1.5 µg RNA was reverse-transcribed with Superscript II reverse transcriptase and oligo dT primers (Invitrogen Corp.). Amplification of cDNAs was performed under the following PCR conditions: 7 min at 94°C for 1 cycle; then 26 cycles of 94°C for 30 s, 59°C for 30 s, and 72°C for 30 s; and a final elongation step at 72°C for 10 min. The primers used for the amplification were as follows. hTERT primer1: sense, 5'- cgg aag agt gtc tgg agc aa-3'; antisense, 5'- gga tga agc gga gtc tgg a -3'. *hTERT* primers1 amplified a 145 bp product (18, 206). hTERT primer2: sense, 5'- act ttg tca agg tgg atg tga cgg -3' (exon6); antisense, 5'- aag aaa tca tcc acc aaa cgc agg -3' (exon10). *hTERT* primers2 amplified a 493 bp product spanning exon 6 to exon 10 (207-209). GAPDH: sense, 5'-gaaggtgaaggtcggagtc-3'; antisense, 5'-caaagttgtcatggatgacc-3'. The amplified GAPDH fragment was used as a positive control. The RT-PCR products were separated by electrophoresis on a 2% agarose gel, stained with ethidium bromide, and viewed by UV. The great majority of oral cancer cell lines contained detectable amounts of hTERT transcript; HSC4 and SAS cells showed especially high levels of mRNA, as indicated by the very dense bands that appear in the relevant lanes. The gels clearly show that Hep2 cells contained no hTERT mRNA. >From [Murakami J et al. (2005) Cancer Chemother Pharmacol. 56(1):22-28.], with permission.

The Control of hTERT Expression

hTERT expression is apparently governed by complicated regulatory pathways. Expression of hTERT is known to be regulated mainly at the transcriptional level [20]. To date, a number of factors have been identified as directly or indirectly affecting hTERT gene activity (i.e., c-Myc, E6, and the estrogen receptor) [21-25]. Among them, c-Myc and Sp1 are the major activators of hTERT transcription, and their binding sites are located within the proximal core promoter [26, 27]. The promoter of hTERT is GC rich, and lacks both TATA and CAAT boxes; however, the hTERT proximal promoter harbors two E-boxes and five GC-boxes, the consensus binding motifs for the Myc network and Sp family proteins, respectively [26, 28]. It has also been demonstrated that the Myc and Mad proteins, respectively, can activate or repress the hTERT promoter through their interaction with the E-boxes [29-31].

Recent reports have revealed that epigenetic modification, such as DNA methylation and histone deacetylation, is involved in repression of the hTERT transcript in human cells. In human cell lines without expression of hTERT transcript and telomerase activity, a DNA demethylating agent 5-aza-2'-deoxycytidine (5-Aza) induces demethylation of the hTERT promoter and expression of the hTERT transcript [32, 33]. Taking these results together, histone modifications including methylation and acetylation have been shown to transcriptionally target the hTERT gene, suggesting the importance of the chromatin environment in the regulation of telomerase gene expression [32, 34-39]. These findings provide insights into the regulatory mechanism underlying telomerase activity in human cells, which may be implicated in the development of therapeutic strategies for telomerase dysregulation-associated human cancers. Histone deacetylase (HDAC) inhibitor, an anticancer drug, is known to modulate transcription and change the expression of hTERT mRNA and telomerase activity in several types of cancer cells; however, it remains incompletely understood how the hTERT gene in normal and cancerous cells is regulated.

This section focuses on the effects of HDAC inhibitor on hTERT gene transcription.

2. Histone Modifications

Histone Acetylation and Deacetylation

Chromatin structure plays an important role in the maintenance of cellular integrity [40]. In eukaryotes, transcriptional regulation largely depends on chromatin remodeling. For the regulation of gene transcription, chromatin alters its structure between an extended, transcriptionally active euchromatin and a compact, transcriptionally silent heterochromatin [41, 42]. These chromatin structures are modulated by various post-translational modifications of histone tails, including acetylation, phosphorylation, methylation and ribosylation [43]. Among those modifications, acetylation and deacetylation of histones have been particularly important and associated with transcriptional activation and repression, respectively [44, 45].

The extent of acetylation of histones is regulated via the balance between acetylation by acetyl transferase (HAT) and deacetylation by histone deacetylase (HDAC). The balance between the two processes defines chromatin structure and accessibility for key cellular proteins to specific target sites [46-48]. Histone acetyltransferases function to acetylate specific lysine residues in the positively charged N-terminal tails that protrude from the octamer. These tails are important for histone-DNA and histone-non-histone protein interactions. Histone acetylation of N-terminal lysine residues triggers the initiation of gene transcription by recruiting chromatin remodeling factors, resulting in a more open, transcriptionally active chromatin structure [49].

In contrast, histone deacetylation, which is mediated by HDACs [50], contributes to the formation of transcriptionally inactive heterochromatin [51]. Histone deacetylation works synergistically with methylation of cytosine residues in DNA CpG islands at the promoter region, both of which result in gene silencing. Alteration of gene expression due to these epigenetic modifications (i.e., HDAC-mediated silencing of specific tumor suppressor genes) is one of main etiologic factors for cancer [52].

Other Modifications: Histone Methylation and Demethylation

Both lysine and arginine residues of histone H3 [53, 54] and H4 [55, 56] are targeted for histone methylation, which plays a pivotal role in the regulation of chromatin structure and gene expression. In histone H3, the lysine methylation is observed on residues 4, 9, 27, 36 and 79, whereas in histone H4, only lys20 is methylated [41, 57, 58]. Methylation of H3-K4, H3-K36 and H3-K79 is predominantly associated with transcriptionally active genes, whereas methylation of H3-K9, H3-K27 and H4-K20 marks silent genes or heterochromatin [57-59].

Almost all lysine histone methyltransferases contain a highly conserved SET domain [60, 61]; thus, histone methylation seems to be mediated through the SET domain. The SET domain proteins, SUV39H1 protein [62], G9a [63] and [64, 65] ESET/SETDB1, have recently been shown to be histone methyltransferases specific for K9 and 27 (G9a only) of H3.

Hypermethylation of Lys9 of histone H3 (H3-K9) leads to the recruitment of the heterochromatin protein 1 (HP1) and the establishment of heterochromatin [51, 66]. This process plays an important role in the maintenance of DNA methylation. In arabidopsis or in the fungus Neurospora crassa, loss of H3-K9 methylation leads to a substantial decrease in DNA methylation [67, 68].

In contrast to H3-K9 methylation, hypermethylation of Lys4 of histone H3 (H3-K4) occurs in transcriptionally active euchromatin [69, 70]. K4 methylation inhibits the association of the HDAC1/2-containing NuRD complex to the chromatin [70]. Additionally, K4 methylation might be required for the establishment of lysine acetylation on H3 [71]. Acetylation together with the H3-K4 methylation impairs the methylation on H3-K9 [72], and could thereby prevent establishment or maintenance of CpG methylation. For H3-K4 methylation, SET7/9 plays a role as a histone methyltransferase that transfers methyl groups to lys4 of histone H3 [72, 73].

The SET and MYND domain-containing protein 3 (SMYD3) has recently been identified as a histone H3-K4–specific methyltransferase. SMYD3 has been shown to dimethylate or trimethylate H3-K4 [74]. SMYD3 specifically binds to a DNA sequence, CCCTCC, in the promoter region of its target genes as a transcription factor [74]. SMYD3 is significantly up-regulated in the great majority of cases of colorectal cancer, hepatocellular cancer, and breast cancer, whereas it is repressed in most normal human cells [74-76]. Recently, SMYD3 has been implicated in the oncogenic process [74, 75]. The recent discovery of multiple histone demethylases reveals that histone methylation, like acetylation and phosphorylation, is a dynamic and reversible process. In 2004, lysine-specific demethylase 1 (LSD1) was identified as a histone demethylase [77]. Interestingly, LSD1, which catalyzes the demethylation of mono- and di-methylated histone H3-K4 or K9, has been shown to either repress or activate target genes by interacting with diverse co-factors [77-81].

Histone Deacetylase Inhibitors

Recently, inhibition of HDAC has been shown to be a novel strategy for the treatment of cancers via the restoration of aberrantly silenced genes. In a clinical setting, HDAC inhibitors represent a novel and promising class of chemotherapeutic agent, modulating the transcription of target genes by disrupting the balance of acetylation/deacetylation in particular regions of chromatin [82, 83]. HDAC inhibitors can inactivate HDAC activity, and then induce hyperacetylation of histones through a relative increase of histone acetyltransferase (HAT) activity. However, the mechanisms involved are still under investigation. A variety of compounds with HDAC inhibitor activities have been chemically synthesized. The structural classifications include [1] short-chain fatty acids (i.e., sodium butyrate), [2] hydroxamic acids [i.e., suberoylanilide hydroxamic acid, oxamflatin, and trichostatin A (TSA), [3] cyclic tetrapeptide (i.e., trapoxin A), [4] cyclic peptides [i.e., depsipeptide (FR901228) and apicidin], and [5] benzamides (i.e., MS-275). To confirm the inhibition of HDAC activity by HDAC inhibitors, we performed immunocytochemical staining against HDAC1 in TSA-treated oral cancer cell line HSC4 cells (Figure 2).

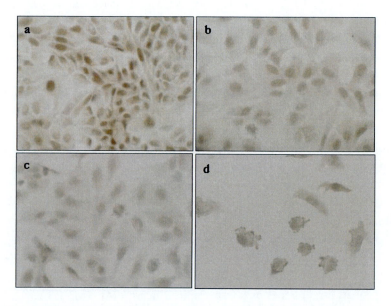

Figure 2. An HDAC inhibitor inhibits histone deacetylase in oral cancer cells. Immunocytochemistry of histone deacetylase 1 (HDAC1) in HSC4 cells treated with varying concentrations of an HDAC inhibitor, TSA, for 24 h. The inhibition of HDAC1 by TSA was monitored with specific antibodies recognizing HDAC1 (Cell Signaling Technology, Beverly, MA). Cells were fixed with 10% formaldehyde and processed with the standard avidin-biotin peroxidase complex. DAB was used as a chromogen. (a) HSC4, non-treated; (b) HSC4, TSA-treated (800 nM); (c) HSC4, TSA-treated (1.6 microM); (d) HSC4, TSA-treated (2.4 microM). Original magnification, x40. HDAC1 was highly expressed in TSA-untreated HSC4 cells, and was downregulated after TSA treatment in a concentration-dependent manner. >From [Murakami J et al. (2008) NOVASCIENCE], with permission.

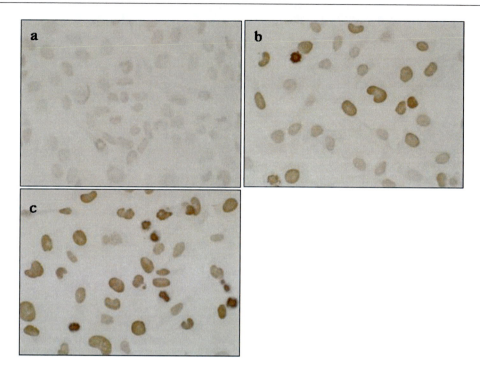

Figure 3. An HDAC inhibitor upregulates histone acetylation in oral cancer cells. Immunocytochemistry of acetyl-histone H3 (Lys9) in HSC4 cells treated with varying concentrations of TSA for 24 h. The TSA-induced hyperacetylation of histone H3 was monitored with specific antibodies recognizing acetyl-histone H3 (Lys9) (Cell Signaling Technology). Cells were fixed with 10% formaldehyde and processed with the standard avidin-biotin peroxidase complex. DAB was used as a chromogen. (a) non-treated; (b) TSA-treated (500 nM); (c) TSA-treated (800 nM). Original magnification, x40. Acetylation of histone tails in HSC4 cells was upregulated after TSA treatment in a concentration-dependent manner. >From [Murakami J et al. (2008) NOVASCIENCE], with permission.

HDAC1-positive cells disappeared after TSA treatment in a concentration-dependent manner. We also examined whether the administration of TSA induced histone H3 acetylation in oral cancer cells (Figures 3 and 4). TSA-induced hyperacetylation of histone H3 was monitored with specific antibodies recognizing acetylated N-terminal lysine residues (Lys9) of histone H3. As shown in Figure 3, TSA mediated an increase in the acetyl histone H3 (Lys9) levels in HSC4 cells after the onset of drug exposure. Acetyl histone H3 (Lys9)-positive HSC4 cells appeared after TSA treatment in a concentration-dependent manner. The histone acetylation was further determined by Western blot analysis. HSC4 and KB cells were treated with TSA (800 nM) for up to 24 h (Figure 4). In a time course experiment, treatment with TSA led to a marked increase in acetyl histone H3 (Lys9) in 4 h; thus, the accumulation of acetylated histone remained elevated to 8 h.

As anti-cancer agents for the treatment of human cancers, HDAC inhibitors are advantageous for their abilities to induce not only cell cycle arrest and differentiation but also apoptosis [43, 84, 85]. We determined the cytotoxicity of TSA in HSC4 (Figure 5a). We found that the treatment of HSC4 cells for 24 h with TSA induced apoptosis in a dose-dependent manner (Figure 5b).

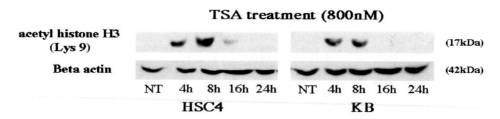

Figure 4. Effect of TSA on histone acetylation. To confirm the upregulation of histone acetylation by HDAC inhibitors, we carried out Western blotting using HSC4 and KB cell lines. The histone acetylation was further determined by Western blot analysis using a specific antibody against acetyl histone H3 (Lys9). The total cell lysate from untreated HSC4, KB cells, and cells treated with TSA (800 nM) for up to 24 hours was electrophoretically separated and transferred to a polyvinylidene difluoride membrane. The membrane was then subsequently probed with specific antibodies recognizing histones acetylated at the following site: lysine 9 of histone H3 (H3-K9). Western blotting was carried out according to the standard method. Representative results are shown. Whereas little acetyl histone H3 (Lys9) expression was detected in non-treated cells, in a time course experiment, the acetylation of the histone tail became evident 4 to 8 hours after treatment with TSA. >From [Murakami J et al. (2008) NOVASCIENCE], with permission.

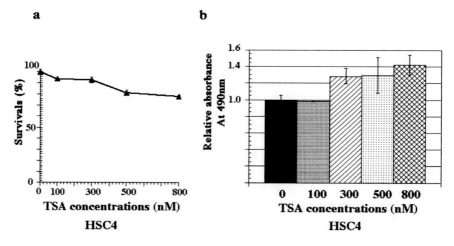

Figure 5. Effect of TSA on the proliferation and apoptosis of HSC4 cells. HSC4 cells were cultured in the presence or absence of varying concentrations of TSA for 24 h. (a) Cellular sensitivity to TSA. The cell survival rates were assayed by an MTT (3-(4,5-dimethylthiazol-2-yl)-2,5-diphenyl tetrasodium bromide) assay. Cells were seeded at a density of 1.5×10^4 cells in 0.1 ml of medium into 8 wells of a 96-well plate and incubated at 37°C for 24 h. The cells were incubated with several concentrations of TSA ranging from 100-800 nM for an additional 24 h. At the end of the TSA exposure, an MTT assay was carried out using an MTT Cell Growth Kit (Chemicon International, Inc., Temecula, CA) according to the manufacturer's instructions. The percentage of cell-growth inhibition was calculated by comparison of the absorbance reading from treated *versus* untreated control cells. The concentration-dependent inhibition of cell proliferation was observed. (b) Apoptotic effects of TSA. We assessed TSA-induced apoptosis using an apoptosis screening kit according to the manufacturer's protocol (Wako Pure Chemical Industries, Ltd., Osaka, Japan). The cells were seeded in 96-well plates, as described above. The cells were then treated with varying doses of TSA ranging from 100-800 nM for 24 h. After drug treatment, the cells were fixed with 10% formaldehyde and assayed for apoptosis. The percentage of apoptosis was calculated by comparing the absorbance at a 490 nm reading from treated cells *versus* that from untreated control cells. The treatment of HSC4 cells for 24 h with the TSA induced apoptosis of cells in a dose-dependent manner. >From [Murakami J et al. (2008) NOVASCIENCE], with permission.

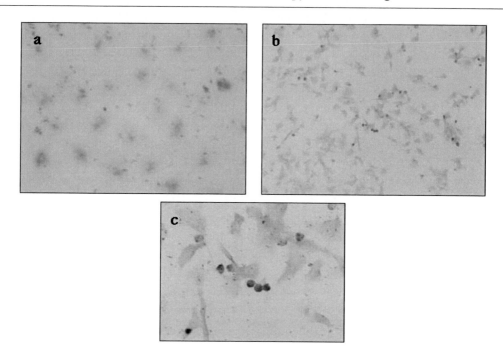

Figure 6. TUNEL detection of apoptosis in HSC4 cells following TSA treatment. TSA-induced apoptosis was assessed by an apoptosis *in situ* detection kit according to the manufacturer's protocol (Wako Pure Chemical Industries, Ltd.). HSC4 cells were cultured in the presence or absence of 800 nM of TSA for 24 h. The cells were fixed with 10% formaldehyde and stained with the TUNEL method. (a) non-treated: original magnification x40; (b) TSA-treated: original magnification x10; (c) TSA-treated: original magnification x40. TUNEL-positive cells appeared after 24 h of culture, indicating that TSA causes apoptosis in HSC4 cells. Apoptotic cells were stained dark brown. >From [Murakami J et al. (2008) NOVASCIENCE], with permission.

As shown in Figure 6, to confirm the induction of apoptosis by the HDAC inhibitor, we performed TUNEL staining for TSA-treated HSC4 cells. TUNEL-positive cells appeared after 24 h of TSA treatment, indicating that TSA causes apoptosis in HSC4. A number of HDAC inhibitors are currently in phase I and II clinical trials, including suberoylanilide hydroxamic acid, MS-275, phenylbutyrate, and depsipeptide (FR901228, FK228, NSC630176) [86-93]. Recently, promising results have been reported for the treatment against a variety of cancers [84, 94], supporting the development of these compounds for clinical use.

Molecular Targets for HDAC Inhibitors

HDAC inhibitors are promising anticancer drugs, but the molecular mechanisms for their activity remain unclear. HDAC inhibitors are thought to induced histone hyperacetylation and chromatin remodeling, leading to selective modulation of the expression of genes related to cell growth, cell cycle and apoptosis [95-99]. Recently, array studies have revealed that only 2% to 5% of genes are affected by HDAC inhibitor treatment [86, 87, 100]. The genes responsible for the inhibition of proliferation and induction of cell differentiation or death by

HDAC inhibitors continue to remain elusive, although some promising candidates have been identified, e.g., the cyclin-dependent kinase inhibitor p21^{WAF1} [101-103], c-Myc [104, 105], and the anti-apoptotic bcl-2 gene [106, 107]. Previous reports have shown HDAC inhibitor-induced upregulation of Fas, FasL, and Bax and downregulation of Bcl-XL [108-114]. Glaser et al. studied the gene profiles of HDAC inhibitor-treated cells [115], and found 13 core genes modulated by several HDAC inhibitors in various cancer cell lines. They compared the gene expression profiles of three different bladder and breast cancer cell lines treated with three HDAC inhibitors: SAHA, TSA, and MS-27-275. They identified a common set of genes that are positively or negatively regulated by all of the HDAC inhibitors in all of the cell lines tested (8 genes were found to be upregulated and 5 genes to be downregulated among the 6,800 genes). Suzuki et al. [116] reported that TSA upregulated 23 genes in the colorectal cancer cell line RKO among 10,814 genes examined using a subtraction microarray. Most of them are classified as genes encoding enzymes and signal transducers, and none are growth-regulatory genes, with the exception of TRADD.

One of the HDAC inhibitor-mediated inducible genes, p21^{WAF1}, which plays a role as a cell cycle inhibitor in the G1 cell cycle arrest of many cancer cells, is commonly induced in response to HDAC inhibitors [101, 117]. HDAC inhibitors accumulate target cells at either the G1 or G2/M phase of the cell cycle, through transcriptional activation of p21^{WAF1} [117, 118]. In Figure 7, the effect of TSA on p21^{WAF1} expression was determined. HSC4 cells were treated with TSA and the induction of p21^{WAF1} proteins was detected using immunofluorescence. TSA strongly induced the expressions of acetyl-histone H3 (Lys9) (Figure 7f) and p21^{WAF1} (Figure 7g).

From the merged image (Figure 7h), the induction of both proteins was observed in the nucleus of the same cell. The induction of p21^{WAF1} by HDAC inhibitors might occur via histone acetylation.

As for apoptosis mediated by HDAC inhibitors, the death receptor-related genes are thought to be involved. In HDAC inhibitor-induced activation of death receptor-mediated apoptosis, the transcriptional activation of proapoptotic genes such as Fas and Bax might mediate HDI-induced apoptosis [109]. In addition, the upregulation of Fas and FasL occurs in cells derived from neuroblastoma, promyelocytic leukemia and uveal melanoma [108-110].

As another HDAC inhibitor-induced apoptosis pathway, the mitochondria-mediated death signals may be crucial for HDAC inhibitor-mediated apoptosis. Some classes of HDAC inhibitors have been reported to induce apoptosis by activating either caspase-8 or caspase-9 in various types of cancer cell lines. For example, Henderson et al. reported that an HDAC inhibitor resulted in the release of cytochrome c and activation of caspase-9 [119]. Amin et al. reported that an HDAC inhibitor modulated the expression of mitochondria-related Bcl-2 family proteins [120].

In Figure 8, we examined the pathway of TSA-stimulated apoptosis. The effectiveness of TSA in activating apoptosis was assessed by the cleavage of the caspases and PARP. The activation of caspases became evident 16 h after treatment with TSA in a time course experiment. Furthermore, TSA treatment induced proteolytic PARP, which is a caspase-3 substrate, indicating that the caspases were actually activated in TSA-treated cells.

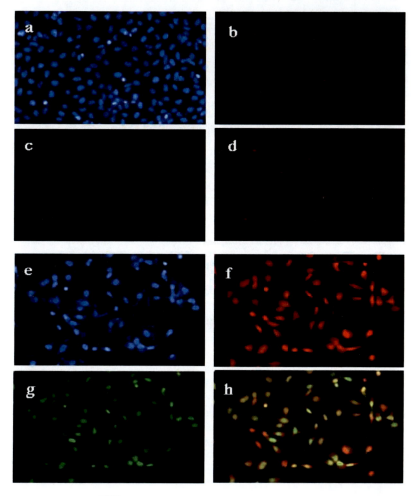

Figure 7. The induction of p21^{WAF1} by HDAC inhibitors might occur via histone acetylation. HSC4 cells were cultured 24 h in the absence or the presence of 800 nM TSA. After TSA treatment, immunofluorescence microscopy was carried out. Briefly, the cells were fixed with 4% formaldehyde in PBS for 10 min at RT, followed by 70% to 100% ethanol. Non-specific binding was blocked with normal goat and swine serum in PBS for 30 min at RT. The cells were probed with rabbit polyclonal anti-acetyl-histone H3 (Lys9) antibody and mouse monoclonal anti-p21^{WAF1} Ab-3 (DCS-60.2) antibody (NeoMarkers, Fremont, CA, USA) overnight at 4°C. The cells were then washed with PBS three times, followed by incubation with tetramethylrhodamine isomer R (TRITC)-conjugated swine anti-rabbit IgG (Dakocytomation, Glostrup, Denmark) and fluorescein-isothiocyanate isomer 1(FITC)-conjugated goat anti-mouse IgG (Dakocytomation) in PBS for 30 min. After the cells were washed with PBS three times, they were counterstained with 4,6-diamidine-2-phenylindole, dihydrochloride (DAPI) and then mounted with GEL/MOUNT (Biomeda Corp., Foster City, CA, USA). They were then examined with a confocal laser scanning microscope (LSM Pascal; Zeiss, Oberkochen, Germany). The panels show representative data of HSC4 cells in the absence (a, b, c, d) or presence of TSA (e, f, g, h). (a) non-treated, indicating DAPI (blue); (b) non-treated, indicating acetyl-histone H3 (Lys9) (red); (c) non-treated, indicating p21^{WAF1} (green); (d) non-treated, merge of green and red; (e) TSA-treated, indicating DAPI (blue); (f) TSA-treated, indicating acetyl-histone H3 (Lys9) (red); (g) TSA-treated, indicating p21^{WAF1} (green); (h) TSA-treated, merge of green and red. TSA strongly induced the expressions of acetyl-histone H3 (Lys9) (f) and p21^{WAF1} (g). From the merged image (h), the induction of both proteins was observed in the nucleus of the same cell. >From [Murakami J et al. (2008) NOVASCIENCE], with permission.

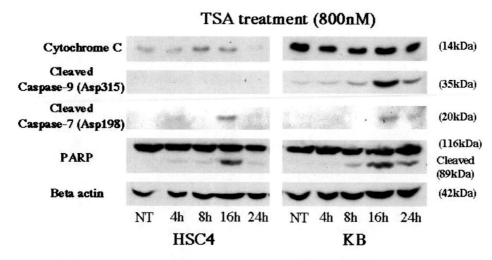

Figure 8. Effect of TSA on the expression level of proteins involved in apoptosis. To confirm the activation of apoptosis by HDAC inhibitors, we carried out Western blotting using the cell lines HSC4 and KB and the following antibodies: anti-cleaved caspase-7, anti-cleaved caspase-9, anti-PARP, and anti-cytochrome c (all purchased from Cell Signaling Technology). The effectiveness of TSA in activating caspase activity was assessed by analysis of the cleavage of the caspases and PARP. The activation of the mitochondrial apoptotic pathway by TSA was assessed by analysis of the expression level of cytochrome c. The total cell lysates were separated from cells untreated and treated with TSA (800 nM) for the indicated time. TSA administration caused cleavage of either PARP or caspase-9. In a time course experiment, the activation of caspases became evident 16 hours after treatment with TSA. The cleaved caspases-7 and -9 appeared in cells after 16 h of culture with TSA, whereas no such changes were detected in the cytochrome c expression. Furthermore, TSA treatment induced proteolytic PARP, which is a caspase-3 substrate, indicating that caspases were really activated in TSA-treated cells. These results indicate that TSA induces apoptosis by facilitating the cleavage of caspases. >From [Murakami J et al. (2008) NOVASCIENCE], with permission.

These results indicated that TSA induces apoptosis by facilitating the cleavage of caspases. Given that the acetylation of the histone tail became evident 4 to 8 hours after treatment with TSA (Figure 4), TSA induced the apoptosis of HSC4 or KB cells in a time-dependent manner, preceded by the accumulation of acetylated histone H3. The activation of the mitochondrial apoptotic pathway by TSA was assessed by analysis of the expression level of cytochrome c. We used whole cell lysates in Western blotting, and did not assess the release of cytochrome c from the mitochondria into the cytosol. However, TSA treatment had no effects on the levels of cytochrome c in the KB and HSC4 cells. The mitochondrial pathway may not play a major role in the activation of apoptosis by TSA in oral cancer cells.

Acquired Resistance to HDAC Inhibitors

Unfortunately, HDAC inhibitor treatment may also induce resistance-related genes in cancers. In a clinical setting, upregulation of resistance-related genes by HDAC inhibitors may become a problem. Acquired drug resistance is known to be mediated by increased efflux of anticancer agents from cancer cells [121]. Some HDAC inhibitors are removed from cancer cells by upregulation of P-glycoprotein (P-gp) or other genes of the ATP-binding cassette (ABC) transporter family. In fact, P-gp-expressing cancer cells are resistant to the HDAC inhibitor depsipeptide [98, 122]. However, the hydroxamic acid-based HDAC inhibitor SAHA could kill P-gp-expressing multi-drug resistant cells [96, 123]. It might be assumed that the larger, natural product HDAC inhibitors like depsipeptide would acquire resistance via the P-gp system, but resistance mechanisms associated with small molecular weight HDAC inhibitors may differ. In addition, Duan et al. [124] suggested that activation of nuclear factor-B (NF-B) is one of the major mechanisms of resistance of head and neck squamous cell carcinoma (HNSCC) to HDAC inhibitors. NF-B is a transcription factor which promotes expression of proliferative and antiapoptotic genes; thus, they reported that limited activity of classic HDAC inhibitors TSA and sodium butyrate (NaB) was associated with enhanced activation of NF-B reporter activity in a panel of six HNSCC cell lines.

HDAC Inhibitors: Synergistic Activity with Existing Cancer Therapy

Recently, several studies have indicated that HDAC inhibitors show synergistic activity in combination with other anticancer agents (*i.e.*, retinoic acids, vitamin D analogues, etc.) or treatments [125-134]. Although the mechanisms remain to be defined, butyrate, an HDAC inhibitor, has been reported to enhance radiation injury in human colon cancer cells and to increase the radiation sensitivity of V79 Chinese hamster lung fibroblasts [135, 136]. In gene therapy, the addition of HDAC inhibitors after adenovirus infection is known to increase viral protein and transgene expression [137, 138]. Lee et al. also reported that the addition of NaB increased the transduction of adenovirus [139].

HDAC Inhibitors as Anti-Angiogenic Agents

Recently, many reports have supported the possible involvement of HDAC in the control of angiogenesis. Angiogenesis is the multi-step process of new blood vessel formation by endothelial cells. To date, several agents have been developed as antiangiogenic agents, including angiostatin [140], endostatin [141], TNP-470 [142], radicicol [143], depudecin [144], and SU6668 [145]. HDAC inhibitors, depudecin and TSA, have also been shown to inhibit angiogenesis [144, 146]. TSA has been shown to inhibit expression of the interleukin-2 (IL-2) gene, which plays a role as an angiogenic amplifier, during T-cell activation [147, 148] and also to inhibit the expression of the angiogenic factor IL-8 in a dose-dependent manner [149].

We investigated the inhibitory effect of an HDAC inhibitor on angiogenesis by means of a co-culture tubule formation assay (Figure 9). To explore the *in vitro* antiangiogenic activity of TSA, HUVECs were co-cultured with human fibroblasts and incubated for 11 days with or without various concentrations of TSA. At the end of the assay, tubule formation was examined by immunochemical staining with anti-CD31 (PECAM-1) antibody according to the manufacturer's protocol.

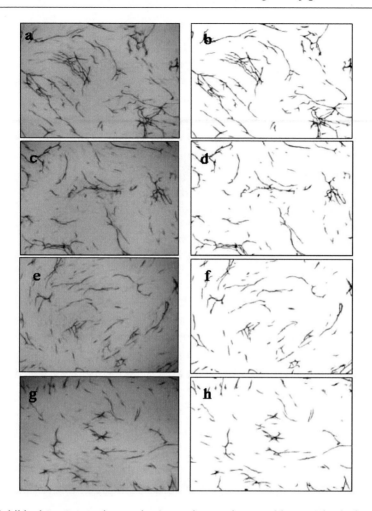

Figure 9. TSA inhibited *in vitro* angiogenesis. An angiogenesis assay kit comprised of co-cultures of human umbilical vein endothelial cells (HUVEC) and human dermal fibroblasts was purchased from KURABO Inc. (Osaka, Japan) and used according to the manufacturer's instructions. Cells were seeded onto a 24-well plate and cultivated in the basal medium with 10 ng/ml vascular endothelial growth factor (VEGF). Various concentrations of TSA were added to the medium of the HUVEC/fibroblast co-culture. The medium was changed every 3 days, and the HUVECs were cultured in the presence or absence of TSA for 11 days. At the end of the assay, the cells were then fixed with 70% ethanol and incubated with diluted anti-human PECAM-1 (CD31) antibody (Kurabo Inc., Osaka, Japan) for 1 hour followed by a 1-hour incubation with a goat anti-mouse IgG alkaline phosphatase-conjugated secondary antibody. Visualization of the tube-like structures formed by HUVECs was achieved with 5-bromo-4-chloro-3-indolyl phosphate/nitro blue tetrazolium (BCIP/NBT), and tubule formation was measured. Left panel: microscopic pictures of tubules stained with CD31. Right panel: images graphically analyzed by KURABO angiogenesis Image Analyzer Ver. 2 (KURABO Inc., Osaka, Japan). (a) non-treated, microscopic picture; (b) non-treated, analyzed image; (c) TSA-treated (1.0 nM), microscopic picture; (d) TSA-treated (1.0 nM), analyzed image; (e) TSA-treated (10 nM), microscopic picture; (f) TSA-treated (10 nM), analyzed image; (g) TSA-treated (100 nM), microscopic picture; (h) TSA-treated (100 nM), analyzed image. The culture of TSA-untreated HUVECs in which VEGF (10 ng/ml) was present from day 0, showed increased capillary-like structures. In contrast, the formation of capillary-like structures in the TSA-treated HUVECs was less than that of the TSA-untreated control. Representative micrographs of capillary structures formed by HUVECs. Original magnification, x4.
>From [Murakami J et al. (2008) NOVASCIENCE], with permission.

As can be seen in Figure 9, the TSA treatment (100 nM) reduced tubule formations and inhibited *in vitro* angiogenesis. On day 11 of the culture of TSA-untreated HUVECs, in which VEGF was present from day 0, increased capillary-like structures were seen. In contrast, the formation of capillary-like structures in the TSA-treated HUVECs was less than that of the TSA-untreated control. In a recent study, hypoxia was indicated to be one of the key factors triggering angiogenesis, and HDAC was shown to be induced under hypoxia. Thus, the increased HDAC activity under hypoxia can suppress the expression of p53 and VHL, the latter of which plays an antiangiogenic role, and downregulate the angiogenic factors VEGF and HIF-1a [146].

Depsipeptide FK228 (FR901228): A Novel HDAC Inhibitor

FK228 formerly FR901288, one of HDAC inhibitors, is under clinical development as a novel drug for cancer therapy [82, 84, 150, 151, 152]. FK228 is a cytotoxic depsipeptide isolated by the Fujisawa Pharmaceutical Company from fermentation broth of Chromobacterium violaceum [152]. FK228 exhibits HDAC inhibitor activity [82, 83, 151] and possesses potent antitumor effects for cancer cells. FK228 had a stronger cytotoxicic activity than TSA. FK228 has been found to inhibit tumor cell proliferation both *in vitro* and *in vivo* at nanomolar concentrations [152-154]. FK228 is currently under clinical trial in patients solid and hematologic malignancies [91, 93]. Phase II trials have been initiated in cutaneous T-cell lymphoma, peripheral T-cell lymphoma, and recurrent or metastatic squamous cell carcinoma (SCC) of the head and neck [91, 155, 156]. These clinical trials have revealed that FK228 is effective as an anti-cancer agent [92, 93, 157, 158]. Moreover, FK228 can be safely administered without any life-threatening toxicities or cardiac toxicities in clinical trials for advanced cancer and leukemia [92, 157].

Molecular Target for FK228

Recently, Yu et al. reported that FK228 acetylated wild-type p53, induced p21 expression, and depleted mutant p53, erbB1, erbB2, and RAF-1 protein levels in cancer cells [97]. $p21^{WAF1}$ expression is induced by FK228, while cyclin D1 and c-Myc are downregulated [95, 99]. Sandor et al. [95] also reported that FK228 diminishes cyclin D1 and phosphorylated-Rb protein in cancer cells.

The results in Figure 10 demonstrate that FK228 is a potent inducer of $p21^{WAF1}$ expression in oral cancer cells (SAS, Hep2). Cells were cultured either continuously in medium containing FK228 or in FK228-free medium for up to 16 h. Western blotting and RT-PCR experiments revealed that FK228 mediated increases in $p21^{WAF1}$ at both the mRNA and protein level after the onset of drug exposure.

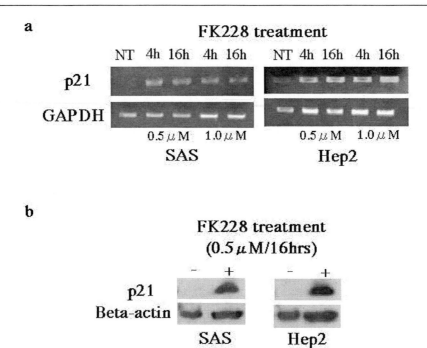

Figure 10. FK228 induced $p21^{WAF1}$ expression in oral cancer cells. SAS and Hep2 cells were treated with FK228 for up to 16 hours. The samples were collected at indicated time points; and assessed by RT-PCR and western blotting. (a) Reverse transcription-PCR analysis for $p21^{WAF1}$ mRNA. Amplification of cDNAs was performed under the following PCR conditions: 7 min at 94□ for 1 cycle; then repeated cycles at 94□ for 30 s, 59□ for 30 s, 72□ for 30 s; and a final elongation step at 72□ for 10 min. The primers used for the amplification were as follows: $p21^{WAF1}$; sense: 5'-CCTCTTCGGCCCGGTGGAC -3', antisense: 5'-CCGTTTTCGACCCTGAGAG -3'. The amplified GAPDH fragment was used as a positive control. (b) Western blot analysis for $p21^{WAF1}$ protein. The cells were cultured either continuously in medium containing FK228 or in FK228-free medium for 16h. Whole cell lysates were prepared and subjected to immunoblot analysis with mouse monoclonal antibody $p21^{WAF1}$. Representative results. Each of the blots shown was demonstrated to have equal protein loading by reprobing with the monoclonal antibody for beta-actin. The amounts of $p21^{WAF1}$ readily increased in the cells after 16h of culture with FK228. >From [Murakami J et al. (2008) NOVASCIENCE], with permission.

FK228 has been shown to strongly inhibit proliferation of tumor cells by inducing a p21-dependent G1 arrest and a p21-independent G2-M arrest [95]. Sandor et al. eported that FK228, while not directly inhibiting kinase activity, causes cyclin D1 downregulation and a p53-independent induction of p21, which in turn lead to the inhibition of CDK and dephosphorylation of Rb, and ultimately result in growth arrest in the early G1 phase [95]. Cell clones lacking p21 were not arrested in G1 phase, but continued DNA synthesis and were arrested in G2/M phase following FK228 treatment [95]. FK228 induced apoptosis with different apoptosis-activating cascades according to the type of malignancy. In small cell lung cancer, Doi et al. reported that FK228 activated caspase-9 and, mainly induced caspase-dependent apoptosis via the mitochondrial pathway [159]. In osteosarcoma and chronic lymphocytic leukemia cells, FK228-induced apoptosis is caspase-dependent and is selectively involved in the death receptor pathway that initiates caspase-8 activation [111, 160]. With respect to the apoptosis-related genes, FK228 increases the pro-apoptotic BH3-

only protein Bim in MKN45 and DLD-1 cells, and inhibits expression of the anti-apoptotic molecules Bcl-2, Bcl-XL and MCl-1 [161, 162]. Furthermore, bortezomib and FK228 can directly up-regulate the expression of TRAIL-R2 (DR5), sensitizing tumor cells to TRAIL-induced apoptosis [163, 164].

Recent studies have suggested that antiangiogenesis may play important roles in FK228-induced cytotoxicity. As mentioned above, HDAC is being treated as a novel target in the development of angiogenic inhibitors. Two HDAC inhibitors, depudecin and TSA, have also recently been reported to inhibit angiogenesis. FK228 regulates the transcription of angiogenesis-responsive gene expression both in cancer and endothelial cells, and inhibits angiogenesis. In a clinical setting, it would be worthwhile to uncover the underlying mechanisms of the antiangiogenic activity of FK228.

In gene therapy, Kitazono et al. demonstrated that in the presence of FK228, much less adenovirus is required *in vitro* [165]. The treatment with low concentrations of FK228 before infection is known to increase viral and transgene expression following adenovirus infection in cancer cell lines [165]. FK228 can increase CAR gene expression in cancer cell lines [165-169], which in turn facilitates virus infection, leading to an increased adenovirus transgene expression [165-167].

Acquired Resistance to FK228

Because FK228 is a relatively large molecular weight natural product, it is excreted by multidrug resistance proteins (MDR1, P-glycoprotein), and cancer cells overexpressing MDR1 show FK228 resistance [98, 122, 156]. Recently, FK228 was reported to be a substrate for MDR1 [95, 170] and multidrug resistance-associated protein 1 (MRP1, ABCC1), both of which mediate FK228 resistance. FK228 was also shown to induce MDR1 when incubated with cancer cells [170]. Based on these reports, MDR1 seems to be a major contributor to FK228 resistance in cancer cells [98, 171, 172].

3. hTERT and Histone Modifications

Histone Modifications on hTERT

Histone modification–mediated chromatin patterns at promoter sequences are associated with repression and activation of hTERT in normal and cancer cell lines. Acetylation of histones, currently the best studied of these modifications, has been shown to transcriptionally target the hTERT gene [32, 34-39]. HDAC inhibitors modulate transcription and change the expression of hTERT mRNA and telomerase activity in several types of cancer cells; however, it remains unclear how the repression or activation of the hTERT gene in normal and cancerous cells is achieved.

In recent years, TSA has been shown to activate hTERT expression in telomerase-negative cells [32, 34, 38, 39]. Even in normal human cells lacking telomerase activity and hTERT, TSA has been reported to induce hTERT expression and activate telomerase [173,

174]. Takakura et al. reported that treatment with TSA induced significant activation of hTERT mRNA expression and telomerase activity in normal cells, but not in cancer cells [38]. Based on their observation, they suggested that TSA activity is specific to normal cells. In cancer cells, Wang et al. reported that TSA induced hTERT transcription and also a general increase in chromatin sensitivity to DNase treatment in telomerase-negative cells (175].

In 2005, we analyzed the epigenetic effects of FK228 on *hTERT* transcription in oral cancer cell lines [176]. Our results showed a clear increase in hTERT expression following FK228 treatment of oral cancer cell lines (Figure 11a).

The hTERT-downregulated cell line Hep2 was incubated with FK228 at a concentration of 0.5 or 1.0 μM. The mRNA expression of hTERT was upregulated after exposure to FK228 in hTERT-negative Hep2 cells, and also in SAS and KB cells, which are known to show high-level expression of hTERT.

Moreover, to determine whether the FK228-mediated induction of hTERT expression was a direct effect or indirect effect, the hTERT-downregulated cell line Hep2 and the hTERT-expressing cell line SAS were incubated with FK228 in the presence of 200 μg/ml of cycloheximide (CHX), a potent protein synthesis inhibitor, and were then analyzed for hTERT mRNA (Figure 11b). Following FK228 treatment, all three cell lines showed inductions of hTERT expression; however, co-treatment with FK228 and CHX resulted in an enhancement of the hTERT expression in SAS and KB cells.

This suggests that the induction of hTERT by FK228 is likely to have a direct rather than an indirect effect on epigenetic changes such as histone acetylation/deacetylation. Furthermore, Wang et al. also reported that TSA-induced hTERT transcription and chromatin alterations were not blocked by CHX, suggesting that this induction does not require *de novo* protein synthesis and that TSA induces hTERT expression through the inhibition of histone deacetylation at the hTERT promoter [175]. Hou et al. also reported that TSA and CHX induced hTERT mRNA [35], suggesting direct induction of the hTERT gene by an HDAC inhibitor.

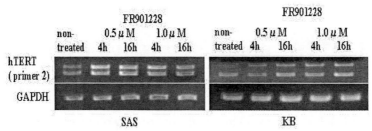

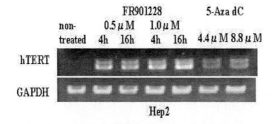

a

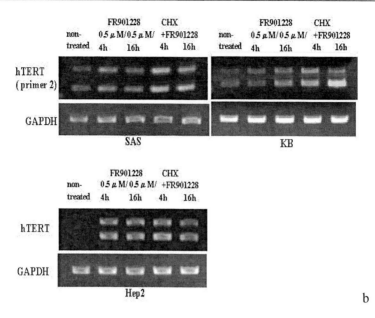

Figure 11. Epigenetic-targeted agent-mediated transactivation of the hTERT gene. (a) Comparison of hTERT expressions before and after FK228 (FR901228) or 5-Aza-2'-deoxycytidine(5-Aza) treatment. FR901228 was a gift from Fujisawa Pharmaceutical Company (Osaka, Japan). FR901288, 5-Aza-2'-deoxycytidine (5-Aza) (Sigma Chemical Co., St Louis, MO) and cycloheximide (CHX)(Wako Pure Chemical Industries, Ltd.) diluted in water were added to MDEM to the final concentration indicated in each treatment. Cells of the hTERT-downregulated cell line Hep2 were incubated with FK228 (FR901228) at a concentration of 0.5 or 1.0 µM. SAS and KB, the cell lines expressing high levels of hTERT, were also analyzed as a control. Substantial amounts of hTERT mRNA were detected by RT-PCR. The gels clearly show that the hTERT expression was induced after 4 -16 h of treatment with FR901228 in all three cell lines. In addition, hTERT-downregulated Hep2 cells were incubated for 7 days with several concentrations (ranging from 4.4 to 8.8 µM) of 5-Aza, an inhibitor that prevents methylation of newly synthesized DNA. The Hep2 cells regained their ability to produce high levels of hTERT mRNA. (b) Effects on hTERT expression of FR901228 treatment combined with cycloheximide (CHX). To determine whether or not FR901228-mediated induction of hTERT expression had a direct effect, Hep2, KB and SAS cells were incubated with FR901228 in the presence/absence of 200 µg/ml of CHX, a potent protein synthesis inhibitor.. CHX was added to cells 30 min before the addition of 0.5 µM FR901228. Cells were harvested at the indicated times (4 or 16 hrs) following FR901228 treatment and then analyzed for hTERT mRNA by RT-PCR. SAS, KB and Hep2 cells showed slight inductions of hTERT expression; however, co-treatment with CHX resulted in an enhancement of the hTERT expression in comparison with FR901228 alone.

The results of several reports suggest that the HDAC inhibitors might exhibit different effects on the hTERT mRNA expression and telomerase activity according to the type of cancer cells. Suenaga et al. reported that TSA and NaB inhibit the expression of hTERT mRNA and telomerase activity in prostate cancer cells *in vitro* [177]. Moreover, CHX blocked the suppression of hTERT mRNA by FK228 in small cell lung carcinoma cells *in vitro*, and the inhibition of hTERT mRNA by FK228 required newly synthesized proteins [178]. In contrast, Zhu et al. reported that TSA or NaB had no effect on the expression of hTERC and hTERT mRNA and on telomerase activity in a human ovarian cancer cell line, SK-OV-3, *in vitro* [179].

Recent Findings on Histone Methylation and hTERT

Interestingly, Atkinson et al. observed that highly trimethylated H3-K4 was associated with the actively transcribed hTERT gene in telomerase-proficient tumor cells [180]. Liu et al. also proposed a potential role for H3-K4 methylation in controlling hTERT transcription [181]. From recent research, SMYD3, a histone methyltransferase (as mentioned before), induced H3-K4 methylation in the hTERT promoter, which recruits histone acetyltransferases and promotes Sp1 and c-myc access to the hTERT promoter [181]. For the recognition of any target site by the c-myc oncoprotein [182], high H3-K4 methylation within appropriate regions of target promoters was one of the strict prerequisites. These sequential events finally lead to transcriptional activation of the hTERT gene.

Furthermore, lack of histone H3-K4 methylation at the hTERT promoter may be required to lock the chromatin in a closed state in telomerase-deficient cells. Zhu et al. reported the effect of a histone demethylase, Lysine-specific demethylase 1 (LSD1), on hTERT transcription. Based on their report, LSD1 represses hTERT transcription via demethylation of H3-K4 in normal and cancer cells, and together with HDACs, participates in the establishment of a stable repression state of the hTERT gene in normal or cancer cells [183]. In normal human fibroblasts with a tight hTERT repression, a pharmacological inhibition of LSD1 led to a weak hTERT expression, and a robust induction of hTERT mRNA was observed when LSD1 and HDACs were both inhibited. In cancer cells, inhibition of LSD1 activity or knock-down of its expression led to significant increases in levels of hTERT mRNA and telomerase activity. Given the fact that the methylated H3-K4 is one of the prerequisites for the c-myc binding [181, 182] and that the hTERT gene is a direct target of c-Myc [28, 184], they considered that H3-K4 demethylation by LSD1 might prevent the c-Myc binding to the hTERT promoter, thereby resulting in transcriptional repression of the hTERT gene.

In addition, some human cancers and immortal cell lines maintain their telomeres via a telomerase-independent mechanism, referred to as alternative lengthening of telomeres (ALT) [185]. Atkinson et al. proposed that specific chromatin modifications of the hTR and hTERT promoters could be used as novel markers for the ALT phenotype [180]. In ALT cell lines, hTR and hTERT expressions are absent due to histone H3 and H4 hypoacetylation and methylation of Lys9 histone H3. Moreover, methylation of Lys20 histone H4 was specific to the hTR and hTERT promoters of ALT cells. Further studies are needed to define the roles of histone methylation in the maintenance of telomeres.

Histone Modifications on c-Myc (A Regulatory Element of hTERT)

Several transcription factors regulating hTERT (i.e., c-Myc, E6, and the estrogen receptor) have been identified [21-25]. Among them, the oncoprotein c-Myc was the first factor shown to be capable of inducing hTERT expression in normal human cells. c-Myc and Sp1 are thought to be major activators of hTERT transcription [26, 27]. The hTERT promoter contains two E-boxes and five GC-boxes, the consensus binding motifs for the myc network and Sp family [26, 28]. On the other hand, overexpression another hTERT regulating factor, p21, was reported to downregulate the level of hTERT mRNA and telomerase activity [186, 187].

One of the oncogenes, c-Myc, is a transcription factor that regulates cell proliferation, cell cycle progression, apoptosis induction, or metabolism [188-196]. Previous studies indicate that c-myc negatively regulates p21 expression [197], thereby overriding a p21-mediated cell cycle checkpoint. The overexpression of c-Myc is known to block terminal differentiation in some cell lines [198-200]. The misregulation of genes by c-Myc leads to cell transformation or tumorigenesis.

HDAC inhibitors, such as TSA and NaB, have been reported to decrease c-Myc mRNA expression and to increase p21 mRNA [104, 105, 201-203]. Although the precise mechanism by which HDAC inhibitors abrogate c-Myc expression remains to be elucidated, it is reasonable to speculate that histone hyper and hypo-acetylation are common underlying features of c-Myc transactivation and repression, respectively. In addition, given the fact that c-Myc and mad1 recruit coactivator hGCN5 with inherent HAT activity and HDAC-containing corepressors to the target promoters, respectively [204, 205], the myc network proteins might play significant roles in the HDAC-mediated regulation of hTERT.

Previously, we examined whether or not FK228 (FR901228) has an effect on the hTERT transcription activator, c-Myc. Figure 12 (panels a and b) shows that FK228 reduced the endogenous expression of c-Myc in SAS, KB and Hep2 cells. We further examined the level of c-Myc mRNA in those cells incubated with FK228 in the presence of CHX (Figure 12c). Compared with the results when cells were exposed to FK228 alone, SAS, KB and Hep2 cells showed a smaller decrease in the levels of c-Myc mRNA when incubated with both FK228 and CHX. Interestingly, in Hep2 cells, CHX treatment even seemed to upregulate the baseline expression of c-Myc, an effect that appeared to depend on the absence of new protein synthesis in the presence of CHX. The rapid reduction of c-Myc mRNA in response to FK228 treatment clearly reflects an indirect role of the HDAC inhibitor in the reduction of this gene, but both direct and indirect effects are at play. In addition, the combined treatment with FR901228 and CHX resulted in the enhancement of hTERT expression independent of c-Myc expression [Figure 11b and 12). It is reasonable to speculate that the decreased expression of c-Myc protein is not required for activation of the hTERT expression by FK228.

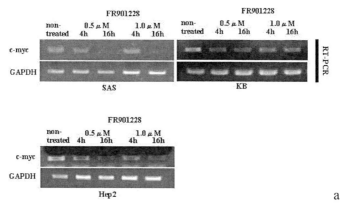

Figure 12. (Continues)

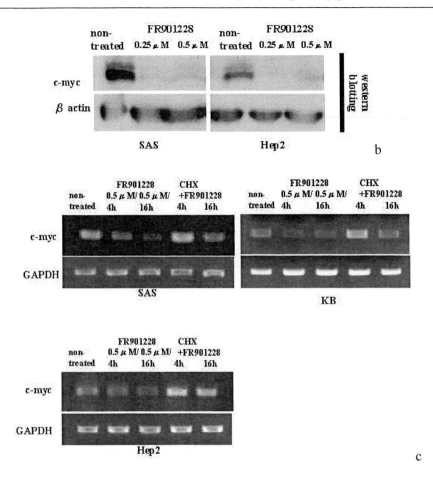

Figure 12. Comparison of c-Myc expression before and after FR901228 treatment. (a) Comparison by RT-PCR of c-Myc mRNA before and after FR901228 treatment. Amplification of cDNAs was performed under the following PCR conditions: 7 min at 94☐ for 1 cycle; then 26 cycles of 94☐ for 30 s, 59☐ for 30 s, and 72☐ for 30 s; and a final elongation step at 72☐ for 10 min. The primers used for the amplification were as follows: c-Myc: sense: 5'-aagtcctgcgcctcgcaa-3', antisense: 5'-gctgtggcctccagcaga-3'. GAPDH:sense: 5'-gaaggtgaaggtcggagtc-3', antisense: 5'-caaagttgtcatggatgacc-3'. The amplified GAPDH fragment was used as a positive control. The RT-PCR products were separated by electrophoresis on a 2% agarose gel, stained with ethidium bromide, and viewed by UV. (b) Comparison by Western blotting of c-Myc protein before and after FR901228 treatment. To examine whether or not FR901228 has an effect on the hTERT transcription activator c-Myc, we analyzed c-Myc expression before and after FR901228 treatment in SAS, KB and Hep2 cells. The protein-blotted membranes were probed for 2 h with mouse anti-c-Myc monoclonal antibody c-Myc Ab-5 (Clone 67P05) (Neomarkers, Fremont, CA) diluted 1:1000. Bound antibody was detected using an ECL + plus kit (Amersham Pharmacia Biotech Inc., Little Chalfont, UK) according to the manufacturer's instructions. To quantify the amount of protein that was loaded per lane, we performed Western blotting for beta actin as a control. FR901228 reduced the endogenous expressions of c-myc in both cell lines. (c) Effects on c-Myc expression of FR901228 treatment combined with CHX. SAS, KB and Hep2 cells were incubated with FR901228 in the presence of CHX and analyzed for c-Myc mRNA by RT-PCR. Compared with the results when cells were exposed to FR901228 alone, SAS, KB and Hep2 cells showed a smaller decrease in the levels of c-Myc mRNA when incubated with both FR901228 and CHX.

This section focused on the critical role played by the endogenous chromatin environment in the regulation of hTERT expression. Taken together, these findings on the telomerase regulatory mechanism may have important clinical implications for the use of HDAC inhibitors in cancer chemotherapy.

References

[1] Hübscher U, Nasheuer HP, Syväoja J. Eukaryotic DNA polymerases, a growing family. (2000) *Trends Biochem. Sci. 25:* 143-147.
[2] Blackburn EH. Telomere states and cell fates. (2000) *Nature. 408:* 53-55.
[3] Greider CW. Telomere length regulation. (1996) *Annu. Rev. Biochem. 65:* 337-365.
[4] Greider CW. Chromosome first aid. (1991) *Cell. 67:* 645-647.
[5] Watson JD. Origin of concatemeric T7 DNA. (1972) *Nature. 239:* 197-201.
[6] Harley CB, Futcher AB, Greider CW. Telomeres shorten during ageing of human fibroblasts. (1990) *Nature. 345:* 458-460.
[7] Kim NW, Piatyszek MA, Prowse KR, Harley CB, West MD, Ho PL, Coviello GM, Wright WE, Weinrich SL, Shay JW. Specific association of human telomerase activity with immortal cells and cancer. (1994) *Science. 266:* 2011-2015.
[8] Chiu CP, Harley CB. Replicative senescence and cell immortality: the role of telomeres and telomerase. (1997) *Proc. Soc. Exp. Biol. Med. 214:* 99-106.
[9] Klapper W, Parwaresch R, Krupp G. Telomere biology in human aging and aging syndromes. (2001) *Mech. Ageing Dev. 122:* 695-712.
[10] Levy MZ, Allsopp RC, Futcher AB, Greider CW, Harley CB. Telomere end-replication problem and cell aging. (1992) *J. Mol. Biol. 225:* 951-960.
[11] Autexier C, Greider CW. Telomerase and cancer: revisiting the telomere hypothesis. (1996) *Trends Biochem. Sci. 21:* 387-391.
[12] Bodnar AG, Ouellette M, Frolkis M, Holt SE, Chiu CP, Morin GB, Harley CB, Shay JW, Lichtsteiner S, Wright WE. Extension of life-span by introduction of telomerase into normal human cells. (1998) *Science (Washington DC). 279:* 349-352.
[13] Zhang X, Mar V, Zhou W, Harrington L, Robinson MO. Telomere shortening and apoptosis in telomerase-inhibited human tumor cells. (1999) *Genes. Dev. 13:* 2388-2399.
[14] Feng J, Funk WD, Wang SS, Weinrich SL, Avilion AA, Chiu CP, Adams RR, Chang E, Allsopp RC, Yu J, Le S, West MD, Harley CB, Andrews WH, Greider CW, Villeponteau B. The RNA component of human telomerase. (1995) *Science (Washington DC). 269:* 1236-1241.
[15] Harrington L, McPhail T, Mar V, Zhou W, Oulton R, Bass MB, Arruda I, Robinson MO. A mammalian telomerase-associated protein. (1997) *Science (Washington DC). 275:* 973-977.
[16] Nakayama J, Saito M, Nakamura H, Matsuura A, Ishikawa F. TLP1: a gene encoding a protein component of mammalian telomerase is a novel member of WD repeats family. (1997) *Cell. 88:* 875-884.

[17] Meyerson M, Counter CM, Eaton EN, Ellisen LW, Steiner P, Caddle SD, Ziaugra L, Beijersbergen RL, Davidoff MJ, Liu Q, Bacchetti S, Haber DA, Weinberg RA. hEST2, the putative human telomerase catalytic subunit gene, is up-regulated in tumor cells and during immortalization. (1997) *Cell. 90:* 785-795.

[18] Nakamura TM, Morin GB, Chapman KB, Weinrich SL, Andrews WH, Lingner J, Harley CB, Cech TR. Telomerase catalytic subunit homologs from fission yeast and human. (1997) *Science (Washington DC). 277:* 955-959.

[19] Takakura M, Kyo S, Kanaya T, Tanaka M, Inoue M. Expression of human telomerase subunits and correlation with telomerase activity in cervical cancer. (1998) *Cancer Res. 58:* 1558-1561.

[20] Ducrest AL, Szutorisz H, Lingner J, Nabholz M. Regulation of the human telomerase reverse transcriptase gene. (2002) *Oncogene. 21:* 541-552.

[21] Kyo S, Takakura M, Kanaya T, Zhuo W, Fujimoto K, Nishio Y, Orimo A, Inoue M. Estrogen activates telomerase. (1999) *Cancer Res. 59:* 5917-5921.

[22] Misiti S, Nanni S, Fontemaggi G, Cong YS, Wen J, Hirte HW, Piaggio G, Sacchi A, Pontecorvi A, Bacchetti S, Farsetti A. Induction of hTERT expression and telomerase activity by estrogens in human ovary epithelium cells. (2000) *Mol. Cell Biol. 20:* 3764-3771.

[23] Klingelhutz AJ, Foster SA, McDougall JK. Telomerase activation by the E6 gene product of human papillomavirus type 16. (1996) *Nature. 380:* 79-82.

[24] Greenberg RA, O'Hagan RC, Deng H, Xiao Q, Hann SR, Adams RR, Lichtsteiner S, Chin L, Morin GB, DePinho RA. Telomerase reverse transcriptase gene is a direct target of c-Myc but is not functionally equivalent in cellular transformation. (1999) *Oncogene. 18:* 1219-1226.

[25] Liu X, Yuan H, Fu B, Disbrow GL, Apolinario T, Tomaic V, Kelley ML, Baker CC, Huibregtse J, Schlegel R.The E6AP ubiquitin ligase is required for transactivation of the hTERT promoter by the human papillomavirus E6 oncoprotein. (2005) *J. Biol. Chem. 280(11):*10807-10816.

[26] Kyo S, Takakura M, Kanaya T, Taira T, Kanaya T, Itoh H, Yutsudo M, ArigaH, Inoue M. Sp1 cooperates with c-Myc to activate transcription of human telomerase reverse transcriptase (hTERT) gene. (2000*) Nucleic Acids Res. 28:* 669-677.

[27] Wu KJ, Grandori C, Amacker M, Simon-Vermot N, Polack A, Lingner J, Dalla-Favera R. Direct activation of TERT transcription by c-MYC. (1999) *Nature Genet. 21:* 220-224.

[28] Takakura M, Kyo S, Kanaya T, Hirano H, Takeda J, Yutsudo M, Inoue M. Cloning of human telomerase catalytic subunit (hTERT) gene promoter and identification of proximal core promoter sequences essential for tanscriptional activation in immortalized and cancer cells. (1999) *Cancer Res. 59:* 551-557.

[29] Wang J, Xie LY, Allan S, Beach D, Hannon GJ. Myc activates telomerase. (1998) *Genes Dev. 12:* 1769-1774.

[30] Oh S, Song YH, Yim J, Kim TK. Identification of Mad as a repressor of the human telomerase (hTERT) gene. (2000) *Oncogene. 19:* 1485-1490.

[31] Günes Ç, Lichtsteiner S, Vasserot AP, Englert C. Expression of the hTERT gene is regulated at the level of transcriptional initiation and repressed by Mad1. (2000) *Cancer Res. 60:* 2116-2121.

[32] Devereux TR, Horikawa I, Anna CH, Annab LA, Afshari CA, Barrett JC. DNA methylation analysis of the promoter region of the human telomerase reverse transcriptase (hTERT) gene. (1999) *Cancer Res. 59:* 6087-6090.

[33] Dessain SK, Yu H, Reddel RR, Beijersbergen RL, Weinberg RA. Methylation of the human telomerase gene CpG island. (2000) *Cancer Res. 60:* 537-541.

[34] Cong YS, Bacchetti S. Histone deacetylation is involved in the transcriptional repression of hTERT in normal human cells. (2000) *J. Biol. Chem. 275:* 35665-35668.

[35] Hou M, Wang X, Popov N, Zhang A, Zhao X, Zhou R, Zetterberg A, Björkholm M, Henriksson M, Gruber A, Xu D. The histone deacetylase inhibitor trichostatin A derepresses the telomerase reverse transcriptase (hTERT) gene in human cells. (2002) *Exp. Cell Res. 274:* 25-34.

[36] Lai SR, Phipps SM, Liu L, Andrews LG, Tollefsbol TO. Epigenetic control of telomerase and modes of telomere maintenance in aging and abnormal systems. (2005) *Front Biosci. 10:* 1779-1796.

[37] Liu L, Saldanha SN, Pate MS, Andrews LG, Tollefsbol TO. Epigenetic regulation of human telomerase reverse transcriptase promoter activity during cellular differentiation. (2004) *Genes Chromosomes Cancer. 41:* 26-37.

[38] Takakura M, Kyo S, Sowa Y, Wang Z, Yatabe N, Maida Y, Tanaka M, Inoue M. Telomerase activation by histone deacetylase inhibitor in normal cells. (2001) *Nucleic Acids Res. 29:* 3006-3011.

[39] Xu D, Popov N, Hou M, Wang Q, Bjorkholm M, Gruber A, Menkel AR, Henriksson M. Switch from Myc/Max to Mad1/Max binding and decrease in histone acetylation at the telomerase reverse transcriptase promoter during differentiation of HL60 cells. (2001) *Proc. Natl. Acad. Sci. USA. 98:* 3826-3831.

[40] Kornberg R, Lorch Y. Twenty-five years of the nucleosome, fundamental particle of the eukaryote chromosome. (1999) *Cell. 98:* 285-294.

[41] van Holde KE. Histone modifications. In: Rich A. (ed), Chromatin, Springer Series in Molecular Biology. (1998) New York: Springer Press. pp.11-148.

[42] Wolffe AP. Transcriptional regulation in the context of chromatin structure. (2001) *Essays Biochem. 37:* 45-57.

[43] de Ruijter AJ, van Gennip AH, Caron HN, Kemp S, van Kuilenburg AB. Histone deacetylases (HDACs): Characterization of the classical HDAC family. (2003) *Biochem. J. 370:* 737-749.

[44] Marmorstein R, Roth SY. Histone acetyltransferases: function, structure, and catalysis. (2001) *Curr. Opin. Genet. Dev. 11:* 155-161.

[45] Ng HH, Bird A. Histone deacetylases: silencers for hire. (2000) *Trends Biochem. Sci. 25:* 121-126.

[46] Kouzarides T. Histone acetylases and deacetylases in cell proliferation. (1999*) Curr. Opin. Genet. Dev. 9:* 40-48.

[47] Jenuwein T, Allis CD. Translating the histone code. (2001) *Science. 293:* 1074-1080.

[48] Johnstone RW, Licht JD. Histone deacetylase inhibitors in cancer therapy: Is transcription the primary target?. (2003) *Cancer Cell. 4:* 13-18.

[49] Agalioti T, Chen G, Thanos D. Deciphering the transcriptional histone acetylation code for a human gene. (2002) *Cell. 111:* 381-392.

[50] Khochbin S, Verdel A, Lemercier C, Seigneurin-Berny D. Functional significance of histone deacetylase diversity. (2001) *Curr. Opin. Genet. Dev. 11:* 162-166.

[51] Nakayama J-I, Rice JC, Strahl BD, Allis CD, Grewal SIS. Role of histone H3 lysine 9 methylation in epigenetic control of heterochromatin assembly. (2001) *Science. 292:* 110-113.

[52] Lund AH, van Lohuizen M. Epigenetics and cancer. (2004) *Genes Dev. 18:* 2315-2335.

[53] Rea S, Eisenhaber F, O'Carroll D, Strahl BD, Sun ZW, Schmid M, Opravil S, Mechtler K, Ponting CP, Allis CD, Jenuwein T. Regulation of chromatin structure by site-specific histone H3 methyltransferases. (2000) *Nature. 406:* 593-599.

[54] Ma H, Baumann CT, Li H, Strahl BD, Rice R, Jelinek MA, Aswad DW, Allis CD, Hager GL, Stallcup MR. Hormone-dependent, CARM1-directed, arginine-specific methylation of histone H3 on a steroid-regulated promoter. (2001) *Curr. Biol. 11:* 1981-1985.

[55] Strahl BD, Briggs SD, Brame CJ, Caldwell JA, Koh SS, Ma H, Cook RG, Shabanowitz J, Hunt DF, Stallcup MR, Allis CD. Methylation of histone H4 at arginine 3 occurs in vivo and is mediated by the nuclear receptor coactivator PRMT1. (2001) *Curr. Biol. 11:* 996-1000.

[56] Wang H, Huang ZQ, Xia L, Feng Q, Erdjument-Bromage H, Strahl BD, Briggs SD, Allis CD, Wong J, Tempst P, Zhang Y. Methylation of histone H4 at arginine 3 facilitating transcriptional activation by nuclear hormone receptor. (2001) *Science. 293:* 853-857.

[57] Ruthenburg AJ, Allis CD, Wysocka J. Methylation of lysine 4 on histone H3: intricacy of writing and reading a single epigenetic mark. (2007) *Mol. Cell. 25:* 15-30.

[58] Shi Y, Whetstine JR. Dynamic regulation of histone lysine methylation by demethylases. (2007) *Mol. Cell. 25:* 1-14.

[59] Berger SL. The complex language of chromatin regulation during transcription. (2007) *Nature. 447:* 407-412.

[60] Jenuwein T, Laible G, Dorn R, Reuter G. SET domain proteins modulate chromatin domains in eu- and heterochromatin. (1998) *Cell Mol. Life Sci. 54:* 80-93.

[61] Kouzarides T. Histone methylation in transcriptional control. (2002) *Curr. Opin. Genet. Dev. 12:* 198-209.

[62] Tschiersch B, Hofmann A, Krauss V, Dorn R, Korge G, Reuter G. The protein encoded by the Drosophila position-effect variegation suppressor gene Su(var)3–9 combines domains of antagonistic regulators of homeotic gene complexes. (1994) *EMBO J. 13:* 3822-3831.

[63] Tachibana M, Sugimoto K, Fukushima T, Shinkai Y. Set domain-containing protein, G9a, is a novel lysine-preferring mammalian histone methyltransferase with hyperactivity and specific selectivity to lysines 9 and 27 of histone H3. (2001) *J. Biol. Chem. 276:* 25309-25317.

[64] Yang L, Xia L, Wu DY, Wang H, Chansky HA, Schubach WH, Hickstein DD, Zhang Y. Molecular cloning of ESET, a novel histone H3-specific methyltransferase that interacts with ERG transcription factor. (2002) *Oncogene. 21:* 148-152.

[65] Schultz DC, Ayyananthan K, Negorev D, Maul GG, Rauscher FJ. SETDB1: a novel KAP-1-associated histone H3, lysine 9-specific methyltransferase that contributes to HP1-mediated silencing of euchromatic genes by krab zinc-finger proteins. (2002) *Genes Dev. 16:* 919-932.

[66] Schotta G, Ebert A, Krauss V, Fischer A, Hoffmann J, Rea S, Jenuwein T, Dorn R, Reuter G. Central role of Drosophila Su(var)3–9 in histone H3-K9 methylation and heterochromatin-dependent gene silencing. (2002) *EMBO J. 21:* 1121-1131.

[67] Gendrel AV, Lippman Z, Yordan C, Colot V, Martienssens R. Dependence of heterochromatic histone H3 methylation patterns on the Arabidopsis gene DDM1. (2002) *Science. 20:* 20.

[68] Tamaru H, Selker EU. A histone H3 methyltransferase controls DNA methylation in Neurospora crassa. (2001) *Nature. 414:* 277-283.

[69] Litt MD, Simpson M, Gaszner M, Allis CD, Felsenfeld G. Correlation between histone lysine methylation and developmental changes at the chicken -globin locus. (2001) *Science. 293:* 2453-2455.

[70] Zegerman P, Canas B, Pappin D, Kouzarides T. Histone H3 lysine 4 methylation disrupts binding of nucleosome remodeling and deacetylase (NuRD) repressor complex. (2002) *J. Biol. Chem. 277:* 11621-11624.

[71] Bernstein BE, Humphrey EL, Erlich RL, Schneider R, Bouman P, Liu JS, Kouzarides T, Schreiber SL. Methylation of histone H3 Lys 4 in coding regions of active genes. (2002) *Proc. Natl. Acad. Sci. U S A. 99:* 8695-8700.

[72] Nishioka K, Chuikov S, Sarma K, Erdjument-Bromage H, Allis D, Tempst P, Reinberg D. Set9, a novel histone H3 methyltransferase that facilitates transcription by precluding histone tail modifications required for heterochromatin formation. (2002) *Genes. Dev. 16:* 479-489.

[73] Wang H, Cao R, Xia L, Erdjument-Bromage H, Borchers C, Tempst P, Zhang Y. Purification and functional characterization of a histone H3-lysine 4-specific methyltransferase. (2001) *Mol. Cell. 8:* 1207-1217.

[74] Hamamoto R, Furukawa Y, Morita M, Iimura Y, Silva FP, Li M, Yagyu R, Nakamura Y. SMYD3 encodes a histone methyltransferase involved in the proliferation of cancer cells. (2004) *Nat. Cell Biol. 6:* 731-740.

[75] Hamamoto R, Silva FP, Tsuge M, Nishidate T, Katagiri T, Nakamura Y, Furukawa Y. Enhanced SMYD3 expression is essential for the growth of breast cancer cells. (2006) *Cancer Sci. 97:* 113-118.

[76] Tsuge M, Hamamoto R, Silva FP, Ohnishi Y, Chayama K, Kamatani N, Furukawa Y, Nakamura Y. A variable number of tandem repeats polymorphism in an E2F-1 binding element in the 5' flanking region of SMYD3 is a risk factor for human cancers. (2005) *Nat. Genet. 37:* 1104-1107.

[77] Shi Y, Lan F, Matson C, Mulligan P, Whetstine JR, Cole PA, Casero RA, Shi Y. Histone demethylation mediated by the nuclear amine oxidase homolog LSD1. (2004) *Cell. 119:* 941-953.

[78] Metzger E, Wissmann M, Yin N, Müller JM, Schneider R, Peters AH, Günther T, Buettner R, Schüle R. LSD1 demethylates repressive histone marks to promote androgen-receptor-dependent transcription. (2005) *Nature. 437:* 436-439.

[79] Lee MG, Wynder C, Cooch N, Shiekhattar R. An essential role for CoREST in nucleosomal histone 3 lysine 4 demethylation. (2005) *Nature. 437:* 432-435.

[80] Wissmann M, Yin N, Müller JM, Greschik H, Fodor BD, Jenuwein T, Vogler C, Schneider R, Günther T, Buettner R, Metzger E, Schüle R. Cooperative demethylation by JMJD2C and LSD1 promotes androgen receptor-dependent gene expression. (2007) *Nat Cell Biol. 9:* 347-353.

[81] Lee MG, Wynder C, Schmidt DM, McCafferty DG, Shiekhattar R. Histone H3 lysine 4 demethylation is a target of nonselective antidepressive medications. (2006) *Chem. Biol. 13:* 563-567.

[82] Ueda H, Nakajima H, Hori Y, Fujita T, Nishimura M, Goto T, Okuhara M. FR901228, a novel antitumor bicyclic depsipeptide produced by Chromobacterium violaceum No. 968: I. Taxonomy, fermentation, isolation, physico-chemical and biological properties, and antitumor activity. (1994) *J. Antibiot. 47:* 301-310.

[83] Nakajima H, Kim YB, Terano H, Yoshida M, Horinouchi S. FR901228, a potent antitumor antibiotic, is a novel histone deacetylase inhibitor. (1998) *Exp. Cell Res. 241:* 126-133.

[84] Johnstone RW. Histone-deacetylase inhibitors: novel drugs for the treatment of cancer. (2002) *Nat. Rev. Drug Discov. 1:* 287-299.

[85] Marks PA, Rifkind RA, Richon VM, Breslow R. Inhibitors of histone deacetylase are potentially effective anticancer agents. (2001) *Clin. Cancer Res. 7:* 759-760.

[86] Munster PN, Troso-Sandoval T, Rosen N, Rifkind R, Marks PA, Richon VM. The histone deacetylase inhibitor suberoylanilide hydroxamic acid induces differentiation of human breast cancer cells. (2001) *Cancer Res. 61:* 8492-8497.

[87] Lee BI, Park SH, Kim JW, Sausville EA, Kim HT, Nakanishi O, Trepel JB, Kim SJ. MS-275, a histone deacetylase inhibitor, selectively induces transforming growth factor ß type II receptor expression in human breast cancer cells. (2001) *Cancer Res. 61:* 931-934.

[88] Kelly WK, O'Connor OA, Krug LM, Chiao JH, Heaney M, Curley T, MacGregore-Cortelli B, Tong W, Secrist JP, Schwartz L, Richardson S, Chu E, Olgac S, Marks PA, Scher H, Richon VM. Phase I study of an oral histone deacetylase inhibitor, suberoylanilide hydroxamic acid, in patients with advanced cancer. (2005) *J. Clin. Oncol. 23:* 3923-3931.

[89] Ryan QC, Headlee D, Acharya M, Sparreboom A, Trepel JB, Ye J, Figg WD, Hwang K, Chung EJ, Murgo A, Melillo G, Elsayed Y, Monga M, Kalnitskiy M, Zwiebel J, Sausville EA. Phase I and pharmacokinetic study of MS-275, a histone deacetylase inhibitor, in patients with advanced and refractory solid tumors or lymphoma. (2005) *J. Clin. Oncol. 23(17):* 3912-3922.

[90] Carducci MA, Gilbert J, Bowling MK, Noe D, Eisenberger MA, Sinibaldi V, Zabelina Y, Chen TL, Grochow LB, Donehower RC. A phase I clinical and pharmacological evaluation of sodium phenylbutyrate on an 120-h infusion schedule. (2001) *Clin. Cancer Res. 7:* 3047-3055.

[91] Sandor V, Bakke S, Robey RW, Kang MH, Blagosklonny MV, Bender J, Brooks R, Piekarz RL, Tucker E, Figg WD, Chan KK, Goldspiel B, Fojo AT, Balcerzak SP, Bates SE. Phase I trial of the histone deacetylase inhibitor, depsipeptide (FR901228, NSC 630176), in patients with refractory neoplasms. (2002) *Clin. Cancer Res. 8:* 718-728.

[92] Byrd JC, Marcucci G, Parthun MR, Xiao JJ, Klisovic RB, Moran M, Lin TS, Liu S, Sklenar AR, Davis ME, Lucas DM, Fischer B, Shank R, Tejaswi SL, Binkley P, Wright J, Chan KK, Grever MR. A phase 1 and pharmacodynamic study of depsipeptide (FK228) in chronic lymphocytic leukemia and acute myeloid leukemia. (2005) *Blood. 105:* 959-967.

[93] Piekarz RL, Robey R, Sandor V, Bakke S, Wilson WH, Dahmoush L, Kingma DM, Turner ML, Altemus R, Bates SE. Inhibitor of histone deacetylation, depsipeptide (FR901228), in the treatment of peripheral and cutaneous T-cell lymphoma: a case report. (2001) *Blood. 98:* 2865-2868.

[94] Marks P, Rifkind RA, Richon VM, Breslow R, Miller T, Kelly WK. Histone deacetylases and cancer: causes and therapies. (2001) *Nat. Rev. Cancer. 1:* 194-202.

[95] Sandor V, Senderowicz A, Mertins S, Sackett D, Sausville E, Blagosklonny MV, Bates SE. P21-dependent G1arrest with downregulation of cyclin D1 and upregulation of cyclin E by the histone deacetylase inhibitor FR901228. (2000) *Br. J. Cancer. 83:* 817-825.

[96] Ruefli AA, Ausserlechner MJ, Bernhard D, Sutton VR, Tainton KM, Kofler R, Smyth MJ, Johnstone RW. The histone deacetylase inhibitor and chemotherapeutic agent suberoylanilide hydroxamic acid (SAHA) induces a cell-death pathway characterized by cleavage of Bid and production of reactive oxygen species. (2001) *Proc. Natl. Acad. Sci. USA. 98:* 10833-10838.

[97] Yu X, Guo ZS, Marcu MG, Neckers L, Nguyen DM, Chen GA, Schrump DS. Modulation of p53, ErbB1, ErbB2, and Raf-1 expression in lung cancer cells by depsipeptide FR901228. (2002) *J. Natl. Cancer Inst. 94:* 504-513.

[98] Peart MJ, Tainton KM, Ruefli AA, Dear AE, Sedelies KA, O'Reilly LA, Waterhouse NJ, Trapani JA, Johnstone RW. Novel mechanisms of apoptosis induced by histone deacetylase inhibitors. (2003) *Cancer Res. 63:* 4460-4471.

[99] Rosato RR, Almenara JA, Grant S. The histone deacetylase inhibitor MS-275 promotes differentiation or apoptosis in human leukemia cells through a process regulated by generation of reactive oxygen species and induction of p21CIP1/WAF1. (2003) *Cancer Res. 63:* 3637-3645.

[100] Marks P, Rifkind RA, Richon VM, Breslow R, Miller T, Kelly WK. Histone deacetylases and cancer: causes and therapies. (2001) *Nat Rev Cancer. 1:* 194-202.

[101] Archer SY, Meng S F, Shei A, Hodin RA. p21WAF1 is required for butyrate-mediated growth inhibition of human colon cancer cells. (1998) *Proc. Natl. Acad. Sci. U S A. 95:* 6791-6796.

[102] Nakano K, Mizuno T, Sowa Y, Orita T, Yoshino T, Okuyama Y, Fujita T, Ohtani-Fujita N, Matsukawa Y, Tokino T, Yamagishi H, Oka T, Nomura H, Sakai T. Butyrate activates the WAF1/Cip1 gene promoter through Sp1 sites in a p53-negative human colon cancer cell line. (1997) *J. Biol. Chem. 272:* 22199-22206.

[103] Janson W, Brandner G, Siegel J. Butyrate modulates DNA-damage-induced p53 response by induction of p53-independent differentiation and apoptosis. (1997) *Oncogene. 15:* 1395-1406.

[104] Heruth DP, Zirnstein GW, Bradley JF, Rothberg PG. Sodium butyrate causes an increase in the block to transcriptional elongation in the c-myc gene in SW837 rectal carcinoma cells. (1993) *J. Biol. Chem. 268:* 20466-20472.

[105] Buckley AR, Leff MA, Buckley DJ, Magnuson NS, De Jong G, Gout PW. Alterations in pim-1 and c-myc expression associated with butyrate-induced growth factor dependence in autonomous rat NB2 lymphoma cells. (1996) *Cell Growth and Differ. 7:* 1713-1721.

[106] Mandal M, Kumar R. Bcl-2 expression regulates sodium butyrate-induced apoptosis in human MCF-7 breast cancer cells. (1996) *Cell Growth and Differ. 7:* 311-318.

[107] Hague A, Díaz GD, Hicks DJ, Krajewski S, Reed JC, Paraskeva C. Bcl-2 and bak may play a pivotal role in sodium butyrate-induced apoptosis in colonic epithelial cells; however, overexpression of bcl-2 does not protect against bak-mediated apoptosis. (1997) *Int. J. Cancer. 72:* 898-905.

[108] Glick RD, Swendeman SL, Coffey DC, Rifkind RA, Marks PA, Richon VM, La Quaglia MP. Hybrid polar histone deacetylase inhibitor induces apoptosis and CD95/CD95 ligand expression in human neuroblastoma. (1999) *Cancer Res. 59:* 4392−4399.

[109] Kwon SH, Ahn SH, Kim YK, Bae GU, Yoon JW, Hong S, Lee HY, Lee YW, Lee HW, Han JW. Apicidin, a histone deacetylase inhibitor, induces apoptosis and Fas/Fas ligand expression in human acute promyelocytic leukemia cells. (2002) *J. Biol. Chem. 277:* 2073-2080.

[110] Klisovic DD, Katz SE, Effron D, Klisovic MI, Wickham J, Parthun MR, Guimond M, Marcucci G. Depsipeptide (FR901228) inhibits proliferation and induces apoptosis in primary and metastatic human uveal melanoma cell lines. (2003) *Invest Ophthalmol Vis. Sci. 44:* 2390-2398.

[111] Imai T, Adachi S, Nishijo K, Ohgushi M, Okada M, Yasumi T, Watanabe K, Nishikomori R, Nakayama T, Yonehara S, Toguchida J, Nakahata T. FR901228 induces tumor regression associated with induction of Fas ligand and activation of Fas signaling in human osteosarcoma cells. (2003) *Oncogene. 22:* 9231-9242.

[112] Suzuki T, Yokozaki H, Kuniyasu H, Hayashi K, Naka K, Ono S, Ishikawa T, Tahara E, Yasui W. Effect of trichostatin A on cell growth and expression of cell cycle- and apoptosis-related molecules in human gastric and oral carcinoma cell lines. (2000) *Int. J. Cancer. 88:* 992-997.

[113] Neuzil J, Swettenham E, Gellert N. Sensitization of mesothelioma to TRAIL apoptosis by inhibition of histone deacetylase: role of Bcl-x(L) down-regulation. (2004) *Biochem. Biophys. Res. Commun. 314:* 186-191.

[114] Cao XX, Mohuiddin I, Ece F, McConkey DJ, Smythe WR. 2001; Histone deacetylase inhibitor downregulation of bcl-xl gene expression leads to apoptotic cell death in mesothelioma. (2001) *Am. J. Respir Cell Mol. Biol. 25:* 562-568.

[115] Glaser KB, Staver MJ, Waring JF, Stender J, Ulrich RG, Davidsen SK. Gene expression profiling of multiple histone deacetylase (HDAC) inhibitors: defining a

common gene set produced by HDAC inhibition in T24 and MDA carcinoma cell lines. (2003) *Mol. Cancer Ther.* 2: 151-163.

[116] Suzuki H, Gabrielson E, Chen W, Anbazhagan R, van Engeland M, Weijenberg MP, Herman JG, Baylin SB. A genomic screen for genes upregulated by demethylation and histone deacetylase inhibition in human colorectal cancer. (2002) *Nature Genet.* 31: 141-149.

[117] Richon VM, Sandhoff TW, Rifkind RA, Marks PA. Histone deacetylase inhibitor selectively induces p21WAF1 expression and gene-associated histone acetylation. (2000) *Proc. Natl. Acad. Sci. U S A.* 97: 10014-10019.

[118] Derjuga A, Richard C, Crosato M, Wright PS, Chalifour L, Valdez J, Barraso A, Crissman HA, Nishioka W, Bradbury EM, Th'ng JP. Expression of p21Waf1/Cip1 and cyclin D1 is increased in butyrate-resistant HeLa cells. (2001) *J. Biol. Chem.* 276: 37815-37820.

[119] Henderson C, Mizzau M, Paroni G, Maestro R, Schneider C, Brancolini C. Role of caspases, Bid, and p53 in the apoptotic response triggered by histone deacetylase inhibitors trichostatin-A (TSA) and suberoylanilide hydroxamic acid (SAHA). (2003) *J. Biol. Chem.* 278: 12579-12589.

[120] Amin HM, Saeed S, Alkan S. Histone deacetylase inhibitors induce caspase-dependent apoptosis and downregulation of daxx in acute promyelocytic leukaemia with t(15;17). (2001) *Br. J. Haematol.* 115: 287-297.

[121] Szakács G, Annereau JP, Lababidi S, Shankavaram U, Arciello A, Bussey KJ, Reinhold W, Guo Y, Kruh GD, Reimers M, Weinstein JN, Gottesman MM. Predicting drug sensitivity and resistance: profiling ABC transporter genes in cancer cells. (2004) *Cancer Cell.* 6: 129-137.

[122] Scala S, Akhmed N, Rao US, Paull K, Lan LB, Dickstein B, Lee JS, Elgemeie GH, Stein WD, Bates SE. P-glycoprotein substrates and antagonists cluster into two distinct groups. (1997) *Mol. Pharmacol.* 51: 1024-1033.

[123] Ruefli AA, Bernhard D, Tainton KM, Kofler R, Smyth MJ, Johnstone RW. Suberoylanilide hydroxamic acid (SAHA) overcomes multidrug resistance and induces cell death in P-glycoprotein-expressing cells. (2002) *Int. J. Cancer.* 99: 292-298.

[124] Duan J, Friedman J, Nottingham L, Chen Z, Ara G, Van Waes C. Nuclear factor-kappaB p65 small interfering RNA or proteasome inhibitor bortezomib sensitizes head and neck squamous cell carcinomas to classic histone deacetylase inhibitors and novel histone deacetylase inhibitor PXD101. (2007) *Mol. Cancer Ther.* 6: 37-50.

[125] Kitamura K, Hoshi S, Koike M, Kiyoi H, Saito H, Naoe T. Histone deacetylase inhibitor but not arsenic trioxide differentiates acute promyelocytic leukaemia cells with t (11;17) in combination with all-trans retinoic acid. (2000) *Br. J. Haematol.* 108: 696-702.

[126] Rashid SF, Moore JS, Walker E, Driver PM, Engel J, Edwards CE, Brown G, Uskokovic MR, Campbell MJ. Synergistic growth inhibition of prostate cancer cells by 1 alpha,25 Dihydroxyvitamin D(3) and its 19-nor-hexafluoride analogs in combination with either sodium butyrate or trichostatin A. (2001) *Oncogene.* 20: 1860-1872.

[127] Coffey DC, Kutko MC, Glick RD, Butler LM, Heller G, Rifkind RA, Marks PA, Richon VM, La Quaglia MP. The histone deacetylase inhibitor, CBHA, inhibits

growth of human neuroblastoma xenografts in vivo, alone and synergistically with all-trans retinoic acid. (2001) *Cancer Res. 61:* 3591-3594.

[128] Nimmanapalli R, Fuino L,Stobaugh C, Richon V, Bhalla K. Cotreatment with the histone deacetylase inhibitor suberoylanilide hydroxamic acid (SAHA) enhances imatinib-induced apoptosis of Bcr-Abl-positive human acute leukemia cells. (2003) *Blood. 101:* 3236-3239.

[129] Yu C, Rahmani M, Almenara J, Subler M, Krystal G, Conrad D, Varticovski L, Dent P, Grant S. Histone deacetylase inhibitors promote STI571-mediated apoptosis in STI571-sensitive and -resistant Bcr/Abl+ human myeloid leukemia cells. (2003) *Cancer Res. 63:* 2118-2126.

[130] Pei XY, Dai Y, Grant S. Synergistic induction of oxidative injury and apoptosis in human multiple myeloma cells by the proteasome inhibitor bortezomib and histone deacetylase inhibitors. (2004) *Clin. Cancer Res. 10:* 3839-3852.

[131] Camphausen K, Scott T, Sproull M, Tofilon PJ. Enhancement of xenograft tumor radiosensitivity by the histone deacetylase inhibitor MS-275 and correlation with histone hyperacetylation. (2004) *Clin. Cancer Res.* 10: 6066-6071.

[132] 132. Adachi M, Zhang Y, Zhao X, Minami T, Kawamura R, Hinoda Y, Imai K. Synergistic effect of histone deacetylase inhibitors FK228 and m-carboxycinnamic acid bis-hydroxamide with proteasome inhibitors PSI and PS-341 against gastrointestinal adenocarcinoma cells. (2004) *Clin. Cancer Res. 10:* 3853-3862.

[133] Kawano T, Horiguchi-Yamada J, Iwase S, Akiyama M, Furukawa Y, Kan Y, Yamada H. Depsipeptide enhances imatinib mesylate-induced apoptosis of Bcr-Abl-positive cells and ectopic expression of cyclin D1, c-Myc or active MEK abrogates this effect. (2004*) Anticancer Res. 24:* 2705-2712.

[134] Weisberg E, Catley L, Kujawa J, Atadja P, Remiszewski S, Fuerst P, Cavazza C, Anderson K, Griffin JD. Histone deacetylase inhibitor NVP-LAQ824 has significant activity against myeloid leukemia cells in vitro and in vivo. (2004) *Leukemia. 18:* 1951-1963.

[135] Arundel CM, Glicksman AS, Leith JT. Enhancement of radiation injury in human colon tumor cells by the maturational agent sodium butyrate (NaB). (1985) *Radiat Res. 104:* 443-448.

[136] Heussen C, Nackerdien Z, Smit BJ, Bohm L. Irradiation damage in chromatin isolated from V-79 Chinese hamster lung fibroblasts. (1987) *Radiat Res. 110:* 84-94.

[137] Dion LD, Goldsmith KT, Tang DC, Engler JA, Yoshida M, Garver RI. Amplification of recombinant adenoviral transgene products occurs by inhibition of histone deacetylase. (1997) *Virology. 231:* 201-209.

[138] Gaetano C, Catalano A, Palumbo R, Illi B, Orlando G, Ventoruzzo G, Serino F, Capogrossi MC. Transcriptionally active drugs improve adenovirus vector performance in vitro and in vivo. (2000) *Gene. Ther. 7:* 1624-1630.

[139] Lee CT, Seol JY, Park KH, Yoo CG, Kim YW, Ahn C, Song YW, Han SK, Han JS, Kim S, Lee JS, Shim YS. Differential effects of adenovirus-p16 on bladder cancer cell lines can be overcome by the addition of butyrate. (2001) *Clin. Cancer Res. 7:* 210-214.

[140] Indraccolo S, Minuzzo S, Gola E, Habeler W, Carrozzino F, Noonan D, Albini A, Santi L, Amadori A, Chieco-Bianchi L. Generation of expression plasmids for angiostatin, endostatin and TIMP-2 for cancer gene therapy. (1999) *Int. J. Biol. Markers. 14:* 251-256.

[141] Cirri L, Donnini S, Morbidelli L, Chiarugi P, Ziche M, Ledda F. Endostatin: a promising drug for antiangiogenic therapy. (1999) *Int. J. Biol. Markers. 24:* 263-267.

[142] Ingber D, Fujita T, Kishimoto S, Sudo K, Kanamaru T, Brem H, Folkman J. Synthetic analogues of fumagillin that inhibit angiogenesis and suppress tumor growth. (1990) *Nature. 348:* 555-557.

[143] Oikawa T, Ito H, Ashino H, Toi M, Tominaga T, Morita I, Murota S. Radicicol, a microbial cell differentiation modulator, inhibits in vivo angiogenesis. (1993) *Eur. J. Pharmacol. 241:* 221-227.

[144] Oikawa T, Onozawa C, Inose M, Sasaki M. Depudecin, a microbial metabolite containing two epoxide groups, exhibits anti-angiogenic activity in vivo. (1995) *Biol. Pharm. Bull. 18:* 1305-1307.

[145] Smolich BD, Yuen HA, West KA, Giles FJ, Albitar M, Cherrington JM. The antiangiogenic protein kinase inhibitors SU5416 and SU6668 inhibit the SCF receptor (c-kit) in a human myeloid leukemia cell line and in acute myeloid leukemia blasts. (2001) *Blood. 97:* 1413-1421.

[146] Kim MS, Kwon HJ, Lee YM, Baek JH, Jang JE, Lee SW, Moon EJ, Kim HS, Lee SK, Chung HY, Kim CW, Kim KW. Histone deacetylase induce angiogenesis by negative regulation of tumor suppressor genes. (2001) *Nat. Med. 7:* 437-443.

[147] Takahashi I, Miyaji H, Yoshida T, Sato S, Mizukami T. Selective inhibition of IL-2 gene expression by trichostatin A, a potent inhibitor of mammalian histone deacetylase. (1996) *J. Antibiot (Tokyo). 49:* 453-457.

[148] Koyama Y, Adachi M, Sekiya M, Takekawa M, Imai K. Histone deacetylase inhibitors suppress IL-2-mediated gene expression before induction of apoptosis. (2000) *Blood. 96:* 1490-1495.

[149] Huang N, Katz JP, Martin DR, Wu GD. Inhibition of IL-8 gene expression in Caco-2 cells by compounds which induce histone hyperacetylation. (1997) *Cytokine. 9:* 27-36.

[150] Shigematsu N, Ueda H, Takase S, Tanaka H, Yamamoto K, Tada T. FR901228, a novel antitumor bicyclic depsipeptide produced by Chromobacterium violaceum No. 968. II. Structure determination. (1994) *J. Antibiot (Tokyo). 47:* 311-314.

[151] Ueda H, Manda T, Matsumoto S, Mukumoto S, Nishigaki F, Kawamura I, Shimomura K. FR901228, a novel antitumor bicyclic depsipeptide produced by Chromobacterium violaceum No. 968. III. Antitumor activities on experimental tumors in mice. (1994) *J. Antibiot (Tokyo). 47:* 315-323.

[152] Ueda H, Nakajima H, Hori Y, Goto T, Okuhara M. Action of FR901228, a novel antitumor bicyclic depsipeptide produced by Chromobacterium violaceum no. 968, on Ha-ras transformed NIH3T3 cells. (1994) *Biosci Biotechnol Biochem. 58:* 1579-1583.

[153] Kosugi H, Ito M, Yamamoto Y, Towatari M, Ito M, Ueda R, Saito H, Naoe T. In vivo effects of a histone deacetylase inhibitor, FK228, on human acute promyelocytic leukemia in NOD/Shi-scid/scid mice. (2001) *Jpn J. Cancer Res. 92:* 529-536.

[154] Fecteau KA, Mei J, Wang HC. Differential modulation of signaling pathways and apoptosis of ras-transformed 10T1/2 cells by the depsipeptide FR901228. (2002) *J Pharmacol Exp. Ther. 300:* 890-899.

[155] Furumai R, Matsuyama A, Kobashi N, Lee KH, Nishiyama M, Nakajima H, Tanaka A, Komatsu Y, Nishino N, Yoshida M, Horinouchi S. FK228 (depsipeptide) as a natural prodrug that inhibits class I histone deacetylases. (2002) *Cancer Res. 62:* 4916-4921.

[156] Piekarz RL, Robey RW, Zhan Z, Kayastha G, Sayah A, Abdeldaim AH, Torrico S, Bates SE. T-cell lymphoma as a model for the use of histone deacetylase inhibitors in cancer therapy: impact of depsipeptide on molecular markers, therapeutic targets, and mechanisms of resistance. (2004) *Blood. 103:* 4636-4643.

[157] Marshall JL, Rizvi N, Kauh J, Dahut W, Figuera M, Kang MH, Figg WD, Wainer I, Chaissang C, Li MZ, Hawkins MJ. A phase I trial of depsipeptide (FR901228) in patients with advanced cancer. (2002) *J. Exp. Ther. Oncol. 2:* 325-332.

[158] Piekarz R, Bates S. A review of depsipeptide and other histone deacetylase inhibitors in clinical trials. (2004) *Curr. Pharm. Des. 10:* 2289-2298.

[159] Doi S, Soda H, Oka M, Tsurutani J, Kitazaki T, Nakamura Y, Fukuda M, Yamada Y, Kamihira S, Kohno S. The histone deacetylase inhibitor FR901228 induces caspase-dependent apoptosis via the mitochondrial pathway in small cell lung cancer cells. (2004) *Mol. Cancer Ther. 3:* 1397-1402.

[160] Aron JL, Parthun MR, Marcucci G, Kitada S, Mone AP, Davis ME, Shen T, Murphy T, Wickham J, Kanakry C, Lucas DM, Reed JC, Grever MR, Byrd JC. Depsipeptide (FR901228) induces histone acetylation and inhibition of histone deacetylase in chronic lymphocytic leukemia cells concurrent with activation of caspase 8-mediated apoptosis and down-regulation of c-FLIP protein.(2003) *Blood. 102(2):* 652-658.

[161] Zhang Y, Adachi M, Zhao X, Kawamura R, Imai K. Histone deacetylase inhibitors FK228, *N*-(2-aminophenyl)-4-[*N*-(pyridin-3-yl- methoxycarbonyl) amino- methyl] benzamide and m-carboxycinnamic acid bis-hydroxamide augment radiation-induced cell death in gastrointestinal adenocarcinoma cells. (2004) *Int. J. Cancer. 110:* 301-308.

[162] Khan SB, Maududi T, Barton K, Ayers J, Alkan S. Analysis of histone deacetylase inhibitor, depsipeptide (FR901228), effect on multiple myeloma. (2004) *Br. J. Haematol. 125:* 156-161.

[163] Vanoosten RL, Moore JM, Karacay B, Griffith TS. Histone deacetylase inhibitors modulate renal cell carcinoma sensitivity to TRAIL/Apo-2L-induced apoptosis by enhancing TRAIL-R2 expression. (2005) *Cancer Biol. Ther. 4:* 1104-1112.

[164] Zhu H, Guo W, Zhang L, Wu S, Teraishi F, Davis JJ, Dong F, Fang B. Proteasome inhibitors-mediated TRAIL resensitization and Bik accumulation. (2005) *Cancer Biol. Ther. 4:* 781-786.

[165] Kitazono M, Goldsmith ME, Aikou T, Bates S, Fojo T. Enhanced adenovirus transgene expression in malignant cells treated with the histone deacetylase inhibitor FR901228. (2001) *Cancer Res. 61:* 6328-6330.

[166] Kitazono M, Rao VK, Robey R, Aikou T, Bates S, Fojo T, Goldsmith ME. Histone deacetylase inhibitor FR901228 enhances adenovirus infection of hematopoietic cells. (2002) *Blood. 99:* 2248-2251.

[167] Goldsmit ME, Kitazono M, Fok P, Aikou T, Bates S, Fojo T. The histone deacetylase inhibitor FK228 preferentially enhances adenovirus transgene expression in malignant cells. (2003) *Clin. Cancer Res. 9:* 5394-5401.

[168] Pong RC, Lai YJ, Chen H, Okegawa T, Frenkel E, Sagalowsky A, Hsieh JT. Epigenetic regulation of coxsackie and adenovirus receptor (CAR) gene promoter in urogenital cancer cells. (2003) *Cancer Res. 63:* 8680-8686.

[169] Hemminki A, Kanerva A, Liu B, Wang M, Alvarez RD, Siegal GP, Curiel DT. Modulation of coxsackie-adenovirus receptor expression for increased adenoviral transgene expression. (2003) *Cancer Res. 63:* 847-853.

[170] Lee JS, Paull K, Alvarez M, Hose C, Monks A, Grever M, Fojo AT, Bates SE. Rhodamine efflux patterns predict P-glycoprotein substrates in the National Cancer Institute Drug Screen. (1994) *Mol. Pharmacol. 46:* 627-638.

[171] Xiao JJ, Huang Y, Dai Z, Sadée W, Chen J, Liu S, Marcucci G, Byrd J, Covey JM, Wright J, Grever M, Chan KK. Chemoresistance to depsipeptide FK228 [(E)-(1S,4S,10S,21R)-7-[(Z)-Ethylidene]-4,21-diisopropyl-2-oxa-12,13-dithia-5,8,20,23-tetraazabicyclo[8,7,6]-tricos-16-ene-3,6,9,22-pentanone] is mediated by reversible MDR1 induction in human cancer cell lines. (2005) *J. Pharmacol Exp. Ther. 314:* 467-475.

[172] Xiao JJ, Foraker AB, Swaan PW, Liu S, Huang Y, Dai Z, Chen J, Sadée W, Byrd J, Marcucci G, Chan KK. Efflux of depsipeptide FK228 (FR901228, NSC-630176) is mediated by P-glycoprotein and multidrug resistance-associated protein 1. (2005) *J. Pharmacol. Exp. Ther. 313:* 268-276.

[173] Jung JW, Cho SD, Ahn NS, Yang SR, Park JS, Jo EH, Hwang JW, Aruoma OI, Lee YS, Kang KS. Effects of the histone deacetylases inhibitors sodium butyrate and trichostatin A on the inhibition of gap junctional intercellular communication by H2O2- and 12-O-tetradecanoylphorbol-13-acetate in rat liver epithelial cells. (2006) *Cancer Lett. 241(2):* 301-308.

[174] Wu Y, Guo SW. Histone deacetylase inhibitors trichostatin A and valproic acid induce cell cycle arrest and p21 expression in immortalized human endometrial stromal cells. (2008) *Eur. J. Obstet. Gynecol. Reprod Biol. 137(2):* 198-203.

[175] Wang S, Zhu J. Evidence for a Relief of Repression Mechanism for Activation of the Human Telomerase Reverse Transcriptase Promoter. (2003) *J. Biol. Chem. 278:* 18842-18850.

[176] Murakami J, Asaumi J, Kawai N, Tsujigiwa H, Yanagi Y, Nagatsuka H, Inoue T, Kokeguchi S, Kawasaki S, Kuroda M, Tanaka N, Matsubara N, Kishi K. Effects of histone deacetylase inhibitor FR901228 on the expression level of telomerase reverse transcriptase in oral cancer. (2005) *Cancer Chemother Pharmacol. 56(1):* 22-28.

[177] Suenaga M, Soda H, Oka M, Yamaguchi A, Nakatomi K, Shiozawa K, Kawabata S, Kasai T, Yamada Y, Kamihira S, Tei C, Kohno S. Histone deacetylase inhibitors suppress telomerase reverse transcriptase mRNA expression in prostate cancer cells. (2002) *Int. J. Cancer. 97:* 621-625.

[178] Tsurutani J, Soda H, Oka M, Suenaga M, Doi S, Nakamura Y, Nakatomi K, Shiozawa K, Yamada Y, Kamihira S, Kohno S. Antiproliferative effects of the histone

deacetylase inhibitor FR901228 on small-cell lung cancer lines and drug-resistant sublines. (2003) *Int. J. Cancer. 104(2):* 238-242.

[179] Zhu K, Qu D, Sakamoto T, Fukasawa I, Hayashi M, Inaba N. Telomerase expression and cell proliferation in ovarian cancer cells induced by histone deacetylase inhibitors. (2008) *Arch. Gynecol. Obstet. 277(1):* 15-19.

[180] Atkinson SP, Hoare SF, Glasspool RM, Keith WN. Lack of telomerase gene expression in alternative lengthening of telomere cells is associated with chromatin remodeling of the hTR and hTERT gene promoters. (2005) *Cancer Res. 65:* 7585-7590.

[181] Liu C, Fang X, Ge Z, Jalink M, Kyo S, Björkholm M, Gruber A, Sjöberg J, Xu D. The telomerase reverse transcriptase (hTERT) gene is a direct target of the histone methyltransferase SMYD3. (2007) *Cancer Res. 67:* 2626-2631.

[182] Guccione E, Martinato F, Finocchiaro G, Luzi L, Tizzoni L, Dall' Olio V, Zardo G, Nervi C, Bernard L, Amati B. Myc-binding-site recognition in the human genome is determined by chromatin context. (2006) *Nat. Cell Biol. 8:* 764-770.

[183] Zhu Q, Liu C, Ge Z, Fang X, Zhang X, Strååt K, Björkholm M, Xu D. Lysine-specific demethylase 1 (LSD1) is required for the transcriptional repression of the telomerase reverse transcriptase (hTERT) gene. (2008) *PLoS ONE. 3(1):* e1446

[184] Horikawa I, Cable PL, Afshari C, Barrett JC. Cloning and characterization of the promoter region of human telomerase reverse transcriptase gene. (1999) *Cancer Res. 59:* 826-830.

[185] Bryan TM, Englezou A, Dalla-Pozza L, Dunham MA, Reddel RR. Evidence for an alternative mechanism for maintaining telomere length in human tumors and tumor-derived cell lines. (1997) *Nat. Med. 3:* 1271-1274.

[186] Wang Z, Kyo S, Takakura M, Tanaka M, Yatabe N, Maida Y, Fujiwara M, Hayakawa J, Ohmichi M, Koike K, Inoue M. Progesterone regulates human telomerase reverse transcriptase gene expression via activation of mitogen-activated protein kinase signaling pathway. (2000) *Cancer Res. 60:* 5367-5381.

[187] Henderson YC, Breau RL, Liu TJ, Clayman GL. Telomerase activity in head and neck tumors after introduction of wild-type p53, p21, p16, and E2F-1 genes by means of recombinant adenovirus. (2000) *Head Neck. 22:* 347-354.

[188] Henriksson M, Lüscher B. Proteins of the Myc network: essential regulators of cell growth and differentiation. (1996) *Adv. Cancer Res. 68:* 109-182.

[189] Ryan KM, Birnie GD. Free in PMC. Myc oncogenes: the enigmatic family. (1996) *Biochem J. 314 (3):* 713-721.

[190] Hoffman B, Liebermann DA, Selvakumaran M, Nguyen HQ. Role of c-myc in myeloid differentiation, growth arrest and apoptosis. (1996) *Curr. Top Microbiol. Immunol. 211:* 17-27.

[191] Dang CV. c-Myc target genes involved in cell growth, apoptosis, and metabolism. (1999) *Mol Cell Biol. 19:* 1-11.

[192] Prendergast GC. Mechanisms of apoptosis by c-Myc. (1999) *Oncogene. 18:* 2967-2987.

[193] Obaya AJ, Mateyak MK, Sedivy JM. Mysterious liaisons: the relationship between c-Myc and the cell cycle. (1999) *Oncogene. 18:* 2934-2941.

[194] Littlewood TD, Evan GI. The role of myc oncogenes in cell growth and differentiation. (1990) *Adv. Dent Res. 4:* 69-79.

[195] Marcu KB, Bossone SA, Patel AJ. myc function and regulation. (1992) *Annu. Rev. Biochem. 61:* 809-860.

[196] Schmid P, Schulz WA, Hameister H. Dynamic expression pattern of the myc protooncogene in midgestation mouse embryos. (1989) *Science. 243:* 226-229.

[197] Coller HA, Grandori C, Tamayo P, Colbert T, Lander ES, Eisenman RN, Golub TR. Expression analysis with oligonucleotide microarrays reveals that MYC regulates genes involved in growth, cell cycle, signaling, and adhesion. (2000) *Proc. Natl. Acad. Sci. USA. 97:* 3260-3265.

[198] Coppola JA, Cole MD. Constitutive c-myc oncogene expression blocks mouse erythroleukaemia cell differentiation but not commitment. (1986) *Nature. 320:* 760-763.

[199] Dmitrovsky E, Kuehl WM, Hollis GF, Kirsch IR, Bender TP, Segal S. Expression of a transfected human c-myc oncogene inhibits differentiation of a mouse erythroleukaemia cell line. (1986) *Nature. 322:* 748-750.

[200] Prochownik EV, Kukowska J. Deregulated expression of c-myc by murine erythroleukaemia cells prevents differentiation. (1986) *Nature. 322:* 848-850.

[201] Weidle UH, Grossmann A. Inhibition of histone deacetylases: a new strategy to target epigenetic modifications for anticancer treatment. (2000) *Anticancer Res. 20:* 1471-1485.

[202] Bernhard D, Ausserlechner MJ, Tonko M, Lofler M, Hartmann BL, Csordas A, Kofler R. Apoptosis induced by the histone deacetylase inhibitor sodium butyrate in human leukemic lymphoblasts. (1999) *FASEB J. 13(14):* 1991-2001.

[203] Krupitza G, Harant H, Dittrich E, Szekeres T, Huber H, Dittrich C. Sodium butyrate inhibits c-myc splicing and interferes with signal transduction in ovarian carcinoma cells. (1995) *Carcinogenesis. 16:* 1199-1205.

[204] Hassig CA, Fleischer TC, Andrew NB, Schreiber SL, Ayer DE. Histone deacetylase activity is required for full transcriptional repression by mSin3A. (1997) *Cell. 89:* 341-347.

[205] Orlando V, Strutt H, Paro R. Analysis of chromatin structure by in vivo formaldehyde cross-linking. (1997) *Methods. 11:* 205-214.

[206] Komata T, Kondo Y, Kanzawa T, Hirohata S, Koga S, Sumiyoshi H, Srinivasula SM, Barna BP, Germano IM, Takakura M, Inoue M, Alnemri ES, Shay JW, Kyo S, Kondo S. Treatment of Malignant Glioma Cells with the Transfer of Constitutively Active Caspase-6 Using the Human Telomerase Catalytic Subunit (Human Telomerase Reverse Transcriptase) Gene Promoter. (2001) *Cancer Res. 61:* 5796-5802.

[207] Cong YS, Wen J, Bacchetti S. The human telomerase catalytic subunit hTERT: organization of the gene and characterization of the promoter. (1999) *Hum. Mol. Genet. 8:* 137-142.

[208] Kim HR, Christensen R, Park NH, Sapp P, Kang MK, Park NH. Elevated Expression of hTERT Is Associated with Dysplastic Cell Transformation during Human Oral Carcinogenesis in Situ. (2001) *Clin. Cancer Res. 7:* 3079-3086.

[209] Nakamura Y, Tahara E, Tahara H, Yasiu W, Tahara E, Ide T. Quantitative reevaluation of telomerase activity in cancerous and noncancerous gastrointestinal tissues. (1999) *Mol. Carcinog. 26:* 312-320.

In: Bacterial DNA, DNA Polymerase and DNA Helicases
Editor: Walter D. Knudsen and Sam S. Bruns

ISBN 978-1-60741-094-2
© 2009 Nova Science Publishers, Inc.

Chapter VII

Dynamics of DNA Polymerase

Ping Xie[*]
Laboratory of Soft Matter Physics, Institute of Physics,
Chinese Academy of Sciences, Beijing 100190, China

Abstract

Replicative DNA polymerase (DNAP) is an enzyme that synthesizes a new DNA strand on a single-stranded template with a high processivity and a high fidelity. The high fidelity is mainly realized via a mechanism of proofreading that is performed at the exonuclease active site spatially separate from the polymerase active site. Here we present a detailed account of our proposed model for the processive nucleotide incorporation and switching transition of DNA between the polymerase and exonuclease active sites by DNAP. Based on the model, we present detailed theoretical studies on its dynamics. The moving time of DNAP to next site after a correct incorporation and that after an incorrect incorporation are analytically studied. The experimentally measured dependence of polymerization rate on tension applied to the template is well simulated. The transfer rates of DNA from the polymerase to exonuclease active sites after a correct incorporation and after incorrect incorporations as well as the transfer rate from the exonuclease to polymerase active sites are analytically studied. Moreover, the backward motion of DNAP when large tensions are applied to the template is also analytically studied. The theoretical results are in good agreement with available experimental data. Some predicted results are presented.

Keywords: DNA replication; Proofreading; Processivity; Transition; Model.

---

[*] E-mail address: pxie@aphy.iphy.ac.cn

1. Introduction

DNA polymerase (DNAP) is an enzyme that catalyzes the synthesis of a new DNA strand on a single-stranded (ss) template (Lodish et al., 2000). Replicative DNAPs elongate the new strand with high rate, high processivity and high fidelity. The high fidelity (e.g., with an error frequency of one in $10^9 - 10^{10}$ bases in bacteriophage T7 DNAP) is realized by using both the base-selection mechanism and the proofreading mechanism (Echols and Goodman, 1991; Johnson, 1993; Joyce and Steitz, 1994; Steitz, 1999; Brautigam and Steitz, 1998; Doublie and Ellenberger, 1998). The DNA synthesis is performed at the polymerase active site (or simply called polymerase site) that selectively incorporates the nucleotide complementary to the template in the 5' – 3' direction, while the proofreading is performed at a separate exonuclease active site (or simply called exonuclease site) that removes mistakes in the 3' – 5' direction.

DNAPs have highly-conserved structure, which means that their overall catalytic subunits vary, on a whole, little from species to species. The structures showed that they possess a hand-like shape with fingers, palm and thumb domains (Ollis et al., 1985). The structural studies of DNAPs complexed to DNA and/or nucleotide substrates revealed how the fingers and thumbs move in response to substrate binding (Korolev et al., 1995; Kim et al., 1995; Eom et al., 1996; Kiefer et al., 1997; 1998; Doublie et al., 1998; Doublie and Ellenberger, 1998; Li et al., 1998). These substrate-induced movements suggested that, during each cycle of nucleotide incorporation, DNAPs undergo a conformational change between an "open finger" state, which allows the polymerase site to sample nucleotides, and a "closed finger" state, which configures the polymerase site for nucleotide incorporation.

Using rapid chemical quench-flow techniques, Johnson and coworkers have obtained that, for correctly base-paired DNA, the rate of forward polymerization is about $300\,\text{s}^{-1}$ in T7 DNAP (Patel et al., 1991); whereas the rate of polymerization on top of a mismatch is reduced to only $0.025\,\text{s}^{-1}$ (Wong et al., 1991). Recently, by using single-molecule techniques, the replication rates of some DNAPs have also been elaborately monitored as it catalyzes the replication of a mechanically stretched DNA template (Wuite et al., 2000; Maier et al., 2000). It was shown interestingly that, as the tension increases, the replication rate increases until a maximum was reached. Then, further increase in tension, however, causes the rate to decrease until polymerization stalled. On the kinetics of transition between the polymerase and exonuclease sites, Donlin et al. (1991) have experimentally obtained that, in T7 DNA polymerase, the rate of transfer of correctly base-paired DNA from the polymerase to exonuclease sites is about $0.16\,\text{s}^{-1}$. In contrast, after incorporation of a mismatched base, the rate of transfer of the DNA to the exonuclease site is increased to about $2.3\,\text{s}^{-1}$. Relative to the slow rate of transfer from the polymerase to exonuclease sites, the rate of reverse transfer (from the exonuclease to polymerase sites) is very high ($\sim 714\,\text{s}^{-1}$).

To explain the observed effect of template tension on the polymerization rate, several kinetic models have been proposed. Originally, the force dependence of DNA replication rates was interpreted by using the global force-extension curves for dsDNA and ssDNA (Wuite et al., 2000; Maier et al., 2000). Later, Goel et al. (2001; 2003) proposed a model that considers only local interactions in the neighborhood of the enzyme to explain the force dependence of DNA replication rates. The effect of the external force on nucleotide incorporation was also studied by using molecular dynamics simulations (Andricioaei et al., 2004; Venkatramani and Radhakrishnan, 2008). The transfer of DNA from the polymerase to exonuclease sites after misincorporation of a nucleotide was explained by assuming that a sheared base pair, caused by a primer-template mismatch, compromises the interaction of dsDNA with the floor of the polymerase, triggering the movement of the dsDNA to the exonuclease site (Doublie and Ellenberger, 1998).

Recently, I have proposed another model for the forward polymerization (Xie, 2007) and DNA transition between the polymerase and exonuxlease sites (Xie, 2009a). In this review, I give a detailed account of the proposed model. Based on the model, I present systematic studies on the dynamics of DNAP, which includes the incorporation rates on top of a correct incorporation and on top of an incorrect incorporation in either absence or presence of the external force, the transition rates between the polymerase and exonuclease sites and the backward motion of DNAP when large tensions are applied to the template. The theoretical results are in agreement with the available experimental data. Moreover, some predicted results are presented.

2. Model Assumptions

Before the presentation of the model for forward polymerization and DNA transition between the polymerase and exonuclease sites by DNAP, two preliminary hypotheses are presented in Sections 2.1 and 2.2.

2.1. Existence of Two ssDNA-Binding Sites and dsDNA-Binding Site

It is assumed that there exist two ssDNA-binding sites in the DNAP, one (called binding site P) locating in the finger domain and having a high affinity for 5' – 3' ssDNA, while the other one (called binding site X) locating in the proofreading domain and having a high affinity for 3' – 5' ssDNA. The existence of binding site P is supported by the experimental results on bacteriophage T4 DNAP and Klenow fragment, showing that the finger domain of DNAP has a high binding affinity for the ssDNA template (Delagoutte and von Hippel, 2003; Turner et al., 2003; Datta et al., 2006); while the existence of binding site X is supported by the recent experiment of Kukreti et al. (2008). Besides these two ssDNA-binding sites, there should also exist the dsDNA-binding site locating in other domain. For example, the interaction of T7 DNAP with thioredoxin orders a proteolytically sensitive loop, located at the tip of the thumb domain, between α helices H and H1 and it is most likely this thioredoxin-binding loop binds to dsDNA (Doublie et al., 1998; Doublie and Ellenberger, 1998), thus increasing the processivity of DNA polymerase activity (Tabor et al., 1987). DNA polymerase III holoenzyme increases processivity of synthesis by associating with a dimeric protein ring termed a sliding clamp, which encircles and binds to the primer–template duplex (Kong et al., 1992). Therefore, the high processivity of polymerization is achieved through the combined effects of both strong binding to the ssDNA template and strong binding to dsDNA.

The DNAP structure shows that the polymerase site lies in the junction between the finger and palm domains (Korolev et al., 1995; Kim et al., 1995; Eom et al., 1996; Kiefer et al., 1997; 1998; Doublie et al., 1998; Doublie and Ellenberger, 1998; Li et al., 1998). Based on this hypothesis of the existence of the ssDNA-binding site P in the finger domain, the induced-fit mechanism, i.e., the polymerase site generally binding correct nucleotides with high affinity while binding incorrect nucleotides with low affinity or incorporating correct nucleotide much more efficiently than incorrect nucleotide (Krahn et al., 2004; Mendelman et al., 1990, Yu and Goodman, 1992), can be understood as follows. The strong interaction of the finger domain with the unpaired base on the DNA template induces the conformational change in the finger domain, which results in the conformational change in the polymerase site that is adjacent to the finger domain. This unpaired-base-related conformational change thus results in the polymerase site having a much higher affinity for the structurally compatible nucleotide than structurally incompatible nucleotides.

2.2. Open Fingers Allow Nucleotide Binding While Closed Fingers Promote Nucleotide Incorporation

From crystal structure studies on DNAP, it was shown that the binding of dNTP to the polymerase site induces the conformational change from the "open finger" state to the "closed finger" state. For example, dNTP binding induces fingers of T7 DNAP and Taq DNAP rotating inwards by about $40°$ and $46°$, respectively (Doublie et al., 1998; Doublie and Ellenberger, 1998; Li et al., 1998; Doublie et al., 1999); dNTP binding to HIV-1 reverse transcriptase induces its p66 fingers rotating inwards by about $20°$ (Huang et al., 1998). In the "open finger" state the polymerase site samples nucleotides, while in the "closed finger" state the chemical reaction of nucleotide incorporation can proceed. Moreover, the release of nucleotide, i.e., PPi, from the polymerase site accompanies the rotation of fingers from the "closed" conformation to the "open" conformation. In other words, the nucleotide-free polymerase site corresponds to the "open" conformation.

Moreover, as shown experimentally (Patel et al., 1991), the PPi release in fact involves two steps, which is described by the pathway

$$EP^* \cdot DNA_n \cdot PPi \to EP \cdot DNA_n \cdot PPi \to EP \cdot DNA_n + PPi,$$

i.e., PPi is released via a transition from the activated $EP^* \cdot DNA_n \cdot PPi$ complex to an unactivated $EP \cdot DNA_n \cdot PPi$ complex and then the release of PPi from the unactivated complex. The first step from the activated to unactivated complexes accompanies the outward rotation of the fingers (Patel et al., 1991). Similarly, the dNTP binding also involves two steps, which is described by the pathway

$$EP \cdot DNA_n + dNTP \rightarrow EP \cdot DNA_n \cdot dNTP \rightarrow EP^* \cdot DNA_n \cdot dNTP$$

The transition from the unactivated ternary complex $EP \cdot DNA_n \cdot dNTP$ to the activated $EP^* \cdot DNA_n \cdot dNTP$ complex accompanies the inward rotation of the fingers.

3. Model

In this Section, the model is presented for forward polymerization and DNA transition between the polymerase and exonuclease sites by DNAP based on the two assumptions presented in Section 2.

3.1. Forward Polymerization

We begin with the finger domain of a DNAP, with no nucleotide in the polymerase site, binding the ssDNA near the replication fork, as shown in Figure 1a. In this nucleotide-free state, either a correct (matched) or an incorrect (mismatched) dNTP can bind to the polymerase site, although a matched dNTP has a much larger probability to bind (see Section 2.1). Thus we consider the two cases separately.

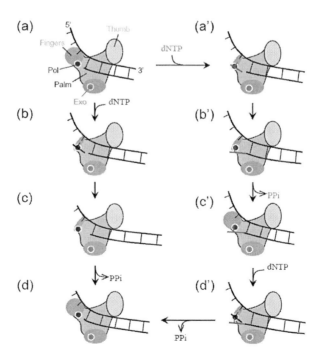

Figure 1. Schematic illustrations of the model for processive nucleotide incorporation by DNAP (see text for the detailed description). The green circles represent open fingers while green ellipses represent closed fingers. "Pol" represents polymerase active site while "Exo" represents exonuclease active site. For clarity, the matched nucleotide or base is drawn in black while the mismatched nucleotide is drawn in red.

3.1.1. Incorporation of a Matched Base

After a matched dNTP binds to the polymerase site ($EP \cdot DNA_n + dNTP \to EP \cdot DNA_n \cdot dNTP$), the ternary complex $EP \cdot DNA_n \cdot dNTP$ is changed to the activated $EP^* \cdot DNA_n \cdot dNTP$ complex ($EP \cdot DNA_n \cdot dNTP \to EP^* \cdot DNA_n \cdot dNTP$), which accompanies the inward rotation of the fingers to the "closed" conformation where the nucleotide incorporation can proceed, as shown in Figure 1*b*. Upon the completion of the nucleotide incorporation ($EP^* \cdot DNA_n \cdot dNTP \to EP^* \cdot DNA_{n+1} \cdot PPi \to EP \cdot DNA_{n+1} \cdot PPi$), the finger domain will bind to new nearest unpaired base (or the sugar-phosphate backbone of the unpaired base) of the ssDNA template because the previous base where the finger domain has just bound has disappeared due to base pairing, as shown in Figure 1*c*. At the same time, accompanying the transition, $EP^* \cdot DNA_{n+1} \cdot PPi \to EP \cdot DNA_{n+1} \cdot PPi$, the fingers change to the "open" conformation and then PPi is released ($EP \cdot DNA_{n+1} \cdot PPi \to EP \cdot DNA_{n+1} + PPi$), as shown in Figure 1*d*.

Note that, during either the conformational change from the "open finger" to "closed finger" states (from Figure 1*a* to *b*) or that from the "closed finger" to "open finger" states (from Figure 1*c* to *d*), DNAP has not moved relative to DNA. However, the binding of ssDNA-binding site *P* to the new base on the ssDNA template (from Figure 1*b* to *c*) makes DNAP move forwards relative to DNA by a distance of *p*, where $p = 0.34$ nm is the distance between two successive bases. The conformational change from Figure 1*a* to *b* induces bending of the ssDNA template while that from Figure 1*c* to *d* induces reverse change of the ssDNA template. In experiments of Wuite et al. (2000) and Maier et al. (2000), the bending of the ssDNA template drives the bead attached to the ssDNA template moved backwards by a distance of l_F, where l_F is the movement distance of binding site *P* induced by the finger rotation; while the reverse change of the ssDNA template makes the bead moved forwards by the same distance of l_F. Therefore, for a catalytic cycle from Figure 1*a* through *b* and *c* to *d*, DNAP has moved forwards relative to DNA by a distance of *p*, while the position of bead has not been changed. For correct incorporations, the above mechanochemical cycle proceeds continuously.

3.1.2. Incorporation of a Mismatched Base

We still begin with the finger domain, with no nucleotide in the polymerase site, binding the ssDNA template near the replication fork, as shown in Figure 1*a*. After a mismatched dNTP binds to the polymerase site ($EP \cdot DNA_n + dNTP \to EP \cdot DNA_n \cdot dNTP$), the chemical transition, $EP \cdot DNA_n \cdot dNTP \to EP^* \cdot DNA_n \cdot dNTP$, accompanies the rotation of the finger domain to the "closed" conformation where the nucleotide incorporation can proceed, as shown in Figure 1*a'*. In the experiments of Wuite et al. (2000) and Maier et al. (2000), this finger rotation drives the bead attached to the ssDNA template moved backwards by a distance of l_F. After the incorporation ($EP^* \cdot DNA_n \cdot dNTP \to EP^* \cdot DNA_{n+1} \cdot PPi \to EP \cdot DNA_{n+1} \cdot PPi$), although the sugar-phosphate backbone of the mismatched dNTP was connected to the backbones of the already formed dsDNA, the mismatched base is not paired with the sterically corresponding base on the ssDNA template. Thus the finger domain is still binding to the same unpaired base (or the sugar-phosphate backbone of the same unpaired base) of the ssDNA template, as shown in Figure 1b'. That means that, from Figure 1a' to b', DNAP is kept unmoved relative to DNA, which is different from the case for the incorporation of a matched base (from Figure 1b to c). Accompanying the transition, $EP^* \cdot DNA_{n+1} \cdot PPi \rightarrow EP \cdot DNA_{n+1} \cdot PPi$, the fingers change to the "open" conformation and then PPi is released ($EP \cdot DNA_{n+1} \cdot PPi \rightarrow EP \cdot DNA_{n+1} + PPi$), as shown in Figure 1$c'$. In the experiments of Wuite et al. (2000) and Maier et al. (2000), this drives the bead attached to the ssDNA template moved forwards by a distance of l_F. Thus, from Figure 1a through a' and b' to c', there is no net movement of DNAP relative to DNA.

After another dNTP binds to the nucleotide-free polymerase site ($EP \cdot DNA_{n+1} + dNTP \rightarrow EP \cdot DNA_{n+1} \cdot dNTP$), accompanying the transition, $EP \cdot DNA_{n+1} \cdot dNTP \rightarrow EP^* \cdot DNA_{n+1} \cdot dNTP$, the fingers change to the "closed" conformation again where the nucleotide incorporation can proceed. However, because of the steric obstacle from the previously incorporated sugar-phosphate backbone, the new dNTP cannot be connected to the backbones of the already formed dsDNA and, thus, the incorporation ($EP^* \cdot DNA_{n+1} \cdot dNTP \rightarrow EP^* \cdot DNA_{n+2} \cdot PPi$) cannot be completed, as shown in Figure 1d'.

Note now that, if the bound dNTP is dissociated from the polymerase site ($EP^* \cdot DNA_{n+1} \cdot dNTP \rightarrow EP \cdot DNA_{n+1} \cdot dNTP \rightarrow EP \cdot DNA_{n+1} + dNTP$) by the thermal noise (the dissociation rate is very low because of the large binding affinity for correct dNTP) in Figure 1d', the fingers rotate back to the "open" conformation. Then the binding of a third dNTP induces the fingers rotating inwards again to the state in Figure 1d'. Thus, even though there may be many times of shuttling rotation of the fingers, the polymerization ($EP^* \cdot DNA_{n+1} \cdot dNTP \rightarrow EP^* \cdot DNA_{n+2} \cdot PPi$) always cannot be completed. In other words, the polymerization becomes stalled, giving sufficient time for the mismatched 3' – 5' DNA primer tail to switch to the binding site X located in the exonuclease domain for proofreading. Now, if the mismatch is not excised for an exonuclease-deficient mutant of DNAP (for example, for Sequensse), only when the polymerase site is moved from position at n-th site to position at (n+1)-th site via overcoming the large binding affinity of DNAP for DNA can the forward polymerization continues (see Section 6.1). Moreover, after the polymerase site is moved to the position at (n+1)-th site, since the mismatched base is not correctly positioned as for the case of the matched base, this impairs the catalysis of nucleotide incorporation. Therefore, due to both the long moving time from n-th to (n+1)-th sites and the long nucleotide-incorporation time at (n+1)-th site, the polymerization rate on top of the mismatch is much reduced (see Section 6.1). This is in agreement with the experimental results that the forward polymerization rate of T7 DNAP is $300 \, \text{s}^{-1}$ for correctly base-paired DNA while it is reduced to only $0.025 \, \text{s}^{-1}$ after incorporation of a mismatch (Dolin et al., 1991).

In Figure 1d', however, if the mismatched base is excised, as will be discussed in the following Section, the new dNTP becomes able to be incorporated into the sugar-phosphate backbone of dsDNA. Then the finger domain, with no nucleotide in the polymerase site, changes to the "open" conformation, as shown in Figure 1d. From Figure 1d the next catalytic cycle will begin and the polymerization will thus proceed continuously again.

3.2. Switching Transition of DNA between Polymerase and Exonuclease Sites

The model of DNA transition between the polymerase and exonuclease sites is illustrated schematically in Figure 2 (Xie, 2009a).

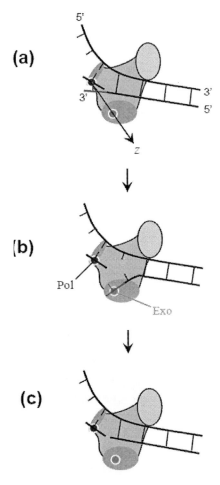

Figure 2. Schematic illustrations of the model for DNA transition between polymerase and exonuclease sites. (a) The sugar-phosphate backbone of the incorrect dNTP is connected to the backbones of the already formed dsDNA, but the incorrect base is not paired with the sterically corresponding base on the ssDNA template. (b) Due to the binding affinity of 3' – 5' ssDNA primer for binding site X and thermal noise, the 3' – 5' ssDNA primer tail diffuses to the exonuclease site, inducing the unwinding of some base pairs. (c) After the mismatched base is excised at the exonuclease site, the 3' – 5' ssDNA primer is switched back rapidly due to the high affinities of the matched base pairs of the dsDNA.

Consider that there is a mismatched base pair, as shown in Figure 2a that is the same as Figure 1d'. The 3' – 5' ssDNA primer tail will bind to the exonuclease domain due to the binding affinity, E_X, of 3' – 5' ssDNA for the binding site X in the exonuclease domain, as shown in Figure 2b. Because the exonuclease site is ~ 3.5 nm away from the polymerase site (Goel et al., 2003), when the 3' – 5' ssDNA primer binds to the exonuclease site several base pairs must be broken or unwound, as shown in Figure 2b, which means that the transfer of the 3' – 5' ssDNA primer from the polymerase to exonuclease sites requires overcoming the free energy for unwinding these base pairs. Once the 3' – 5' ssDNA reaching the exonuclease site, the base is excised with a very high rate k_x (Donlin et al., 1991). Due to the interaction potential between the matched base pairs, the 3' – 5' ssDNA primer will transfer back to the polymerase site with a high rate $K_{X \to P}$ by overcoming the energy barrier E_X of the 3' – 5' ssDNA primer binding to the exonuclease domain, as shown in Figure 2c. Note that, even if there is no mismatched base pair and if the thermal noise induces a matched base pair to unwind, the unwound 3' – 5' ssDNA primer tail can also bind to the exonuclease domain by overcoming the total free energy of base pairs that are required to unwind by the transfer of the 3' – 5' ssDNA primer to the exonuclease site.

4. Pathway of DNAP

Based on the model presented in Section 3, the chemical reaction pathway of a DNA polymerase is shown in Figure 3, which is similar to that given in the literature (Johnson, 1993; Wuite et al., 2000).

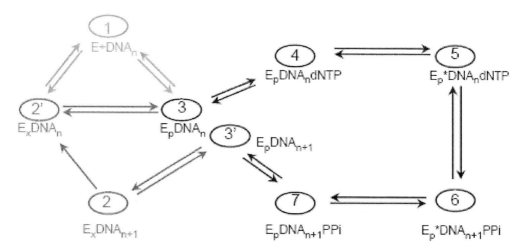

Figure 3. Kinetic pathway of DNAP. The transition 4 → 5 involve the conformational change from "open finger" to "closed finger" states and the transition 6 → 7 involve the conformational changes from "closed finger" to "open finger" states.

The pathway consists mainly of three parts: the chemical reactions related to the polymerization, i.e., the transitions 3□ 4□ 5□ 6□ 7□ 3', the reactions related to the switching transition between the polymerase and exonuclease sites and the exonucleolysis, i.e. the transitions 3'□ 2→2' □ 3, and the reactions related to the association to and dissociation from DNA, i.e. the transitions 1□ 3 and 1□ 2'.

5. Equations for Movement of DNAP along DNA

Without loss of generality, we consider that the ssDNA-binding site P covers four bases on the ssDNA template (Turner et al., 2003). Consider DNA segment near the replication fork, as schematically shown in Figure 4a (upper diagram). Then, the interaction potential, $V_{ssDNA}(x_1)$, between the ssDNA-binding site P and the ssDNA template can be approximately shown in the middle diagram of Figure 4a, where E_1 is the binding affinity for all four bases of the ssDNA template that the binding site P can cover, E_1' is the binding affinity for three bases that only part of the binding site P can cover, and x_1 represents the coordinate of the leftmost point on the binding site P along the DNA. As structure shows, the polymerase site should be located near the rightmost point on the binding site P. If we assume that the dsDNA-binding site also covers four base pairs on the dsDNA duplex, the interaction potential, $V_{dsDNA}(x_2)$, between the dsDNA-binding site and the dsDNA can be approximately shown in the lower diagram of Figure 4a, where E_2 is the binding affinity for the sugar-phosphate backbones connecting all four base pairs on the dsDNA, E_2' is the binding affinity for the backbones connecting only three base pairs on the dsDNA, and x_2 represents the coordinate of the rightmost point on the dsDNA-binding site along the DNA. Note that, since the ssDNA-binding site P at position of (n+2)-th base can bind only three bases, the binding affinity E_1' is smaller than E_1 that corresponds to binding all four bases. It should be also noted that different points on the binding site P should have different binding affinities for unpaired bases. Due to the same reason, the binding affinity E_2' is smaller than E_2. Moreover, considering that the interaction between the enzyme and DNA is via electrostatic force, with the interaction distance which is approximately equal to the Debye length (~ 1 nm) in solution larger than the distance p = 0.34 nm between two successive base pairs, it is expected that E_{10} is larger than E_1 and E_1'.

In our analysis, we represent the position, x, of DNAP by that of the leftmost point on the ssDNA-binding site P. Assuming that the leftmost point of dsDNA-binding site that locates in the thumb domain is m base pairs (more than four base pairs) away from the rightmost point of ssDNA-binding site P that locates in the finger domain, the potential $V_{dsDNA}(x)$ can be represented by the lower diagram of Figure 4b (i.e., $V_{dsDNA}(x_2)$ in the lower diagram of Figure 4a is shifted towards $-x$ direction by m+7 base pairs). As our model shows, before the incorporation of n-th base, $V_{ssDNA}(x)$ and $V_{dsDNA}(x)$ are shown in Figure 4b; while after the incorporation of an incorrect n-th base, $V_{ssDNA}(x)$ and $V_{dsDNA}(x)$ are still shown in

Figure 4b. However, after the incorporation of a correct n-th base, $V_{ssDNA}(x)$ is changed to that shown in upper diagram of Figure 4c, while $V_{dsDNA}(x)$ is shown in lower diagram of Figure 4c.

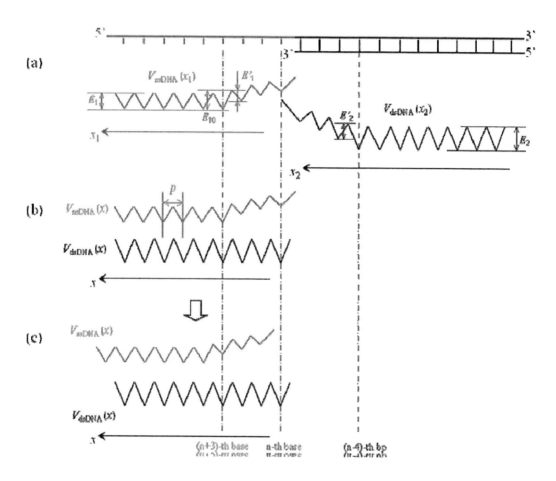

Figure 4. Interaction potentials V_{ssDNA} between the ssDNA-binding site P and the ssDNA template and V_{dsDNA} between the dsDNA-binding site and the dsDNA (see text for the detailed description). Without loss of generality, it is considered that the ssDNA-binding site P covers four bases and the dsDNA-binding site covers four base pairs. $p = 0.34$ nm is the distance between two successive bases (or between two successive base pairs) along ssDNA (or dsDNA).

In the case that the 3' – 5' ssDNA primer is not switched to the exonuclease site, the movement of DNAP along DNA in the over-damped environment can be described by the following Langevin equation

$$\Gamma \frac{dx}{dt} = -\frac{\partial V(x,t)}{\partial x} + \xi(t), \tag{1}$$

where $V(x)$ is the potential of DNAP interacting with DNA, which can be written as $V(x) = V_{ssDNA}(x) + V_{dsDNA}(x)$. Γ is the frictional drag coefficient on DNAP and $\xi(t)$ is the fluctuating Langevin force with $\langle \xi(t) \rangle = 0$ and $\langle \xi(t)\xi(t') \rangle = 2k_B T \Gamma \delta(t-t')$. The drag coefficient $\Gamma = 6\pi\eta R = 9.4 \times 10^{-11}$ kg·s^{-1}, where the viscosity of the aqueous medium is $\eta = 0.01$ g·cm^{-1}·s^{-1} and the DNAP is considered as a sphere with radius $R = 5$ nm.

The Fokker-Planck equation corresponding to Langevin equation (1) has the form

$$\frac{\partial P(x,t)}{\partial t} = \frac{1}{\Gamma} \frac{\partial}{\partial x}\left[\frac{\partial V(x)}{\partial x} P(x,t)\right] + D \frac{\partial^2 P(x,t)}{\partial^2 x}, \qquad (2)$$

where $D = k_B T / \Gamma$.

6. Forward Polymerization

6.1. Moving time

As just shown in Section 5, after the incorporation of a correct n-th base, potentials $V_{ssDNA}(x)$ and $V_{dsDNA}(x)$ are changed from those shown in Figure 4b to those shown in Figure 4c. With $V(x) = V_{ssDNA}(x) + V_{dsDNA}(x)$ given in Figure 4c, from Eq. (2), the mean first-passage time, T_c, for the DNAP to move from (n+3)-th site at position $x = 0$, with the polymerase site positioned at n-th site, to the next (n+4)-th site at position $x = p = 2l$, with the polymerase site positioned at (n+1)-th site, can be approximately calculated by (Gardiner, 1983)

$$T_c = \frac{1}{D} \int_0^{2l} \exp\left(\frac{V(y)}{\Gamma D}\right) dy \int_0^y \exp\left(-\frac{V(z)}{\Gamma D}\right) dz, \qquad (3)$$

where $p = 2l$ is the distance between two successive bases. For simplicity, in Eq. (3) we have neglected the small contribution of the movement to the range of $x < 0$, because, for the form of $V_{ssDNA}(x)$ shown in Figure 4b, the probability for the DNAP to jump from the potential well at (n+3)-th base to the backward shallower potential well at (n+2)-th base is very small.

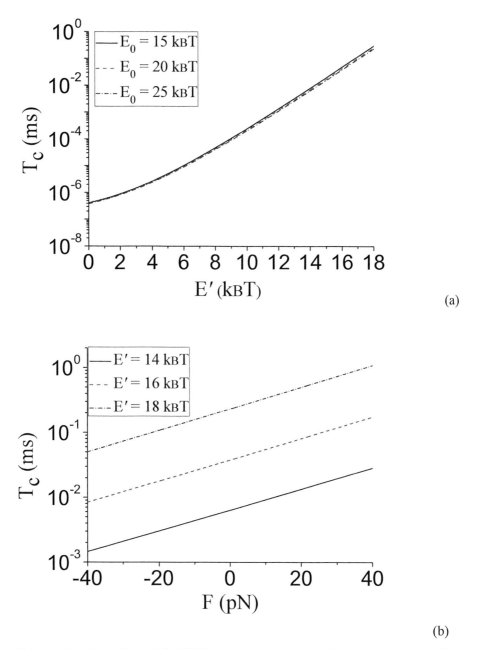

Figure 5. Calculated results of time T_c for DNAP to move to next site after the incorporation of a correct n-th base. (a) T_c versus E' for different values of E_0 under $F = 0$. (b) T_c versus backward load F for different values of E', with $E_0 = 25\, k_B T$.

From Eq. (3), the moving time T_c is obtained as follows

$$T_c = \frac{(l\Gamma)^2 D}{(E'+Fl)}\left[\exp\left(\frac{E'+Fl}{\Gamma D}\right)-1\right] - \frac{l^2\Gamma}{E'+Fl}$$

$$+ \frac{(l\Gamma)^2 D}{(E'+Fl)(E_0-Fl)} \left[\exp\left(\frac{E'+Fl}{\Gamma D}\right) - \exp\left(\frac{E'-E_0+2Fl}{\Gamma D}\right) \right]$$

$$\cdot \left[1 - \exp\left(-\frac{E'+Fl}{\Gamma D}\right)\right]\left(1 + \frac{E'+Fl}{E_0-Fl}\right) + \frac{l^2\Gamma}{E_0-Fl}, \qquad (4)$$

where $E' = E_2 + E_1'$, $E_0 = E_2 + E_{10}$ and F is an external load acted on DNAP if DNA template is fixed. Here it is defined that the backward load has the positive value of F while the forward load has the negative value.

Using Eq. (4), the calculated results of the moving time T_c versus $E' = E_2 + E_1'$ for different values of E_0 under no load are shown in Figure 5a. As anticipated, T_c is insensitive to the variation of E_0. It is seen that, even for a very large value of $E' = 18\,k_B T$, the moving time ($T_c < 0.3$ ms) is much shorter than the dwell time, T_d, between two successive nucleotide incorporations, which is about $T_d = 3.33$ ms that is obtained from the incorporation rate of about $300\,\text{s}^{-1}$ for T7 DNAP (Patel et al., 1991). The results of T_c versus backward load F for different values of E' are shown in Figure 5b. It is seen that, even for a large backward load of 40 pN, the moving time T_c is increased by only about 4.4 folds. As it will be inferred from the calculation in the following Section 6.2, a large backward load acted on the DNAP greatly reduces the nucleotide-incorporation rate. Thus, we conclude that that the moving time is much shorter than the time of nucleotide incorporation under the external load that is not very large and, consequently, the polymerization rate is approximately only dependent on the chemical reaction rates. This is a characteristic similar to that for other polymerase enzymes such as RNA polymerase (Xie, 2008; 2009b) that catalyzes the synthesis of RNA upon a DNA template.

After the incorporation of an incorrect n-th base, potentials $V_{ssDNA}(x)$ and $V_{dsDNA}(x)$ shown in Figure 4b are kept unchanged. Thus, similar to the derivation of Eq. (4), the mean first-passage time, T_i, for the DNAP to move from (n+3)-th site, with the polymerase site positioned at n-th site, to the next (n+4)-th site, with the polymerase site positioned at (n+1)-th site, is obtained as follows

$$T_i = \frac{(l\Gamma)^2 D}{(E+Fl)} \left[\exp\left(\frac{E+Fl}{\Gamma D}\right) - 1 \right] - \frac{l^2\Gamma}{E+Fl}$$

$$+ \frac{(l\Gamma)^2 D}{(E+Fl)(E-Fl)}\left[\exp\left(\frac{E+Fl}{\Gamma D}\right) - \exp\left(\frac{2Fl}{\Gamma D}\right)\right]\cdot\left[1 - \exp\left(-\frac{E+Fl}{\Gamma D}\right)\right]\left(1 + \frac{E+Fl}{E-Fl}\right)$$

$$+ \frac{l^2\Gamma}{E-Fl}, \qquad (5)$$

where $E = E_1 + E_2$. Because $E + Fl \gg \Gamma D = k_B T$, Eq. (5) can be approximately rewritten as

$$T_i = \frac{(l\Gamma)^2 D}{(E+Fl)}\left[\exp\left(\frac{E+Fl}{\Gamma D}\right) - 1\right] - \frac{l^2\Gamma}{E+Fl}$$
$$+ \frac{(l\Gamma)^2 D}{(E+Fl)(E-Fl)}\left[\exp\left(\frac{E+Fl}{\Gamma D}\right) - \exp\left(\frac{2Fl}{\Gamma D}\right)\right] + \frac{l^2\Gamma}{E-Fl} . \quad (6)$$

This time T_i also corresponds to the time that the DNAP stays at (n+3)-th site before the incorporation of correct n-the base.

Using Eq. (6), the calculated results of T_i versus E under no load, i.e., $F = 0$, is shown in Figure 6a. In order to ensure the completion of the nucleotide incorporation, it is required that $T_i \gg T_d = 3.33$ ms for T7 DNAP (e.g., $T_i \geq 10\ T_d = 33.3$ ms). From Figure 6a, it is seen that, for $T_i \geq 33.3$ ms, $E = E_1 + E_2$ should be equal to or larger than $23.34\ k_B T$. From Eq. (6), the calculated results of T_i versus F for different values of E are shown in Figure 6b. It is seen that T_i does not increases largely with the increase of the external load. For example, under a forward load of 40 pN, T_i decreases only about 5 folds compared to that under no load; while under a backward load of 40 pN, T_i increases only about 5 folds compared to that under no load.

Based on our model, after the incorporation of a mismatch, if the mismatch is not excised for an exonuclease-deficient mutant of DNAP or a reverse transcriptase that lacks a proofreading function (Xie 2009c), only when the polymerase site is moved from the position at n-th site to position at (n+1)-th site can the forward polymerization continue. Thus, the polymerase rate on top of a mismatch can be calculated by

$$K_{mismatch} \approx (T_i + T_{inc})^{-1} , \quad (7)$$

where T_{inc} is the time for incorporating a nucleotide complementary to the (n+1)-base on the template strand. Note here that, if the nucleotide dNTP that has been bound to the polymerase site when the polymerase site was located at the n-th base is not complimentary to the (n+1)-th base of the template, the dNTP will be released rapidly due to the low binding affinity (see the induced-fit mechanism as discussed in Section 2.1). Thus, we have neglected the short release time of the incorrect nucleotide dNTP from the polymerase site in Eq. (7).

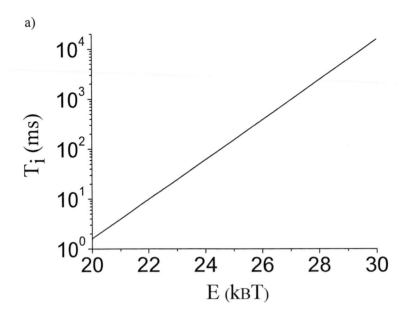

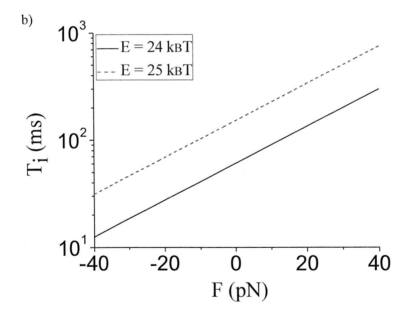

Figure 6. Calculated results of time T_i for DNAP to move to next site after the incorporation of an incorrect n-th base. (a) T_i versus E under $F = 0$. (b) T_i versus F for different values of E.

Structures of DNAP with mismatches at the primer-terminus (Johnson and Beese, 2004) showed that, for some structures, the mismatch is placed near the polymerase site, which corresponds to the case that the polymerase site is at the position of n-th site, as shown in Figure 1c'. For other structures, the polymerase site is moved to the position of next (n+1)-th site. In the latter case, since the mismatched base pair is not in the correct orientation as for the case of matched base pair, this impairs the catalysis of nucleotide incorporation, i.e., increases the incorporation time T_{inc} on top of the incorrect incorporation. The increased incorporation time T_{inc} should be dependent of the identity of the mismatch, thus resulting in the mispair extension efficiency identity-dependent, which is consistent with the experimental results (Mendelman et al., 1990; Johnson et al., 2000; Washington et al., 2001; Picher et al., 2006). It is also noted that different types of DNAP should have different values of binding affinity E, giving very different values of T_i (Figure 6a). Moreover, different types of DNAP should have different values of T_{inc}. Therefore, different types of DNAP, which have very different values of $K_{mismatch} = (T_i + T_{inc})^{-1}$, show very different mispair extension efficiencies. This is also consistent with the available experimental results (Mendelman et al., 1990; Johnson et al., 2000; Washington et al., 2001; Picher et al., 2006; Yu and Goodman, 1992).

6.2. Dependence of Polymerization Rate on Tension Applied to Template

As in the experiment of Wuite et al. (2000) and Maier et al. (2000), in this Section we consider that a mechanical tension, F, is applied to the DNA template, instead of an external force acted on the DNAP as studied above.

In the pathway shown in Figure 3, the reactions related to the polymerization are the transitions $3 \to 4 \to 5 \to 6 \to 7 \to 3'$, where $4 \to 5$ and $6 \to 7$ involve the conformational changes from "open finger" to "closed finger" states and vice verse, respectively. Here, for simplicity, we have neglected the backward transitions because each backward transition rate is much smaller then the corresponding forward transition rate.

For the case that the mechanical force F is acted on the bead attached to the ssDNA template (Wuite et al., 2000; Maier et al., 2000), only during the processes of the conformational change from the "open finger" to "closed finger" states and vice versa, which drives the bead attached to the ssDNA template moved backwards to the chemical reaction active position and then forwards, can the DNAP experience this mechanical force F. Therefore, the applied mechanical force F mainly has the effect on the transition rates of $4 \to 5$ and $6 \to 7$, while has little effect on other transitions. Moreover, since the dsDNA end is fixed and the 5' end of the ssDNA can be moved, the tension on the DNA template should resist the transition of $4 \to 5$ while assist the transition of $6 \to 7$. In addition, since the transition rates k_{56} and k_{73} are much larger than other transition rates k_{45} and k_{67}, we approximately have the polymerization rate k_{pol} as follows

$$\frac{1}{k_{pol}} = \frac{1}{k_{34}} + \frac{1}{k_{45}} + \frac{1}{k_{67}}, \qquad (8)$$

where $k_{34} = k_b[\text{dNTP}]$ is tension independent. According to Arrhenius-Eyring kinetics, k_{45} and k_{67} are written as

$$k_{45} = k_{45}^{(0)} \exp\left(-\frac{Fl_F}{k_B T}\right), \qquad (9a)$$

$$k_{67} = k_{67}^{(0)} \exp\left(\frac{Fl_F}{k_B T}\right), \qquad (9b)$$

where l_F is the movement distance of the binding site P induced by the inward (outward) rotation of the fingers, and $k_{45}^{(0)}$ and $k_{67}^{(0)}$ are the transition rates of $4 \to 5$ and $6 \to 7$ under no mechanical tension, respectively.

From the crystal structure of T7 DNAP (Doublie et al., 1998; Doublie and Ellenberger, 1998), it is estimated that the movement distance of the binding site P is $l_F \approx 0.5$ nm. Taking this value for l_F and the measured values of $k_{45}^{(0)} = 300\,\text{s}^{-1}$ and $k_{67}^{(0)} = 1200\,\text{s}^{-1}$ from kinetic experiment for T7 DNAP (Patel et al., 1991), using Eqs. (8) and (9) we calculate k_{pol} versus F at saturating [dNTP]. The result is shown by dashed line in Figure 7a and Figure 8a. It is noted that, without any adjustable parameter, the calculated results are in agreement with the experimental ones for T7 DNAP in Wuite et al. (2000) and the experimental results for Sequensse (an exonuclease-deficient mutant T7 DNAP) in Maier et al. (2000).

In fact, by adjusting the values of parameters for $k_{45}^{(0)}$, $k_{67}^{(0)}$ and l_F, we can fit the experimental results for T7 DNAP better. The result is shown by solid line in Figure 7a. Note that the fitted value of $l_F = 0.5$ nm is consistent with the estimate from the crystal structure. The fitted value of $k_{45}^{(0)} = 500\,\text{s}^{-1}$ is close to the kinetically measured value of about $300\,\text{s}^{-1}$. However, the fitted value of $k_{67}^{(0)} = 180\,\text{s}^{-1}$ is far from the kinetically measured value of about $1200\,\text{s}^{-1}$ (the discussion on the difference will be given in Section 9). Taking $k_b = 70$ $\mu\text{M}^{-1}\text{s}^{-1}$ (Johnson, 1993) and $k_{45}^{(0)} = 500\,\text{s}^{-1}$ and $k_{67}^{(0)} = 180\,\text{s}^{-1}$ that are the same as those used for solid line in Figure 7a, we show the calculated results of k_{pol} versus F for different concentrations of dNTP in Figure 7b.

Dynamics of DNA Polymerase

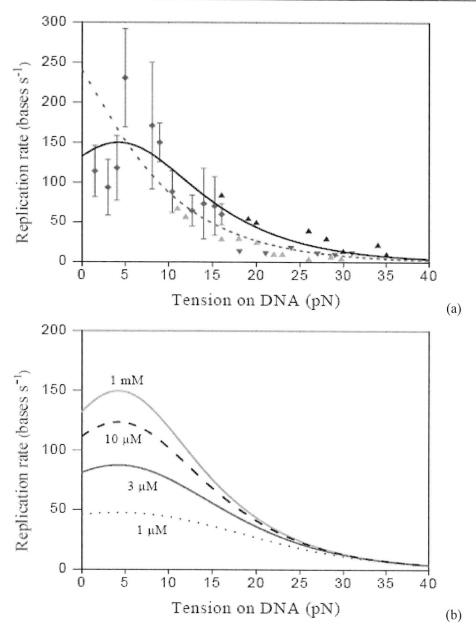

Figure 7. Polymerase rate of T7 DNAP versus tension applied to DNA template calculated by using Eqs. (8) and (9), with $l_F = 0.5$ nm that is estimated from the crystal structure (Doublie et al., 1998). (a) At saturating dNTP concentration. The dashed line is calculated by using $k_{45}^{(0)} = 300$ s^{-1} and $k_{67}^{(0)} = 1200$ s^{-1} taken from kinetic experimental results (Patel et al., 1991). The solid line is fitted to the experimental data of Wuite et al. (2000), with fitted values $k_{45}^{(0)} = 500$ s^{-1} and $k_{67}^{(0)} = 180$ s^{-1}. The data points are taken from Wuite et al. (2000), with diamonds representing 50 polymerization bursts taken at 11 different tensions and triangles representing three traces fitted through a succession of replication bursts measured at constant end-to-end distances until a stalling force in reached. (b) At different dNTP concentrations, with $k_{45}^{(0)} = 500$ s^{-1} and $k_{67}^{(0)} = 180$ s^{-1}.

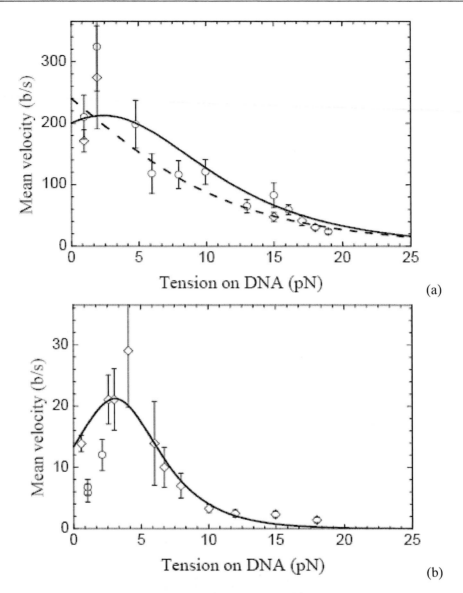

Figure 8. Polymerase rate of Sequensse (a) and Klenow fragment (b) versus tension applied to DNA template calculated by using Eqs. (8) and (9). The dashed line in (a) is calculated by using $l_F = 0.5$ nm that is estimated from the crystal structure (Doublie et al., 1998), $k_{45}^{(0)} = 300\,\mathrm{s}^{-1}$ and $k_{67}^{(0)} = 1200\,\mathrm{s}^{-1}$ taken from kinetic experimental results (Patel et al., 1991). The solid line in (a) is fitted to the experimental data of Maier et al. (2000), with fitted values $k_{45}^{(0)} = 600\,\mathrm{s}^{-1}$, $k_{67}^{(0)} = 300\,\mathrm{s}^{-1}$ and $l_F = 0.6$ nm. The solid line in (b) is fitted to the experimental data of Maier et al. (2000), with fitted values $k_{45}^{(0)} = 120\,\mathrm{s}^{-1}$, $k_{67}^{(0)} = 15\,\mathrm{s}^{-1}$ and $l_F = 1.4$ nm. The data points are taken from Maier et al. (2000), with circles and diamonds in (a) measured in high stringency primer hybridization and random priming, respectively, and circles and diamonds in (b) measured in high and low stringency primer hybridizations, respectively.

Similar to the fitting of the experimental results of Wuite et al. (2000) in Figure 7a, the experimental results for Sequensse and Klenow fragment given by Maier et al. (2000) can also be fitted well by using Eqs. (8) and (9). The results are shown in Figure 8a (for Sequensse) and Figure 8b (for Klenow fragment), where the movement distances of ssDNA driven by the rotation of the fingers are $l_F = 0.6$ nm for Sequensse and $l_F = 1.4$ nm for Klenow fragment. The fitted values of $k_{45}^{(0)} = 600\,\mathrm{s}^{-1}$, $k_{67}^{(0)} = 300\,\mathrm{s}^{-1}$ and $l_F = 0.6$ nm for Sequensse are close to those fitted to the experimental results by Wuite et al. (2000). Note also that the fitted values of both $k_{45}^{(0)} = 120\,\mathrm{s}^{-1}$ and $k_{67}^{(0)} = 15\,\mathrm{s}^{-1}$ for Klenow fragment are close to the kinetically measured values of ~50 s^{-1} and ~15 s^{-1} (Dahlberg and Benkovic, 1991), respectively.

7. DNA Transitions between Polymerase and Exonuclease Sites

In this Section, we study the transition rate of DNA from the polymerase to exonuclease sites and that from the exonuclease to polymerase sites. In the pathway shown in Figure 3, these involve the transitions 3'☐ 2 →2' ☐ 3.

7.1. Equations for Transfer Rates between Polymerase and Exonuclease Sites

Similar to Eq. (1) that describes the movement of DNAP along DNA, the movement of the 3' end of the 3' – 5' ssDNA primer can be described by the following Langevin equation

$$\gamma \frac{dz}{dt} = -\frac{\partial U(z)}{\partial z} + \zeta(t) \tag{10}$$

Here z represents position of the 3' end of the 3' – 5' ssDNA primer along z axis that connects the polymerase and exonuclease sites, as shown in Figure 2a. For convenience, the origin of the z axis is taken at the polymerase site. γ is the drag coefficient on the 3' – 5' ssDNA primer when its 3' end moves between the polymerase and exonuclease sites, i.e., along the z axis. The potential is written as $U(z) = U_{DNA}(z) + U_X(z)$, where $U_{DNA}(z)$ is the potential required to break some base pairs as the 3' end of the 3' – 5' ssDNA primer moves away from the polymerase site, in the vicinity of which the ssDNA template is bound by the ssDNA-binding site P, and $U_X(z)$ is the potential of the 3' – 5' ssDNA primer interacting with the binding site X. It is noted here that an implicit approximation we have adopted is that the palm domain has a very weak interaction with dsDNA composed of several nascent base pairs so that it has negligible effect on the unwinding of these base pairs as the 3' end of the 3' – 5' ssDNA primer moves from the polymerase to exonuclease sites (see discussions in

Section 9). The last term $\zeta(t)$ in Eq. (10) is the fluctuating Langevin force, with $\langle\zeta(t)\rangle=0$ and $\langle\zeta(t)\zeta(t')\rangle = 2k_B T\gamma\delta(t-t')$.

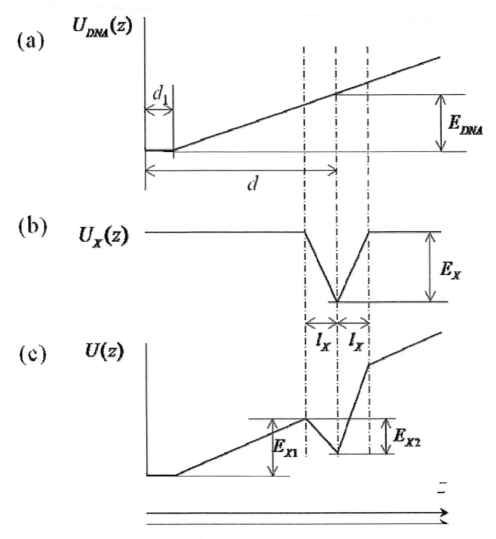

Figure 9. Potentials related to DNA transition between the polymerase and exonuclease sites. (a) Form of potential $U_{DNA}(z)$ to break some base pairs as 3' end of the 3' – 5' ssDNA primer moves away from the polymerase site. (b) Form of potential $U_X(z)$ of 3' – 5' ssDNA interacting with the binding site X. (c) Form of potential $U(z)=U_{DNA}(z)+U_X(z)$.

Consider that m base pairs are needed to break when the 3' end of the 3' – 5' ssDNA primer is moved to the exonuclease site. Since when the 3' end of the 3' – 5' ssDNA primer is located near the polymerase site no base pair is needed to break, it is thus inferred that, as the 3' end is moved from $z = 0$ (the polymerase site) to $z = d$ (the exonuclease site), the free energy required to break base pairs increases from 0 to E_{DNA}. Here $E_{DNA} = m(E_{bp} + E_{BS})$, where E_{bp} is the hydrogen-bonding free energy of each base pair and E_{BS} is the free energy of base-stacking interaction between adjacent base pairs. For simplicity, it is reasonable to take that the free energy increases linearly along the z axis. Note that, if there are n misincorporated bases at the 3' end of the 3' – 5' ssDNA primer, then $(m-n)$ base pairs are needed to break when the 3' end of the 3' – 5' ssDNA primer is bound to the binding site X and, in the range from $z = 0$ to $z = d_1$, the potential $U_{DNA}(z)$ is unchanged, where $d_1 = dn/m$. As a result, for the case that there are n misincorporated bases at the 3' end of the 3' – 5' ssDNA primer, the potential $U_{DNA}(z)$ approximately has the form shown in Figure 9a, where $E_{DNA} = (m-n)(E_{bp} + E_{BS})$. To precisely determine E_{BS} is a rather complicated subject and, for calculation, we take $E_{BS} = \varepsilon E_{bp}$, where ε is the base stacking intensity. As it is known, in general, the stacking of the two bases often contributes as much as, or more than, half of the free energy of the total base pair (Kool, 2001). Thus, in the calculation, we take ε in between 0.5 and 0.7.

The interaction potential $U_X(z)$ between 3'–5' ssDNA and the binding site X can be approximately shown in Figure 9b, where E_X is the binding affinity. With $U_{DNA}(z)$ in Figure 9a and $U_X(z)$ in Figure 9b, the potential $U(x) = U_{DNA}(z) + U_X(z)$ is thus given in Figure 9c.

Similar to Eq. (3), the mean first-passage time for the 3' end of the 3' – 5' ssDNA primer to move from the polymerase to exonuclease sites can be calculated by

$$T = \frac{1}{D_T}\int_0^d dx \exp[U(x)/(\gamma D_T)]\int_0^x \exp[-U(y)/(\gamma D_T)]dy \quad \text{(Gardiner, 1983), where}$$

$D_T = k_B T/\gamma$. By integration of it with the form of $U(z)$ given in Figure 9c, we obtain the mean first-passage time as

$$T_{P\to X} = \frac{d_1^2}{2D_T} + \frac{(d-d_1-l_X)\gamma}{E_{X1}}\left\{\left[d_1 + \frac{(d-d_1-l_X)k_B T}{E_{X1}}\right]\left[\exp\left(\frac{E_{X1}}{k_B T}\right)-1\right]-(d-d_1-l_X)\right\}$$
$$+ \frac{l_X\gamma(d-d_1-l_X)k_B T}{E_{X1}E_{X2}}\exp\left(\frac{E_{X1}}{k_B T}\right)\left[1-\exp\left(-\frac{E_{X1}}{k_B T}\right)\right]\left[1-\exp\left(-\frac{E_{X2}}{k_B T}\right)\right]$$
$$+ \frac{l_X^2\gamma}{E_{X2}} - \frac{l_X^2\gamma k_B T}{(E_{X2})^2}\left[1-\exp\left(-\frac{E_{X2}}{k_B T}\right)\right] + \frac{l_X\gamma d_1}{E_{X2}}\exp\left(\frac{E_{X1}}{k_B T}\right)\left[1-\exp\left(-\frac{E_{X2}}{k_B T}\right)\right],$$

where $E_{X1} = E_{DNA}\left(1 - \dfrac{l_X}{d - d_1}\right)$, $E_{X2} = E_X - E_{DNA}\dfrac{l_X}{d - d_1}$, $d_1 = \dfrac{n}{m}d$, $E_{DNA} = (m - n)E_{bp}(1 + \varepsilon)$. Considering E_{X1} or $E_{X2} \gg k_B T$, the above equation can be approximately written as

$$T_{P \to X} = \dfrac{d_1^2}{2D_T} + \dfrac{(d - d_1 - l_X)\gamma}{E_{X1}}\left\{\left[d_1 + \dfrac{(d - d_1 - l_X)k_B T}{E_{X1}}\right]\left[\exp\left(\dfrac{E_{X1}}{k_B T}\right) - 1\right] - (d - d_1 - l_X)\right\}$$
$$+ \dfrac{l_X \gamma}{E_{X2}}\left\{\left[d_1 + \dfrac{(d - d_1 - l_X)k_B T}{E_{X1}}\right]\exp\left(\dfrac{E_{X1}}{k_B T}\right) + l_X\right\}. \tag{11}$$

Similarly, we obtain the mean first-passage time for the 3' end of the 3' – 5' ssDNA primer to move from the exonuclease to polymerase sites as follows

$$T_{X \to P} = \dfrac{l_X \gamma}{E_{X2}}\left\{\dfrac{l_X k_B T}{E_{X2}}\left[\exp\left(\dfrac{E_{X2}}{k_B T}\right) - 1\right] - l_X\right\}$$
$$+ \dfrac{l_X(d - d_1 - l_X)\gamma k_B T}{E_{X2} E_{X1}}\left[1 - \exp\left(-\dfrac{E_{X1}}{k_B T}\right)\right]\left[\exp\left(\dfrac{E_{X2}}{k_B T}\right) - 1\right]$$
$$+ \dfrac{(d - d_1 - l_X)^2 \gamma}{E_{X1}} - \dfrac{(d - d_1 - l_X)^2 \gamma k_B T}{(E_{X1})^2}\left[1 - \exp\left(-\dfrac{E_{X1}}{k_B T}\right)\right]$$
$$+ \dfrac{l_X \gamma}{E_{X2}}\exp\left(\dfrac{E_{X2} - E_{X1}}{k_B T}\right)\left[1 - \exp\left(-\dfrac{E_{X2}}{k_B T}\right)\right]d_1$$
$$+ \dfrac{(d - d_1 - l_X)\gamma}{E_{X1}}\left[1 - \exp\left(-\dfrac{E_{X1}}{k_B T}\right)\right]d_1 + \dfrac{d_1^2}{2D_T}.$$

Considering E_{X1} or $E_{X2} \gg k_B T$, the above equation can be approximately written as

$$T_{X \to P} = \dfrac{l_X \gamma}{E_{X2}}\left\{\dfrac{l_X k_B T}{E_{X2}}\left[\exp\left(\dfrac{E_{X2}}{k_B T}\right) - 1\right] - l_X\right\}$$
$$+ \dfrac{l_X(d - d_1 - l_X)\gamma k_B T}{E_{X2} E_{X1}}\exp\left(\dfrac{E_{X2}}{k_B T}\right) + \dfrac{(d - d_1 - l_X)^2 \gamma}{E_{X1}}$$
$$+ \left[\dfrac{l_X \gamma}{E_{X2}}\exp\left(\dfrac{E_{X2} - E_{X1}}{k_B T}\right) + \dfrac{(d - d_1 - l_X)\gamma}{E_{X1}}\right]d_1 + \dfrac{d_1^2}{2D_T}. \tag{12}$$

The mean transfer rate of DNA from the polymerase to exonuclease sites is $K_{P \to X} = 1/T_{P \to X}$, where $T_{P \to X}$ is calculated from Eq. (11). Similarly, the mean transfer rate from the exonuclease to polymerase sites is $K_{X \to P} = 1/T_{X \to P}$, where $T_{X \to P}$ is calculated from Eq. (12).

7.2. Results for Transfer Rates between Polymerase and Exonuclease Sites

To calculate the mean transfer rates by using Eqs. (11) and (12), we firstly determine values of relevant parameters from available crystal structures and other experimental data.

From the crystal structure, it is noted that the exonuclease site is separated from the polymerase site by about 35 Å (Goel et al., 2003). Thus, we take $d = 3.5$ nm in our calculation. The structure also suggested that the DNA melts at least 8 or 9 base pairs to bring a single-stranded segment of DNA into the exonuclease site (Johnson, 1993). Thus, we take $m = 9$. It is known that the free energy to break an A:T base pair is $2 k_B T$ while the free energy to break a G:C base pair is $3 k_B T$. Thus, we take the mean free energy to break one base pair to be $E_{bp} = 2.5 k_B T$. As mentioned above, we take the base stacking intensity ε in between 0.5 and 0.7. The interaction between the 3' – 5' ssDNA primer and the binding site X can be considered to be via electrostatic. Thus, the interaction distance is equal about to the Debye length in solution. Inside cell the Debye length is about 1nm. Thus we take $l_X = 1$ nm. From Stokes law, the drag coefficient on a moving sphere of radius $r = 1$nm is $\gamma = 6\pi\eta r \approx 2\times 10^{-11}$ kg s^{-1}, where the viscosity η of the aqueous medium inside cell is 0.01 g cm^{-1}s^{-1}. It is expected that the drag coefficient on the 3' – 5' ssDNA primer is around this value. Thus, we take γ in the range between 0.5×10^{-11} kg s^{-1} and 3×10^{-11} kg s^{-1}.

Since $d - d_1 - l_X > l_X$, the third term in the right-hand side of Eq. (11) is usually smaller than the sum of the first and second terms. For an approximation, we firstly neglect the third term in Eq. (11) to calculate the mean transfer rate of DNA from the polymerase to exonuclease sites $K_{P \to X}$. The results of $K_{P \to X}$ versus ε with fixed $\gamma = 2\times 10^{-11}$ kg s^{-1} and $K_{P \to X}$ versus γ with fixed $\varepsilon = 0.6$ are shown in Figure 10a and Figure 10b, respectively. From Figure 10a, it is seen that in the range of ε between 0.5 and 0.7, $K_{P \to X}$ is in the range between 0.65 s^{-1} and 0.034 s^{-1} in the absence of misincorporation (i.e., when $n = 0$), while $K_{P \to X}$ is in the range between 7.5 s^{-1} and 0.55 s^{-1} in the presence of one misincorporation (i.e., when $n = 1$). From Figure 10b, we see that, in the range of γ between 0.5×10^{-11} kg s^{-1} and 3×10^{-11} kg s^{-1}, $K_{P \to X}$ has the value in between 0.6 s^{-1} and 0.06 s^{-1} when $n = 0$, while $K_{P \to X}$ has the value in between 8.2 s^{-1} and 0.82 s^{-1} when $n = 1$. These results of $K_{P \to X}$ are close to the experimental value of about $0.16\,\mathrm{s}^{-1}$ in the absence of misincorporation and of about $2.3\,\mathrm{s}^{-1}$ in the presence of one misincorporation (Donlin et al., 1991). In particular, from Figure 10a it is seen that, at $\varepsilon = 0.591$ and $\gamma = 2\times 10^{-11}\,\mathrm{kg\,s}^{-1}$ (indicated by dotted line), the calculated $K_{P\to X} = 0.17\,\mathrm{s}^{-1}$ when $n = 0$ and $K_{P\to X} = 2.3\,\mathrm{s}^{-1}$ when $n = 1$ are consistent with the measured values of about $0.16\,\mathrm{s}^{-1}$ and $2.3\,\mathrm{s}^{-1}$, respectively. Alternatively, from Figure 10b we see that, at $\varepsilon = 0.6$ and $\gamma = 1.78\times 10^{-11}\,\mathrm{kg\,s}^{-1}$ (indicated by dotted line), the calculated $K_{P\to X} = 0.167\,\mathrm{s}^{-1}$ when $n = 0$ and $K_{P\to X} = 2.3\,\mathrm{s}^{-1}$ when $n = 1$ are also consistent with the measured values of about $0.16\,\mathrm{s}^{-1}$ and $2.3\,\mathrm{s}^{-1}$, respectively. The former values ($\varepsilon = 0.591$ and $\gamma = 2\times 10^{-11}\,\mathrm{kg\,s}^{-1}$) are very close to the latter ones ($\varepsilon = 0.6$ and $\gamma = 1.78\times 10^{-11}\,\mathrm{kg\,s}^{-1}$). Considering that, in the above calculation, we have neglected the third term in the right-hand side of Eq. (11), which implies an overestimation of $K_{P\to X}$, in the following calculation we thus take $\varepsilon = 0.6$ and $\gamma = 1.2\times 10^{-11}\,\mathrm{kg\,s}^{-1}$ that is slightly smaller than $1.78\times 10^{-11}\,\mathrm{kg\,s}^{-1}$.

Now, we study the mean transfer rate, $T_{X\to P}$, from the exonuclease to polymerase sites. Since the value of the binding affinity E_X is unavailable, we thus take it as a variable parameter. Using Eq. (12), we calculate the mean transfer rate from the exonuclease to polymerase sites, $K_{X\to P}$. The results of $K_{X\to P}$ versus E_X for any value of n are shown in Figure 11, provided that n satisfies $d - d_1 = \left(1 - \dfrac{n}{m}\right)d \geq l_X$. It is seen that $K_{X\to P}$ has a large value even for the very large value of E_X, which is consistent with the experimental data (Donlin et al., 1991). From the experimentally determined rate of about $714\,\mathrm{s}^{-1}$ (Donlin et al., 1991), we estimate $E_X \approx 28\,k_B T$. With this value of E_X and $\varepsilon = 0.6$, from the definitions of E_1 and E_2 given in Eqs. (11) and (12), respectively, we can easily obtain that the depths of the two potential wells shown in Figure 9c are $E_{X1} \approx 25.7\,k_B T$ and $E_{X2} \approx 17.7\,k_B T$ for the matched DNA primer/template ($n = 0$) and $E_{X1} \approx 17.7\,k_B T$ and $E_{X2} \approx 17.7\,k_B T$ for the doubly mismatched primer/template ($n = 2$). From these values of E_1 and E_2 it is thus inferred that, for the matched DNA primer/template, the equilibrium association constant for the polymerase site is much larger than that for the exonuclease site; while for the doubly mismatched primer/template, the equilibrium association constant for the polymerase site is nearly equal to that for the exonuclease site. These are consistent with the experimental data of Bailey et al. (2004).

Then, with $E_X = 28\,k_B T$, $\varepsilon = 0.6$ and $\gamma = 1.2\times 10^{-11}\,\mathrm{kg\,s}^{-1}$, we calculate transfer rate $K_{P\to X}$ by using Eq. (11). The results are $K_{P\to X} = 0.157\,\mathrm{s}^{-1}$ when $n = 1$ and $K_{P\to X} = 2.3\,\mathrm{s}^{-1}$ when $n = 2$. These are in good agreement with the measured values of about $0.16\,\mathrm{s}^{-1}$ and

$2.3\,s^{-1}$, respectively. In addition, we also obatin $K_{P\to X} = 32.5\,s^{-1}$ when $n = 2$ and $K_{P\to X} = 439\,s^{-1}$ when $n = 3$. The results are also consistent with the experiment of Carver et al. (1994), showing that two or more consecutive mismatches caused the transition to the exonuclease site by at least a 250-fold increase in the equilibrium partitioning constant relative to the matched sequence.

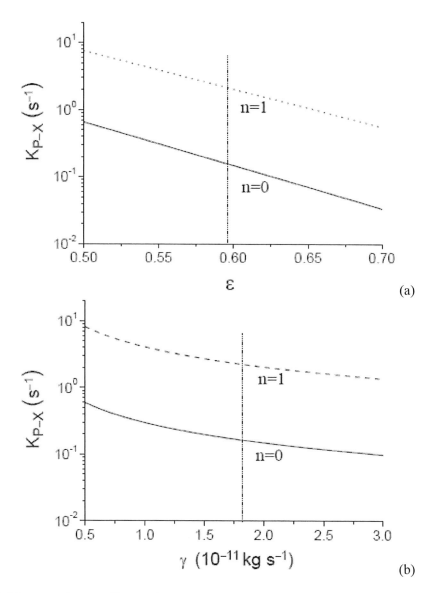

Figure 10. Mean transfer rate $K_{P\to X}$ of the 3' – 5' ssDNA primer from polymerase to exonuclease sites calculated by neglecting the third term in the right-hand side of Eq. (11). (a) $K_{P\to X}$ versus base-stacking intensity ε with fixed drag coefficient $\gamma = 2\times 10^{-11}\,kg\,s^{-1}$. (b) $K_{P\to X}$ versus γ with fixed $\varepsilon = 0.6$.

Moreover, we give some predictive results for the rate $K_{P \to X}$ of transfer of DNA containing only A:T base pairs and that of DNA containing only G:C base pairs. For the former DNA, we take $E_{bp} = 2 k_B T$. Then using Eq. (11), we obtain $K_{P \to X} = 19.2 \text{ s}^{-1}$ when $n = 0$ and $K_{P \to X} = 156 \text{ s}^{-1}$ when $n = 1$. These are much larger than 0.16 s^{-1} (when $n = 0$) and 2.3 s^{-1} (when $n = 1$) for the DNA containing both A:T and G:C base pairs. For the DNA containing only G:C base pairs, by taking $E_{bp} = 3 k_B T$ we obtain $K_{P \to X} = 1.17 \times 10^{-3} \text{ s}^{-1}$ when $n = 0$ and $K_{P \to X} = 0.03 \text{ s}^{-1}$ when $n = 1$. These are much smaller than 0.16 s^{-1} (when $n = 0$) and 2.3 s^{-1} (when $n = 1$).

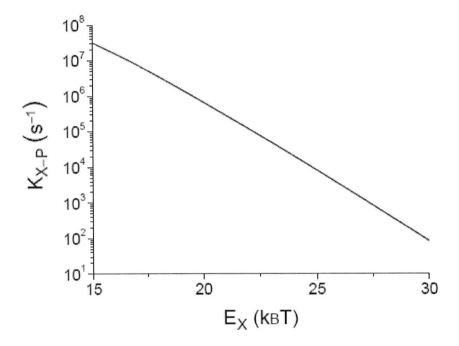

Figure 11. Calculated results of mean transfer rate $K_{X \to P}$ of 3'−5' ssDNA primer from exonuclease to polymerase sites versus binding affinity E_X of 3'−5' ssDNA for binding site X.

8. Backward Motion of DNAP under Template Tension above 40 pN

An interesting experimental result in Wuite et al. (2000) is that, when tension applied to the template is increased above 40 pN, a fast exonuclease rate (~ 30 bases s^{-1}) was initiated with or without dNTPs present and the exonuclease rate becomes force independent. Since it has been known that forces greater than 40 pN may promote melting or fraying in dsDNA, it was suggested that the exonuclease rate simply reflects the rate of fraying of DNA at the tensions applied to the template (Wuite et al., 2000). However, the experiment on the effect of template tension on the activity of *Escherichia coli* exonuclease I, which can only attack ssDNA at its 3' end, demonstrated that under a tension of 50 pN the bases were removed at a rate of ~ 200 s^{-1} (Wuite et al., 2000). This implies that the fraying rate of DNA is at least 200 s^{-1}, thus excluding fraying in DNA induced by tension as the rate-limiting step for the exonucleolysis studied in Wuite et al. (2000). In addition, from the exonuclease rate $k_x =$ 900 s^{-1} measured biochemically (Donlin et al., 1991), the exonucleolysis after the 3' – 5' ssDNA primer is transferred to the exonuclease domain is also excluded to be rate limiting. In the following, based on our model we present a quantitative study of the exonuclease rate under the template tension larger than 40 pN.

When the tension applied to the template is above 40 pN, it deforms the dsDNA geometry at its 3' end, which greatly reduces the interacting intensity E_{bp} between some complementary base pairs near the 3' end. For example, for $E_{bp} < 1 k_B T$, with other parameter $\varepsilon = 0.6$, $\gamma = 1.2 \times 10^{-11}$ kg s^{-1} and $E_X = 28 k_B T$ determined in Section 7.2, we obtain $K_{P \to X} > 1.7 \times 10^5$ s^{-1} by using Eq. (11) and obtain $K_{X \to P} < 1.2$ s^{-1} by using Eq. (12). The very large value of $K_{P \to X}$ but the small value of $K_{X \to P}$ imply that, when the tension on the template is larger than 40 pN, the 3' – 5' ssDNA primer is considered to be nearly always positioned at exonuclease domain, which is different from the case when a small tension is applied to the template (the case studied in Section 6.2). Thus, the ssDNA-binding site P in the finger domain, the ssDNA-binding site X in the exonuclease domain and the dsDNA-binding site in the thumb domain always binds DNA simultaneously.

Without loss of generality, we consider that the ssDNA-binding site X covers four bases on the 3' – 5' ssDNA primer. We represent the position of the ssDNA-binding site X by that of the rightmost point on the binding site X. Before the excision of (n-3)-th base on the 3' – 5' ssDNA primer, the interaction potential $V_X(x)$ of the binding site X with 3' – 5' ssDNA has the form shown in the upper diagram of Figure 12a; while after the excision of the *n*-th base, $V_X(x)$ becomes the form shown in the lower diagram of Figure 12a. Since the ssDNA-binding site X at position of (n-1)-th base can only bind three bases, the binding affinity E_X' is smaller than E_X that corresponds to binding all four bases. Considering that the interaction between the enzyme and DNA is via electrostatic force, with the interaction distance which is approximately equal to the Debye length (~ 1 nm) in solution larger than the distance $p =$ 0.34 nm between two successive base pairs, it is expected that E_{X0} is larger than E_X and E_X'. The potential $V_{ssDNA}(x)$ of the binding site P interacting with the ssDNA template and the potential $V_{dsDNA}(x)$ of the dsDNA-binding site with dsDNA have the same forms before and after the excision of (n-3)-th base, with the form of $V_{ssDNA}(x) + V_{dsDNA}(x)$ being shown in Figure 12b. Note that, because the dsDNA segment is partially deformed under the tension above 40 pN, it is expected that the potential depth of $V_{dsDNA}(x)$, E_2^*, should be smaller than E_2, thus giving $E^* = E_1 + E_2^* < E = E_1 + E_2$.

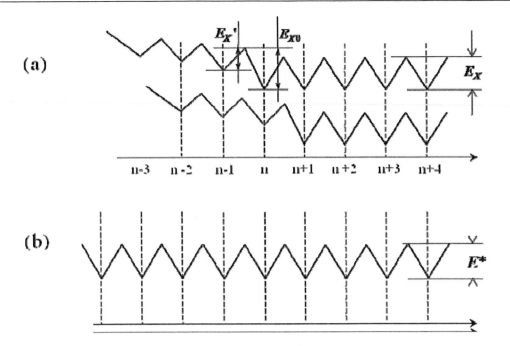

Figure 12. Interactional potential of DNAP with DNA under the case that the tension applied to the template is above 40 pN. (a) Interaction potential $V_X(x)$ of binding site X with 3' – 5' ssDNA primer. Upper diagram represents $V_X(x)$ before the excision of (n-3)-th base, while lower diagram represents $V_X(x)$ after the excision. (b) Sum of the potential of binding site P interacting with 5' – 3' ssDNA and that of dsDNA-binding sites with dsDNA.

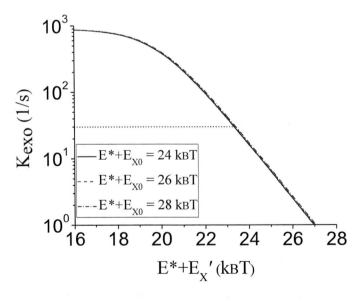

Figure 13. Calculated results of exonuclease rate K_{exo} versus $E^* + E_X'$ under template tension larger than 40 pN.

The backward movement of DNAP along DNA is still described by Eq. (1) but with $V(x) = V_{ssDNA}(x) + V_{dsDNA}(x) + V_X(x)$. Similar to the derivation of Eqs. (4) and (5), the moving time T_X is obtained as follows

$$T_X = \frac{(l\Gamma)^2 D}{E^* + E_X'}\left[\exp\left(\frac{E^* + E_X'}{\Gamma D}\right) - 1\right] - \frac{l^2\Gamma}{E^* + E_X'}$$

$$+ \frac{(l\Gamma)^2 D}{(E^* + E_X')(E^* + E_{X0})}\left[\exp\left(\frac{E^* + E_X'}{\Gamma D}\right) - \exp\left(\frac{(E^* + E_X') - (E^* + E_{X0})}{\Gamma D}\right)\right]$$

$$\cdot \left[1 - \exp\left(-\frac{E^* + E_X'}{\Gamma D}\right)\left(1 + \frac{E^* + E_X'}{E^* + E_{X0}}\right)\right] + \frac{l^2\Gamma}{E^* + E_{X0}} \quad . \tag{13}$$

The backward moving rate or the exonuclease rate is calculated by $K_{exo} \approx (T_X + k_x^{-1})^{-1}$, where $k_x = 900 \text{ s}^{-1}$ (Donlin et al., 1991). In the case of $E^* = E_1 + E_2^* = 0$, the moving mechanism and equation (13) becomes those for exonuclease I and RecJ (Xie, 2009d).

Using Eq. (13), the calculated results of K_{exo} versus $E^* + E_X'$ for different values of $E^* + E_{X0}$ are shown in Figure 13. As anticipated, the exonuclease rate K_{exo} is insensitive to the variation of $E^* + E_{X0}$. It is seen that $K_{exo} = 30$ bases s^{-1} (indicated by dotted line) occurs at $E^* + E_X' \approx 23.4 k_B T$. Considering that $E^* = E_1 + E_2^* < E = E_1 + E_2$ and $E = E_1 + E_2 \geq 23.34 k_B T$ (see Section 6.1), the estimated value of $E^* + E_X' \approx 23.4 k_B T$ is reasonable. It is noted that the backward moving time T_X is much longer than the exonucleolysis time and, thus, the backward motion is rate limiting. Since the backward moving time T_X is only determined by the three binding affinities, it is thus concluded that the exonuclease rate is independent of the presence or absence of dNTP, which is consistent the experimental observation (Wuite et al., 2000). Moreover, it is predicted that the template tension at which the polymerase reverses direction should be independent of dNTP concentration. However, for the case of $E^* = E_1 + E_2^* = 0$, which corresponds to exonuclease I and RecJ, the small value of E_X' gives the moving time much shorter than the exonucleolysis time and, thus, the exonucleolysis is rate limiting (Xie, 2009d).

9. Discussion

In our model, after the incorporation of a mismatched base, the polymerase site is still positioned at the n-th base for a long time (Fig. 1d'). This gives sufficient time for the mismatched 3' – 5' ssDNA primer tail to switch to the binding site X for proofreading. After the mismatched base is excised and then the 3' – 5' ssDNA primer is switched back to the polymerase site, the bound dNTP becomes readily able to be incorporated. This implies that, after the proofreading, the DNAP rapidly resumes continuous forward polymerization. However, consider that the polymerase site is moved to the position at the next (n+1)-th site after the incorporation of a mismatched base, as it is generally believed. Then, after the mismatched base is excised and the 3' – 5' ssDNA primer is switched back to the polymerase site, the polymerase site has to move backwards to n-th site in order for the nucleotide dNTP to be incorporated. Because of the strong binding affinity of DNAP for DNA, it will take a long time for the DNAP to move backwards by one step. This implies that, after the proofreading, the DNAP will take a long time to resume continuous forward polymerization. Therefore, it is reasonable to consider that, during the processive forward polymerization, after a mismatched incorporation at n-th site, the polymerase site is still positioned at the n-th site for a long time rather than rapidly moved to the next (n+1)-th site. Consequently, the much reduced rate on top of the mismatch pair is due to both the long moving time of the polymerase site from the n-th to (n+1)-th sites and the long nucleotide-incorporation time at the (n+1)-th site (see Eq. (7)).

The kinetic experimental results for bacteriophage T7 DNAP (Patel et al., 1991) showed that, under no mechanical tension on the DNA template, the rate-limiting step during the polymerization is the structural change from the "open" to "closed" conformations. However, from the best fitting to the experimental results for T7 DNAP obtained by using single-molecule method (Wuite et al., 2000; Maier et al., 2000), it was indicated that both the structural change from the "open" to "closed" conformations and that from the "closed" to "open" conformations are rate limiting under no mechanical tension on DNA template (see Section 6.2). This is not unreasonable, because the conformational change generally involves a large free-energy change for the protein, which would thus induce the transition rates of the two steps becoming slower than rates of other transitions that involve only chemical reaction while involve no obvious conformational change. This is indeed in agreement with the kinetic experimental results for Klenow fragment (Dahlberg and Benkovic, 1991), where it was shown that both steps that involve the large conformational changes are rate limiting for DNA polymerization.

In the calculation of transfer rates $K_{P \to X}$ and $K_{X \to P}$ (see Section 7), an approximation is that the palm domain has a very weak interaction with dsDNA of 9 nascent base pairs so that it has negligible effect on the unwinding of these 9 base pairs. This approximation can be justified by experimental evidences from following biochemical and structural studies. The T7 DNAP adopts the host *Escherichia coli* thioredoxin that binds tightly to the enzyme to greatly increase its processivity of polymerization (Tabor et al., 1987; Huber et al., 1987). This interaction with thioredoxin orders a proteolytically sensitive loop, located at the tip of the thumb domain, between α helices H and H1 and it is most likely this thioredoxin-binding loop binds to dsDNA (Doublie et al., 1998; Doublie and Ellenberger, 1998). Mutations affecting three lysines in the thioredoxin-binding loop (Lys300, Lys302 and Lys304) decrease the affinity of the enzyme for DNA and lower its processivity (Yang and Richardson, 1997). The addition of the thioredoxin-binding domain to the Klenow fragment confers a thioredoxin-dependent increase in its processivity (Bedford et al., 1997), suggesting that most of the processivity function is located within this small domain. In particular, recent experiment by Turner et al. (2003) and experiment by Datta et al. (2006) directly showed that

Klenow fragment binds dsDNA very weakly and, thus, specific interactions between the polymerase enzyme and the single-stranded portion of the substrate DNA clearly contribute significantly to the overall stability of the complex. Therefore, it is reasonable to consider that the T7 DNAP binds tightly to dsDNA mainly through this thioredoxin-binding loop and, thus, the high processivity of polymerization is achieved through the combined effects of both strong binding to the ssDNA template by the finger domain and the strong binding to dsDNA by the thioredoxin-binding loop.

Instead, if we assume that the binding affinity of palm domain for dsDNA of the 9 nascent base pairs has a low value of $E_I = 1 k_B T$ per base pair, then $E_{DNA} = (m-n)E_{bp}(1+\varepsilon)$ in Eqs. (11) and (12) should be replaced by $E_{DNA} = (m-n)E_{bp}(1+\varepsilon) + (m-n)E_I$. Now, using Eq. (11), the calculated results of $K_{P \to X}$ versus ε (when $n = 1$) with fixed $\gamma = 1.2 \times 10^{-11}$ kg s^{-1} (see Section 7.2) and $\gamma = 0.3 \times 10^{-11}$ kg s^{-1} are shown in Figure 14. It is seen that, even at $\varepsilon = 0.5$ and $\gamma = 1.2 \times 10^{-11}$ kg s^{-1}, $K_{P \to X} = 0.04$ s^{-1}, which is much smaller than the experimental value (~ 2.3 s^{-1}). Even for a very small value of $\gamma = 0.3 \times 10^{-11}$ kg s^{-1}, $K_{P \to X} = 0.16$ s^{-1} at $\varepsilon = 0.5$, which is smaller than one tenth of the experimental value. Thus, our results indicate that at least the dsDNA of 9 nascent base pairs have a very weak interaction with the palm domain.

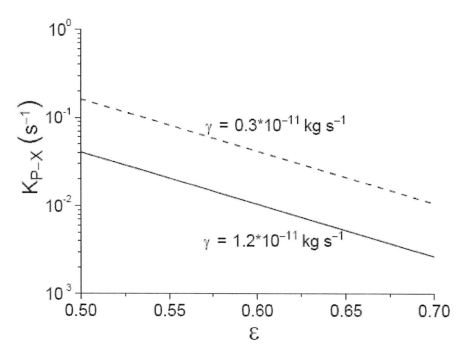

Figure 14. Mean transfer rate $K_{P \to X}$ of the 3' – 5' ssDNA primer from the polymerase to exonuclease sites for the case that there exists a binding affinity $E_I = 1 k_B T$ per base pair between palm domain and dsDNA.

Another approximation is that, in the calculation of $U_{DNA}(z)$ in Section 7, we use a mean base-pair free energy E_{bp} and a mean intensity to take into account the effect of base-stacking interaction. For precise calculation, one should use $E_{bp} = 2 k_B T$ for a A:T base pair and $E_{bp} = 3 k_B T$ for a G:C base pair. Moreover, the base-stacking interaction is also base-sequence dependent (Kool, 2001). Thus, the form of $U_{DNA}(z)$ should be sequence dependent, resulting in that different sequences should cause different transfer rates of DNA to the exonuclease site. Another interesting point to note is that, since an internal base has two neighboring bases while a terminal base has only one neighboring base, the decrease in the base-stacking free energy caused by an internal mismatch should be larger than that caused by a terminal mismatch. Thus, the transfer rate of DNA with internal mismatches is usually larger than that of DNA with the same mismatch at the primer terminus. These are consistent with the experiment of Carver et al. (1994), showed that internal mismatches have larger equilibrium partitioning constant than the same mismatch at the primer terminus.

Distinct differences of our present model from the previous model discussed in Doublie and Ellenberger (1998) are as follows. First, the stall of the polymerization after incorporation of a mismatched base occurs naturally in the present model. However, in the previous model, it was hypothesized that the presence of mismatches at the 3'-end of the template immediately downstream of the polymerase site compromises the interaction of dsDNA with the floor of the polymerase enzyme, causing the polymerization to stall. Second, in the present model the transfer of DNA from the polymerase to exonuclease sites also occurs naturally, due only to thermal noise, and, when a mismatched base is incorporated, only the 3' – 5' ssDNA primer is transferred to the exonuclease site by unwinding several base pairs; whereas in the previous model the whole dsDNA is transferred to the exonuclease site. For the present model, even there is no mismatched base the 3' – 5' ssDNA primer can still transfer to the exonuclease site but with a much low transfer rate because an extra base pair is required to unwind. For the previous model, it was assumed that, a sheared base pair, caused by a primer-template mismatch, compromises the interaction of dsDNA with the floor of the enzyme, perhaps triggering the movement of the dsDNA to the exonuclease site (Doublie and Ellenberger, 1998). Here, using the parameter values that are determined from the available crystal structure and other experimental data, we quantitatively show that the rates of transfer of only the 3' – 5' ssDNA primer to the exonuclease site are in good agreement with the experimental data for both the case that no mismatched base is present and the case that one mismatched base is present. This thus gives a support to our present model.

Conclusion

Based on previous structural, biochemical and single-molecule studies, a model is presented for the processive nucleotide incorporation and switching transition of DNA between the polymerase and exonuclease sites by DNAP. Based on the model, we present detailed theoretical studies on its dynamics such as forward polymerization for both correct and incorrect incorporations, transfer rates of DNA between the two active sites and backward motion of DNAP when tension applied to the template is increased above 40 pN. The theoretical results are consistent with available experimental data. Moreover, some predicted results are presented such as the dependence of polymerase rate on tension applied to the template under different dNTP concentrations (Fig. 7b), the transfer rate of DNA containing only A:T base pairs and that of DNA containing only G:C base pairs (see Section 7.2) and the independence of the template tension at which the polymerase reverses direction on dNTP concentration (see Section 8).

References

Andricioaei I., Goel A., Herschbach D., and Karplus M. (2004). Dependence of DNA polymerase replication rate on external forces: a model based on molecular dynamics simulations. *Biophys. J. 87,* 1478–1497.

Bailey M.F., van der Schans E.J.C., Millar D.P., (2004). Thermodynamic dissection of the polymerizing and editing modes of a DNA polymerase. *J. Mol. Biol. 336,* 673–693.

Bedford E., Tabor S., and Richardson C. C. (1997). The thioredoxin binding domain of bacteriophage T7 DNA polymerase confers processivity on *Escherichia coil* DNA polymerase I. *Proc. Natl. Acad. Sci. USA 94,* 479–484.

Brautigam C. A., and Steitz T. A. (1998). Structural and functional insights provided by crystal structures of DNA polymerases and their substrate complexes. *Curr. Opin. Struct. Biol. 8,* 54–63.

Carver, Jr T. E., Hochstrasser R. A., and Millar D. P. (1994). Proofreading DNA: recognition of aberrant DNA termini by the Klenow fragment of DNA polymerase I. *Proc. Natl. Acad. Sci. USA 91,* 10670–10674.

Dahlberg M.E., and Benkovic S.J. (1991). Kinetic mechanism of DNA polymerase I (Klenow fragment): Identification of a second conformational change and evaluation of the internal equilibrium constant. *Biochemistry 30,* 4835–4843.

Datta K., Wowor A. J., Richard A. J., and LiCata V. J. (2006). Temperature dependence and thermodynamics of Klenow polymerase binding to primed-template DNA. *Biophys. J. 90,* 1739–1751.

Delagoutte E., and von Hippel P.H. (2003). Function and assembly of the Bacteriophage T4 DNA replication complex. *J. Biol. Chem. 278,* 25435–25447.

Donlin M. J., Patel S. S., and Johnson K. A. (1991). Kinetic partitioning between the exonuclease and polymerase sites in DNA error correction. *Biochemistry 30,* 538–546.

Doublie S., and Ellenberger T. (1998). The mechanism of action of T7 DNA polymerase. *Curr. Opin. Struct. Biol. 8,* 704–712.

Doublie S., Tabor S., Long A.M., Richardson C. C., and Ellenberger T. (1998). Crystal structure of a bacteriophage T7 DNA replication complex at 2.2 Å resolution. *Nature 391,* 251–258.

Doublie S., Sawaya M. R., and Ellenberger T. (1999). An open and closed case for all polymerases. *Structure 7,* R31–R35.

Echols H., and Goodman M. F. (1991). Fidelity mechanisms in DNA replication. *Annu. Rev. Biochem. 60*, 477–511.

Eom S. H., Wang J., and Steitz T. A. (1996). Structure of Taq polymerase with DNA at the polymerase active site. *Nature 382*, 278–281.

Gardiner C. W. (1983). *Handbook of stochastic methods for physics, chemistry and the natural sciences.* Berlin:Springer-Verlag.

Goel A., Frank-Kamenetskii M. D., Ellenberger T., and Herschbach D. (2001). Tuning DNA "strings": Modulating the rate of DNA replication with mechanical tension. *Proc. Natl. Acad. Sci. USA 98*, 8485–8489.

Goel A., Astumian R. D., and Herschbach D. (2003). Tuning and stretching a DNA polymerase motor with mechanical tension. *Proc. Natl. Acad. Sci. USA 100*, 9699–9704.

Huber H. E., Tabor S., and Richardson C. C. (1987). *Escherichia coli* thioredoxin stabilizes complexes of bacteriophage T7 DNA polymerase and primed templates. *J. Biol. Chem. 262*, 16224–16232.

Huang H., Chopra R., Verdine G. L., and Harrison S. C. (1998). Structure of a covalently trapped catalytic complex of HIV-1 reverse transcriptase: implications for drug resistance. *Science 282*, 1669–1675.

Johnson K. A. (1993). Conformational coupling in DNA polymerase fidelity. *Annu. Rev. Biochem. 62*, 685–713.

Johnson R. E., Washington M. T., Haracska L., Prakash S., and Prakash, L. (2000). Eukaryotic polymerases ι and ζ act sequentially to bypass DNA lesions. *Nature 406*, 1015–1019.

Johnson,S. J., and Beese,L. S. (2004). Structures of mismatch replication errors observed in a DNA polymerase. *Cell 116*, 803–816.

Joyce C. M., and Steitz T. A. (1994). Function and structure relationships in DNA polymerase. *Ann. Rev. Biochem. 63*, 777–822.

Kiefer J. R., Mao C., Hansen C. J., Basehore S. L., Hogrefe H. H., Braman J. C., and Beese L. S. (1997). Crystal structure of a thermostable *Bacillus* DNA polymerase I large fragment at 2.1 Å resolution. *Structure 5*, 95–108.

Kiefer J. R., Mao C., Braman J. C., and Beese L. S. (1998). Visualizing DNA replication in a catalytically active *Bacillus* DNA polymerase crystal. *Nature 391*, 304–307.

Kim Y., Eom S. H., Wang J., Lee D. S., Suh S. W., and Steitz T. A. (1995). Crystal structure of *Thermus aquaticus* DNA polymerase. *Nature 376*, 612–616.

Kong X.-P., Onrust R., O'Donnell M., and Kuriyan J. (1992). Three-dimensional structure of the β subunit of E. coli DNA polymerase III holoenzyme: A sliding DNA clamp. *Cell 69*, 425–437.

Kool, E. T. (2001). Hydrogen bonding, base stacking, and steric effects in DNA replication. *Annu. Rev. Biophys. Biomol. Struct. 30*, 1–22.

Korolev S., Nayal M., Barnes W. M., Di Cera E., and Waksman G. (1995). Crystal structure of the large fragment of *Thermus aquaticus* DNA polymerase I at 2.5- Å resolution: structural basis for thermostability. *Proc Natl Acad Sci USA 92*, 9264–9268.

Krahn J. M., Beard W. A., and Wilson S. H. (2004). Structural insights into DNA polymerase β deterrents for misincorporation support an induced-fit mechanism for fidelity. *Structure 12*, 1823–1832.

Kukreti P., Singh K., Ketkar A., and Modak M. J. (2008). Identification of a new motif required for the 3'–5' exonuclease activity of *Escherichia coli* DNA polymerase I (Klenow fragment). *J. Biol. Chem.* 283, 17979–17990.

Li Y., Korolev S., and Waksman G. (1998). Crystal structures of open and closed forms of binary and ternary complexes of the large fragment of thermus aquaticus DNA polymerase I: structural basis for nucleotide incorporation. *EMBO J.* 17, 7514–7525.

Lodish H., Berk A., Zipursky S. L., Matsudaira P., Baltimore D., and Darnell J. (2000). *Molecular Cell Biology,* fourth ed. W. H. Freeman and Company, New York (Chapter 12).

Maier B., Bensimon D., and Croquette V. (2000). Replication by a single DNA polymerase of a stretched single-stranded DNA. *Proc. Natl. Acad. Sci. USA* 97, 12002–12007.

Mendelman L. V., Petruska J., and Goodman M. F. (1990). Base mispair extension kinetics. Comparison of DNA polymerase alpha and reverse transcriptase. *J. Biol. Chem.* 265, 2338–2346.

Ollis D. L., Brick, P., Hamlin R., Xuong N. G., and Steitz T. A. (1985). Structure of large fragment of *Escherichia coli* DNA polymerase I complexed with dTMP. *Nature* 313, 762–766.

Patel S. S., Wong I., and Johnson K. A. (1991). Pre-steady-state kinetic analysis of processive DNA replication including complete characterization of an exonuclease-deficient mutant. *Biochemistry* 30, 511–525.

Picher A. J., Garcia-Diaz M., Bebenek K., Pedersen L. C., Kunkel T. A., and Blanco L. (2006). Promiscuous mismatch extension by human DNA polymerase lambda. *Nucleic Acids Res.* 34, 3259–3266.

Steitz T. A. (1999). DNA polymerase: Structural diversity and common mechanism. *J. Biol. Chem.* 274, 17395–17398.

Tabor S., Huber H. E., and Richardson C. C. (1987). *Escherichia coli* thioredoxin confers processivity on the DNA polymerase activity of the gene 5 protein of bacteriophage T7. *J. Biol. Chem.* 262, 16212–16223.

Turner R. M., Grindley N. D. F, and Joyce C. M. (2003). Interaction of DNA polymerase I (Klenow fragment) with the single-stranded template beyond the site of Synthesis, *Biochemistry* 42, 2373–2385.

Venkatramani R., and Radhakrishnan R. (2008). Computational study of the force dependence of phosphoryl transfer during DNA synthesis by a high fidelity polymerase. *Phys. Rev. Lett.* 100, 088102.

Washington, M. T., Johnson, R. E., Prakash, S. and Prakash, L. (2001) Mismatch extension ability of yeast and human DNA polymerase η. *J. Biol. Chem.* 276, 2263–2266.

Wong I., Patel S. S., and Johnson K. A. (1991). An induced-fit kinetic mechanism for DNA replication fidelity: Direct measurement by single-turnover kinetics. *Biochemistry* 30, 526–537.

Wuite G. J. L., Smith S. B., Young M., Keller D., and Bustamante C. (2000). Single-molecule studies of the effect of template tension on T7 DNA polymerase activity. *Nature* 404, 103–106.

Xie P. (2007). Model for forward polymerization and switching transition between polymerase and exonuclease sites by DNA polymerase molecular motors. *Arch. Biochem. Biophys. 457,* 73–84.

Xie P. (2008). A dynamic model for transcription elongation and sequence-dependent short pauses by RNA polymerase, *BioSystem 93,* 199–210.

Xie P., (2009a). A possible mechanism for the dynamics of transition between polymerase and exonuclease sites in a high-fidelity DNA polymerase. *J. Theor. Biol.* doi:10.1016/j.jtbi.2009.04.009

Xie P., (2009b). Dynamics of backtracking long pauses of RNA polymerase. *Biochim. Biophys. Acta 1789,* 212–219.

Xie P., (2009c). A polymerase-site-jumping model for strand transfer during DNA synthesis by reverse transcriptase. *Virus Res.* doi:10.1016/j.virusres.2009.03.022

Xie P., (2009d). Molecular motors that digest their track to rectify Brownian motion: processive movement of exonuclease enzymes. (to be published)

Yang X. M., and Richardson C. C. (1997). Amino acid changes in a unique sequence of bacteriophage T7 DNA polymerase alter the processivity of nucleotide polymerization. *J. Biol. Chem. 272,* 6599–6606.

Yu H., and Goodman M. F., (1992). Comparison of HIV-1 and avian myeloblastosis virus reverse transcriptase fidelity on RNA and DNA templates. *J. Biol. Chem. 267,* 10888–10896.

Chapter VIII

Inhibiting Viral DNA Polymerase for HBV Therapy

Min Jiang and Yuanhao Li
Vertex Pharmaceutical Inc., Cambridge, MA, USA

Abstract

Hepatitis B virus infection is a major public health problem worldwide especially in East Asia. The viral infection, if persistent, may lead to chronic hepatitis, cirrhosis, and hepatocellular carcinoma. In1998, FDA approved GlascoSmithKline's Epivir-HBV (Lamivudine), a small molecule drug that inhibits HBV DNA polymerase and interferes with viral replication. Since then, a few similar drugs have been approved and many candidates are currently at preclinical and clinical development. The use of small molecular drugs that target HBV DNA polymerase is a milestone in the treatment of chronic hepatitis B. In this review, we will focus on HBV DNA polymerase and discuss the preclinical and clinical development process for the HBV polymerase inhibitors. The DNA polymerase mutants associated with drug resistance will also be discussed. The mechanism of the drug resistance and further understanding of these DNA polymerase inhibitors may help the determination of better clinical regimen for HBV therapy and patient care.

1). HBV Biology and Epidemiology

Hepatitis B virus (HBV), the causative agent of B-type hepatitis in humans, belongs to a family of viruses known as *Hepadnaviridae*. This viral family has many members that can cause liver disease in animal species other than humans, such as ducks (DHBV), woodchucks (WHV) and ground squirrels (GSHV).

HBV is an enveloped DNA virus. The viral genomic DNA is about 3.2 kb and appears in a partial double stranded manner referred to as relaxed circular (RC) DNA (Figure 1). The double stranded DNAs are not equal in size and the structure characteristics are associated to the special replication process of HBV DNA. The 5' end of the long (-)-DNA is covalently linked to the terminal domain of the DNA polymerase, and the 5' end of the short (+)-DNA consists of an RNA oligonucleotide, which is derived from the pregenomic RNA (pgRNA) and used as the primer for (+)-DNA synthesis. Based on the genome difference, there are 8 genotypes, A, B, C, D, E, F, G and H in HBV (genomes differ by more than 8%) (Okamoto, Tsuda et al. 1988; Norder, Courouce et al. 2004; Schaefer 2007). These genotypes and further subgenotypes (about 4% diversity) show a distinct geographic distribution and may also contribute to disease development.

During viral replication, HBV gains entry to host cells, probably via endocytosis, however the mechanisms are yet to be further defined (Nassal 2008). The nucleocapsids are transported to the nucleus and the RC-DNA genome is released into the nucleoplasm and converts to covalently closed circular DNA (cccDNA) (Figure 1A). This is a stable form of the viral genome and is not lost during cell division. From the cccDNA template, different mRNAs can be transcribed by host cell RNA polymerase II to generate pgRNA and other subgenomic mRNAs. The pgRNA is a 3.5 kb mRNA containing all viral genetic information and is the RNA intermediate for viral genome replication (Figure 1B). Other RNA transcripts are used for viral gene expression. There are four partially overlapped regions that express viral products. The pre S/S region encodes the large, middle and small surface antigens. The pre C/C region encodes HBe Ag ("e" antigen) and HBc Ag (core antigen). The P and X regions encode HBV polymerase (P protein) and X protein respectively. The pgRNA can be selectively packaged into progeny nucleocapsids and DNA replication is facilitated by the co-packaged viral DNA polymerase. This process includes the reverse transcription of the first (-)-DNA strand to form the immature nucleocapsid and synthesis of the (+)-DNA strand to become mature nucleocapsid (detail description in later section). After forming the partial double-stranded DNA genome, the infectious progeny HBV virus can be released.

The HBV virions, also called as Dane particles, are spherical or tubular in shape. These particles are mainly assembled by hepatitis B surface antigens (HBsAg) that form the outer lipoprotein coat. Some viral particles envelope the icosahedral nucleocapsid protein (HBcAg) cores containing the viral DNA and DNA polymerase. These viruses are the infectious HBV virions. On the other hand, there are numerous core-less particles that are HBsAg-only and are not infectious. These particles may have various sizes and shapes and may outnumber the infectious viral particles in blood. Other than the structural proteins, hepatitis 'e' antigen (HBeAg) is a non-structural peptide that is produced when the viruses are actively reproducing and therefore used as an indicator of HBV viremia. There are some HBV strains that do not produce HBeAg and the infected patients are often grouped in the HBeAg-negative category.

Although effective HBV vaccines have been around for about thirty years, the social and economical issues in different nations prohibit successful universal vaccination. Furthermore, the established infections still result in blood-borne transmissions from infected mothers to newborn babies even after vaccination (Stevens, Toy et al. 1985). Even in normal vaccinated populations, there is approximately 10% who cannot induce protective levels of anti-HBs antibodies. Moreover, the vaccines are not effective in inducing antiviral immunity in a considerable proportion of immunosuppressed individuals such as patients infected with human immunedeficiency virus (HIV) or patients receiving immunosuppressant drugs.

Therefore, three decades after the first launch of HBV vaccine, global HBV infections are still prevalent. Overall, there are about 350 million people who are infected chronically with the disease and approximately 1 million people that die each year as a result of complications from HBV infection (Parkin, Pisani et al. 1999; Seeger and Mason 2000; Strader, Wright et al. 2004). In one of the HBV hot-spots, China, HBsAg carrier rate among the vaccinated population is 4.51% and 9.51% among the unvaccinated (Liang, Chen et al. 2005).

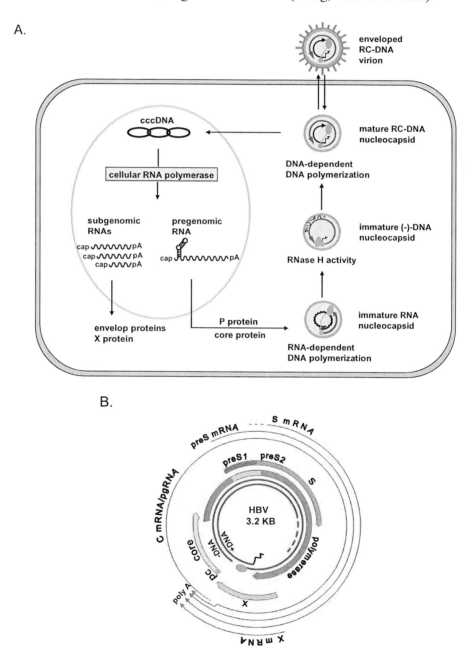

Figure 1. Schematic viral replication cycle (A) and genome structure (B) of human hepatitis B virus.

HBV is transmitted through contact of blood and bodily fluids. Vertical transmission from mother to newborn baby is the most common mode of transmission. In adults, horizontal transmission through sex is the major route. HBV causes transient and chronic infections in human liver. Most acute infections are mild and the virus is cleared in three months. Only about 0.5% of transient infections become serious and fatal. For those who are unable to get rid of the viruses after acute infection, liver diseases become chronic and may cause serious problems. The risk for acute infection becoming chronic has strong ties to age; about 90% of vertical infected infants will progress to chronic HBV, whereas less than 10% of adults become chronically infected (Lok 2000). For a lengthy period time, the chronic hepatitis B is often asymptomatic. However, a significant portion of these patients eventually develop chronic hepatitis, progressive liver fibrosis, liver cirrhosis and hepatocellular carcinoma. Patients who develop cirrhosis may develop decompensated liver disease, which may present itself as jaundice, abdominal distention from ascites, peripheral edema, renal insufficiency, easy bruising and bleeding, confusion, and gastrointestinal varices.

The diagnosis for chronic hepatitis B is difficult due to lack of specific symptoms. However, early diagnosis of chronic HBV infection is essential to ensure favorable treatment outcomes. There are different assay formats to target on viral antigens, antibodies, viral DNA and patient blood chemistry. The liver function test and histology are also important for disease assessment. Table 1 summarizes the clinical tests for HBV diagnosis. Based on comprehensive evaluation of the events in patients, HBV infection can be staged differently and is used to guide clinical care.

1) *Chronic Hepatitis B:* Patient serum is HBsAg positive for over 6 months. The serum HBV DNA is present and HBeAg is usually positive. There are also HBeAg-negative patients due to precore and core promoter mutants suppressing HBeAg. These HBeAg-negative patients are important population of chronic hepatitis B. In chronic hepatitis B patients, alanine aminotransferase and aspartate aminotransferase (ALT/AST) levels are often elevated and chronic hepatitis can be identified by liver biopsy.
2) *Inactive HBsAg Carrier State:* This group is HBsAg positive for over 6 months and has HBeAg clearance (HBeAg seroconversion). The serum DNA level is low and ALT/AST levels are normal. No significant hepatitis.
3) *Resolved Hepatitis B:* This group has previous known history of acute or chronic hepatitis B infection. Serum presence of anti-HBc or anti-HBs antibodies is common. No detectable HBsAg and viral DNA in serum. ALT/AST levels are normal.

The classical serological and blood tests are common in clinical assessment for HBV infection. However, the virological and histological tests are important for evaluating serious liver diseases. For example, HBV circulating virus level has a strong correlation to the progression of cirrhosis. The risk for cirrhosis is independent of HBeAg status and serum ALT level (Iloeje, Yang et al. 2006). The accurate diagnosis of HBV-infected patient is essential for determination of treatment and prognosis.

Table 1. Clinical tests for chronic HBV infection

Test	Diagnosis
Serological	
HBs Ag	General marker of HBV infection
HBs Ab	Marker of recovery and immunity to HBV infection. HBs seroconversion: HBs Ag loss and gain of HBs Ab
HBc Ab	Marker of past or chronic HBV infection.
HBc IgM Ab	Marker of acute HBV infection
HBe Ag	Indicator of active HBV replication and infectious. HBe Ag is generally detectable at same time as HBs Ag and disappears before HBs Ag
HBe Ab	Indicator of low HBV replication. HBe seroconversion: HBe Ag loss and gain of HBe Ab
Biochemical	
ALT/AST	To assess liver function and damage
Virologic	
HBV DNA assay	HBV DNA level can be used to measure viral replication
Histologic	
Histological Activity Index (HAI)	Necroinflammatory score
Fibrosis	Histological evaluation of liver damage
Cirrhosis	Histological evaluation of liver damage

2). Current Preventive and Therapeutic Approaches for HBV Infection

HBV prophylactic and therapeutic interventions were revolutionized in the last three decades. In 1981, the FDA approved the first plasma-derived hepatitis B vaccine for human use (Miller 1982). This vaccine, Heptavax, was manufactured from pooled HBsAg-positive blood from donors. After inactivation, the plasma vaccine was used for immunization. Although it was the first approved product for HBV, Merck discontinued this vaccine in 1990 due to the emergence of the recombinant vaccine, which is both safer and easier to produce.

The vaccine approach relies on host antiviral immunity for HBV prevention. For people who had previously been infected with HBV and subsequently reached seroconversion to clear the viral infection, the naturally produced anti-HBV immunity protected them from reinfection with HBV. The serum from these individuals has a strong presence of antibodies against HBV antigens such as HBsAg. These antibodies can even be purified and passively administered to others in order to induce antiviral immunity. Such hepatitis B immune globulin product has also been recently approved for preventing HBV recurrence following liver transplantation in HBsAg-positive patients (http://www.fda.gov/bbs/topics/NEWS/2007/ NEW01602.html).

For therapeutic application in HBV patient, biological products targeting patients' immunity were among the earliest developed. Interferon has been an important class of antiviral therapeutics and was approved for HBV immunotherapy in 1992. In a subset of HBV carriers, especially those with the most active diseases, interferon is effective and causes seroconversion. However, it remains ineffective in about 50% of HBV carriers (Hoofnagle and di Bisceglie 1997). In order to improve the application of this protein-based therapeutic, pegylated products were added to the market in 2005. These injectable agents have antiviral and immunomodulatory effects in chronic HBV patients and are used as the first line in therapy. However, their therapeutic efficacy and tolerability profiles are not favorable for all patients, and they are also very expensive.

Another major class of HBV therapeutics is the antiviral nucleotide or nucleoside analogues. This class of agents interferes with viral DNA synthesis, which inhibits HBV replication. Currently, there are total of five approved products in the market (Table 2). These are all oral available drugs and are much more affordable than using the injectable interferon products in HBV therapy.

Table 2. FDA approved drugs for HBV therapy

Drug	Commercial name	Mechanism of Action	Developer	FDA approval
Plasma inactivated vaccine	Heptavax	Inactivated viruses as vaccine	Merck	1981 Discontinued in 1990
Recombinant Heptitis B vaccine	RECOMBIVAX HB®	HBsAg vaccine	Merck	1986
Recombinant Heptitis B vaccine	Engerix-B	HBsAg vaccine	GlaxoSmithKline Biologicals	1989
Combination HAV/HBV vaccine	Twinrix (Havrix and Engerix-B)	Inactivated HAV and HBsAg vaccine	GlaxoSmithKline Biologicals	2001
Hepatitis B immune globulin	HepaGam B	Hepatitis B immune globulin	Cangene	2006 (orphan drug)
Pegylated IFN-α-2a /Pegylated IFN-α-2b	Pegasys	Immunotherapy	Roche/ Schering Plough	2005
IFN-α-2a /IFN-α-2b	Intron A	Immunotherapy	Roche/ Schering Plough	1992
Thymosin-α	Zadaxin	Immunotherapy	SciClone Pharmaceuticals	2000
Lamivudine	Epivir, 3TC	Nucleoside analogue, targeted RT as viral DNA terminator	GlaxoSmithKline	1998
Adefovir	Hepsera	Nucleotide analogue	Gilead Sciences	2002
Entecavir	Baraclude	Nucleoside analogue	Bristol Myer Squibb	2005
Telbivudine	Tyzeka, Sebivo	Nucleoside analogue	Novartis/ Idenix	2006
Tenofovir Desoproxil Fumarate	Viread	Nucleoside analogue	Gilead Sciences	2008

The goal for antiviral therapy in chronic HBV infected patients is to inhibit HBV replication and to achieve remission of liver disease. This is done by reducing serum HBV

DNA levels as low as possible for as long as possible and to induce HBeAg/HBsAg to seroconversion for a long term cure. The different products in HBV patient care may need to use in optimized way to specific patient groups in order to reach best antiviral outcome. However, the current pool of anti-HBV artilleries is still too poor to support the mission for HBV patient care. There are different weaknesses associated with these therapeutic products. The call for new products and optimized therapeutic regimen for HBV patients is urgent. In-depth research to understand the HBV biology, comprehensive diagnosis of disease progression, careful assessment of viral and host relationship in liver disease development and the precise determination of therapeutic regimen using available drugs are all critical in managing chronic HBV infection. Better application of this knowledge will improve the clinical outcome in reducing the risk of liver fibrosis, cirrhosis, liver failure and hepatocellular carcinoma. Based on the complex nature of HBV-induced liver diseases, the priority of treating HBV infection is clearance of HBV DNA. Other goals include antiviral seroconversion and improvement in inflammation on liver histology and liver damage.

3). HBV DNA Polymerase as Target for Drug Development

In last decade, the oral antiviral agents that block HBV DNA synthesis are playing a more advanced role in HBV therapy, and the development for new drugs in this class are greatly boosted. These drugs all target on HBV DNA polymerase which has multi-functions including RNA-dependent DNA polymerase (reverse transcriptase, RT), DNA-dependent DNA polymerase, RNase H, and protein priming activities. All of these functions are important for the unique HBV DNA replication process. About 90 kDa in size, HBV DNA polymerase is much larger than retroviral RT. However, the HBV RT and RNase H (RH) domains share sequence motifs similar to retroviral RTs such as HIV-1 RT. Only the terminal protein (TP) domain at the N terminus is HBV-specific and the spacer separating from the RT domain is also variable and dispensable (Figure 2).

The unique structure of HBV DNA polymerase is required for the replication of this DNA virus through the RNA intermediate. During HBV replication, viral cccDNA can be transcribed by host cell RNA polymerase to form the pregenomic RNAs. The pgRNA serves as the mRNA for the synthesis of core proteins and polymerase proteins. P protein binds to the 5' end of its own mRNA and the complex is then packaged into HBV nucleocapsids (Figure 1). The viral DNA synthesis starts in the immature nucleocapsids where the pgRNA is reverse transcribed by the P protein into the negative-sense DNA strand facilitated by the protein-priming mechanism. In this process, a specific tyrosine residue (Y63) on TP is covalently linked to the first nucleotide of the (-)-DNA to initiate the DNA synthesis (Bartenschlager and Schaller 1988; Lanford, Notvall et al. 1997). Therefore, the 5' end of the long (-)-DNA is linked to the terminal domain of the P protein. After the TP-linked DNA oligonucleotide (3-4nts) translocates to the complementary motif at 3' end of pgRNA for template switch, the (-)-DNA elongation can be carried out (Beck and Nassal 2007; Nassal 2008). While the (-)-DNA is being synthesized, the pgRNA template that has already been copied is degraded concomitantly by the RNase H domain of the P protein. When the polymerase reaches the 5' end of the pgRNA, (-)-DNA synthesis is complete. The short pgRNA oligonucleotide that remains undigested is then used to prime (+)-DNA synthesis. Using the newly synthesized (-)-DNA as template, the (+)-DNA is synthesized by DNA polymerase. In the end, the 5' terminus of the short (+)-DNA remains this undigested RNA oligonucleotide from the original pgRNA. When the partial double-stranded DNA genome is formed, the HBV particle can be released as an infectious progeny virus. Alternatively, RC-DNA can enter the cell nucleus to form cccDNA once again.

Based on the complicated role of HBV DNA polymerase in the process of HBV DNA replication, it is possible to block the synthesis of progeny HBV DNA from different avenues. First, the protein-mediated priming step can be a viable process for drug intervention. More importantly, two of the HBV DNA polymerase's functions, the RNA dependent-DNA polymerase (RT) and DNA-dependent DNA polymerase (DNA pol), provide opportunities for antiviral therapy. Nucleotides are the building blocks of DNA synthesis and have been widely used in therapeutic applications for viral diseases such as for human immunodeficiency virus (HIV) and herpes simplex virus (HSV). To inhibit viral DNA synthesis, the modified nucleotide analogues are used to compete with the natural nucleotides and interfere with the viral replication process as chain terminators. Similarly, the unphosphorylated precursors, nucleosides, can also be modified for therapeutic intervention.

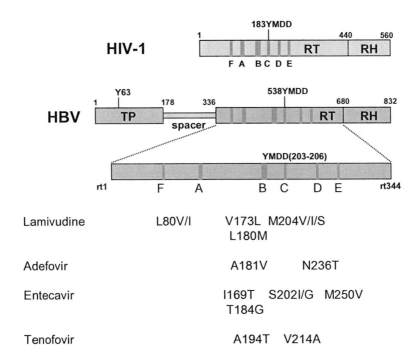

Figure 2. Schematic structure of HBV DNA polymerase and comparison of reverse transcriptase to HIV. Representative drug induced resistant mutants are listed.

At present, the crystal structure of HBV DNA polymerase has not been solved (Das, Xiong et al. 2001; Nassal 2008). There is no direct structural information on any hepadnaviral P protein. However, homology-based models for HBV RT and RH domain have been proposed based on the comparison with retroviral RTs (Allen, Deslauriers et al. 1998; Das, Xiong et al. 2001). Furthermore, several nucleoside analogues originally developed against HIV-RT including lamivudine have shown potent anti-HBV activity, suggesting a similar structure and function relation. Therefore, the similar molecules derived from anti-HIV or anti-HSV therapy have been further studied and modified for HBV therapy. Other newly designed nucleotide analogues that unlike naturally present nucleotides can also be studied for anti-HBV activity.

Currently, there are five nucleotide or nucleoside analogues approved for HBV therapy from last ten years.

Lamivudine

Lamivudine was the first oral nucleoside to be approved for HBV treatment. It is a cytidine nucleoside analog and can inhibit the reverse transcribed 1st (-)-strand DNA synthesis during HBV replication. In fact, lamivudine was originally developed for HIV therapy. This drug has a good safety profile in massive clinical applications across wide dose ranges. It has quick antiviral effects, lowering the amount of viral DNA and ALT (Tassopoulos, Volpes et al. 1999). One year of treatment leads about 18% of patients to reach HBe seroconversion (Schalm, Heathcote et al. 2000; Leung, Lai et al. 2001). Long term use of lamivudine can improve HBV seroconversion and liver histology, even the risk of HCC. However, the high resistance rate is a major problem.

Adefovir

Adefovir is an adenosine nucleotide analog. It exhibits potent antiviral activity at lower doses and has a good safety profile, even to patients with renal problem. Adefovir can interfere with priming and DNA elongation. It is effective against lamivudine-resistant mutants. One year of treatment in HBeAg positive patients has about 14% HBe seroconversion (Marcellin, Chang et al. 2003). In HBeAg negative patients, one year drug treatment results in about 51% patients having reduced serum DNA below 400 copies/mL (Hadziyannis, Tassopoulos et al. 2003). The resistant rate is relatively low, however, the lamivudine resistant mutants may also become adefovir resistant.

Entecavir

Entecavir is an oral deoxyguanine nucleoside analog with extremely high potency to HBV DNA replication. It inhibits the priming of HBV DNA polymerase and the reverse transcription of (-)-DNA strand from pregenomic RNA, and synthesis of (+)-DNA strand. Compared to lamivudine, entecavir had superior efficacy that one year treatment has higher DNA reduction and ALT normalization in both HBeAg positive or negative patients (Chang, Gish et al. 2006; Lai, Shouval et al. 2006). The rate of resistance is also very low that only less than 1% of treatment-naïve patients developed resistance after four years treatment by entecavir (Colonno, Rose et al. 2006).

Telbivudine

Telbivudine is a deoxythymidine nucleotide analog. This drug has good safety and has no treatment-related adverse events at high dose (800mg/day). It has no embroyonic or fetal toxic effects and can be considered for young patients. In clinical application, telbivudine has better viral load reduction and liver histology improvement than lamivudine (Lai, Gane et al. 2007) and probably adefovir (Chan, Heathcote et al. 2007). Telbivudine and lamivudine are similar nucleoside analogues and have cross-resistance. However, telbivudine gets only half resistant rate than lamivudine and two years treatment with telbivudine had about 7-17% resistance observed (Lai, Gane et al. 2007; Hou, Yin et al. 2008).

Tenofovir

Tenofovir is an acyclic nucleoside prodrug and structurally related to adefovir. It is a potent inhibitor for both HBV and HIV. In clinical studies, tenofovir was more potent than adefovir in reaching antiviral responses in different patients including the patients of lamivudine resistance and the patients with HBV and HIV co-infection (van Bommel, Wunsche et al. 2004; Lacombe, Gozlan et al. 2008).

4). Preclinical snd Clinical Development of More HBV DNA Polymerase Inhibitors

The biggest limitation for developing therapeutic products against HBV infection is the absence of cell culture technology and appropriate animal models that allow HBV replication. Currently, there is also a lack of crystal structure information on HBV DNA polymerase. Despite the difficulties in developing more HBV DNA polymerase inhibitors, the high demands for HBV patient care still result in increased R&D efforts in the pharmaceutical industry. Human hepatoma cell lines stably transfected with HBV genome are being used for in vitro study for new drug development. In vivo studies focus on using surrogate animal models that have played a significant role in understanding the biology of hepadnaviruses.

The human hepatocellular carcinoma cell line HepG2 was transfected with HBV DNA and HepG2.2.15 or HepAD38 cells were established to express all HBV RNAs and proteins. These cells produce viral genomes and secrete virus-like particles which mimic natural HBV replication (Sells, Zelent et al. 1988; Ladner, Otto et al. 1997). Upon incubation with the antiviral agent, the derived cell assay can be used in evaluating the anti-HBV DNA replication as an antiviral readout. The nucleotides or nucleosides are therefore screened to select leads for further pharmacokinetics and pharmacology studies. From the comparison of in vitro cell assay data, the current FDA approved HBV DNA polymerase drugs have a wide range of activity against HBV replication. The EC50 in HepG2.2.15 cells varies from low nM (entecavir 3.75nM) to low µM (tenofovir 1.5µM) (Hui and Lau 2005). Based on the in vitro cytotoxicity data, these drugs have a very high selectivity index, suggesting large room for improvements on drug potency with more novel nucleotides or nucleosides.

Human HBV can use chimpanzee and higher primates as permissive hosts and tree shrews as a less permissive host. These animal species are definitely not appropriate for preclinical research for economical and ethical reasons. Due to the similar nature of HBV to other hepadnaviruses, the naturally occurring HBV-like viruses and their animal hosts are currently used for preclinical purposes. Among them, the duck model and woodchuck model have been widely used to assess the antiviral efficacy for different nucleotides or nucleosides. The DHBV model does not develop HCC and the ducks appear to be less sensitive in evaluating the safety profile of each compound tested. In comparison to this avian model, the mammalian woodchuck model may be preferred for preclinical use.

Table 3. HBV drugs in clinical development

Drug	Commercial name	Mechanism of Action	Developer	Clinical status
Famciclovir	FAMVIR	Prodrug of penciclovir	Novartis/ TEVA Pharms	Approved for HSV. Tested in HBV and discontinued
Abacavir	Ziagen	Nucleoside analogue	GlaxosmithKline	Approved for HIV. Phase III for HBV/HIV
Emtricitabine	Emtriva FTC Coviracil	Nucleoside analogue	Gilead Sciences	Approved for HIV. Phase III for HBV or HBV/HIV in combination with other drugs
Truvada (Emtricitabine + tenofovir)		Nucleoside analogue	Gilead Sciences	Approved for HIV. Phase III for HBV
Clevudine	L-FMAU, Levovir	Nucleoside analogue	Pharmasset/ Bukwang	Approved in S. Korea. Phase III for HBV
Valtorcitabine	Valyl-L-dC	Nucleoside analogue	Novartis/ Idenix	Phase II
MIV-210 (FLG-prodrug)	FLG, Medivir	Prodrug	Medivir/Glaxo SmithKline	Phase II for HIV, phase I for HBV in UK
LB80380	ANA 380	Prodrug	Anadys/LG Life Sciences,	Phase II
Amdoxovir	DAPD	Nucleoside analogue	RFS Pharma	Phase II
Pradefovir	Remofovir	Prodrug for adefovir	Schering-Plough	Phase II (discontinued)
Racivir	RCV		Pharmasset	Phase II
Elvucitabine	L-Fd4C	Nucleoside analogue	Achillion Pharmaceuticals	Phase II for HBV, HIV
NOV-205	BAM-205	Non-nucleoside compound	Novelos	Approved in Russia for HBV. Phase I trial for HCV
NUC B1000		Expressed RNAi	Nucleonics	Phase I

Lamivudine, the first nucleoside drug for HBV, has been tested in duck and woodchuck models (Mason, Cullen et al. 1998; Korba, Cote et al. 2000; Marion, Salazar et al. 2002). It was shown that the treatment of lamivudine in the animals had reduced virus titre more than 300-fold. Similar result was also demonstrated when adefovir was tested in duck or woodchuck models (Delmas, Schorr et al. 2002; Menne, Butler et al. 2008). The woodchuck model was also successfully used in demonstrating the efficacy of adefovir in lamivudine-resistant mutants (Jacob, Korba et al. 2004). For another FDA approved HBV DNA polymerase drug, entecavir, both duck and woodchuck models were tested in preclinical studies (Colonno, Genovesi et al. 2001; Marion, Salazar et al. 2002). The experience accumulated from the preclinical studies of these approved drugs has greatly helped the further development of new DNA polymerase inhibitors. For example, clevudine has demonstrated antiviral efficacy in both duck and woodchuck models. In chronic WHV carrier woodchucks, daily oral dosing of clevudine at 10mg/kg for 4 weeks can significantly reduce viraemia, antigenaemia, intrahepatic WHV replication (Peek, Cote et al. 2001). These preclinical pharmacology studies and further toxicology evaluation pave the road for product clinical development. Currently, some of the newly developed HBV DNA polymerase inhibitors have reached clinical stages as summarized in Table 3.

5). A Decade's Application of Lamivudine and Drug Resistance of DNA Polymerase Inhibitors

HBV DNA replicates in a special way that involves the multiple functions of its DNA polymerase. Without the proofreading ability, the viral reverse transcriptase gives a high spontaneous error rate to HBV DNA replication than other DNA viruses. In general, random genomic mutations are common in HBV due to high viral production and a high mutation rate. These naturally occurring mutations will be balanced to their replication fidelities in determining their overall survival and proportion in the HBV pool. In chronic HBV infections, some mutations can result in reduced HBeAg expression. These patients can have mutants that affect translation of a full-length precore proteins by introduction of a stop-codon in the precore gene (Lok, Akarca et al. 1994) or have mutations that affect the basal core promoter regions (Hunt, McGill et al. 2000) to eventually become HBeAg-negative carriers.

In comparison to the naturally occurred mutations, the HBV DNA polymerase inhibitors may expedite and change the landscape of HBV mutants during drug treatment. The drug selective pressure causes mutation at the antiviral target site to be amplified. These drug-selected mutants will result in failed treatment and need to be managed by an alternative therapeutic regimen. Otherwise, these drug-resistant HBV mutants can cause severe clinical problems (Liaw, Chien et al. 1999; Ayres, Bartholomeusz et al. 2003).

The application of oral antiviral agents in HBV therapy has over a decade's history now. These oral drugs are easier to use and more affordable than interferon products. They have minimal side effects and clear clinical benefits for many HBV patients. For example, a landmark study in chronic HBV patients with significant hepatic fibrosis and high virus load demonstrated that lamivudine treatment could prevent the progression of liver disease and reduce HCC risk (Liaw, Sung et al. 2004). However, these nucleotides or nucleosides require a long treatment duration, and the resistant mutants may emerge to cause clinical problems. For lamivudine, the drug resistant mutants can increase very quickly in more than 70% patients after treatment for four years (Lai, Dienstag et al. 2003; Chang, Lai et al. 2004). Patients with mutants showed a significant increase in the HBV DNA and ALT levels and had progression of liver diseases. The antiviral efficacy and liver histological improvement from the initial drug treatment can be progressively lost with the accumulation of drug resistant mutants, requiring the patients to seek additional antiviral therapy.

In contrast to lamivudine, other recently approved HBV DNA polymerase inhibitors have a smaller mutation rate. Adefovir showed a lower resistant rate of about 22-28% after 2-5 years (Angus, Vaughan et al. 2003; Fung, Chae et al. 2006; Hadziyannis, Tassopoulos et al. 2006). Telbivudine has cross-resistance to lamivudine as a similar nucleoside analogue but with only half as much risk. The two years of telbivudine treatment had about 7 to 17% resistance (Lai, Gane et al. 2007). Resistance to entecavir is even lower that in treatment naïve patients; 4 years entecavir treatment had less than 1% resistance (Colonno, Rose et al. 2006). However, the cross-resistance of lamivudine with entecavir results in entecavir having higher resistance in lamivudine resistant mutants (Tenney, Rose et al. 2007).

The detail analysis of the mutants reveals that the mutations are concentrated in the reverse transcriptase around the YMDD motif in the C domain (Figure 2). This motif is the catalytic domain of HBV RT and the mutations may cause structural alterations and interfere with the binding of drug to RT, resulting in the lower sensitivity of drug treatment. Other mutations are also selected from drug induced structural and functional changes in RT. There are examples of different drugs selecting similar mutants and becoming cross-resistant. Further studies in understanding the emergence of each particular mutant and the strategies for eliminating their presence clinically are important for HBV management.

Table 4. CDC recommended treatment for viral resistance

Resistance	Treatment regimen
Lamivudine-resistance	Add adefovir or tenofovir Stop lamivudine, switch to Truvada[1,3] Stop lamivudine, switch to entecavir (preexisting lamivudine-resistant mutation predisposes to entecavir resistance)[2]
Adefovir-resistance	Add lamivudine[2] Stop adefovir, switch to Truvada[1,3] Switch to or add entecavir[2,3]
Entecavir-resistance	Switch to or add adefovir or tenofovir[3]
Telbivudine-resistance[4]	Add adefovir or tenofovir Stop telbivudine, switch to Truvada Stop telbivudine, switch to entecavir (preexisting telbivudine-resistant mutation predisposes to entecavir resistance)

[1] Truvada = combination pill with emtricitabine 200 mg and tenofovir 300 mg
[2] Durability of viral suppression unknown, especially in patient with prior lamivudine resistance
[3] In HIV coinfected persons; scanty in vivo data in non-HIV infected persons
[4] Clinical data not available

In the HBV genome structure, the coding region of HBV genes and the transcriptional regulatory elements are greatly overlapped. The drug selected HBV mutants can differ from wild-type virus not only in the replication competence, but also the virion structure. Therefore, it is important to manage these mutants before they become a serious health problem. So far, the main lamivudine resistant mutants remain sensitive to other drugs and further treatment with a new antiviral drug may be sufficient to manage the mutants. Other treatment regimen can also be considered in order to better manage the mutants. Table 4 is the recommendation from CDC for some guidelines in managing the HBV resistance during treatment with HBV DNA polymerase inhibitors. Other guidelines from various institutions and research groups are also widely available.

6). Perspective for Newer Regimens using DNA Polymerase Inhibitors

When to treat and what to treat are the fundamental questions for physicians who manage chronically infected HBV patients. The timing for HBV patient treatment is debated because unwise treatment may not have clinical benefits, but rather provide chances for unwanted mutants to occur. In general, one preventive principal is to avoid unnecessary treatment and to initiate treatment with potent antiviral agents that have a low rate of drug resistance or with combination therapy.

Chronic HBV associated liver diseases are very complicated. Given the development of more sensitive and sophisticated diagnostic approaches, HBV infections will need an early and appropriate diagnosis to determine the best treatment regimen. Based on the study of lamivudine mutants, it was suggested that the risk factor for generating mutants include pre-therapy HBV DNA and ALT level, inadequate HBV DNA suppression and duration of the treatment (Lai, Dienstag et al. 2003). It would be critical to determine the best time and duration to apply the drug treatment.

Obviously, for chronic HBV management, the treatment goal is to achieve sustained suppression of viral replication and remission of liver disease. The clinical readouts would be to reduce serum HBV DNA, to normalize ALT and to improve liver histological conditions. However, early treatment for some patients may not be enough to achieve this goal. For example, patients in an immunotolerant phase who have a high HBV DNA load and persistently normal ALTn levels will have no chance of HBeAg seroconversion after receiving antiviral treatment. It is not recommended to treat these patients with current oral drugs. A similar situation is with the HBsAg inactive carriers who are HBsAg positive and have low DNA levels and normal ALT.

Although the oral antiviral agents for HBV therapy are becoming the dominant treatment for chronic HBV infection, there is still a shortage of optimized treatment regimens for eradicating HBV. There is a continuous need for clinical research in finding the best use of the HBV DNA polymerase inhibitors currently approved by FDA. In the meantime, it is important to develop more potent and less resistant antiviral drugs against HBV DNA polymerase or other HBV-specific targets.

In regards to currently available drugs, the best way to manage drug resistance is still a hard question for all clinical researchers. The common consensus is to apply an add-on therapy rather than switching to second monotherapy during drug resistance. The combination therapy would be also a valuable alternative in treating drug-resistant patients. The drugs in the combination should be the most potent ones with no overlap in resistant mutations. Currently, combination therapy is being tested actively. For example, lamivudine plus adefovir dipivoxil treatment has shown to have lower serum HBV DNA and higher rates of ALT normalization than lamivudine alone. This combination therapy can also decelerate resistance development (Sung, Lai et al. 2008). Truvada's trials (emtricitabine and tenofovir) in HBV therapy are aiming to extend its success in fighting HIV in its combination format. The comparison of adefovir monotherapy versus adefovir plus emtricitabine in HBeAg-positive chronic hepatitis B also demonstrated higher efficacy in combination groups, and no resistance was found after a two year treatment (Hui, Zhang et al. 2008). Furthermore, combinations of entecavir with others are being actively tested and even combinations between interferon and antiviral oral agent are also generating valuable clinical finding.

Another issue associated with the antiviral treatment of chronic HBV infection is relapse after treatment discontinuation. Half of the patients with lamivudine-induced HBeAg seroconversion had a relapse at two or three years after therapy (Song, Suh et al. 2000). Therefore, lamivudine-induced HBeAg seroconversion may not be durable. It was found that the duration of additional lamivudine treatment after HBeAg seroconversion is the major risk factor for posttreatment relapse. Additional risk factors will need to be clarified and similar research is also needed for other antiviral drugs.

The HBV DNA polymerase inhibitors are certainly the most important drugs for chronic HBV therapy and are playing an integral role in managing liver diseases. With the help of all preclinical and clinical studies and a better understanding of the knowledge collected, we may reach the optimal regimen for fighting HBV infection. This will direct us to use our current antiviral assets to control HBV infection and deal with drug resistance and relapse problems. It may also provide clear guidance for new drug development for the benefit of the HBV infected patients.

References

Allen, M.I., Deslauriers, M., Andrews, C.W., Tipples, G.A., Walters, K.A., Tyrrell, D.L., Brown, N. and Condreay, L.D. (1998). Identification and characterization of mutations in hepatitis B virus resistant to lamivudine. Lamivudine Clinical Investigation Group. *Hepatology* 27, 1670-7.

Angus, P., Vaughan, R., Xiong, S., Yang, H., Delaney, W., Gibbs, C., Brosgart, C., Colledge, D., Edwards, R., Ayres, A., Bartholomeusz, A. and Locarnini, S. (2003). Resistance to adefovir dipivoxil therapy associated with the selection of a novel mutation in the HBV polymerase. *Gastroenterology* 125, 292-7.

Ayres, A., Bartholomeusz, A., Lau, G., Lam, K.C., Lee, J.Y. and Locarnini, S. (2003). Lamivudine and Famciclovir resistant hepatitis B virus associated with fatal hepatic failure. *J Clin Virol* 27, 111-6.

Bartenschlager, R. and Schaller, H. (1988). The amino-terminal domain of the hepadnaviral P-gene encodes the terminal protein (genome-linked protein) believed to prime reverse transcription. *Embo J* 7, 4185-92.

Beck, J. and Nassal, M. (2007). Hepatitis B virus replication. *World J Gastroenterol* 13, 48-64.

Chan, H.L., Heathcote, E.J., Marcellin, P., Lai, C.L., Cho, M., Moon, Y.M., Chao, Y.C., Myers, R.P., Minuk, G.Y., Jeffers, L., Sievert, W., Bzowej, N., Harb, G., Kaiser, R., Qiao, X.J. and Brown, N.A. (2007). Treatment of hepatitis B e antigen positive chronic hepatitis with telbivudine or adefovir: a randomized trial. *Ann Intern Med* 147, 745-54.

Chang, T.T., Gish, R.G., de Man, R., Gadano, A., Sollano, J., Chao, Y.C., Lok, A.S., Han, K.H., Goodman, Z., Zhu, J., Cross, A., DeHertogh, D., Wilber, R., Colonno, R. and Apelian, D. (2006). A comparison of entecavir and lamivudine for HBeAg-positive chronic hepatitis B. *N Engl J Med* 354, 1001-10.

Chang, T.T., Lai, C.L., Chien, R.N., Guan, R., Lim, S.G., Lee, C.M., Ng, K.Y., Nicholls, G.J., Dent, J.C. and Leung, N.W. (2004). Four years of lamivudine treatment in Chinese patients with chronic hepatitis B. *J Gastroenterol Hepatol* 19, 1276-82.

Colonno, R.J., Genovesi, E.V., Medina, I., Lamb, L., Durham, S.K., Huang, M.L., Corey, L., Littlejohn, M., Locarnini, S., Tennant, B.C., Rose, B. and Clark, J.M. (2001). Long-term entecavir treatment results in sustained antiviral efficacy and prolonged life span in the woodchuck model of chronic hepatitis infection. *J Infect Dis* 184, 1236-45.

Colonno, R.J., Rose, R., Baldick, C.J., Levine, S., Pokornowski, K., Yu, C.F., Walsh, A., Fang, J., Hsu, M., Mazzucco, C., Eggers, B., Zhang, S., Plym, M., Klesczewski, K. and Tenney, D.J. (2006). Entecavir resistance is rare in nucleoside naive patients with hepatitis B. *Hepatology* 44, 1656-65.

Das, K., Xiong, X., Yang, H., Westland, C.E., Gibbs, C.S., Sarafianos, S.G. and Arnold, E. (2001). Molecular modeling and biochemical characterization reveal the mechanism of hepatitis B virus polymerase resistance to lamivudine (3TC) and emtricitabine (FTC). *J Virol* 75, 4771-9.

Delmas, J., Schorr, O., Jamard, C., Gibbs, C., Trepo, C., Hantz, O. and Zoulim, F. (2002). Inhibitory effect of adefovir on viral DNA synthesis and covalently closed circular DNA formation in duck hepatitis B virus-infected hepatocytes in vivo and in vitro. Antimicrob Agents *Chemother* 46, 425-33.

Fung, S.K., Chae, H.B., Fontana, R.J., Conjeevaram, H., Marrero, J., Oberhelman, K., Hussain, M. and Lok, A.S. (2006). Virologic response and resistance to adefovir in patients with chronic hepatitis B. *J Hepatol* 44, 283-90.

Hadziyannis, S.J., Tassopoulos, N.C., Heathcote, E.J., Chang, T.T., Kitis, G., Rizzetto, M., Marcellin, P., Lim, S.G., Goodman, Z., Ma, J., Brosgart, C.L., Borroto-Esoda, K., Arterburn, S. and Chuck, S.L. (2006). Long-term therapy with adefovir dipivoxil for HBeAg-negative chronic hepatitis B for up to 5 years. *Gastroenterology* 131, 1743-51.

Hadziyannis, S.J., Tassopoulos, N.C., Heathcote, E.J., Chang, T.T., Kitis, G., Rizzetto, M., Marcellin, P., Lim, S.G., Goodman, Z., Wulfsohn, M.S., Xiong, S., Fry, J. and Brosgart, C.L. (2003). Adefovir dipivoxil for the treatment of hepatitis B e antigen-negative chronic hepatitis B. *N Engl J Med* 348, 800-7.

Hoofnagle, J.H. and di Bisceglie, A.M. (1997). The treatment of chronic viral hepatitis. *N Engl J Med* 336, 347-56.

Hou, J., Yin, Y.K., Xu, D., Tan, D., Niu, J., Zhou, X., Wang, Y., Zhu, L., He, Y., Ren, H., Wan, M., Chen, C., Wu, S., Chen, Y., Xu, J., Wang, Q., Wei, L., Chao, G., Constance, B.F., Harb, G., Brown, N.A. and Jia, J. (2008). Telbivudine versus lamivudine in Chinese patients with chronic hepatitis B: Results at 1 year of a randomized, double-blind trial. *Hepatology* 47, 447-54.

Hui, C.K. and Lau, G.K. (2005). Clevudine for the treatment of chronic hepatitis B virus infection. *Expert Opin Investig Drugs* 14, 1277-84.

Hui, C.K., Zhang, H.Y., Bowden, S., Locarnini, S., Luk, J.M., Leung, K.W., Yueng, Y.H., Wong, A., Rousseau, F., Yuen, K.Y., Naoumov, N.N. and Lau, G.K. (2008). 96 weeks combination of adefovir dipivoxil plus emtricitabine vs. adefovir dipivoxil monotherapy in the treatment of chronic hepatitis B. *J Hepatol* 48, 714-20.

Hunt, C.M., McGill, J.M., Allen, M.I. and Condreay, L.D. (2000). Clinical relevance of hepatitis B viral mutations. *Hepatology* 31, 1037-44.

Iloeje, U.H., Yang, H.I., Su, J., Jen, C.L., You, S.L. and Chen, C.J. (2006). Predicting cirrhosis risk based on the level of circulating hepatitis B viral load. *Gastroenterology* 130, 678-86.

Jacob, J.R., Korba, B.E., Cote, P.J., Toshkov, I., Delaney, W.E.t., Gerin, J.L. and Tennant, B.C. (2004). Suppression of lamivudine-resistant B-domain mutants by adefovir dipivoxil in the woodchuck hepatitis virus model. *Antiviral Res* 63, 115-21.

Korba, B.E., Cote, P., Hornbuckle, W., Schinazi, R., Gangemi, J.D., Tennant, B.C. and Gerin, J.L. (2000). Enhanced antiviral benefit of combination therapy with lamivudine and alpha interferon against WHV replication in chronic carrier woodchucks. *Antivir Ther* 5, 95-104.

Lacombe, K., Gozlan, J., Boyd, A., Boelle, P.Y., Bonnard, P., Molina, J.M., Miailhes, P., Lascoux-Combe, C., Serfaty, L., Zoulim, F. and Girard, P.M. (2008). Comparison of the antiviral activity of adefovir and tenofovir on hepatitis B virus in HIV-HBV-coinfected patients. *Antivir Ther* 13, 705-13.

Ladner, S.K., Otto, M.J., Barker, C.S., Zaifert, K., Wang, G.H., Guo, J.T., Seeger, C. and King, R.W. (1997). Inducible expression of human hepatitis B virus (HBV) in stably transfected hepatoblastoma cells: a novel system for screening potential inhibitors of HBV replication. *Antimicrob Agents Chemother* 41, 1715-20.

Lai, C.L., Dienstag, J., Schiff, E., Leung, N.W., Atkins, M., Hunt, C., Brown, N., Woessner, M., Boehme, R. and Condreay, L. (2003). Prevalence and clinical correlates of YMDD variants during lamivudine therapy for patients with chronic hepatitis B. *Clin Infect Dis* 36, 687-96.

Lai, C.L., Gane, E., Liaw, Y.F., Hsu, C.W., Thongsawat, S., Wang, Y., Chen, Y., Heathcote, E.J., Rasenack, J., Bzowej, N., Naoumov, N.V., Di Bisceglie, A.M., Zeuzem, S., Moon, Y.M., Goodman, Z., Chao, G., Constance, B.F. and Brown, N.A. (2007). Telbivudine versus lamivudine in patients with chronic hepatitis B. *N Engl J Med* 357, 2576-88.

Lai, C.L., Shouval, D., Lok, A.S., Chang, T.T., Cheinquer, H., Goodman, Z., DeHertogh, D., Wilber, R., Zink, R.C., Cross, A., Colonno, R. and Fernandes, L. (2006). Entecavir versus lamivudine for patients with HBeAg-negative chronic hepatitis B. *N Engl J Med* 354, 1011-20.

Lanford, R.E., Notvall, L., Lee, H. and Beames, B. (1997). Transcomplementation of nucleotide priming and reverse transcription between independently expressed TP and RT domains of the hepatitis B virus reverse transcriptase. *J Virol* 71, 2996-3004.

Leung, N.W., Lai, C.L., Chang, T.T., Guan, R., Lee, C.M., Ng, K.Y., Lim, S.G., Wu, P.C., Dent, J.C., Edmundson, S., Condreay, L.D. and Chien, R.N. (2001). Extended lamivudine treatment in patients with chronic hepatitis B enhances hepatitis B e antigen seroconversion rates: results after 3 years of therapy. *Hepatology* 33, 1527-32.

Liang, X.F., Chen, Y.S., Wang, X.J., He, X., Chen, L.J., Wang, J., Lin, C.Y., Bai, H.Q., Yan, J., Cui, G. and Yu, J.J. (2005). *[A study on the sero-epidemiology of hepatitis B in Chinese population aged over 3-years old]*. Zhonghua Liu Xing Bing Xue Za Zhi 26, 655-8.

Liaw, Y.F., Chien, R.N., Yeh, C.T., Tsai, S.L. and Chu, C.M. (1999). Acute exacerbation and hepatitis B virus clearance after emergence of YMDD motif mutation during lamivudine therapy. *Hepatology* 30, 567-72.

Liaw, Y.F., Sung, J.J., Chow, W.C., Farrell, G., Lee, C.Z., Yuen, H., Tanwandee, T., Tao, Q.M., Shue, K., Keene, O.N., Dixon, J.S., Gray, D.F. and Sabbat, J. (2004). Lamivudine for patients with chronic hepatitis B and advanced liver disease. *N Engl J Med* 351, 1521-31.

Lok, A.S. (2000). Hepatitis B infection: pathogenesis and management. J Hepatol 32, 89-97.

Lok, A.S., Akarca, U. and Greene, S. (1994). Mutations in the pre-core region of hepatitis B virus serve to enhance the stability of the secondary structure of the pre-genome encapsidation signal. *Proc Natl Acad Sci* U S A 91, 4077-81.

Marcellin, P., Chang, T.T., Lim, S.G., Tong, M.J., Sievert, W., Shiffman, M.L., Jeffers, L., Goodman, Z., Wulfsohn, M.S., Xiong, S., Fry, J. and Brosgart, C.L. (2003). Adefovir dipivoxil for the treatment of hepatitis B e antigen-positive chronic hepatitis B. *N Engl J Med* 348, 808-16.

Marion, P.L., Salazar, F.H., Winters, M.A. and Colonno, R.J. (2002). Potent efficacy of entecavir (BMS-200475) in a duck model of hepatitis B virus replication. *Antimicrob Agents Chemother* 46, 82-8.

Mason, W.S., Cullen, J., Moraleda, G., Saputelli, J., Aldrich, C.E., Miller, D.S., Tennant, B., Frick, L., Averett, D., Condreay, L.D. and Jilbert, A.R. (1998). Lamivudine therapy of WHV-infected woodchucks. *Virology* 245, 18-32.

Menne, S., Butler, S.D., George, A.L., Tochkov, I.A., Zhu, Y., Xiong, S., Gerin, J.L., Cote, P.J. and Tennant, B.C. (2008). Antiviral effects of lamivudine, emtricitabine, adefovir dipivoxil, and tenofovir disoproxil fumarate administered orally alone and in combination to woodchucks with chronic woodchuck hepatitis virus infection. *Antimicrob Agents Chemother* 52, 3617-32.

Miller, P.J. (1982). Update: hepatitis B vaccine "the FDA approved the new vaccine on November 17, 1981". *Infect Control* 3, 76-7.

Nassal, M. (2008). Hepatitis B viruses: reverse transcription a different way. *Virus Res* 134, 235-49.

Norder, H., Courouce, A.M., Coursaget, P., Echevarria, J.M., Lee, S.D., Mushahwar, I.K., Robertson, B.H., Locarnini, S. and Magnius, L.O. (2004). Genetic diversity of hepatitis B virus strains derived worldwide: genotypes, subgenotypes, and HBsAg subtypes. *Intervirology* 47, 289-309.

Okamoto, H., Tsuda, F., Sakugawa, H., Sastrosoewignjo, R.I., Imai, M., Miyakawa, Y. and Mayumi, M. (1988). Typing hepatitis B virus by homology in nucleotide sequence: comparison of surface antigen subtypes. *J Gen Virol* 69 (Pt 10), 2575-83.

Parkin, D.M., Pisani, P. and Ferlay, J. (1999). Estimates of the worldwide incidence of 25 major cancers in 1990. *Int J Cancer* 80, 827-41.

Peek, S.F., Cote, P.J., Jacob, J.R., Toshkov, I.A., Hornbuckle, W.E., Baldwin, B.H., Wells, F.V., Chu, C.K., Gerin, J.L., Tennant, B.C. and Korba, B.E. (2001). Antiviral activity of clevudine [L-FMAU, (1-(2-fluoro-5-methyl-beta, L-arabinofuranosyl) uracil)] against woodchuck hepatitis virus replication and gene expression in chronically infected woodchucks (Marmota monax). *Hepatology* 33, 254-66.

Schaefer, S. (2007). Hepatitis B virus taxonomy and hepatitis B virus genotypes. *World J Gastroenterol* 13, 14-21.

Schalm, S.W., Heathcote, J., Cianciara, J., Farrell, G., Sherman, M., Willems, B., Dhillon, A., Moorat, A., Barber, J. and Gray, D.F. (2000). Lamivudine and alpha interferon combination treatment of patients with chronic hepatitis B infection: a randomised trial. *Gut* 46, 562-8.

Seeger, C. and Mason, W.S. (2000). Hepatitis B virus biology. *Microbiol Mol Biol Rev* 64, 51-68.

Sells, M.A., Zelent, A.Z., Shvartsman, M. and Acs, G. (1988). Replicative intermediates of hepatitis B virus in HepG2 cells that produce infectious virions. *J Virol* 62, 2836-44.

Song, B.C., Suh, D.J., Lee, H.C., Chung, Y.H. and Lee, Y.S. (2000). Hepatitis B e antigen seroconversion after lamivudine therapy is not durable in patients with chronic hepatitis B in Korea. *Hepatology* 32, 803-6.

Stevens, C.E., Toy, P.T., Tong, M.J., Taylor, P.E., Vyas, G.N., Nair, P.V., Gudavalli, M. and Krugman, S. (1985). Perinatal hepatitis B virus transmission in the United States. Prevention by passive-active immunization. *Jama* 253, 1740-5.

Strader, D.B., Wright, T., Thomas, D.L. and Seeff, L.B. (2004). Diagnosis, management, and treatment of hepatitis C. *Hepatology* 39, 1147-71.

Sung, J.J., Lai, J.Y., Zeuzem, S., Chow, W.C., Heathcote, E.J., Perrillo, R.P., Brosgart, C.L., Woessner, M.A., Scott, S.A., Gray, D.F. and Gardner, S.D. (2008). Lamivudine compared with lamivudine and adefovir dipivoxil for the treatment of HBeAg-positive chronic hepatitis B. *J Hepatol* 48, 728-35.

Tassopoulos, N.C., Volpes, R., Pastore, G., Heathcote, J., Buti, M., Goldin, R.D., Hawley, S., Barber, J., Condreay, L. and Gray, D.F. (1999). Efficacy of lamivudine in patients with hepatitis B e antigen-negative/hepatitis B virus DNA-positive (precore mutant) chronic hepatitis B.Lamivudine Precore Mutant Study Group. *Hepatology* 29, 889-96.

Tenney, D.J., Rose, R.E., Baldick, C.J., Levine, S.M., Pokornowski, K.A., Walsh, A.W., Fang, J., Yu, C.F., Zhang, S., Mazzucco, C.E., Eggers, B., Hsu, M., Plym, M.J., Poundstone, P., Yang, J. and Colonno, R.J. (2007). Two-year assessment of entecavir resistance in Lamivudine-refractory hepatitis B virus patients reveals different clinical outcomes depending on the resistance substitutions present. *Antimicrob Agents Chemother* 51, 902-11.

van Bommel, F., Wunsche, T., Mauss, S., Reinke, P., Bergk, A., Schurmann, D., Wiedenmann, B. and Berg, T. (2004). Comparison of adefovir and tenofovir in the treatment of lamivudine-resistant hepatitis B virus infection. *Hepatology* 40, 1421-5.

In: Bacterial DNA, DNA Polymerase and DNA Helicases
Editor: Walter D. Knudsen and Sam S. Bruns

ISBN 978-1-60741-094-2
© 2009 Nova Science Publishers, Inc.

Chapter IX

Modeling DNA Translocation and Unwinding by Helicase RecG

*Ping Xie**

Laboratory of Soft Matter Physics, Institute of Physics,
Chinese Academy of Sciences, Beijing, China

Abstract

RecG is a DNA helicase involved in the repair of damage at a replication fork by catalyzing the reversal of the fork to create a Holliday junction. Here, based on previous structural and biochemical studies a model is presented on how RecG catalyzes the processive translocation and unwinding of DNA so as to realize the fork reversal. In the model, a power stroke induces the DNA translocation and unwinding. Moreover, in order to effectively unwind DNA of a helical structure, it is assumed that the residues (i.e., the long α helix) connecting the wedge domain and the helicase domains in which the dsDNA-binding loop is located behave elastically along the torsional direction. Thus, accompanying the power stroke, the wedge domain is forced elastically to rotate relative to the dsDNA-binding loop that binds dsDNA strongly. Using the model, the calculated DNA translocation size per ATP hydrolysis is consistent with previous experimental data. Moreover, it is showed that RecG has a strong preference for negatively supercoiled DNA, i.e., RecG shows a much higher ATPase activity and longer processivity when interacting with the negatively supercoiled DNA than with the relaxed and/or positively supercoiled DNAs. This is also in agreement with the previous experimental data. In addition, the rotational rate or the rotational angle of RecG relative to DNA per ATP hydrolysis and the torsional elastic coefficient of the residues connecting the wedge domain and the helicase domains are predicted.

Keywords: RecG; DNA helicase; DNA repair; recombination; molecular motor.

---

* E-mail address: pxie@aphy.iphy.ac.cn

1. Introduction

Helicases are ubiquitous enzymes that play an essential role in nearly all aspects of DNA metabolism including DNA replication, repair, recombination, and transcription. They are motor proteins that can processively translocate along and unwind DNA by using the energy derived from NTP (generally ATP) hydrolysis [1–7]. Previous studies showed that some helicases such as Dda, HCV NS3 and RecQ work as monomers [8–11] while others such as *Escherichia coli* DnaB and bacteriophage T7 helicases function as oligomers [12,13].

RecG is a member of the superfamily 2 (SF2) helicases. It was first identified in *Escherichia coli* as a protein involved in DNA recombination and repair [14]. Biochemical studies revealed that RecG is active as a monomer [15] and catalyzes the interconversion of a replication stalled fork and a Holliday junction via the reversal of the fork to a point beyond the damage in the template strand [16–18]. The conversion of a replication fork into a Holliday junction requires the simultaneous unwinding of the leading and lagging strands followed by the reannealing of the two parental strands and the annealing of the two nascent strands [19]. The Holliday junction formed by this process may then migrate by the action of RuvAB or RecG [20–22], to re-establish the replication fork with the DNA lesion bypassed for later repair.

The crystal structure of *Thermatoga maritima* RecG in complex with a replication fork revealed that RecG comprises three structural domains [23]. The largest one (domain 1) is at the N terminus and consists of about half of the protein. Within this domain there is a greek key motif and thus it is referred to as the "wedge" domain that provides specificity for binding a three-way branched DNA structure. It has been proposed that the translocation of the dsDNA portion of the junction pulls the parental strands of a replication fork structure through two separate channels flanking the wedge domain [23]. The other two C-terminal domains (domain 2 and domain 3) of RecG contain the characteristic motifs that identify RecG as an SF2 helicase and thus they are referred to as "helicase domains". Experiments by Mahdi et al. [24] identified a motif containing a helical hairpin structure in the helicase domains and showed that changes to this structure interfere with the unwinding of branched DNA molecules. It was thus suggested that the motif mediates translocation of RecG on dsDNA by coupling ATP hydrolysis to movement at the dsDNA-binding site [24].

In this work, based on these structural and biochemical studies we propose a model on how RecG drives the processive translocation and unwinding of DNA so as to realize the fork reversal. The movement of the wedge domain induced by the power stroke results in the DNA translocation and unwinding, similar to that proposed previously [23,24]. Moreover, in order to effectively unwind DNA that has a helical structure, it is considered that the residues (i.e., the long α helix) connecting the wedge domain and the helicase domains in which the dsDNA-binding loop is located behave elastically along the torsional direction. As a result, accompanying the DNA translocation and unwinding, the wedge domain is also forced elastically to rotate relative to the dsDNA-binding loop that binds dsDNA strongly. Based on this model, we study the translocation step size per ATP hydrolysis and the dependences of ATPase activity and processivity on the twist change of supercoiled DNA. The results are in good agreement with previous experimental results [25,26]. Moreover, the rotational rate or the rotational angle of RecG relative to DNA per ATP hydrolysis and the torsional elastic coefficient of the residues connecting the wedge domain and the helicase domains are predicted.

2. Model

Before the presentation of the model for the DNA translocation and unwinding by RecG, we present some preliminary hypotheses and the supporting experimental evidences to the hypotheses in the following sections 2.1 – 2.3.

2.1. Nucleotide-State-Dependent Binding Affinity for dsDNA

The nucleotide-dependent binding affinity of RecG for dsDNA has not been experimentally determined yet. Here, without loss of generality, we assume that the affinity is similar to other motor proteins that can interact with dsDNA such as RuvB protein, another motor protein in conjunction with RuvA that can also interact with Holliday junctions, and bacteriophage $\phi 29$ DNA packaging motor that can translocate dsDNA of the *Bacillus subtilis* bacteriophage $\phi 29$ into a precursor capsid. Previous observations showed that both RuvB protein and $\phi 29$ DNA packaging motor bind dsDNA strongly in the presence of ATP while their affinities for dsDNA are low in the presence of ADP or in the absence of nucleotide [27,28]. Based on these evidences, we make the similar minimal hypothesis for RecG. The dsDNA-binding site located in the helicase domains has a strong binding affinity for the phosphodiester backbone of dsDNA in ATP and ADP.Pi states, while has a weak binding affinity in ADP and nucleotide-free states. This characteristic of nucleotide-dependent binding affinity for dsDNA is also consistent with the experimental results for SF1 helicases such as Rep that shows a high affinity for dsDNA in ATP state while a low affinity for dsDNA in ADP state [29,30].

2.2. Interaction between RecG and ssDNA

The crystal structure of RecG from *Thermatoga maritima* in a complex with a DNA fork substrate [23] reveals that the splitting of the duplex arm into two single strands is through the N-terminal wedge domain as it is moved forwards relative to DNA. The RecG-DNA complex is stabilized mainly by the interaction between an aromatic residue Phe204 from the RecG protein and the first unpaired base from the fork on the leading strand and the interaction between another aromatic residue Tyr208 from the RecG protein and a base on the lagging strand [23]. Based on this structural evidence, we make the following hypothesis. As the wedge domain is moved forwards, it forces the duplex to split into two single strands (see the following section 2.3). The ssDNA-binding site (residue Phe204) on one side of the wedge domain has the interaction with a nucleic acid base on the leading strand, while the ssDNA-binding site (residue Tyr208) on the other side of the wedge domain has the interaction with a base on the lagging strand. The interaction strength between the aromatic residue and nucleic acid base is dependent on the distance and relative orientation between them while independent of the nucleotide state of the protein, which is in contrast to the nucleotide-dependent interaction between dsDNA-binding site and the phosphodiester backbone of dsDNA as assumed in section 2.1.

2.3. Movement of dsDNA-Binding Site Induced by ATP Hydrolysis

Experiments by Mahdi et al. [24] demonstrated that a highly conserved helical hairpin motif in helicase domains plays a crucial role for RecG function. Two arginines (R609 and R630) are placed in opposing positions within this motif where they are stabilized by a network of hydrogen bonds involving a glutamate from helicase motif VI. Based on this observation, it was suggested that disruption of this feature, triggered by ATP hydrolysis, causes the movement of an adjacent loop in the dsDNA-binding channel. Here we also adopt the similar assumption. ATP hydrolysis induces the movement of dsDNA-binding site or loop in the helicase domains by a distance, $L \approx np$, towards the wedge domain, where n is the number of base pairs unwound by the wedge domain as it is moved forwards and $p = 0.34$ nm. Moreover, to enable the effective unwinding of DNA that has a helical structure, the following complementary hypothesis is necessary. *The residues (i.e., the long α helix) that connect the wedge domain and the helicase domains in which the dsDNA-binding site is located behave elastically along the torsional direction.*

Note here that, in other motor proteins such as bacteriophage T7 helicase [31] and ϕ 12 P4 RNA packaging motor [32], the movements of DNA-binding site (power strokes) are also induced by NTP hydrolysis. In addition, without loss of generality, we assume that the reverse movement of the dsDNA-binding site of RecG occurs following ADP release.

2.4. Mechanism of DNA Translocation and Unwinding by RecG

Based on the hypotheses in above sections 2.1−2.3, we describe the unidirectional dsDNA translocation and unwinding of DNA by RecG as follows. Here, for convenience of writing, DNA is considered in the relaxed form.

We begin with RecG in ADP state, with the dsDNA-binding site having weak interaction with the phosphodiester backbone of dsDNA (see section 2.1). As structure shows [23], the ssDNA-binding site (residue Phe204) on one side of the wedge domain binds the first unpaired base (base 1) from the duplex fork on the leading strand; while the ssDNA-binding site (residue Tyr208) on the other side of the wedge domain binds the second base (base 2') from the fork on the lagging strand (see section 2.2), as shown in Figure 1a. After ADP is released, the dsDNA-binding site restores to its pre-power-stroke position (see section 2.3). Because now the binding affinity of dsDNA-binding site for dsDNA is weak, this movement has no effect on the interactions between ssDNA-binding sites and the two single strands and thus the relative position between the wedge domain and DNA, as shown in Figure 1b.

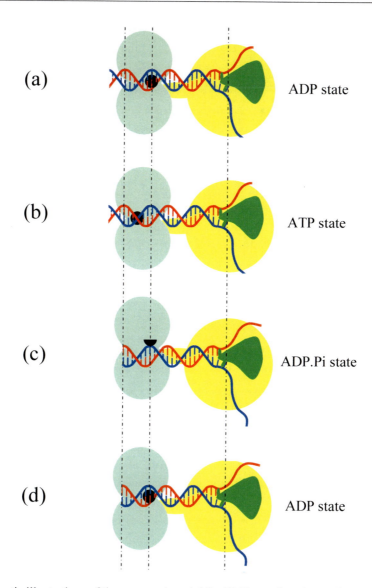

Figure 1. Schematic illustrations of the proposed model for DNA translocation and unwinding by helicase RecG (see text for details). Yellow sphere represents domain 1, with green part in domain 1 representing the wedge domain. Two light green spheres represent helicase domains composed of domain 2 and domain 3. Small dark sphere represents RecG dsDNA-binding site, with the complete sphere and half sphere represent different orientations of the dsDNA-binding site that is located solidly in the helicase domains.

After ATP binds, the dsDNA-binding site becomes bound strongly to the nearest sugar and phosphate residues on the dsDNA backbone (see section 2.1). Then ATP hydrolysis induces the power stroke. Because now dsDNA-binding site binds dsDNA strongly, the power stroke firstly disrupts the relatively weaker interaction between residue Phe204 and base 1 and that between residue Tyr208 and base 2'. Then the power stroke forces DNA to be unwound by n base pairs, where n satisfies $L \approx np$. During the unwinding process, due to the elastic connection between the wedge domain and the helicase domains in which the dsDNA-binding site is located, the wedge domain flanked by two nascent single strands is forced to rotate clockwisely by an angle ϕ relative to the dsDNA-binding site (see from the wedge domain). Simultaneously, the segment of duplex DNA in between the position to which the dsDNA-binding site is binding and the position at the duplex fork is forced to rotate by an angle of $\Theta = n\theta_0 - \phi$ in counterclockwise direction. Here $\theta_0 = 36°$ is the angle between two successive base pairs of B-form DNA. After the power stroke, the interaction of residue Phe204 with the base on the leading strand makes Phe204 bind its nearest base, i.e., the first nascent unpaired base (base 1) from the duplex fork on the leading strand; similarly, residue Tyr208 binds its nearest base, i.e., the second nascent unpaired base (base 2') from the duplex fork on the lagging strand, as shown in Figure 1c (noting that the viewing direction of Figure 1c is different from that of Figure 1a or 1b).

Then after Pi release, the binding of RecG to dsDNA becomes weak. This allows the dsDNA rotating by the angle $\Theta = n\theta_0 - \phi$ relative to the dsDNA-binding site in counterclockwise direction (see from duplex fork) and thus the torsional torque in DNA vanishes. Moreover, the induced elastic torque in RecG makes the dsDNA-binding site further rotate by the angle ϕ relative to the wedge domain towards the equilibrium position, as shown in Figure 1d. Thus, from Figure 1a to Figure 1d, RecG is translocated by a distance $L = np$ and is rotated by an angle $\phi + \Theta = n\theta_0$ relative to DNA in one ATPase cycle. Then the next ATPase cycle and the induced mechanical step will proceed.

It is noted that, in the present model, the interactions between RecG and DNA substrate during different periods of an ATPase cycle can be described briefly as follows: (i) In ADP and nucleotide-free states, RecG binds two single strands. (ii) In ATP state, RecG binds both the duplex region and two single strands. (iii) In ADP.Pi state, except during the very short period of power stroke when RecG binds only the duplex region, it binds both the duplex region and two single strands. Therefore, RecG always has strong interactions with DNA substrate, thus ensuring a high processivity. As shown in Briggs et al. [33], the wedge domain in the present model acts as both a strand separation module and a processivity factor via its binding to the two single strands.

3. Determination of RecG step Size and Rotation Rate

In this section, based on the model presented in above section we determine the translocation distance and rotational angle of RecG relative to DNA per ATP hydrolysis. Moreover, we give an estimate of the torsional elastic coefficient of the residues connecting the wedge domain and the helicase domains. Here we considered that DNA is in the relaxed form and the effect of DNA twist change on RecG activity will be studied in the next section.

First, we determine the torsional energy within the segment of dsDNA in between the position to which the dsDNA-binding site is binding and the position at the duplex fork after the power stroke. The torsional torque, τ_D, acting on the dsDNA segment is related to the rational angle θ by the following equation

$$\tau_D = k_B T C_D \frac{\theta}{l}, \qquad (1)$$

where $C_D \approx 75$ nm is the twist persistence length of dsDNA [34] and l is the length of the duplex DNA segment. From the crystal structure [23] we approximately have $l = 15$ bp $= 5.1$ nm. The torsional energy of the dsDNA segment can be calculated by using $E_{DNA} = \int_0^\Theta \tau_D d\theta$, where $\Theta = n\theta_0 - \phi$. From Eq. (1) we have

$$E_{DNA} = \frac{k_B T C_D}{2l} (n\theta_0 - \phi)^2. \qquad (2)$$

Then, we determine the torsional energy within RecG protein after the power stroke. Similar to Eq. (1), the torsional torque, τ_G, on RecG induced by the rotation of the wedge domain relative to the dsDNA-binding site by an angle ϕ can be written as

$$\tau_G = k_B T \frac{C_G}{l} \phi, \qquad (3)$$

where C_G/l is the elastic coefficient for the relative rotation between RecG helicase domains and wedge domain or the torsional elastic coefficient of the residues (i.e., the long α helix) that connect them, with C_G equivalent to C_D for DNA in Eq. (1). From the equilibrium condition $\tau_G = \tau_D$, we have $\Theta = C_G/C_D \phi$, from which, together with $\phi + \Theta = n\theta_0$, we can easily obtain

$$\phi = \frac{C_D}{C_G + C_D} n\theta_0, \qquad (4a)$$

$$\Theta = \frac{C_G}{C_G + C_D} n\theta_0. \qquad (4b)$$

From Eq. (4a), the torsional energy of RecG is obtained as

$$E_{RecG} = \frac{k_B T C_G}{2l} \frac{C_D^2}{(C_G + C_D)^2} (n\theta_0)^2. \qquad (5)$$

Using Eq. (4b), Eq. (2) can be rewritten as

$$E_{DNA} = \frac{k_B T C_D}{2l} \frac{C_G^2}{(C_G + C_D)^2} (n\theta_0)^2 . \qquad (6)$$

From Eqs. (5) and (6), the total torsional energy, $E_{Torsion} = E_{\mathrm{Re}cG} + E_{DNA}$, induced by the power stroke is

$$E_{Torsion} = \frac{k_B T}{2l} \frac{C_D C_G}{C_G + C_D} (n\theta_0)^2 . \qquad (7)$$

Based on the model, the power stroke also induces the unwinding of n base pairs. The free energy required to unwind these n base pairs can be calculated by $E_{Unwind} = n(E_{bp} + E_{BS})$. Here E_{bp} = 2.5 $k_B T$ is the average hydrogen-bonding free energy of one Watson-Crick base pair and E_{BS} is the free energy of base-stacking interaction between adjacent base pairs, which originates mainly from noncovalent van der Waals interactions between adjacent base pairs [35]. To precisely determine E_{BS} is a rather complicated subject and, for calculation, we take $E_{BS} = \varepsilon E_{bp}$, where ε is the base stacking intensity.

Under physiological conditions, the free-energy change of ATP hydrolysis is about 12 kcal/mol or $\Delta G \approx 20\ k_B T$ for the hydrolysis of an ATP molecule at room temperature. In this model, the free energy ΔG released by ATP hydrolysis is converted into both the unwinding energy and the torsional energy, i.e.,

$$\Delta G = E_{Torsion} + E_{Unwind} . \qquad (8)$$

From Eq. (8) we can obtain C_G as a function of n for different values of ε. The results are shown in Figure 2. As it is known, in general, the stacking of the two bases often contributes as much as, or more than, half of the free energy of the total base pair [36]. Thus in Figure 2 we take ε in between 0.5 and 1.

From Figure 2 it is seen that, for the value of C_G being in the reasonable range from 10 to 100 nm, n is larger than 2 but smaller than 4. As the model requires that the number of base pairs unwound should be approximately an integer, we thus have n = 3, meaning that the step size is 3 bp. This is in good agreement with the previous biochemical results showed that RecG on average translocates 3 bp per ATP molecule hydrolyzed [25]. Accordingly, based on the model the mean rotational angle of RecG relative to DNA per ATP molecule hydrolyzed is $n\theta_0 = 3\theta_0 = 108°$. From the measured ATPase rate of about $7.6\,\mathrm{s}^{-1}$ during translocation [25], it is predicted that the mean rotation rate of RecG relative to DNA is about $821°$ per second.

Moreover, as seen from Figure 2, at n = 3, C_G has the value of about 27 nm for ε = 0.75 which is close to the average base stacking intensity determined experimentally [36]. We thus predict that the torsional elastic coefficient of the residues connecting the helicase domains and the wedge domain is around $k_B T C_G / l = 5.3\,k_B T$ per radian.

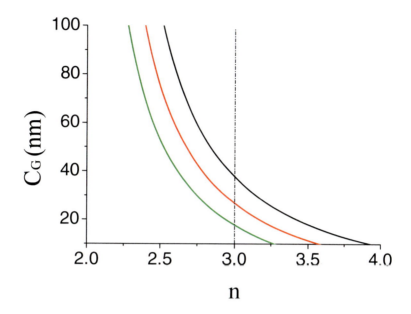

Figure 2. Calculated results of C_G as a function of n for $\varepsilon = 0.5$ (black line), 0.75 (red line) and 1 (green line).

4. Dependence of RecG Activity on DNA Topology

In section 3 we used relaxed DNA. In this section we will study the effect of the twist change in supercoiled DNA on RecG activity.

For the supercoiled DNA, after the power stroke, the rotational angle of the wedge domain relative to the dsDNA-binding site and that of DNA segment are written, instead of Eqs. (4a) and (4b), as follows

$$\phi = \frac{C_D}{C_G + C_D}\left(3\theta_0 + \frac{2\pi}{h_0}\cdot\frac{\Delta Tw}{Tw_0}l\right), \qquad (9a)$$

$$\Theta = \frac{C_G}{C_G + C_D}\left(3\theta_0 + \frac{2\pi}{h_0}\cdot\frac{\Delta Tw}{Tw_0}l\right), \qquad (9b)$$

where $h_0 = 3.4$ nm and $\Delta Tw = Tw - Tw_0$ is the change in twist of the DNA, with Tw and Tw_0 being the numbers of helical repeats of the DNA in supercoiled and relaxed states, respectively. From Eqs. (9a) and (9b), the total torsional energy, Eq. (7), is replaced by

$$E_{Torsion} = \frac{k_B T}{2l} \frac{C_G C_D}{C_G + C_D} \left(3\theta_0 + \frac{2\pi}{h_0} \cdot \frac{\Delta Tw}{Tw_0} l\right)^2. \qquad (10)$$

As it is known, a supercoiled DNA is easier to be locally unwound than a relaxed DNA. Thus, during a power stroke, RecG will overcome a smaller energy barrier to unwind the three base pairs of the supercoiled DNA than the relaxed DNA. To quantitatively calculate the free energy, $E_{Unwind}(\Delta Tw)$, that is required to unwind the three base pairs of a supercoiled DNA is a very difficult task. Here, it can be phenomenologically written as $E_{Unwind}(\Delta Tw) = \left(E_{Unwind}(0) - \alpha |\Delta Tw/Tw_0|\right)^x$, where $E_{Unwind}(0)$ is the free energy required to unwind the three base pairs of the relaxed DNA, and $\alpha > 0$ and $x > 0$ are parameters independent of $\Delta Tw/Tw_0$ [Note 1]. Considering $E_{Unwind}(0) \gg \alpha |\Delta Tw/Tw_0|$ in the case of very small $|\Delta Tw/Tw_0|$, we obtain the difference between the free energy required to unwind the supercoiled DNA and that to unwind the relaxed DNA as follows

$$\Delta E_{Unwind} \approx -\beta \left|\frac{\Delta Tw}{Tw_0}\right|, \qquad (11)$$

where $\beta = \alpha x \left[E_{Unwind}(0)\right]^{x-1}$ is a parameter independent of $\Delta Tw/Tw_0$.

Now we use Eqs. (10) and (11) to determine the ATPase rate K. From our model, the ATPase cycling of RecG is described by the following pathway

$$\text{Empty} \xrightarrow{k_b[\text{ATP}]} \text{ATP} \xrightarrow[\text{powerstroke}]{k_2} \text{ADP.Pi} \xrightarrow{k_3} \text{ADP} \xrightarrow{k_4} \text{Empty}, \qquad (12)$$

where, for simplicity, we have neglected the backward transitions since each backward-transition rate is much smaller than the corresponding forward-transition rate under the condition of no ADP and Pi. The transition ATP $\xrightarrow{k_2}$ ADP.Pi results in the translocation of DNA relative to RecG due to the power stroke. The twisted free energy $E_{Torsion}$ has the effect on the transitions ATP $\xrightarrow{k_2}$ ADP.Pi and ADP.Pi $\xrightarrow{k_3}$ ADP while has no effect on other transitions. The energy $E_{Unwind}(\Delta Tw)$ only has the effect on the transition ATP $\xrightarrow{k_2}$ ADP.Pi while has no effect on other transitions. Thus, based on Arrhenius equation, we have the following relation

$$k_i = k_i^{(0)}, (i = b, 4) \qquad (13a)$$

$$k_2 = k_2^{(0)} \exp\left(-\frac{E_{Torsion} + \Delta E_{Unwind}}{k_B T}\right), \qquad (13b)$$

$$k_3 = k_3^{(0)} \exp\left(-\frac{E_{Torsion}}{k_B T}\right), \tag{13c}$$

where $k_i^{(0)}$ is the transition rate with $E_{Torsion} = 0$ (i.e., without the torsional torque acting on RecG) and $\Delta E_{Unwind} = 0$.

From Scheme (8) the mean ATPase rate, K, at saturating ATP concentration can be written as $K^{-1} = k_2^{-1} + k_3^{-1} + k_4^{-1}$, which is rewritten by using Eq. (13) as follows

$$K = K^{(0)} \frac{(1+A+B)}{1+\left[A\exp\left(\frac{\Delta E_{Unwind}}{k_B T}\right) + B\right]\exp\left(\frac{E_{Torsion}}{k_B T}\right)}, \tag{14}$$

where $\left(K^{(0)}\right)^{-1} = \left(k_2^{(0)}\right)^{-1} + \left(k_3^{(0)}\right)^{-1} + \left(k_4^{(0)}\right)^{-1}$ is the inverse of mean ATPase rate with $E_{Torsion} = 0$, $\Delta E_{Unwind} = 0$, $A = k_4^{(0)}/k_2^{(0)}$ and $B = k_4^{(0)}/k_3^{(0)}$. It is expected that the transition ATP $\xrightarrow{k_2}$ ADP.Pi that induces the power stroke is the rate-limiting step and thus A should be a large value. Therefore, Eq. (14) can be approximately rewritten as

$$K \approx K^{(0)} \exp\left(-\frac{E_{Torsion} + \Delta E_{Unwind}}{k_B T}\right). \tag{15}$$

As determined in section 3, C_G has the value of about 27 nm for $\varepsilon = 0.75$ which is close to the average base stacking intensity determined experimentally [36]. Thus we take $C_G = 27$ nm in the following calculation. In addition, since β is not available, we take it as an adjustable parameter. In Figure 3 we show the calculated K versus $\Delta Tw/Tw_0$ for different values of β by using Eqs. (10), (11) and (15), where, to be consistent with the experimental value [25], we take $K = 7.6\,\text{s}^{-1}$ under $\Delta Tw/Tw_0 = 0$. It is seen that, for any value of β, K is larger for the negatively supercoiled DNA than for the relaxed and positively supercoiled DNA. For the large value of β, K has the largest value in the presence of negatively supercoiled DNA and has the smallest value in the presence of relaxed DNA and, moreover, K is much larger with negatively supercoiled DNA than with relaxed and/or positively supercoiled DNA. These are in agreement with the experimental results by Slocum et al. [26].

From Eqs. (1) and (9b) or Eqs. (3) and (9a), we obtain that the torsional torque acting on the dsDNA segament or on RecG is

$$\tau_D = \tau_G = \frac{k_B T}{l} \frac{C_G C_D}{C_G + C_D}\left(3\theta_0 + \frac{2\pi}{h_0} \cdot \frac{\Delta Tw}{Tw_0} l\right). \tag{16}$$

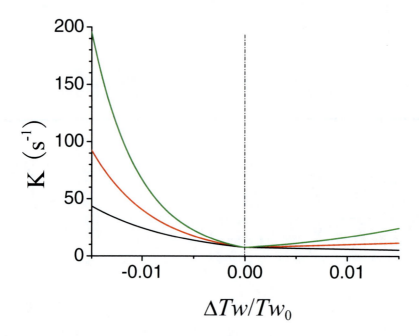

Figure 3. Calculated results of ATPase rate K versus $\Delta Tw/Tw_0$ for β = 50 k_BT (black line), 100 k_BT (red line) and 150 k_BT (green line). To be consistent with the experimental value [25], we take K = 7.6 s^{-1} under $\Delta Tw/Tw_0$ = 0.

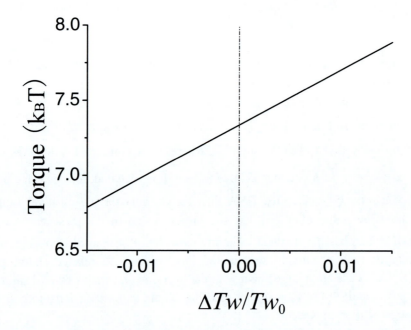

Figure 4. Calculated results of torque acting on RecG or DNA as a function of $\Delta Tw/Tw_0$.

Using Eq. (16), the calculated torque versus $\Delta Tw/Tw_0$ is shown in Figure 4. It is seen that, for $\Delta Tw/Tw_0 < 0$, τ_D or τ_G has a smaller value than that for $\Delta Tw/Tw_0 = 0$, which is smaller than that for $\Delta Tw/Tw_0 > 0$. On the other hand, as theoretically shown by Sarkar and Marko [37], the local torsional stress in DNA can eject a DNA-bound protein. Similarly, the torsional stress in the protein can also eject the protein from DNA. Thus the processivity of RecG in the presence of negatively supercoiled DNA is longer than in the presence of relaxed DNA, which is longer than in the presence of positively supercoiled DNA. This is also in agreement with the experimental result by Slocum et al. [26].

5. Comparison with RuvAB

Since both the monomeric RecG and hexameric RuvB in conjunction with RuvA tetramer can drive DNA translocation and unwinding and catalyze the branch migration of Holliday junctions [20–22], it is interesting to make comparisons between the working mechanism by RecG presented here and that by RuvAB presented in the previous work [38].

The activity carried out by monomeric RecG is via the relative movement between the dsDNA-binding site and the wedge domain. To effectively unwind DNA that has a helical structure, it is required that, as the wedge domain is moved towards the dsDNA-binding site that binds dsDNA strongly, the wedge domain is forced elastically to rotate relative to the dsDNA-binding site. By contrast, during the branch migration of Holliday junctions mediated by RuvAB, the translocation of DNA is driven by RuvB hexamer while the unwinding of DNA is via RuvA tetramer as it is moved relative to DNA. Moreover, the outward straight translocations of the two opposite arms of dsDNA can make the other two arms in perpendicular direction simultaneously translocate inwards and rotate in a right-handed manner. Thus the straight movement of the RuvB dsDNA-binding site is enough to effectively unwind DNA that has a helical structure.

Since the dsDNA-binding site in RecG is far away from two ssDNA-binding sites, the three sites can simultaneously bind DNA by inducing a torsional torque due to the mismatch in angles between RecG and DNA. The binding to DNA by three binding sites ensure the high processivity for RecG. However, since the power-stroke size of RuvB is only few base pairs and the twist persistence length of dsDNA is long, the dsDNA-binding sites of one RuvB subunit and its adjacent RuvB subunit are unable to simultaneously bind DNA due to the mismatch in angles between RuvB hexamer and DNA. The configuration arrangement that DNA is surrounded by the ring-shaped RuvB ensures the high processivity for RuvAB.

For both cases, the rotations of motor proteins relative to DNA during one ATPase cycle are via two steps. For RecG, one rotation is through the rotation of the wedge domain relative to the dsDNA-binding site that binds dsDNA during the power stroke, while the other one is following the release of dsDNA-site from DNA after Pi release, with the ssDNA-binding sites binding to two single strands near the fork. For RuvAB, one rotation is through the relative rotation between RuvB hexamer and RuvA trtamer, while the other one is through the passage of the strong DNA binding from the previous RuvB subunits that face each other to the next adjacent two facing subunits.

Conclusion

In this work, a power-stroke model is presented for the DNA translocation and unwinding by RecG helicase, which takes into account the torsional elasticity of the residues (i.e., the long α helix) connecting the wedge domain and the helicase domains in which the dsDNA-binding loop is located. Based on the model, from the available free-energy change of ATP hydrolysis we determine that the step size is 3 bp, the rotational angle of RecG relative to DNA per step is 108° and the torsional elastic coefficient of the residues that connect the wedge domain and the helicase domains is around $5.3\,k_B T$ per radian. Using these determined parameters and one adjustable parameter β we study the ATPase rate versus twist change $\Delta Tw/Tw_0$ of the supercoiled DNA. The theoretical results are in agreement with available experimental data.

Finally, we mention that, in order to test the model for RecG presented here, it is expected to measure the rotational rate or the rotational angle of RecG relative to DNA per ATP hydrolysis and the torsional elastic coefficient of the residues that connect the wedge domain and the helicase domains. Measure the ATPase rate as a function of $\Delta Tw/Tw_0$ so that the adjustable parameter β can be determined.

This research was supported by the National Natural Science Foundation of China.

Note 1. Precisely speaking, the value of α for $\Delta Tw/Tw_0 < 0$ should be larger than that for $\Delta Tw/Tw_0 > 0$, because a negatively supercoiled DNA is easier to be locally unwound than a positively supercoiled DNA. Here, for simplicity, we take α having the same value for both negative and positive values of $\Delta Tw/Tw_0$.

References

[1] Matson, S. W. and Kaiser-Rogers, K. A. (1990). DNA helicases. *Annu. Rev. Biochem., 59,* 289–329.

[2] Lohman, T. M. and Bjornson, K. P. (1996). Mechanisms of helicase-catalyzed DNA unwinding. *Annu. Rev. Biochem., 65,* 169–214.

[3] Patel, S. S. and Picha, K. M. (2000). Structure and function of hexameric helicases. *Annu. Rev. Biochem., 69,* 651–697.

[4] Marians, K. J. (2000). Crawling and wiggling on DNA: structural insights to the mechanism of DNA unwinding by helicases. *Structure, 8,* R227–R235.

[5] Bianco, P. R.; Brewer, L. R.; Corzett, M.; Balhorn, R.; Yeh, Y.; Kowalczykowski, S. C.; and Baskin, R. J. (2001). Processive translocation and DNA unwinding by individual RecBCD enzyme molecules. *Nature, 409,* 374–378.

[6] Xie, P. (2006). Model for helicase translocating along single-stranded DNA and unwinding double-stranded DNA. *Biochim. Biophys. Acta, 1764,* 1719–1729.

[7] Xie, P. (2007). On translocation mechanism of ring-shaped helicase along single-stranded DNA. *Biochim. Biophys. Acta, 1774*, 737–748.

[8] Morris, P. D.; Tackett, A. J.; Babb, K.;, Nanduri, B.; Chick, C.; Scott, J.; and Raney, K. D. (2001). Evidence for a functional monomeric form of the bacteriophage T4 Dda helicase. *J. Biol. Chem. 276*, 19691–19698.

[9] Nanduri, B.; Byrd, A. K.; Eoff, R. L.; Tackett, A. J.; and Raney, K. D. (2002). Pre-steady-state DNA unwinding by bacteriophage T4 Dda helicase reveals a monomeric molecular motor. *Proc. Natl Acad. Sci. USA 99*, 14722–14727.

[10] Levin, M. K.; Gurjar, M. M.; and Patel, S. S. (2005). A Brownian motor mechanism of translocation and strand separation by hepatits C virus helicase. *Nature Struc. Mol. Biol. 12*, 429–435.

[11] Zhang, X.-D.; Dou, S.-X.; Xie, P.; Hu, J.-S.; Wang, P.-Y.; and Xi, X. G. (2006). E. coli RecQ is a rapid, efficient and monomeric helicase. *J. Biol.Chem. 281*, 12655-12663.

[12] Kaplan, D. L. and O'Donnell, M. (2002). DnaB drives DNA branch migration and dislodges proteins while encircling two DNA strands. *Mol. Cell, 10*, 647–657.

[13] Kim, D.-E.; Narayan, M.; and Patel, S. S. (2002). T7 DNA helicase: A molecular motor that processively and unidirectionally translocates along single-stranded DNA. *J. Mol. Biol., 321*, 807–819.

[14] Lloyd, R. G. and Buckman, C. (1991). Genetic analysis of the *recG* locus of *Escherichia coli* K-12 and of its role in recombination and DNA repair. *J. Bacteriol., 173*, 1004–1011.

[15] McGlynn, P.; Mahdi, A. A.; and Lloyd, R. G. (2000). Characterisation of the catalytically active form of RecG helicase. *Nucleic Acids Res., 28*, 2324–2332.

[16] McGlynn, P. and Lloyd, R. G. (2000). Modulation of RNA polymerase by (p)ppGpp reveals a RecG-dependent mechanism for replication fork progression. *Cell, 101*, 35–45.

[17] McGlynn, P.; Lloyd, R. G.; and Marians, K. J. (2001). Formation of Holliday junctions by regression of nascent DNA in intermediates containing stalled replication forks: RecG stimulates regression even when the DNA is negatively supercoiled. *Proc. Natl. Acad. Sci. U.S.A., 98*, 8235–8240.

[18] Gregg, A. V.; McGlynn, P.; Jaktaji, R. P.; and Lloyd, R. G. (2002). Direct rescue of stalled DNA replication forks via the combined action of PriA and RecG helicase activities. *Mol. Cell, 9*, 241–251.

[19] McGlynn, P. and Lloyd, R. G. (2001). Rescue of stalled replication forks by RecG: Simultaneous translocation on the leading and lagging strand templates supports an active DNA unwinding model of fork reversal and Holliday junction formation. *Proc. Natl. Acad. Sci. U.S.A., 98*, 8227–8234.

[20] West, S. C. (1998). RuvA gets X-rayed on Holliday. *Cell, 94*, 699–701.

[21] Whitby, M. C. and Lloyd, R. G. (1998). Targeting Holliday junctions by the RecG branch migration protein of *Escherichia coli*. *J. Biol. Chem., 273*, 19729–19739.

[22] Sharples, G. J.; Ingleston, S. M.; and Lloyd, R. G. (1999). Holliday junction processing in bacteria: insights from the evolutionary conservation of RuvABC, RecG, and RusA. *J. Bacteriol., 181*, 5543–5550.

[23] Singleton, M. R.; Scaife, S.; and Wigley, D. B. (2001). Structural analysis of DNA replication fork reversal by RecG. *Cell, 107,* 79–89.

[24] Mahdi, A. A.; Briggs, G. S.; Sharples, G. J.; Wen, Q.; and Lloyd, R. G. (2003). A model for dsDNA translocation revealed by a structural motif common to RecG and Mfd proteins. *EMBO J., 22,* 724–734.

[25] Martinez-Senac, M. M. and Webb, M. R. (2005). Mechanism of translocation and kinetics of DNA unwinding by the helicase RecG. *Biochemistry, 44,* 16967–16976.

[26] Slocum, S. L.; Buss, J. A.; Kimura, Y.; and Bianco, P. R. (2007). Characterization of the ATPase activity of the *Escherichia coli* RecG protein reveals that the preferred cofactor is negatively supercoiled DNA. *J. Mol. Biol., 367,* 647–664.

[27] Muller, B.; Tsaneva, I. R.; and West, S.C. (1993). Branch migration of Holliday junctions promoted by the *Escherichia coli* RuvA and RuvB proteins: II. Interaction of RuvB with DNA. *J. Biol. Chem., 268,* 17185–17189.

[28] Chemla, Y. R.; Aathavan, K.; Michaelis, J.; Grimes, S.; Jardine, P. J.; Anderson, D. L.; and Bustamante, C. (2005). Mechanism of force generation of a viral DNA packaging motor. *Cell, 122,* 683–692.

[29] Wong, I. and Lohman, T. M. (1992). Allosteric effects of nucleotide cofactors on *Escherichia coli* Rep helicase-DNA binding. *Science, 256,* 350–355.

[30] Wong, I.; Chao, K. L.; Bujalowski, W.; and Lohman, T. M. (1992). DNA-induced dimerisation of the *Escherichia coli* Rep helicase: allosteric effects of single stranded and duplex DNA. *J. Biol. Chem., 267,* 7596–7610.

[31] Singleton, M.; Sawaya, M.; Ellenberger, T.; and Wigley, D. B. (2000). Crystal structure of T7 gene 4 ring helicase indicates a mechanism for sequential hydrolysis of nucleotides. *Cell, 101,* 589–600.

[32] Mancini, E. J.; Kainov, D. E.; Grimes, J. M.; Tuma, R.; Bamford, D. H.; and Stuart, D. I. (2004). Atomic snapshots of an RNA packaging motor reveal conformational changes linking ATP hydrolysis to RNA translocation. *Cell, 118,* 743–755.

[33] Briggs, G. S.; Mahdi, A. A.; Wen, Q.; and Lloyd, R. G. (2005). DNA binding by the substrate specificity (wedge) domain of RecG helicase suggests a role in processivity. *J. Biol. Chem., 280,* 13921–13927.

[34] Strick, T. R.; Allemand, J.-F.; Bensimon, D.; and Croquette, V. (1998). Behavior of Supercoiled DNA. *Biophys. J., 74,* 2016–2028.

[35] Saenger, W. (1984). *Principles of Nucleic Acid Structure.* New York, Springer-Verlag.

[36] Kool, E. T. (2001). Hydrogen bonding, base stacking, and steric effects in DNA replication. *Annu. Rev. Biophys. Biomol. Struct., 30,* 1–22.

[37] Sarkar, A. and Marko, J. F. (2001). Removal of DNA-bound proteins by DNA twisting. *Phys. Rev. E, 64,* 061909.

[38] Xie, P. (2007). Model for RuvAB-mediated branch migration of Holliday junctions. *J. Theor. Biol., 249,* 566–573.

In: Bacterial DNA, DNA Polymerase and DNA Helicases
Editor: Walter D. Knudsen and Sam S. Bruns
ISBN 978-1-60741-094-2
© 2009 Nova Science Publishers, Inc.

Chapter X

De Novo DNA Synthesis by DNA Polymerase

Xingguo Liang,* Tomohiro Kato and Hiroyuki Asanuma
Department of Molecular Design and Engineering, Graduate School of Engineering, Nagoya University, Chikusa, Nagoya, Japan

Abstract

In this chapter, the *de novo* DNA synthesis by DNA polymerase in the absence of any added template and/or primer is described. As early as the 1960s, Kornberg et al. reported that DNA polymerase I could *de novo* synthesize DNA, although it was later pointed out that the contaminated DNA or RNA may provide the seeds for DNA synthesis and the contaminated transferase that can polymerize dNDPs might play the role of template-independent DNA synthesis. The synthesized DNA polymers were characterized as homopolymer pairs such as poly(dA)/poly(dT) and poly(dI)/poly(dC), or the alternating copolymers poly(dA-dT), poly(dG-dC) and poly(dI-dC). In late 1990s, after a long silent time, Ogata et al. found that the highly purified thermophilic DNA polymerase could carry out *de novo* DNA synthesis at higher temperatures (> 65°C) and the synthesized DNA polymers were composed of repetitive palindromic sequences such as $(TACATGTA)_n$, and $(TAAT)_n$. Surprisingly, Liang et al. reported that thermophilic DNA polymerases could carry out *de novo* DNA synthesis even at room temperature. Basically, *de novo* DNA synthesis is considered to occur in two stages, although the detail mechanism is not clear. In the initial stage, dNTPs are polymerized to short DNA oligos, and then the selected repetitive sequences are elongated to long DNA in the elongation stage through self-priming and primer extension. More interestingly, the *de novo* DNA synthesis can be greatly accelerated by adding endonuclease such as restriction enzymes, which digest DNA at their recognition sites. A Cut-grow mechanism was proposed to explain both the paradox of endonuclease and the extremely high efficiency: the synthesized long DNA is cut to shorter seeds for the elongation of the next cycle so that DNA can be exponentially amplified. The possible biological and evolutionary roles of *de novo* DNA synthesis are also discussed.

---

* Correspondence should be addressed to liang@mol.nagoya-u.ac.jp

Keywords: *de novo* DNA synthesis, DNA polymerase, hairpin formation, elongation, primer extension, molecular evolution, homopolymer, endonuclease, restriction enzyme, exonuclease.

1. Introduction

The *de novo* DNA synthesis described here is designated as the synthesis of DNA from dNTPs directly by DNA polymerase (or the enzymes of DNA polymerase family) in the total absence of any added DNA or RNA as the template/primer. Although *de novo* DNA synthesis was well described by Kornberg et al. in as early as the 1960s [1–6], it had been a controversy for a long time because it was hard to guarantee that the DNA synthesis did not come from tiny DNA or RNA contaminant [6–8]. Another reason for the doubt of *de novo* synthesis is the failure to find its persuadable biological meaning. The *de novo* DNA synthesis seems to make no sense from the viewpoint of molecular biology because DNA is only utilized as the carrier of genetic information and shows no other function. In addition, DNA polymerases usually have extremely high fidelity and mutation of DNA is highly suppressed. Even at present, most of the biochemists and biologists believe that the DNA synthesis catalyzed by DNA polymerases definitely requires a primer/template complex as the substrate, and DNA polymerases are not capable of *de novo* DNA synthesis [9–10]. In almost all the textbooks, we are taught that DNA synthesis requires at least a primer synthesized by the primase during DNA replication.

However, experimental results in recent years supported the occurrence of *de novo* synthesis, and it has shed light again on one of the most remarkable and unanticipated events in the history of DNA polymerase studies [6]. Not only on the level of describing the facts of *de novo* DNA synthesis, developments in both biochemistry and biotechnology show us new opportunities to go further to clarify the mechanism and significance underlying these miraculous phenomena. New findings and hypothesizes have been accumulating to make us think again about the nature of life. For example, most of the newly found DNA polymerases that can carry out *de novo* DNA synthesis are obtained from thermophilic archaebacteria, which have been widely considered to be close to "the origins" on the evolutionary tree model. Most of the DNAs synthesized *de novo* have repetitive sequences that are ubiquitous in nature [11-13], and the abnormal expansion of these sequences have been found to cause hereditary neuromuscular diseases [14–17]. The paradox of endonuclease that greatly accelerates *de novo* DNA synthesis according to Cut-grow mechanism may provide a model of simplest "life" [18]. Furthermore, in theory, the *de novo* DNA synthesis is considered to be possible at the active center of DNA polymerase starting with the synthesis of 2-nt-long deoxyribonucleotide (pppNpN) from two dNTPs, although the efficiency is supposed to be extremely low. Based on the above exciting findings, development of this promising research field of *de novo* DNA synthesis is highly expected in the following decades. In this chapter, the short history, synthesis by various DNA polymerases, mechanism, and proposed biological meanings of *de novo* DNA synthesis are demonstrated.

2. *De Novo* DNA Synthesis by DNA Polymerase I – The Early History

In 1960, only 4 years after Authur Kornberg found the first DNA polymerase (Polymerase I) from *E. coli* [19], it was reported that the DNA polymerase could carry out *de novo* DNA synthesis. Unexpectedly, a huge DNA-like polymer was obtained after several hours of incubation of dATP, dTTP, and polymerase I without adding any oligonucleotide [2].

The product was poly(dA-dT) in which the deoxyadenylate and thymidylate residues are arranged in perfectly ordered alternation. Figure 1a shows the typical time course of *de novo* synthesis, evaluated by viscometric and radioisotope incorporation methods. After a lag period of several hours, DNA started to be synthesized with an exponential rate. The synthesized DNA became degraded by the exonuclease activity of Polymerase I after 60–80% of the dATP and dTTP were consumed. Similar results were also obtained by using sedimentation and spectrophotometric studies for characterization. This finding surprised all the members of the research group, although work several years later by Arthur Kornberg and Gobind Khorana showed that the trace amounts of DNA present in the DNA polymerase preparation probably initiated the DNA synthesis [4,9]. However, these researches proposed the interesting question of "Can DNA polymerase carry out *de novo* DNA synthesis?" and caused a long debate simultaneously [6,8].

In the subsequent years, the *de novo* synthesis of a large variety of homopolymers and copolymers such as poly(dA-dT), poly(dG-dC), poly(dA)/poly(dT), and poly(dG)/poly(dC), were observed under various conditions [2–4, 20–23]. Interestingly, the *de novo* synthesis of DNA polymers involving non-natural base Inosine (I) and ^{Br}C (5-bromide cytosine), such as poly(dG-d^{Br}C) [2], poly(dI)/poly(dC) [21,22] and poly(dI-dC) [21,23] were also observed (Figure 1b). Factors and conditions on both the sequences and kinetics of the *de novo* DNA synthesis were also investigated, although the detail mechanism remained largely unexplained, especially for the case of synthesizing homopolymer.

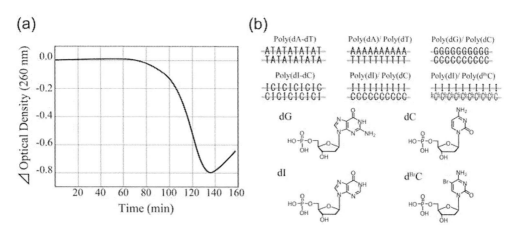

Figure 1. Typical time course of *de novo* DNA synthesis of Poly(dA-dT) (a) and structures of *de novo* synthesized DNA polymers (b). Structures of base I (Inosine) and ^{Br}C are also shown.

2.1. Factors and Conditions Influencing the Lag Time of *De Novo* DNA Synthesis

During the initiation of *de novo* DNA synthesis, a period of time is usually required for enough DNA to be detected. It is designated as the lag time, ranging from 1 to 100 hours. The longer the lag time indicates the more difficult *de novo* DNA synthesis. In the first case of *de novo* DNA synthesis (when dTTP and dATP were used), the lag time was 1–5 hours, depending on the reaction conditions. For example, increasing the enzyme concentration shortens the lag period, but increasing the ionic strength lengthens the lag period [1]. Furthermore, with increasing purity and modified separation procedure, the capacity of DNA polymerase to catalyze the *de novo* reaction decreases. It was even hard to observe *de novo* DNA synthesis after 48 hours when only dCTP and dGTP were used [21]. By using the partially purified polymerase I, however, poly(dG)/poly(dC) could be detected after 20 hours under the same conditions. Change of pH also influenced the *de novo* synthesis. For example, the incubation with dGTP and dCTP gave no reaction in Tris-HCl at pH 8.2, but did react at pH 7.4 by using partially purified polymerase I [21].

Obviously, poly(dG)/poly(dC) was much more difficult to be synthesized as compared with poly(dA-dT), due to the high stability of GC rich duplex. Similarly, in the case of *de novo* synthesizing poly(dI-dC) and poly(dI)/poly(dC), the lag time varied from 5 to 25 hours in the reaction with the partially purified Polymerase I, whereas longer lag time (25 to 40 hours) was found with the highly purified polymerase [21]. Interestingly, the 5′→3′ exonuclease activity is found to be necessary for the *de novo* DNA synthesis of poly(dA-dT): the Klenow Large Fragment lacking of 5′→3′ exonuclease activity (but fully active in catalyzing DNA synthesis at a nick without degrading the 5′-strand ahead of the growing point) failed to *de novo* synthesize poly(dA-dT) even after 84 hours [25]. Note that the function of nucleases is to destroy DNA from 5′-end, and it is hard to imagine that it can help to synthesize it (the possible mechanism will be discussed later).

In 1978 and 1979, a group from the former USSR reported that the deoxynucleoside diphosphate: oligonucleotide deoxynucleotidyl transferase (dNDP-deoxynucleotidyl transferase) separated from *E.coli* DNA polymerase I preparation might contribute to the short oligo DNA synthesis during the lag time [7,8]. They pointed out that the dNDPs present in the dNTPs solution are sufficient for oligo(dG) and oligo(dC) synthesis by the mixed dNDP-deoxynucleotidyl transferase. They also found that the lag time of *de novo* synthesis of poly(dA-dT) was abolished by preincubation of DNA-polymerase I preparations with dADP and dTDP. One of the evidence of above claim is dithiothreitol and *N*-ethylmaleimide, the inhibitors of dNDP-transferase, can suppress the *de novo* DNA synthesis but not the primed synthesis of poly(dA-dT) [9]. However, it is hard to conclude that DNA polymerase cannot carry out *de novo* DNA synthesis from above results.

Another factor influencing the lag time is the presence of contaminant DNA. The lag time for synthesizing poly(dA-dT) was greatly decreased or abolished by adding $(AT)_n$ (n=3-7). Adding AT rich crab DNA also caused the reduction of lag time [4]. Thus, we have to consider the possible variation of contaminant when other conditions such as concentration and purity of polymerase are changed. For example, one of the possibilities is that the less contaminant of highly purified DNA polymerase cause the longer lag time. However, one may argue that it cannot deny the *de novo* DNA synthesis because the presence of some DNA may just accelerate the *de novo* synthesis. The effects by other polymerase will be discussed separately.

2.2. Factors and Conditions on the Sequence of *De Novo* DNA Synthesis

Sequences of *de novo* synthesized DNA depended greatly on the purity of polymerase and reaction conditions such as pH, substrates (dNTPs), buffer, and metal ions (ion strength). When dATP and dTTP were present, only poly(dA-dT) was obtained under most of the conditions. However, when proflavin [26] or nogalamycin [27] was added as an intercalator agent, poly(dA)/poly(dT) was synthesized instead of poly(dA-dT). Even when four deoxyribonucleoside triphosphates (dNTPs) were all present, dCTP and dGTP were excluded and poly(dA-dT) was obtained [1]. When only dCTP and dGTP were present, however, homopolymer poly(dG)/poly(dC) was synthesized by the partially purified polymerase I [4,21].

Table 1. Effect of reaction conditions on the sequences of *de novo* synthesized DNA from dITP and dCTP by DNA polymerase I

DNA polymerase	Buffer type	Buffer conc. (mM)	Ionic strength	pH	Metal ion	Products (%) Poly(dI) / Poly(dC)	Poly(dI-dC)
E. coli	Tris-HCl	50	0.042	7.4	Mg	0	100
		50	0.042	7.4	Mn	0	100
		100	0.082	7.4	Mg	87	13
		150	0.128	7.4	Mg	100	0
		200	0.166	7.4	Mg	0	100
	K₂HPO₄/ KH₂PO₄	25	0.047	7.4	Mg	0	100
		50	Not cal.	7.3	Mg	100	0
		50	0.095	7.4	Mg	100	0
		50	0.095	7.4	Mn	0	100
		50	Not cal.	7.7	Mg	58	42
		50	Not cal.	7.8	Mg	68	32
		50	Not cal.	7.9	Mg	0	100
		150	0.288	7.4	Mn	100	0
		200	0.383	7.4	Mn	100	0
M. luteus	K₂HPO₄/ KH₂PO₄	50	0.095	6.8	Mg	0	100
		50	0.095	7.0	Mg	0	100
		50	0.095	7.4	Mg	60	40
		100	0.190	7.4	Mg	40	60

Not like poly(dA-dT), poly(dG-dC) was hard to be obtained under all the conditions that had been tried. Interestingly, when dITP was used instead of dCTP, both poly(dG-dI) and poly(dG-dC) could be *de novo* synthesized depended on the reaction conditions. For example, when 50 mM of potassium phosphate was used as the buffer (pH 7.4) in the presence of MnCl$_2$ gave 100% of poly(dI-dC), but MgCl$_2$ gave poly(dI)/poly(dC) [21]. In the presence of 5.0 mM MgCl$_2$, 50 mM of Tris-HCl (pH7.4), poly(dI-dC) was synthesized, but 50 mM of potassium phosphate (pH 7.4) gave poly(dI)/poly(dC). More interestingly, only by increasing the concentration of potassium phosphate from 50 mM to 150 mM, poly(dI)/poly(dC) was obtained even when Mn^{2+} was used. Furthermore, when pH changes from 7.3 to 7.9, the product of *de novo* DNA synthesis changed gradually from poly(dI)/poly(dC) to poly(dI-dC) (Table 1) [21].

Seize and Kornberg pointed out that these seemingly slight perturbations may produce profound changes in the kinetics and specificity of *de novo* polymer synthesis [6]. For example, the consequence of DNA polymerase conformation might change during the polymerizing, as well in the conformation of DNA molecules for elongation. Another possibility is that the condition change may imitate different mechanisms of *de novo* DNA synthesis. When poly(dG)/poly(dC) is used as the template/primer, poly(dG)/poly(d^{Br}C) can be obtained in the presence of dGTP and d^{Br}CTP, and poly(dI)/poly(d^{Br}C) is synthesized from d^{Br}CTP and dITP (Figure 1b) [2, 28-29]. However, when poly(dI-dC) or poly(dG-dC) was used as the template/primer, only the natural dGTP and dCTP can be incorporated and poly(dG-dC) is synthesized [29,30].

3. *De Novo* DNA Synthesis by Other Non-Thermophilic DNA Polymerase

By the end of 1960s, it had been reported that *de novo* DNA synthesis could also be carried out by DNA polymerase from B. *subtilis* [24], M. *lysodeicticus* [23,31], M. *luteus* [32], bacteriophage T4 [33], and *calf thymus* [34,35]. Basically, poly(dA-dT) was synthesized in most of the cases, which is similar as polymerase I. Burd and Wells reported that with the partially purified M. *Luteus* DNA polymerase in Tris-HCl buffer and the substrate dATP and dTTP, poly(dA)/poly(dT) was formed only under highly restricted conditions of pH (7.9-8.3); the alternating copolymer was found at all other pH ranges tested (7.3-7.9, 8.3-8.7) [21]. At low concentrations of Tris-HCl buffer (pH 8.1, 15 mM), the product is a mixture of poly(dA)/poly(dT) and poly(dA-dT). After the DNA polymerase was highly purified, only poly(dA-dT) was synthesized in principle. Again, the highly purified DNA polymerase from M. *luteus* could not carry out *de novo* DNA synthesis when dGTP and dCTP were used [32]. Grant et al. reported that the DNA polymerase from *Micrococcus lysodeikticus* could *de novo* synthesize poly(dI-dC) after a lag period of many hours, then the reaction exhibited exponential kinetics and was complete in 34.5 h. [23].

In 1975, Henner and Furth reported that calf thymus DNA Polymerase α (DNA nucleotidyltransferase) could *de novo* synthesize poly(dA-dT) with a lag time of longer than 6 hours [34]. In another paper, Henner and Furth found that the unwinding protein is favorable for *de novo* synthesis by increasing both the exponential rate constant and the rate of linear synthesis [35]. Combined with the results that the rate of synthesis is markedly affected by the Mg^{2+} concentration and has a higher temperature optimum than replication of activated DNA, they pointed out that the "strand slippage" may be related to the mechanism of synthesis. They also mentioned that the exponential amplification may come from the endonuclease activity, although they could not detect it [34]. At last, it is significant to note that the polyribonucleotide with alternating inosinic acid (I) and cytidylic acid (C) units is synthesized *de novo* by the *Azotobacter vinelandii* RNA polymerase [36].

4. *De Novo* DNA Synthesis by Thermophilic DNA Polymerase

Non-requirement of template/primer during *de novo* synthesis of poly(dA-dT) caused a long debate till 1979. However, the description of this activity appears to be omitted after that, and more and more researchers began to doubt the *de novo* DNA synthesis. The studies in the late 1970s indicated that the reaction reported earlier was at least partly related to the contaminating nucleotide oligomers or other enzymes, but the nature of the reaction remained obscure. One of the confusing questions is "Even in the presence of tiny DNA contaminants, how can a large amount of DNA molecules be synthesized in the absence of endonuclease?" After a quiet period of 18 years, the *de novo* DNA synthesis by thermophilic DNA polymerase aroused researchers' interest again because of the breakthrough that all the four dNTPs can also be incorporated by highly purified DNA polymerase. Moreover, it may help to explain the non-specific amplification that happens accidentally during modern approaches such as RCA (rolling circle amplification) and PCR (polymerase chain reaction). Another interest is the underlined relationship between *de novo* DNA synthesis and the molecular evolution of nucleic acids, noting that most of the thermophilic DNA polymerases are usually attained from archaea that live in extreme environments such as deep vent on the bottom of ocean or hot spring close to a volcano [37,38]. From the viewpoint of origin of life, both the fact that DNA polymerases from thermophilic archaebacteria can carry out *de novo* DNA synthesis efficiently and the obtained repetitive sequences may tell us some hint about the evolution of nucleic acids [39–42].

4.1. *De Novo* DNA Synthesis by Thermophilic DNA Polymerase at High Temperatures

In 1997 and 1998, Ogata and Miura reported that 10–50 kb DNA could be *de novo* synthesized at 69–94°C in the total absence of added nucleic acid by *Tli* DNA polymerase (a thermophilic DNA polymerase from archaea *Thermococcus litoralis*) and *Tth* DNA polymerase (a thermophilic DNA polymerase from *Thermus thermophilus*) [43–45]. Figure 2 shows the gel patterns of synthesized DNA at various temperatures. For the first time, long DNA involving all four bases (dA, dG, dC, and dT) such as (AGATATCT)$_n$ and (TAGATATCTATC)$_n$ was *de novo* synthesized. They claimed that genetic information is 'created' by the protein, DNA polymerase. The possibility of DNA contamination in the reaction mixtures was vigorously excluded because the *de novo* DNA synthesis could not be abolished by using the highly purified DNA polymerase and dNTPs.

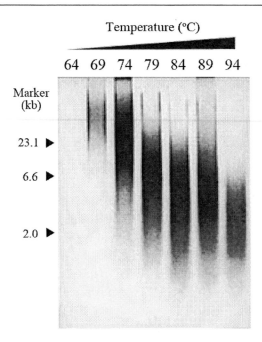

Figure 2. *de novo* DNA synthesis by Vent DNA polymerase at various temperatures [44]. Other reaction conditions: 20 μL of 1X ThermoPol buffer containing 10 mM KCl, 10 mM $(NH_4)_2SO_4$, 6 mM $MgSO_4$, 20 mM Tris-HCl (pH 7.3 at 74°C) and 0.1% Triton X-100; 0.2 mM each of dNTPs; 0.4 U of Vent; 3 h. Electrophoresis was carried out on a 0.8% agarose gel stained afterwards with ethidium bromide.

Moreover, the DNase I and RNase A treatment for removing tiny contaminated DNA or RNA did not influence the DNA synthesis either. Another interesting point is that the 5'→3' exonuclease activity which is required for *de novo* DNA synthesis by non-thermophilic DNA polymerase is not necessary here. Both *Tli* and *Tth* are lack of 5'→3' exonuclease activity, although they hold the 3'→5' exonuclease activity which is responsible for removing the mis-added nucleotide at 3'-terminus. However, after the 3'→5' exonuclease activity of these two DNA polymerases is silenced through the mutation of only two amino acid residues (designated as Vent(exo⁻) and *Tth*(exo⁻) DNA polymerase), *de novo* DNA synthesis could not be detected even after their incubation with dNTPs for several days at higher temperatures than 70°C [18]. These results support the idea that DNA is really *de novo* synthesized because both the natural and mutated DNA polymerases are prepared using the same approach and the polymerase activity of Vent(exo⁻) and *Tth*(exo⁻) is not influenced by the mutation. Unexpectedly, it can be concluded that 3'→5' exonuclease activity is favorable for *de novo* DNA synthesis, although no persuadable explanation is available.

4.2. Sequences of the *De Novo* DNA Synthesized by Thermophilic DNA Polymerase

The DNA sequences *de novo* synthesized depend on the reaction temperatures and ionic strength (Table 2). For instance, $(TAAT)_n$, $(ATCCGGAT)_n$, and $(TATCGCGATAGCGATCGC)_n$ were synthesized by *Tli* DNA polymerase at 69°C, 84°C, and 89°C, respectively

[44,45]. At 74°C, the reaction products were (ATCTAGAT)$_n$ with 0 mM KCl, (GTATATAC)$_n$ with 50 mM KCl, and (TAGTTATAACTA)$_n$ with 100 mM KCl. All the synthesized DNAs have the tandem repetitive sequences and most of them are palindromic.

Table 2. *de novo* **synthesized DNA sequences at various conditions by thermophilic DNA polymerases**

DNA polymerase	Temp. (°C)	KCl conc. (mM)	pH (at 25 °C)	Sequence
Tli	69	10	8.8	(TAAT)$_n$
	74	10	8.8	(TATCTAGA)$_n$
	79	10	8.8	(TATAGC)$_n$
	84	10	8.8	(TATCCGGA)$_n$
	89	10	8.8	(TATCGCGATAGCGATCGC)$_n$
	94	10	8.8	(GATCGC)$_n$
	74	0	8.8	(TATCTAGA)$_n$
	74	10	8.8	(TATCTAGA)$_n$
	74	50	8.8	(TATCTAGA)$_n$
	74	75	8.8	(TATCTAGA)$_n$
	74	100	8.8	(TATAGTTATAAC)$_n$
	74	10	7.8	(TATCTAGA)$_n$
	74	10	8.8	(TATCTAGA)$_n$
	74	10	9.8	(TATCTAGA)$_n$
	74	10	10.8	(TATCTAGA)$_n$
Tth	74	50	9.0	(CATGTATA)$_n$ (TGTATGTATACATACATA)$_n$ (TATACGTA)$_n$ (TGTATGTATATACATACA)$_n$ (TGTACATATA)$_n$ (TATACGTATA)$_n$ (TGTATACATATA)$_n$ (TGTATGTATACATACACGTATACATACATA)$_n$

It can be seen that the GC content increased when *de novo* DNA synthesis was carried out at higher temperatures. The size of synthesized DNA decreased with increasing pH, and *de novo* DNA synthesis can even be carried out at pH 10.8. In the case of *Tth* DNA polymerase, the similar sequences were also obtained such as (TACATGTA)$_n$, (TGTATGTATACATACATA)$_n$, (ATACGTAT)$_n$, (TGTATGTATATACATACA)$_n$, (TATGTACATA)$_n$, (TATGTATACATA)$_n$, (TATACGTATA)$_n$, and (TGTATGTATACATACATACGTATGTATACATACATA)$_n$. The *de novo* DNA synthesis could also be carried out by DNA polymerase from *Pyrococcus sp.* strain KOD1 (KOD DNA polymerase) and *Thermus flavus ubiquitos* (*Tub* DNA polymerawse), indicating that *de novo* DNA synthesis is a fairly common phenomenon at high temperatures.

4.3. *De Novo* DNA Synthesis by Thermophilic DNA Polymerase at Relatively Lower Temperatures

Interestingly, Liang et al. found that DNA can be *de novo* synthesized more efficiently at relatively lower temperatures by Vent DNA polymerase [46]. For example, the *de novo* DNA synthesis shows higher efficiency at about 50°C than that at 65-75°C, which is optimal for normal DNA synthesis using DNA template-primer. The *de novo* DNA synthesis can also be synthesized even at 25°C by the same thermophilic DNA polymerase after a long lag time of several days. Even for Vent(exo⁻) DNA polymerase in the presence of four dNTPs, DNA can be *de novo* synthesized with a lag time of 20 h at 50°C, but no DNA can be detected at 70°C after several days' incubation (Figure 3).

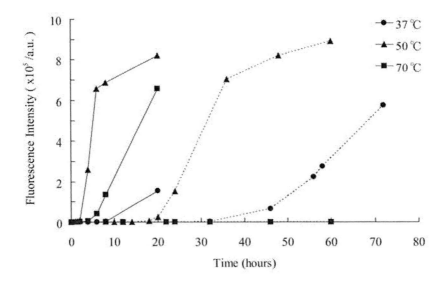

Figure 3. Time course of *de novo* DNA synthesis by Vent (solid lines) or Vent(exo-) (dotted lines) at 37°C (circles), 50°C (triangles), and 70°C (squares) [46]. Other reaction conditions: 100 µL of 1X ThermoPol buffer containing 10 mM KCl, 10 mM $(NH_4)_2SO_4$, 6 mM $MgSO_4$, 20 mM Tris-HCl (pH8.8 at 25°C) and 0.1% Triton X-100; 0.5 mM each of dNTPs; 1 U of DNA polymerase. Fluorescence was measured after stained with SYBR Green I.

4.4. Effect of Deoxynucleotidyl Transferase Activity on *De Novo* DNA Synthesis

Some DNA polymerases such as human DNA polymerase λ and DNA polymerase µ show terminal deoxynucleotidyl transferase (TdT) activity which catalyses the addition of nucleotides (from dNTPs) to the 3' terminus of a DNA molecule in the absence of a template. One possibility is that the TdT-like activity is responsible for the elongation of very short DNAs during *de novo* DNA synthesis. In 2004, Ramadan et al. reported that human DNA polymerase λ, DNA polymerase µ, and TdT can carry out *de novo* DNA synthesis. However, only short DNA fragments (2-21 nt) are synthesized and the synthesized DNA cannot be digested completely by nuclease, although the synthesized DNA has a canonical 3'-5' phosphodiester bond [47]. In 1997 and 1998, Hanaki et al. investigated dozens of thermophilic DNA polymerases for their template/primer-independent synthesis of Poly(dA-dT), and found that non-modified DNA polymerases from *genus Thermus* (*T. acquaticus, T. flavus, T. icelandicus,* and *T. thermophilus*) showed the capacity of *de novo* DNA synthesis after several hours' incubation [48-49]. On the other hand, the thermostable DNA polymerases derived from *Bacillus cardotenax,* three species belonging to genus *Pyrococcus*, and *Thermotoga maritime* could not show that activity under similar conditions. They claimed that there were not obvious relationship between *de novo* DNA synthesis and TdT-like activity, indicating that the presence of TdT-like activity is not essential. However, the research work by Ramadan et al. showed us the possibility of real *de novo* DNA synthesis by DNA polymerase [47]. In some cases, the TdT-like activity is hard to detect probably because it is too weak.

5. Cut-Grow *De Novo* DNA Synthesis

As described in section 2.1, as well as reported by many groups, short oligonucleotides with proper sequences can be elongated by DNA polymerase [50-58]. Accordingly, one can easily consider that the addition of endonuclease that can digest the elongated DNA to short ones with proper length should accelerate the DNA amplification by an exponential amplification mechanism. It is designated as "Cut-grow" involving digestion-elongation cycles: long DNA is digested by endonuclease to many short ones as seeds; and then the short oligos are elongated to long DNA again followed by digestion of next cycle (Figure 4).

Kornberg et al. mentioned that the mixed endonuclease I favored the *de novo* synthesis of poly(dG)/poly(dC). However, the prerequisite that the short seeds obtained by digestion have the sequence to be elongated efficiently should be satisfied for smooth amplification.

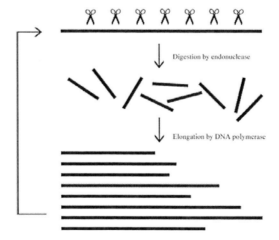

Figure 4. Illustration of Cut-grow mechanism [18]. The synthesized long duplex DNA is digested by endonuclease to short fragments; then the short fragments are elongated by DNA polymerase to give long DNA. The cycles are repeated many times so the synthesized DNA is amplified with extremely high efficiency. All reactions proceed isothermally.

5.1. Typical Cut-Grow *De Novo* DNA Synthesis by the Combination of Thermophilic DNA Polymerase and Thermophilic DNA Endonuclease

In 2004, Liang et al. reported that the *de novo* DNA synthesis was greatly accelerated by using the combination of thermophilic restriction enzyme with thermophilic DNA polymerase. In each unit of the synthesized DNA with repetitive sequences, there is one or two recognition sites of the restriction enzyme used. For example, DNA sequences involving (AAAAATTTTT)$_n$ and (AAGAATTCTT)$_n$ are synthesized when *Tsp*509I (recognition site: ↓AATT, the arrow indicates the digestion position) is used. In the case of *Tsp*RI (recognition site: NNCA(G/C)TGNN↓), (ATACACTGTATATACAGTGTAT)$_n$ is synthesized. It is interesting that the acceleration can be explained well with the Cut-grow mechanism: a longer DNA with numerous recognition sites for the restriction enzyme is digested to short fragments, and the short fragments are used as seeds for elongation to synthesize long DNA. More importantly, if the reaction is really *de novo*, the information of recognition sequence flows from restriction enzyme to DNA with the aid of DNA polymerase.

The gel patterns of the DNA synthesized in the presence of dNTPs, Vent(exo-) DNA polymerase, and *Tsp*RI restriction enzyme are shown in Figure 5.

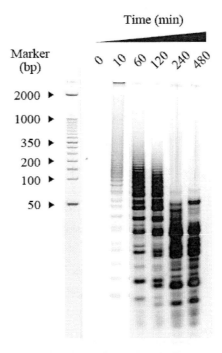

Figure 5. Time course of Cut-grow DNA synthesis by using the combination of *Tsp*RI restriction enzyme and Vent(exo⁻) DNA polymerase [18]. Reaction conditions: 50 μL of mixed buffer containing 0.5X ThermoPol (5 mM KCl, 5 mM (NH$_4$)$_2$SO$_4$, 3 mM MgSO$_4$, 10 mM Tris-HCl (pH 8.8 at 25°C), and 0.05% Triton X-100) and 0.5X NEB4 (10 mM potassium acetate, 10 mM Tris-acetate, 5 mM magnesium acetate (pH 7.9 at 25°C), and 0.5 mM dithiothreitol); 0.4 mM of each dNTP; 1 U of DNA polymerase; 0.5 U of *Tsp*RI.; 70°C. Electrophoresis was carried out on a 1% agarose gel stained with SYBR Green I.

Not like the *de novo* DNA synthesis described in 4.1., which has a smear pattern of DNA longer than 2 kb, the synthesized DNA showed ladder-like bands of relatively short DNA. It can be seen that the *de novo* DNA synthesis is extremely efficient: a large amount of DNA is detected after only 10 min of incubation at 70°C; and the DNA synthesis is almost complete after 1 h, indicating that all the dNTPs are consumed. After that the synthesized DNA is digested by *TspR*I to shorter than 50 bp, showing that the synthesized DNA contains many recognition sites of *TspR*I. In the case of restriction enzyme *Tsp*509I, DNA was even detected within 5 min when only dATP and dTTP were used [18]. Again, no DNA can be detected in the absence of *TspR*I restriction enzyme even after incubating the Vent(exo-) DNA polymerase and dNTPs under the same conditions for 3 days. Thus the effect of endonuclease in the reaction solution is extremely remarkable. It is worth noting that no special conditions are needed in most of the cases. The concentration of DNA polymerase can be used in a range of 0.02-0.1 U/μL, restriction enzyme in a range of 0.002-0.05 U/μL. Usually the mixed buffer involving 0.5× Polymerase buffer and 0.5× restriction enzyme buffer is used when the buffers required for DNA polymerization and digestion are different.

5.2. Factors and Conditions on the Efficiency of Cut-Grow *De Novo* DNA Synthesis

The efficiency and the length of *de novo* synthesized DNA by Cut-grow depends greatly on the reaction conditions. Basically, it depends on the equilibrium between the DNA synthesis activity of DNA polymerase and the DNA cleavage activity. For example, when smaller amount of restriction enzyme is used, longer DNA is obtained because the elongated DNA cannot be digested in time. On the contrary, the synthesis efficiency becomes lower when too much restriction enzyme is present because the synthesized DNA is digested to too short to be elongated efficiently. However, when a much higher concentration of restriction enzyme is used, the Cut-grow can be recovered due to the digestion of different sequence with its star activity so that a different gel pattern is observed [18]. At lower temperatures and when thermophilic DNA restriction enzyme is used, the Cut-grow becomes less efficient and longer DNA is obtained in some cases, probably the activity of restriction enzyme decreased greatly at low temperatures.

For various restriction enzymes, the optimal conditions for cut-grow *de novo* DNA synthesis are different. For example, *fli*I (recognition site: G↓A(A/T)TC) requires longer reaction time (6 h), *Psp*GI (recognition site: ↓CC(A/T)GG) requires higher reaction temperature (85°C), *Bsa*BI (recognition site: GATNN↓NNATC) needs higher concentration of *Bsa*BI and lower temperature (60°C). All the following thermophilic restriction enzymes tried by Liang et al. can also carry out cut-grow *de novo* DNA synthesis at a temperature higher than 60°C: *Tsp*45I (↓GT(G/C)AC), *Tli* (C↓TCGAG), *Tth*111I (GACN↓NNGTC), *Taq*^{α}I (T↓CGA), *Apo*I ((A/G)↓AATT(T/C)), and *Tas*I (↓). The endonuclease that can carry out Cut-grow was also expanded to nicking enzyme. Zyrina et al. reported recently that nickase N.BspD6I (recognition site: GAGTCNNNN↓) from thermophilic strain *Bacillus* species D6 could accelerate the *de novo* DNA synthesis by *Bst* DNA polymerase (large fragment) and the synthesized DNA contains non-palindromic tandem repeats of hexanucleotides [59].

5.3. Cut-Grow *De Novo* DNA Synthesis by Using Non-Thermophilic Endonucleases and Other DNA Polymerases

Liang et al. further showed that the Cut-grow *de novo* DNA synthesis is a common characteristic of many restriction enzymes and DNA polymerases under a wide range of temperature [60]. Several non-thermophilic restriction enzymes such as *Eco*RI (G↓AATTC), *Pst*I (CTGCA↓G), *Mlu*I (A↓CGCGT), *Ssp*I (AAT↓ATT), and *Bsr*BI (CCG↓CTC), can carry out Cut-grow in the presence of Vent(exo⁻) DNA polymerase at 37°C. Figure 6 shows the gel pattern of the synthesized DNA in the case of *Eco*RI. Again, the efficiency becomes lower when too much *Eco*RI is added, and the gel pattern changed due to the star activity (Lane 4-6, 10-12, 16-18 in Figure 6.).

It was found that the synthesis efficiency depends greatly on the GC content of the recognition sites of restriction enzymes: more GC content causes less efficiency. According to the Cut-grow mechanism, the sequences of synthesized DNA ought to contain the recognition sites of the corresponding restriction enzymes.

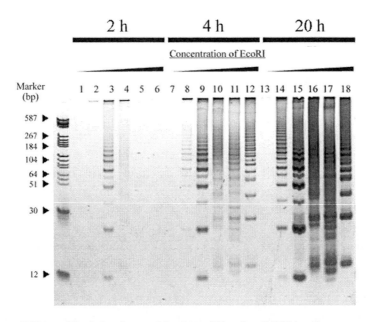

Figure 6. Cut-grow DNA synthesis by the combination of Vent(exo⁻) DNA polymerase and *Eco*RI restriction enzyme at various concentrations in the absence of any added nucleic acids [60]. Reaction conditions: 50 µL of 1X ThermoPol buffer (10 mM KCl, 10 mM (NH$_4$)$_2$SO$_4$, 6 mM MgSO$_4$, 20 mM Tris-HCl (pH 8.8 at 25°C), and 0.1% Triton X-100); 0.4 mM each of dNTPs; 1 U of DNA polymerase; 37°C. *Eco*RI was added at various concentrations: Lane 1,7,13 (0 U/µL), Lane 2,8,14 (0.01 U/µL), Lane 3,9,15 (0.05 U/µL), Lane 4,10,16 (0.2 U/µL), Lane 5,11,17 (0.5 U/µL), Lane 6,12,18 (1.0 U/µL). The DNA products were analyzed by 10% nondenaturing PAGE (stained with ethidium bromide).

Thus, DNA containing more GC should be synthesized in the case of higher GC content in the recognition site. Because the reaction temperature was 37°C or lower, the synthesized DNA duplex became too stable to slip or dissociate partly for elongation. Here, it is amazing that the information of protein structure can flow back to DNA under the ambient conditions, noting that the recognition site of restriction enzyme determined basically the sequences of synthesized DNA.

Unexpectedly, Cut-grow can even be carried out at 4°C in the presence of DNase I and Vent(exo⁻), although the efficiency is very low and DNA can only be detected only after several weeks. Both Vent and Vent(exo⁻) DNA polymerase can carry out Cut-grow and no obvious difference was observed, although, as described previously in Section 4, the *de novo* DNA synthesis by Vent is much faster than that of Vent(exo⁻) in the absence of restriction enzymes. Several thermophilic DNA polymerases such as 9°N$_m$ and Bst (large fragment) DNA polymerase have also shown the activity of Cut-grow DNA synthesis. The DNA polymerase with very strong nuclease activity such as T7 DNA polymerase and phi29 DNA polymerase failed to *de novo* synthesize DNA under the conditions being tried (unpublished results by Liang et al).

5.4. Theoretical Analysis of the Potential Power of Cut-Grow DNA Synthesis

All the above results can be explained by the Cut-grow mechanism as shown in Figure 4. Moreover, the mechanism has been further experimentally proved: the synthesized DNA can be elongated again with DNA polymerase, and the elongated DNA can be digested to short oligonucleotides back again [18]. It can be seen that the DNA is capable to be amplified very efficiently once the proper DNA sequence involving the recognition site of present endonuclease is synthesized. The exponential amplification can be expressed with the following formula:

$$N = N_0 \exp(nt/l)$$

wherein, N is the number of DNA molecules at reaction time t (min), N_0 is the number of initial molecules, n (bp/min) is the speed of DNA synthesis by DNA polymerase, and l (bp) is average length of the short ODNs during synthesis. Supposing that elongated DNA can be immediately digested to short ODNs with the length of l. If n = 1000 bp/min, l = 50 bp, the DNA can be amplified by 10^{26} folds after 3 minutes according to above formula. Actually, the extremely efficient amplification cannot be kept long because the DNA polymerase and restriction enzyme molecules are not enough for remaining this exponential amplification. However, it displays the extremely high potential of this model for amplifying DNA.

6. Mechanism of *De Novo* DNA Synthesis

6.1. Steps of *De Novo* DNA Synthesis

Based on the time course of the apparent *de novo* synthesis of poly(dA-dT) and the effect of $(AT)_n$ to shorten its lag time, Kornberg et al. proposed a mechanism involving four stages (Figure 7) [4-6].

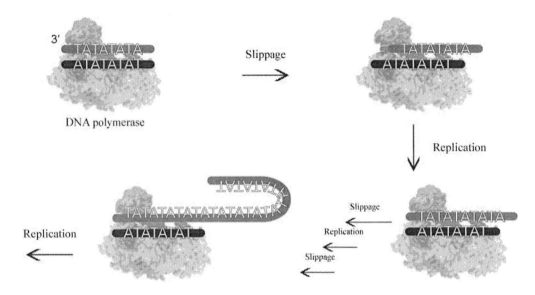

Figure 7. Proposed mechanism of *de novo* synthesis of poly(dA-dT) by Kornberg et al. [6]

At the initiation stage, short oligonucleotide of alternating A and T residues (e.g. $(AT)_n$, 6 nt or longer) was synthesized. At the second stage, the synthesized short oligo is bound and held fixed by the polymerase in the template site, another DNA strand may then serve as the primer for extensive chain growth. Obviously, successive slippage and polymerization can happen to allow gradual elongation until the primer strand has grown to a size sufficient to fold back on itself and form a hairpin structure (reiterative replication). The synthesized DNA at the first two stages is not enough for detection. At the third stage, short DNA is elongated to long one and amplified exponentially. When dATP and dTTP substrates have be exhausted, the synthesized DNA is digested gradually by the exonuclease activity of DNA polymerase at the last stage.

For *de novo* DNA synthesis in 1960s, three principal possibilities for the initiation of the *de novo* reaction were suggested [6]. One is that the short DNA as seeds for growing is *de novo* synthesized from dNTPs. The second alternative is that DNA fragments such as $(AT)_n$ strongly bind in DNA polymerase as impurities. The third one is that such fragments are made by a contaminating deoxynucleotidyl transferase that can polymerize dNDPs without primer or template direction. It should be noted that in some cases, two or three approaches may happen in the same reaction solution, which makes it more complicated.

Kornberg et al. also pointed that the reiterative replication can become exponential only when new DNA molecules are generated to serve as template-primers for growing by the excess of idle polymerase molecules [6]. The nuclease cut of DNA might supply the new 3'-OH for extension and afford rapid and exponential amplification. As described previously, it fails to obtain *de novo* poly(dG)/poly(dC) synthesis when pol I preparations are free of traces of endonuclease I, and the polymer synthesis is restored when small amounts of endonuclease are added to the highly purified polymerase reaction mixture. Another evidence is that no *de novo* synthesis of poly(dA-dT) was detected by the large fragment of pol I (losing 5'→3' exonuclease activity). They claimed that the 5'→3' exonuclease activity may help to generate fragments that possess priming activity. However, the 5'→3' exonuclease activity could not produce free 3'-OH because 5'→3' exonuclease digest DNA from 5'-end one nucleotide by one nucleotide. The highly purified DNA polymerases such as Polymerase I and Vent DNA polymerase do not show detectable endonuclease activity. It is obvious that the above explanation is obscure and subjective. New reasonable mechanism should be proposed and evidenced.

Based on the time course, the *de novo* DNA synthesis can largely be divided into two steps: initiation and exponential amplification. In initiation step (lag time), the short oligo which can be used as the seeds for quick elongation was produced. In the exponential amplification stage, the short oligo was elongated and exponentially amplified. Under the premise that DNA is really *de novo* synthesized, the authors proposed a more detail mechanism described as follows (Figure 8).

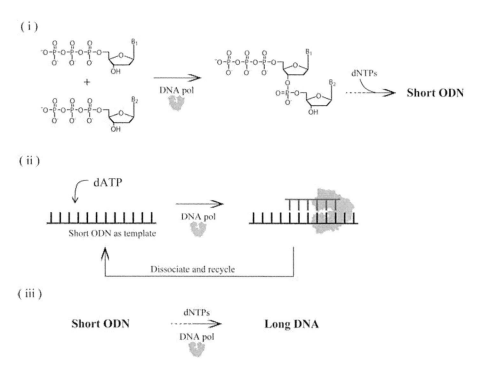

Figure 8. Proposed mechanism of the *de novo* DNA synthesis proposed by the authors. (i) At first dimers are synthesized from dNTPs at the active site of DNA polymerase, and then the short oligonucleotides (ODN) are synthesized gradually. (ii) The complementary strand is further synthesized using the *de novo* synthesized short ODN as the template. (iii) The short ODN with a proper sequence is elongated to long DNA and DNA synthesis goes to the exponential stage.

At first, the dimer was synthesized from dNTPs at the active site of DNA polymerase which has the intrinsic ability for forming the phosphodiester bond from a 5'-triphosphate and a free 3'-OH, although the efficiency might be extremely low. Then the trimer and short DNA was synthesized gradually. At the following stage, the synthesized short DNA can be used as the template to synthesize the complementary DNA (actually having the same sequence for palindromic sequence) with improved efficiency as compared with the synthesis of the first seeds without template. Once the proper sequence that can be elongated is synthesized, the obtained long DNA can be used as the template for DNA synthesis at many sites by the excess DNA polymerase so that the synthesis goes to the exponential stage.

For this model, the exponential amplification is possible even in the absence of endonuclease, although the presence of endonuclease will definitely accelerate the DNA amplification with an extremely high efficiency according to the Cut-grow mechanism. However, it can be seen that provement of the mechanism is difficult because a single short molecule with the proper sequence for elongation present in the solution may stimulate the DNA synthesis. Another difficult point is that synthesized DNA can only be detected when the DNA reaches a certain concentration. Although the authors do not have the evidence that short DNA can be really *de novo* synthesized from dNTPs directly, it was found that poly(A) can be *de novo* synthesized from dATP in the presence of 12-nt-long poly(dT) (TTTTTTTTTTTT) as the template (unpublished data), indicating that the DNA can be *de novo* synthesized at least in the absence of primer.

Kong et al. reported that the ssDNA 3'→5' exonuclease activity of Vent DNA polymerase required binding of a 12–16 nt ssDNA, and the oligo shorter than 12 nt was hard to be digested [38]. As described in section 4.1, the 3'→5' exonuclease activity domain of Vent favored the *de novo* DNA synthesis greatly as compared with Vent(exo⁻). This can be explained as follows: the *de novo* synthesized short oligos can be elongated at the polymerization activity domain more efficiently because the exonuclease domain can capture them and avoid their complete *diffusion* to the solution. This may be very important at the initiation stage because the concentration of short DNAs is extremely low. The concentrating effect of the 3'→5' exonuclease domain can improve greatly the efficiency of the elongation. The improved *de novo* DNA synthesis by proper 5'→3' exonuclease activity can also be explained by the similar reason, because many DNA polymerases with 5'→3' exonuclease activity degrade long ssDNA to about 20 nt long and the digestion is hard to go further [49].

6.2. Mechanisms for Elongation during *De Novo* DNA Synthesis

After the lag time, long DNA is detected during *de novo* DNA synthesis, and an exponential increase of DNA is usually observed. Obviously, the synthesized long DNA should be obtained from the elongation of short ODNs. On the other hand, elongation of short DNA to very long DNA does not only arouse interest for explaining the mechanism of *de novo* DNA synthesis, but also repetitive DNA sequences, especially the triplet repeat such as $(CAG)_n/(CTG)_n$, are known to elongate *in vivo* through expansion of repeat number and cause hereditary neuromuscular diseases [58,60]. Many kinds of short oligonucleotides with repetitive sequences such as $(TA)_n$, $(TG)_n$, $(CAG)_n$, $(TAGG)_n$, $(TTAGGG)_n$, $(TACATGTA)_n$,

(AGATATCT)$_n$ have been found to be elongated to very long DNA by DNA polymerase [50–57]. Furthermore, the palindromic sequences are probably important for the regulation of cellular metabolism [62,63]. For example, the palindromic sequences that are repeated in tandem exist frequently in the genome [64–66] and have been found near promoters [66,67], reporter binding sites [68,69], and replication origins [70,71]. Several mechanisms of DNA elongation, as shown in Figure 9, have been proposed up to now [4,18,56–58,72–74], although the detail mechanism and reasonable explanation for many cases remain largely unknown.

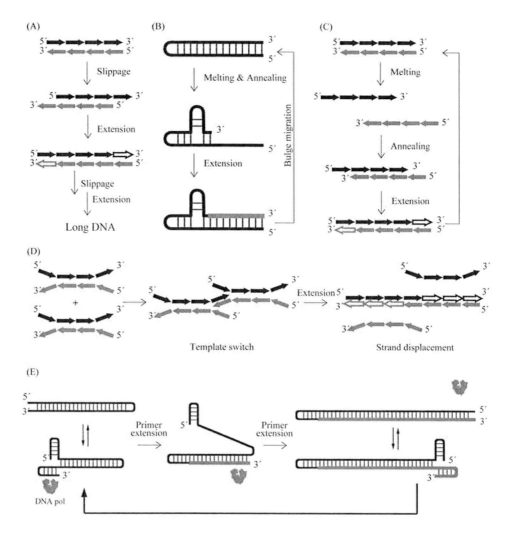

Figure 9. The proposed mechanisms of DNA elongation. (A) Strand slippage model by Kornberg et al. [4-6] (B) Hairpin-elongation model by Ogata and Miura [57]. (C) DEME (Duplex Elongation at Melting Equilibrium) model by Ogata [56, 73]. (D) Template switching and strand displacement model by Ramadan et al. [47] (E) THF-SPE (Terminal Hairpin Formation and Self-Priming Extension) model by Liang et al. [46].

When discussing the DNA elongation mechanism during *de novo* DNA synthesis, the concentration of DNA seeds, DNA sequence, and temperature factors have to be considered. In some cases, two or more mechanisms can also occur at the same time. For Model A (Figure 9A), slippage is favorable to happen for very short repetitive units such as (AT)$_n$, (ATG)$_n$, for repetitive units longer than 5 nt, however, it is difficult to slip, especially at lower temperatures. In addition, the slippage becomes difficult for very long duplex (e.g. > 1000 bp) so that the efficiency should slow down greatly. This slippage mechanism can explain the *de novo* synthesis of mono-polymers such as poly(dG)/poly(dC) and poly(dA)/poly(dT), although the efficiency may be not very high. For Model B, the hairpin is hard to form if it is too short (< 8 nt), and only several nucleotides are elongated for each motion. After several cycles, the hairpin will be easily formed at the ends but not internal so that the efficiency of DNA elongation by this mechanism decreases greatly. At high temperatures and high concentrations of the seeds, Model C may happen with a high efficiency. However, the complete dissociation at lower temperature, e.g. at 37°C, is impossible. Model D can explain the amplification of non-palindromic seeds, but it only happens at very high concentration (e.g. > 1 µM) [58].

Although, THF-SPE model (Model E) cannot explain the expansion of non-palindromic sequence such as (ATG)$_n$, it shows several merits for explaining the *de novo* synthesis: 1) It can happen at single molecule level once the polymerase can bind to the hairpin complex to start the primer extension; 2) At temperatures lower than the T_m, dissociation of DNA duplex at ends to form a hairpin structure does not cause much free energy loss; 3) The DNA length is almost doubled for each cycle so the elongation and amplification are very efficient; 4) During elongation, short oligos can be used as primers by binding to the strand which is peeling off. Accordingly, THF-SPE should be the main mechanism during *de novo* DNA synthesis, especially during the lag time, when the short DNAs are at a very low concentration.

Based on the THF-SPE model, Liang et al. designed several short ODNs that can form hairpin structure for elongation. For a 28-nt-long ODN (Seq28) with the sequence 5'-AATTCTTAAGAATT GTTAGTGCACTAAC-3' that can form two hairpin structures at both ends (Figure 10a), it can be efficiently elongated by Vent(exo-) DNA polymerase at 70°C in the presence of dNTPs. The DNA longer than 20000 nt was obtained within 30 min (Figure 10b).

The elongation process can be described as follows. At first, a bigger hairpin is formed after primer extension, and then two hairpin structures forms at the open end and the DNA can be elongated according to THF-SPE. Although the hairpin structure is not so stable in the reaction buffer at 70°C, which is 17°C higher than the melting temperature (T_m) of Seq28 (T_m = 53.0°C), this structure can temporarily form and be captured by Vent(exo-) because the DNA polymerase shows a high affinity to primer-template complex (K_m= 0.1 nM). Actually, hundreds folds of elongation has been detected for even 1 nM of Seq28 within 10 min [74].

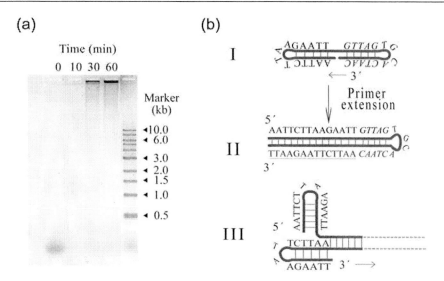

Figure 10. Efficient elongation of short ODN S28 (5'-AATTCTTAAGAATT GTTAGTGCACTAAC-3') by Vent(exo⁻) DNA polymerase [74]. (a) Gel patterns of the elongated products analyzed on 1% agarose gel stained with SYBR Green I. (b) Scheme of the elongation process. Conditions: 100 μL of 1X ThermoPol buffer (10 mM KCl, 10 mM $(NH_4)_2SO_4$, 6 mM $MgSO_4$, 20 mM Tris-HCl (pH 8.8 at 25°C) and 0.1% Triton X-100); 0.5 mM each of dNTPs; 2 U of Vent; 0.1 μM S28, at 70°C. The reaction time is shown in the above panel.

7. Biological Implications of *De Novo* DNA Synthesis

If *de novo* DNA synthesis happens in an efficient way to some extent, one may ask an interesting question: why does it happen? Is it just a by-production because the chemical nature of polymerase cannot avoid this? We would like to believe that it had special meaning either on nucleic acid evolution or during biological processes in a cell. At least, the mechanisms of Cut-grow and DNA elongation may help us to explain some happenings *in vivo* or *in vitro*. However, the possible biological significance of *de novo* DNA synthesis remains obscure although several speculations have been described [8,43-47,72].

From the research on *de novo* DNA synthesis and the investigation of repetitive DNA sequences in organisms, the following facts are clear: 1) The sequences of *de novo* synthesized are palindromic repetitive sequences (e.g. $(TAAT)_n$ and $(ATCCGGAT)_n$) and simple alternating copolymer such as poly(dA-dT) and poly(dG)/poly(dC); 2) These kinds of sequences can be efficiently elongated by self-priming; 3) In the presence of endonucleases, DNA can be amplified by Cut-grow mechanism with an extremely high efficiency; 4) Repetitive tracts including the simple microsatellite repetitive sequence such as di-, tri-, tetra-, and penta-nucleotide tandem repeats constitute significant fractions of most genomes from bacteria to human being (e.g. at least more than 25% of human genome are repetitive sequences); 5) The repetitive sequences hold functions either in a DNA or in a transcript state, e.g. the repetitive DNA sequences of the animal genome are extensively represented in cellar RNA [75-78]. Thus, it is easy to hit upon the idea that *de novo* DNA synthesis of the repetitive DNA sequences may relate to the origin and evolution of genome DNA [78-80]. Another point is that the complex genomic DNA has to originate from simple DNA sequences if we believe the theory of molecular evolution.

Ohno proposed that the first set of coding sequences that arose in the prebiotic world were repeats of nucleotide oligomers [41,42]. He claimed the three advantages of repeats of base oligomers as the primordial coding sequences: 1) When compared with randomly generated base sequences in general, the repeats are more likely to have long open reading frames; 2) Periodical polypeptide chains specified by such repeats are more likely to assume either α–helical or β–sheet secondary structures, which are more important than random ones; 3) The internally repetitious coding sequences can be impervious to mutations such as substitutions, deletions, and insertions [41]. In the light of this idea, it may be plausible that a primitive polypeptide or a catalytic surface having polymerase-like activity synthesized long stretches of DNAs with simple repetitive sequences and that they gradually 'evolved' into information-rich sequences by accumulating mutations. Note that the replication of these DNA sequences are not necessarily required complicate DNA polymerase. The sequences itself hold the internal ability to be amplified by self-priming, and the catalyst can be either nucleic acids having special sequence or even the surface of minerals.

For further evolution to genome-like DNA, large amounts of DNA materials with various repetitive DNA sequences should have been required. The cut-grow *de novo* DNA synthesis could have held the competency through the combination of polymerization and digestion activity. Here, the cleavage of DNA can be carried out by any inorganic or organic molecule having nuclease-like activity. The exponential amplification potential of Cut-grow suggests that the digestion of nucleic acids may play an important role in the evolution of genetic materials for procreating the diversification of genetic information on the early earth. Using a similar mechanism like Cut-grow, the machinery consisting of polymerization and digestion could not only provide materials for nucleic acid evolution but also supply the simplest style of "life": polymerization utilizes the energy and monomers to grow and cleavage activity gives birth of new seeds. In this primordial system, the polymerization and digestion activity could be carried out by protein, RNA, or other functional molecules such as metal ions or their complexes [18,60]. Figure 11 shows the outline of a speculated model about the origin and evolution of nucleic acids.

Another question related to *de novo* DNA synthesis is "Does it happen in a modern cell?" Although we cannot exclude the possibility that *de novo* DNA synthesis is only the trace of original bioprocess, one may doubt that it may occur during the bioprocess in the cell now, either in a positive or a negative direction. Unfortunately, we have few ideas at present and consider that most people prefer to the negative answer. One possibility is that the present DNA and RNA in the cell can suppress the *de novo* DNA synthesis due to their binding to DNA polymerase. We believe that the quick development of biotechnology will help us to find the answer in the near future.

If we can clarify the detail mechanism of this reaction, it may help us to develop new methods for DNA amplification and detection, sharing a big contribution to either enzymology or biochemistry.

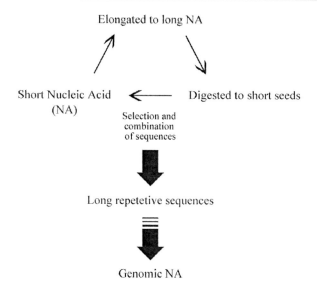

Figure 11. A speculated model about the origin and evolution of nucleic acids [60].

Conclusion

In conclusion, DNA polymerases, especially some thermophilic polymerases, probably hold the ability of *de novo* DNA synthesis from dNTPs, although no experimental data can clearly prove this at present. Investigation of *de novo* DNA synthesis, especially at the initiation stage, requires special discretion to exclude all the possible impurities (enzymes, oligonucleotides, or some other small molecules) affecting DNA synthesis. For clarifying the mechanism at the initiation stage, new robust experiments have to be designed.

By the combination of DNA polymerase and endonuclease, DNA with repetitive sequences can be amplified with an extremely high efficiency according to the Cut-grow mechanism. It is amazing but doubtless that the endonuclease destroying DNA can be used to accelerate the *de novo* DNA synthesis. From the researches described in this chapter, we are certain that DNA sequences easily amplified can be selected by the *de novo* DNA synthesis. Especially when restriction enzyme is present, we can obtain various sequences to be efficiently amplified by DNA polymerase under certain conditions. Furthermore, the genetic information can be considered to flow from protein to DNA because the synthesized DNA contains repetitive sequence of the recognition site of the added restriction enzyme. The internal characteristics of these sequences may possibly give us a hint of the origin and evolution of nucleic acids. The biological significance of *de novo* DNA synthesis remains unknown. Interestingly, however, the elongation of short oligonucleotides has already been used for preparing DNA materials in the nanotechnology field [81–85]. On the road we are traveling to find the nature of the world, we can always find the techniques we need by accident, which may be unrelated totally to our original goals.

Acknowledgments

We thank M.D. Professor Frank-Kamenetskii for his kind discussion about the research. Support from the Japan Society for the Promotion of Science and Nagoya University is appreciated.

References

[1] Schachman, HK; Adler, J; Radding, CM; Lehman, IR; Kornberg, A. Enzymatic synthesis of deoxyribonucleic acid VII: Synthesis of a polymer of deoxyadenylate and deoxythymidylate. *J. Biol. Chem.* 235, 3242-3249, 1960.

[2] Radding, CM; Josse, J; Kornberg, A. Enzymatic synthesis of deoxyribonucleic acid XII: A polymer of deoxyguanylate and deoxycytidylate. *J. Biol. Chem.* 237, 2869-2876, 1962.

[3] Radding, CM; Kornberg, A. Enzymatic synthesis of deoxyribonucleic acid XIII: Kinetics of primed and *de novo* synthesis of deoxynucleotide polymers. *J. Biol. Chem.* 237, 2877-2882, 1962.

[4] Kornberg, A; Jackson, JF; Khorana, HG; Bertsch, LL. Enzymatic synthesis of deoxyribonucleic acid XVI: Oligonucleotides as templates and mechanism of their replication. *Proc. Natl. Acad. Sci. USA.* 51, 315-323, 1964.

[5] Kornberg, A. Synthesis of DNA-like polymers *de novo* or by reiterative replication. In *Evolving Genes and Proteins.* Edited by Bryson, V; Vogel, HJ. Academic Press, New York, 403-417, 1965.

[6] Kornberg, A; Baker, T. Apparent *de novo* synthesis of repetitive DNA. In *DNA replication.* 2nd Edition, W. H. Freeman and Co., New York, 144-150, 1992.

[7] Nazarenko, IA; Potapov, VA; Romaschenko, AG; Salganik, RI. Separation of the enzyme catalyzing polymerization of deoxyribonucleoside diphosphates from the preparations of *Escherichia coli* DNA polymerase I. *FEBS lett.* 86, 201-204, 1978.

[8] Nazarenko, IA; Bobko, LE; Romaschenko, AG; Khripin, YL; Salganik, RI. A study on the unprimed poly(dA-dT) synthesis catalyzed by preparations of *E.coli* DNA polymerase I. *Nucleic Acid Res.* 6, 2545-2560, 1979.

[9] Lehman, IR. Discovery of DNA polymerase. *J. Biol. Chem.* 278, 34733-34738, 2003.

[10] Friedberg, EC. The eureka enzyme: the discovery of DNA polymerase. *Nat. Rew. Mol. Cell Biol.* 7, 143-147, 2006.

[11] Jurka, J; Kapitonov, W; Kohany, O; Jurka, MV. Repetitive sequences in complex genomes: structure and evolution. *Annu. Rev. Genom. Hum. Genet.* 8, 241-259, 2007.

[12] Elder, JF; Turner, BJ. Concerted evolution of repetitive DNA-sequences in eukaryotes. *Quart. Rev. Biol.* 70, 297-320, 1995.

[13] Lupski, JR; Weinstock, GM. Short, interspersed repetitive DNA-sequences in prokaryotic genomes. *J. Bact.* 174, 4525-4529, 1992.

[14] Charlesworth, B; Sniegowski, P; Stephan, W. The evolutionary dynamics of repetitive DNA in eukaryotes. *Nature* 371, 215-220, 1994.

[15] Lovett, ST. Encoded errors: mutation and rearrangements mediated by misalignment at repetitive DNA sequences. *Mol. microbiol.* 52, 1243-1253, 2004.

[16] Imarisio, S; Carmichael, J; Korolchuk, V; Chen, CW; Saiki, S; Rose, C; Krishna, G; Davies, JE; Ttofi, E; Underwood, BR; Rubinsztein, DC. Huntington's disease: from pathology and genetics to potential therapies. *Biochem. J.* 412, 191-209, 2008.

[17] Heidenfelder, BL; Makhov AM; Topal, MD. Hairpin formation in Friedreich's ataxia triplet repeat expansion. *J. Biol. Chem.* 278, 2425-2431, 2003.

[18] Liang, XG; Jensen, K; Frank-Kamenetskii, MD. Very efficient template/primer-independent DNA synthesis by thermophilic DNA polymerase in the presence of a thermophilic restriction endonuclease. *Biochemistry* 43, 13459-13466, 2004.

[19] Kornberg, A; Lehman, IR; Bessman, MJ; Simms, ES. Enzymatic synthesis of deoxyribonucleic acid. *Biochim, Biophys. Acta* 21, 197-198, 1956.

[20] Jovin, TM; Englund, PT; Bertsch, LL. Enzymatic synthesis of deoxyribonucleic Acid XXVI: physical and chemical studies of a homogeneous deoxyribonucleic acid polymerase. *J. Biol. Chem.* 244, 2996-3008, 1969.

[21] Burd, JF; Wells, RD. Effect of incubation conditions on the nucleotide sequence of DNA products of unprimed DNA polymerase reactions. *J. Mol. Biol.* 53, 435-459, 1970.

[22] Inman, RB; Baldwin, RL. Helix-random coil transitions in DNA homopolymer pairs. *J. Mol. Biol.* 8, 452-469, 1964.

[23] Grant, RC; Harwood, SJ; Wells, RD. The synthesis and characterization of poly d(I-C)·poly d(I-C). *J. Am. Chem. Soc.* 90, 4474-4476, 1968.

[24] Okazaki, T; Kornberg, A. Enzymatic synthesis of deoxyribonuleic acid XV: purification and properties of a polymerase from *Bacillus subtilis*. *J. Biol. Chem.* 239, 259-268, 1964.

[25] Setlow, P; Brutlag, D; Kornberg, A. Deoxyribonucleic acid polymerase: two distinct Enzymes in one polypeptide. *J. Biol. Chem.* 247, 224-231, 1972.

[26] McCarter, JA; Kadohama, N; Tsiapalis, C. Proflavine and fidelity of DNA polymerase. *Can. J. Biochem.* 47, 391-399, 1969.

[27] Olson, K; Luk, D; Harvey, CL. Formation of poly(dA)-poly(dT) by *Escherichia coli* DNA polymerase in presence of anthracycline antibiotics. *Biochim. Biophys. Acta.* 269-275, 1972.

[28] Inman, RB; Baldwin, RL. Helix-random coil transitions in synthetic DNAs of alternating sequence. *J. Mol. Biol.* 5, 172-184, 1962.

[29] Inman, RB; Baldwin, RL. Formation of hybrid molecules from two alternating DNA copolymers. *J. Mol. Biol.* 5, 185-200, 1962.

[30] Grant, RC; Kodama, M; Wells, RD. Enzymatic and physical studies on (dI-dC)$_n$·(dI-dC)$_n$ and (dG-dC)$_n$·(dG-dC)$_n$. *Biochemsitry* 11, 805-815, 1972.

[31] Zimmerman, BK. Purification and properties of deoxyribonucleic acid polymerase from *micrococcus lysodeikticus*. *J. Biol. Chem.* 241, 2035-2041, 1966.

[32] Harwood, SJ; Schendel, PF; Wells, RD. *Micrococcus luteus* deoxyribonucleic acid polymerase – Studies of enzymic reaction and properties of deoxyribonucleic acid product. *J. Biol. Chem.* 245, 5614-5624, 1970.

[33] Nossal, NG. DNA-synthesis on double-stranded DNA template by T4-bacteriophage DNA-polymerase and T4 gene 32 DNA unwinding protein. *J. Biol. Chem.* 249, 5668-5676, 1974.

[34] Henner, D; Furth, JJ. *de novo* Synthesis of a polymer of deoxyadenylate and deoxythymidylate by Calf Thymus DNA polymerse α. *Proc. Natl. Acad. Sci. USA.* 72, 3944-3946, 1975.

[35] Henner, D; Furth, JJ; Primed and unprimed synthesis of poly(dA-dT) by Calf Thymus DNA polymerse α. *J. Biol. Chem.* 252, 1932-1937, 1977.

[36] Krakow JS, Karstadt, M. *Azotobacter vinelandii* ribonucleic acid polymerase IV: unprimed synthesis of rIC copolymer. *Proc. Natl. Acad. Sci. USA.* 58, 2094-2101, 1967.

[37] Chien, A; Edgar, DB; Trela, JM. Deoxyribonucleic acid polymerase from the extreme thermophile thermus aquaticus. *J. Bact.* 127, 1550-1557, 1976.

[38] Kong, H; Kucera, RB; Jack, WE. Characterization of a DNA polymerase from the hyperthermophile archaea thermococcus litoralis – Vent DNA polymerase, steady-state kinetics, thermal-stability, processivity, strand displacement, and exonuclease activities. *J. Biol. Chem.* 268, 1965-1975, 1993.

[39] Glansdorff N. About the last common ancestor, the universal life-tree and lateral gene transfer: reappraisal. *Mol. Microbiol.* 38, 177-185, 2000.

[40] Wolf, YI; Rogozin, IB; Grishin, NV; Koonin, EV. Genome trees and the Tree of Life. *Trends in Gene.* 18, 472-479, 2002.

[41] Ohno, S. Repeats of base oligomers as the primordial coding sequences of the primeval earth and their vestiges in modern genes. *J. Mol. Evol.* 20, 313-321, 1984.

[42] Ohno, S. Early genes that were oligomeric repeats generated a number of divergent domains on their own. *Proc. Natl. Acad. Sci. USA.* 84, 6486-6490, 1987.

[43] Ogata, N; Miura, T. Genetic information 'created' by archaebacterial DNA polymerase, *Biochem. J.* 324, 667-671, 1997.

[44] Ogata, N; Miura, T. Creation of genetic information by DNA polymerase of the archaeon *Thermococcus litoralis*: influences of temperature and ionic strength. *Nucleic Acids Res.* 26, 4652-4656, 1998.

[45] Ogata, N; Miura, T. Creation of genetic information by DNA polymerase of the thermophilic bacterium *Thermus thermophilus*. *Nucleic Acids Res.* 26, 4657-4661, 1998.

[46] Liang, XG; Kato, T; Asanuma, H. Unexpected efficient *ab initio* DNA synthesis at low temperatures by using thermophilic DNA polymerase. *Nucleic Acids Symp. Ser.* 51, 351-352, 2007.

[47] Ramadan, K; Shevelev, IV; Maga, G; Hübscher, U. *de novo* DNA synthesis by human DNA polymerase λ, DNA polymerase μ, and terminal deoxyribonucleotidyl transferase. *J. Mol. Biol.* 339, 395-404, 2004.

[48] Hanaki, K; Odawara, T; Muramatsu, T; Kuchino, Y.; Masuda, M; Yamamoto, K; Nozaki, C.; Mizuno, K.; Yoshikura, H. Primer/template-independent synthesis of poly d(A-T) by Taq DNA polymerase. *Biochem. Biophys. Res. Commun.* 238, 113-118, 1997.

[49] Hanaki, K; Odawara, T; Nakajima, N; Shimizu, YK; Nozaki, C; Mizuno, K; Muramatsu, T; Kuchino, Y; Yoshikura, H. Two different reactions involved in the primer/template-independent polymerization of dATP and dTTP by Taq DNA polymerase. *Biochem. Biophys. Res. Commun.* 244, 210-219, 1998.

[50] Wells, RD; Ohtsuka, E; Khorana, HG. Studies on polynucleotides. L. Synthetic deoxyribopolynucleotides as templates for the DNA polymerase of *Escherichia coli*: A new double-stranded DNA-like polymer containing repeating dinucleotide sequences. *J. Mol. Biol.* 14, 221-240, 1965.

[51] Wells, RD; Jocob, TM; Narang, SA; Khorana, HG. Studies on polynucleotides 69. Synthetic deoxyribopolynucleotides as templates for the DNA polymerase of *Escherichia coli*: DNA-like polymers containing repeating trinucleotide sequences. *J. Mol. Biol.* 27, 237-263, 1967.

[52] Wells, RD; Büchi, H; Kössel, H; Ohtsuka, E; Khorana, HG. Studies on polynucleotides 70. Synthetic deoxyribopolynucleotides as templates for the DNA polymerase of *Escherichia coli*: DNA-like polymers containing repeating tetranucleotide sequences. *J. Mol. Biol.* 27, 265-272, 1967.

[53] Schlötterer, C; Tautz, D. Slippage synthesis of simple sequence DNA. *Nucleic Acids Res.* 20, 211-215, 1992.

[54] Ji, JP; Clegg, NJ; Peterson, KR; Jackson, AL; Laird, CD; Loeb, LA. In vitro expansion of GGC:GCC repeats: identification of the preferred strand of expansion. *Nucleic Acids Res*, 24, 2835-2840, 1996.

[55] Lyons-Darden, T; Topal, MD. Effects of temperature, Mg^{2+} concentration and mismatches on triplet-repeat expansion during DNA replication *in vitro*. *Nucleic Acids Res.* 27, 2235-2240, 1999.

[56] Ogata, N; Morino, H. Elongation of repetitive DNA by DNA polymerase from a hyperthermophilic bacterium *thermus thermophilus*. *Nucleic Acids Res.* 28, 3999-4004, 2000.

[57] Ogata, N; Miura, T. Elongation of tandem repetitive DNA by the DNA polymerase of the hyperthermophilic archaeon *Thermococcus litoralis* at a hairpin-coil transitional state: A model of amplification of a primordial simple DNA sequence. *Biochemistry* 39, 13993-14001, 2000.

[58] Tuntiwechapikul, W; Salazar, M. Mechanism of in vitro expansion of long DNA repeats: effect of temperature, repeat length, repeat sequence, and DNA polymerases. *Biochemistry* 41, 854-860, 2002.

[59] Zyrina, NY; Zheleznaya, LA; Dvoretsky, EV; Vasiliev, VD; Chernov, A; Matvienko, NI. N.BstD6I DNA nickase strongly stimulates template-independent synthesis of non-palindromic repetitive DNA by Bst DNA polymerase. *Biol. Chem.* 388, 367-372, 2007.

[60] Liang, XG; Li, BC; Jensen, K; Frank-Kamenetskii, MD. Ab initio DNA synthesis accelerated by endonuclease. *Nucleic Acids Symp. Ser.* 50, 95-96, 2006.

[61] Cummings, CJ; Zoghbi, HY. Fourteen and counting: unraveling trinucleotide repeat diseases. *Hum. Mol. Genet.* 9, 906-916, 2000.

[62] Horng, JT; Huang, HD; Jin, MH; Wu, LC; Huang, SL. The repetitive sequences database and mining putative regulatory elements in gene promoter regions. *J. Comput. Biol.* 9, 621-640, 2002.

[63] Hui, JY; Bindereif, A. Alternative pre-mRNA splicing in the human system: unexpected role of repetitive sequences as regulatory elements. *Biol. Chem.* 386, 1265-1271, 2005.

[64] Gusella, JF; MacDonald, ME. Trinucleotide instability: A repeating theme in human inherited disorders. *Annu. Rev. Med.* 47, 201-209, 1996.

[65] Warren, ST. The expanding world of trinucleotide repeats. *Science* 271, 1374-1375, 1996.

[66] Phillips, SEV; Manfield, I; Parsons, I; Davidson, BE; Rafferty, JB; Somers, WS; Margarita, D; Cohen, GN; Saint Girons, I. Stockley, PG. Cooperative tandem binding of Met repressor of *Eschierichia-Coli. Nature* 341, 711-715, 1989.

[67] Dai, X; Greizerstein, MB; Nadas-Chinni, K; Rothman-Denes, LB. Supercoil-induced extrusion of a regulatory DNA hairpin. *Proc. Natl. Acad. Sci. USA.* 94, 2174-2179, 1997.

[68] Zazopoulos, E; Lalli, E; Stocco, DM; Sassone-Corsi, P. DNA binding and transcriptional repression by DAX-1 blocks steroidogenesis. *Nature* 390, 311-315, 1997.

[69] Spiro, C; Richards, JP; Chandrasekaran, S; Brennan, RG; McMurray, CT. Secondary structure creates mismatched base-pairs required for high-affinity binding of camp response element-binding protein to the human enkephalin enhancer. *Proc. Natl. Acad. Sci. USA.* 90, 4606-4610, 1993.

[70] Grosschedl, R; Hobom, G. DNA sequences and structural homologies of the replication origins of lambdoid bacteriophages. *Nature* 277, 621-627, 1979.

[71] Willwand, K; Mumtsidu, E; Kuntz-Simon, G; Rommelaere, J. Initiation of DNA replication at palindromic telomeres is mediated by a duplex-to-hairpin transition induced by the minute virus of mice nonstructural protein NS1. *J. Biol. Chem.* 273, 1165-1174, 1998.

[72] Miura, T. Origin of genomic DNA: Discussion from reverse-transcription and expansion of repetitive oligonucleotides. *Viva Origino* 31, 46-61, 2003.

[73] Ogata, N. Elongation of palindromic repetitive DNA by DNA polymerase from hyperthermophilic archaea: A mechanism of DNA elongation and diversification. *Biochimie* 89, 702-712, 2007.

[74] Liang, XG; Kato, T; Asanuma, H. Mechanism of DNA elongation during *de novo* DNA synthesis. *Nucleic Acids Symp. Ser.* 52, 411-412, 2008.

[75] Davidson, EH; Posakony, JW. Repetitive sequence transcripts in development. *Nature* 297, 633-635, 1982.

[76] Willard, HF. The genomics of long tandem arrays of satellite DNA in the human genome. *Genome* 31, 737-744, 1989.

[77] Hiatt, EN; Kentner, EK; Dawe, RK. Independently regulated neocentromere activity of two classes of tandem repeat arrays. *Plant Cell* 14, 407-420, 2002.

[78] Eichler, EE; Sankoff, D. Structural dynamics of eukaryotic chromosome evolution. *Science* 301, 793-797, 2003.

[79] McAllister, BF; Werren, JH. Evolution of tandemly repeated sequences: What happens at the end of an array? *J. Mol. Evol.* 48, 469-481, 1999.

[80] Charlesworth, B; Langley, CH; Stephan, W. The evolution of restricted recombination and the accumulation of repeated DNA sequences. *Genetics* 112, 947-962, 1986.

[81] Chi, YS; Jung, YH; Choi, IS; Kim, YG. Surface-initiated growth of poly d(A-T) by Taq DNA polymerase. *Langmuir* 21, 4669-4673, 2005.

[82] Kotlyar, AB; Borovok, N; Molotsky, T; Fadeev, L; Gozin, M. In vitro synthesis of uniform poly(dG)-poly(dC) by Klenow exo⁻ fragment of polymerase I. *Nucleic Acids Res.* 33, 525-535, 2005.

[83] Kotlyar, AB; Borovok, N; Molotsky, T; Cohen, H; Shapir, E; Porath, D. Long, monomolecular guanine-based nanowires. *Adv. Mat.* 17, 1901-1905, 2005.

[84] Shapir, E; Cohen, H; Calzolari, A; Cavazzoni, C; Ryndyk, DA; Cuniberti, G; Kotlyar, A; Di Felice, R; Porath, D. Electronic structure of single DNA molecules resolved by transverse scanning tunnelling spectroscopy. *Nat. Mat.* 7, 68-74, 2008.

[85] Borovok, N; Molotsky, T; Ghabboun, J; Porath, D; Kotlyar, A. Efficient procedure of preparation and properties of long uniform G4-DNA nanowires. *Anal. Biochem.* 374, 71-78, 2008.

In: Bacterial DNA, DNA Polymerase and DNA Helicases   ISBN 978-1-60741-094-2
Editor: Walter D. Knudsen and Sam S. Bruns   © 2009 Nova Science Publishers, Inc.

Chapter XI

Bacteriophage φ29 DNA Polymerase: An Outstanding Replicase

Miguel de Vega and Margarita Salas*

Instituto de Biología Molecular "Eladio Viñuela" (CSIC), Centro de Biología Molecular "Severo Ochoa" (CSIC-UAM), Campus Universidad Autónoma, Cantoblanco, Madrid, Spain

Abstract

Due to the limited processivity of replicative DNA polymerases (replicases), as well as to their incapacity to unwind parental duplex DNA to allow replication fork progression, their replication efficiency depends on the functional assistance of accessory proteins as processivity factors and helicases. In addition, the inability of DNA polymerases to start de novo DNA synthesis requires the use of a short RNA/DNA molecule to provide the 3'-OH group required to initiate DNA replication. This requisite for a primer creates a dilemma to replicate the ends of linear genomes: once the last primer for the lagging strand synthesis is removed, a portion of ssDNA at the end of the genome will remain uncopied. Bacteriophage φ29 has overcome these issues by means of the unique catalytic features of an outstanding enzyme, the φ29 DNA polymerase. This replicase belongs to the family B (eukaryotic-type) of DNA-dependent DNA polymerases and has served as model to understand the enzymology of these polymerases. As most of the family B members, φ29 DNA polymerase contains both 3'-5' exonuclease and polymerization activities residing in two structurally independent domains. During two decades, site-directed mutagenesis studies of individual residues contained in regions of high amino acid similarity have provided the functional insights of this enzyme, extrapolative to other family B members. However, φ29 DNA polymerase is endowed with two distinctive features: high processivity and strand displacement capacity that allow it to replicate the viral genome from a single binding event, without requiring the assistance of unwinding and processivity factors. Recent crystallographic resolution of the structure of the apo and binary/ternary complexes of

---

* Correspondence should be addressed to mdevega@cbm.uam.es and msalas@cbm.uam.es

ϕ29 DNA polymerase, together with the biochemical studies of site-directed mutants, have given insights into the structural basis responsible for the coordination of the processive polymerization and strand displacement. In addition, such structures have provided the mechanism of translocation of family B DNA polymerases. Another difference with respect to the rest of replicases is the ability of ϕ29 DNA polymerase to use a protein (terminal protein, TP) as primer, circumventing the end replication problem. Recent resolution of the structure of the ϕ29 DNA polymerase/TP heterodimer, together with the biochemical analysis of chimerical DNA polymerases and TPs, have given the clues of the specificity of the interaction between both proteins, suggesting a model for the transition from initiation to elongation. We will also discuss how the basic research on the ϕ29 DNA polymerase properties have led to the development of DNA amplification technologies based on this outstanding enzyme.

To accomplish efficient and fast genome replication, replicative DNA polymerases (also called replicases) rely on their functional association to other replicative proteins as helicases, that unwind de double-stranded DNA (dsDNA) allowing progression of the replication fork, and to processivity factors to held the DNA polymerase bound to the template strand (Kornberg and Baker, 1992, Watson et al., 2004). In addition, DNA polymerases are unable to start *de novo* DNA synthesis and they need to use a 3´-OH group, provided in most cases by a short RNA molecule synthesized by a primase activity. This fact creates the so-called end replication problem of linear chromosomes: once the most terminal primer is removed, a short region of unreplicated single-stranded DNA (ssDNA) will remain at the end of the chromosome that would lead to a continuous shortening of the daughter DNA molecule after successive rounds of DNA replication. Thus, it is essential to guarantee replication of the chromosome ends, that otherwise would lead to cell death. Many mechanisms have evolved to prevent such a shortening, most of them making use of the presence of repetitive sequences at the ends of the chromosomes that allow them to form long concatemers, to circularize, or to form hairpin loops to fill the incomplete 5´-ends (Kornberg and Baker, 1992). In higher eukaryotes, telomerase produces an overhanged ssDNA end by direct elongation of the 3´-OH group (Kornberg and Baker, 1992) that finally can invade homologous double-stranded telomeric tracts, protecting chromosome ends (Verdun and Karlseder, 2007).

Bacteriophage ϕ29 has circumvented the end replication problem by means of the specific and distinctive features displayed by the ϕ29 DNA polymerase. This small (66 kDa) single subunit enzyme, the product of ϕ29 gene 2, has been characterized as the viral DNA replicase (Blanco and Salas, 1984, 1985b, Salas, 1991). Based on the sequence similarity and the sensitivity to specific inhibitors of eukaryotic DNA polymerases, as aphidicolin, phosphonoacetic acid, butylanilino-dATP and butylphenyl-dGTP, this enzyme was proposed to belong to the B family (eukaryotic-type) of DNA-dependent DNA polymerases (Blanco and Salas, 1986, Bernad et al., 1987). As any other DNA polymerase, ϕ29 DNA polymerase accomplishes sequential template-directed addition of dNMP units onto the 3´-OH group of a growing DNA chain, with insertion discrimination values ranging from 10^4 to 10^6, and with an efficiency of mismatch elongation 10^5–10^6-fold lower than that of a properly paired primer terminus (Esteban et al., 1993). In addition, ϕ29 DNA polymerase catalyzes two degradative reactions: 1) pyrophosphorolysis, the polymerization reversal, a reaction consisting in the release of dNTPs from the 3´ end of a primer/template structure by addition of PPi as substrate, in the presence of divalent metal ions, probably playing some role in fidelity (Blasco et al., 1991). The fact that φ29 DNA polymerase mutants at the catalytic amino acid residues involved in the DNA polymerization activity were also deprived of pyrophosphorolytic activity indicated that both activities share a common polymerization active site (Blasco et al., 1991). 2) 3′-5′ exonuclease, a reaction usually found in the same polypeptide chain of DNA replicases, requiring divalent metal ions to release dNMP units from the 3′ end of a DNA strand. The 3′-5′ exonuclease of φ29 DNA polymerase plays a role in proofreading of DNA insertion errors, as it degrades preferentially mismatched primer termini rather than correctly paired ones (Blanco and Salas, 1985a, Garmendia et al., 1992). In addition, this activity degrades processively (without dissociation) DNA substrates longer than six nucleotides, the catalytic constant being 500 s^{-1} (Esteban et al., 1994). Shorter substrates are degraded distributively, φ29 DNA polymerase/DNA complex dissociating at a rate of 1 s^{-1} (Esteban et al., 1994).

In addition, φ29 DNA polymerase shows three distinctive features compared to most replicases. First, it initiates DNA replication at the origins located at both ends of the double-stranded linear genome by catalyzing the addition of the initial dAMP onto the hydroxyl group of Ser232 of the φ29 terminal protein (TP), which acts as primer (Salas, 1991, Salas et al., 1996, Salas, 1999), bypassing the need for a primase, and overcoming the end replication problem. After a transition stage in which a sequential switch from TP-priming to DNA-priming occurs, the same polymerase molecule replicates the entire genome processively without dissociating from the DNA, being the only DNA polymerase involved in the viral DNA replication (Blanco et al., 1989). Second, unlike most replicases, φ29 DNA polymerase performs DNA synthesis without the assistance of processivity factors, displaying the highest processivity described for a DNA polymerase [>70 kb; (Blanco et al., 1989)]. A third distinctive property of φ29 DNA polymerase is the efficient coupling of processive DNA polymerization to strand displacement. This capacity enables the enzyme to replicate the φ29 double-strand genome (19285 bp) without the need of a helicase (Blanco et al., 1989).

The lack of the crystallographic structure of φ29 DNA polymerase until recently led us to structurally map the enzymatic activities of this enzyme by site-directed mutagenesis of individual residues contained in regions of high amino acid similarity, as well as by the construction of deletion mutants (Blanco and Salas, 1996) (see Table I). Those studies allowed us to propose a bimodular organization for φ29 DNA polymerase, with a N-terminal domain containing the critical residues responsible for the exonucleolytic activity, whereas both synthetic activities, protein-primed initiation and DNA polymerization, as well as pyrophosphorolysis, were contained in the C-terminal domain (Blanco and Salas, 1996).

The recent crystallographic resolution of the φ29 DNA polymerase structure confirmed our previous mutational results. Thus, φ29 DNA polymerase consists of a N-terminal exonuclease domain (residues 1-190), containing the 3′-5′ exonuclease active site, and a C-terminal polymerization domain (residues 191-575) that, like in other DNA polymerases, is subdivided into the universally conserved palm (containing the catalytic and DNA ligand residues), fingers (containing the dNTP ligands) and thumb (which confers stability to the primer) subdomains ((Kamtekar et al., 2004); Figure 1). In addition, the structure of φ29 DNA polymerase provided for the first time a topological basis for its intrinsic strand displacement capacity and processivity.

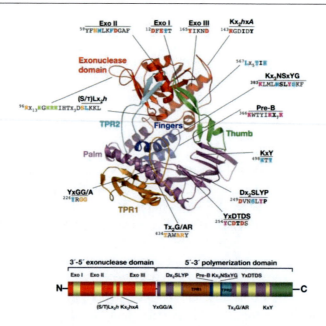

Figure 1. Ribbon representation of φ29 DNA polymerase. The different subdomains of the C-terminal polymerization domain as well as the N-terminal 3′-5′ exonuclease domain of φ29 DNA polymerase (PDB code 1XHX) are coloured as in the original article (Kamtekar et al., 2004). Thus, the 3′-5′ exonuclease domain is shown in red, the palm in pink, the fingers in dark blue, and the thumb in green. The specific insertions TPR1 and TPR2 of protein-primed DNA polymerases are coloured in orange and cyan, respectively. The amino acid sequences corresponding to conserved motifs present in family B DNA polymerases are shown, as well as their spatial placement. Catalytic amino acids responsible for the exonuclease and the polymerization activities are coloured in red, DNA ligand residues in blue, those interacting with both DNA and TP substrates are in orange, incoming nucleotide ligands in magenta, and residues predicted to make contacts with TP-DNA in green. The linear arrangement of these sequence motifs is shown at the bottom.

Table 1.

	Region[a]	Motif	Metal binding and catalysis	DNA binding	TP binding	dNTP binding	Strand displacement
N-terminal	ExoI	"DxE"	D^{12}, E^{14}	T^{15}	F^{65}, Y^{69}, F^{69}, H^{61}		D^{12}, E^{14}
	ExoII	"$Nx_{2-3}F/YD$"	D^{66}	Y^{59}, H^{61}, N^{62}, F^{65}, F^{69}	R^{96}, K^{114}		D^{66}
	pre-"(S/T)Lx$_2$h"			R^{96}, K^{110}, K^{112}, K^{113}	S^{122}, F^{128}		
	"(S/T)Lx$_2$h"			S^{122}, L^{123}, F^{128}			
	"Kx_2h"		K^{143}				K^{143}
	ExoIII	"Yx_3D"	Y^{165}, D^{169}				Y^{165}, D^{169}
	ct	"YxGG"		Y^{226}, F^{230}, R^{223}	R^{223}, G^{228}, G^{229}, F^{230}		
C-terminal	A/I/II	"Dx_2SLYP"	D^{249}	S^{252}		Y^{254}	
	TPR1			K^{305}, Y^{315}	D^{332}, K^{305}, Y^{315}		
	Pre-B					K^{371}, K^{379}	
	B/2a/III	"Kx_3NSxYG"		N^{387}, G^{391}, F^{393}, L^{384}		K^{383}, Y^{390}, K^{392}	
	2b/III	"Tx_2GR"	R^{438}	T^{434}, R^{438}	T^{434}, R^{438}		
	C/3/I	"YxDTDS"	D^{456}, D^{458}			Y^{454}	
	4/VII	"KxY"		K^{498}, Y^{500}			
	Thumb subdomain			[b] K^{538}, K^{555}, L^{567}, T^{573}, K^{575}			

The 3´-5´ Exonuclease Domain

Most of the known replicases contain in the same polypeptidic chain a 3´-5´ exonuclease activity responsible for editing (proofreading) the insertion errors to guarantee a faithful DNA synthesis. The 3´-5´ exonuclease activity resides in an independent domain structurally conserved among the DNA polymerases from families A, B and C that are endowed with a proofreading activity (Bernad et al., 1989), and consists in a central core of β-sheets surrounded by six α-helices (Ollis et al., 1985, Freemont et al., 1988, Beese et al., 1993, Wang et al., 1996, Wang et al., 1997, Doublié et al., 1998, Hopfner et al., 1999, Zhao et al., 1999, Rodriguez et al., 2000, Hashimoto et al., 2001). The 3´-5´ exonuclease active site of prokaryotic and eukaryotic DNA polymerases is evolutionarily conserved, formed by three N-terminal amino acid motifs (Exo I, Exo II and Exo III) containing the invariant four carboxylate groups that bind two metal ions, and a tyrosine residue involved in orienting the attacking water molecule (Bernad et al., 1989, Joyce and Steitz, 1994). The structure of φ29 DNA polymerase complexed with a 5-mer oligonucleotide bound to its exonuclease domain (Kamtekar et al., 2004) supported the mutational data that showed the catalytic carboxylates to be D12 and E14 (Exo I motif), D66 (Exo II motif) and D169 (Exo III motif) (Bernad et al., 1989, Soengas et al., 1992, Esteban et al., 1994) (see Figure 2A). The 10^5-fold reduction of the exonuclease activity displayed by φ29 DNA polymerase mutants lacking the carboxylic group of these residues (Esteban et al., 1994), demonstrated their role in catalysis, allowing to extrapolate the two-metal ion mechanism (Beese and Steitz, 1991) to φ29 DNA polymerase. In addition to the catalytic residues mentioned above, the exonuclease active site of family B DNA polymerases also contains two conserved residues, a Lys belonging to motif Kx_2h (de Vega et al., 1997), and a Tyr from motif Exo III (Bernad et al., 1989, Soengas et al., 1992). Mutations introduced at the corresponding φ29 DNA polymerase residues K143 and Y165 negatively affected catalysis without compromising the stability of the ssDNA binding at the 3´-5´ exonuclease active site (Soengas et al., 1992, de Vega et al., 1997). Interestingly, the crystal structure of φ29 DNA polymerase containing ssDNA at its exonuclease active site has allowed to identify two different conformations for both residues, K143 and Y165 (Berman et al., 2007). Thus, in one conformation, Y165 and K143 are facing away from the active site, being solvent exposed, whereas in the other conformation the hydroxyl group of Y165 would interact with the water nucleophile, as well as with K143, whose ε-amine group interacts with the hydroxyl of Y165, with the catalytic aspartate D169, and with the scissile phosphate through a water-mediated hydrogen bond (Berman et al., 2007) (see Figure 2A). The last conformation most likely corresponds to the chemically and biologically relevant one for exonuclease activity, explaining why mutants at both residues are defective in the exonuclease activity (Soengas et al., 1992, de Vega et al., 1997). Thus, it is tempting to speculate that during normal polymerization the exonuclease active site would exist in its non-catalytically competent state, whereas the presence of a mismatched 3´ terminus would promote a concerted movement of the conserved Tyr and Lys residues into the active site to set it up for the exonucleolysis reaction.

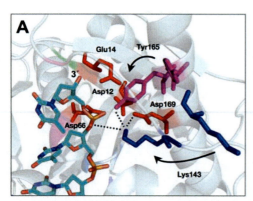 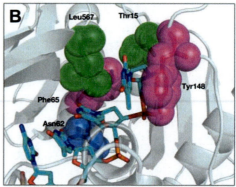

Figure 2. The 3'-5' exonuclease active site of φ29 DNA polymerase. (A) *Amino acid residues responsible for the exonucleolytic activity.* The structure of the ssDNA bound to the exonuclease active site was obtained from the coordinates PDB 2PY5 (Berman et al., 2007). The metal ligands D12 and E14 (Exo I motif), D66 (Exo II motif) and D169 (Exo III motif) are coloured in red. Residues Y165 (Exo III motif), involved in orienting the attacking water molecule, and K143 (Kx2h motif), which plays an auxiliary role in catalysis, are coloured in magenta and dark blue, respectively. The black dashed lines are the hydrogen bonds that residue K143 establishes with residues Y165, D169 and the scissile phosphodiester bond in the catalytically competent state. Black arrows represent the predicted movement of residues Y165 and K143 from an open (inactive) to a close (active) conformation suitable for catalysis (Berman et al., 2007). Adapted from (Berman et al., 2007). (B) *Binding of the ssDNA at the 3'-5' exonuclease active site of φ29 DNA polymerase.* φ29 DNA polymerase ssDNA ligands T15 (Exo I motif), N62 and F65 (Exo II motif), Y148 (Kx2h motif) and L567 (thumb subdomain) are represented as semitransparent spheres.

ssDNA Binding Residues at the 3´-5´ Exonuclease Active Site

Stabilization and correct orientation of the primer-terminus at the exonuclease active site of DNA polymerases is a prerequisite to ensure an efficient editing activity. As in other crystallized proofreading DNA polymerases, the three last nucleotides of a bound 5-mer oligonucleotide at the exonuclease domain of φ29 DNA polymerase are embedded in a groove whose dimensions are designed to bind ssDNA (Kamtekar et al., 2004). Due to the orientation of the exonuclease domain respect to the polymerization one, the ssDNA binding cleft adopts a tunnel conformation [as in other family B DNA polymerases; (Wang et al., 1997, Hopfner et al., 1999)] burying the DNA, instead of the channel shape showed in prokaryotic-type DNA polymerases (Beese et al., 1993, Doublié et al., 1998). The ssDNA binding cleft of φ29 DNA polymerase is formed by residues T15 (Exo I motif) that contacts the 3' nucleotide of the ssDNA, and N62 and F65 (Exo II motif) that directly interact with the penultimate nucleotide, as it was previously anticipated by analysis of φ29 DNA polymerase mutants at these residues (de Vega et al., 1996, 1998b) (see Figure 2B). However, other amino acids initially described also as ssDNA ligands, as Y59 and F69 (Exo II motif; (de Vega et al., 2000)) and S122 ((S/T)Lx$_2$h motif; (de Vega et al., 1998b)) could be playing a more structural role, since they are buried in the protein interior (Kamtekar et al., 2004). In addition, the impairment of ssDNA binding due to the mutation of residues H61 and F128 (de Vega et al., 2000, Rodríguez et al., 2003) could be indirect since they are forming hydrogen bonds with each other and with residue S122. Interestingly, the resolution of the φ29 DNA polymerase structure has allowed to identify two other residues involved in ssDNA stabilization at the catalytic site: Y148, highly conserved in the Kx$_2$h motif (de Vega et al., 1997), which is stacking on the 3´ terminal base, and L567 from the thumb subdomain which is interacting with the two 3´ terminal bases in the exonuclease active site (Kamtekar et al., 2004), and whose role as ssDNA ligand in the stabilization of the frayed primer-terminus has been recently demonstrated by site-directed mutagenesis (Pérez-Arnaiz et al., 2006) (see Figure 2B).

Polymerization Domain

φ29 DNA polymerase C-terminal domain contains the residues involved in binding the φ29 DNA polymerase substrates [DNA, dNTP and TP (Blanco and Salas, 1996, Kamtekar et al., 2004, 2006, Berman et al., 2007)]. As in other eukaryotic-type (family B) DNA polymerases, as those from RB69 (Wang et al., 1997), *Thermococcus gorgonarius* (Hopfner et al., 1999), *Pyrococcus kodakaraensis* (Hashimoto et al., 2001), *E. coli* DNA polymerase II (Protein Data Bank ID code 1Q8I), *Thermococcus* sp.9°N-7 (Rodriguez et al., 2000), and *Desulfurococcus tok* (Zhao et al., 1999), the polymerization domain of φ29 DNA polymerase is structured as a semiopen right hand containing the universal palm, fingers, and thumb subdomains, which form a groove in which the primer-template DNA is bound [(Berman et al., 2007), see Figure 1].

The palm subdomain constitutes the base of the above mentioned groove. It contains the catalytic metal ligands and it is formed by a central core of antiparallel β-sheets flanked by α-helices packed against it (Blanco and Salas, 1996, Kamtekar et al., 2004, Berman et al., 2007). The fingers subdomain is formed by two antiparallel α-helices (like in other family B DNA polymerases) including the residues responsible for binding the incoming nucleotide (Truniger et al., 2002a, Truniger et al., 2002b, Truniger et al., 2003, Kamtekar et al., 2004, Berman et al., 2007). The thumb subdomain is very small and, contrarily to the helical character in other crystallized DNA polymerases, in φ29 DNA polymerase is mainly made by a loop and a long β-turn-β element. It includes most of the residues involved in binding the dsDNA product (Pérez-Arnaiz et al., 2006, Berman et al., 2007). The main difference between φ29 DNA polymerase and the above-mentioned family B DNA polymerases is the presence in the φ29 enzyme of two additional subdomains, both corresponding to sequence insertions specifically conserved in the protein-primed subgroup of DNA polymerases, and named Terminal Protein Region 1 and 2 [(TPR1 and TPR2; Blasco et al., 1990; Blanco et al., 1991; Dufour et al., 2000), see Figure 1]. TPR1 insertion (residues 261 to 358) forms a well-defined subdomain composed by α-helices and β-sheets. It lies at the edge of the palm, and contains an internal and flexible β-turn-β structure (residues 302-316) just in the entrance of the dsDNA binding groove (Figure 1). TPR2 is formed by residues 394-427 and is structured as a long β-hairpin emerging between the fingers and palm subdomains, opposite to the thumb. The spatial arrangement of the five subdomains mentioned above results in the presence of three electropositive paths leading into the active site of φ29 DNA polymerase (Kamtekar et al., 2004), described in detail below.

Binding of the dsDNA

Crystallographic resolution of the φ29 DNA polymerase binary and ternary complexes have shown that the dsDNA portion of the substrate is completely encircled by the toroidal structure formed by the TPR2, thumb, palm and fingers subdomains (upstream duplex DNA tunnel) (Berman et al., 2007).

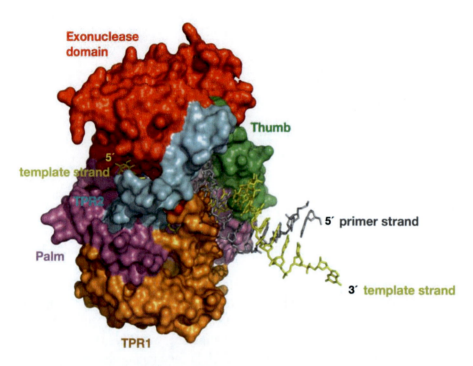

Figure 3. φ29 DNA polymerase/dsDNA complex. Crystallographic data are from (Berman et al., 2007). φ29 DNA polymerase is space-filling represented, and coloured as in Figure 1; exonuclease in red, palm in pink, TPR1 in orange, fingers in dark blue, TPR2 in cyan and thumb in green. The structure of the φ29 DNA polymerase binary complex was obtained from the coordinates PDB 2PZS (Berman et al., 2007).

Residues belonging to such subdomains make extensive contacts with the eight base pairs upstream from the catalytic site. In addition, the protruding 5´ template strand is threaded through a 10Å narrow tunnel (downstream template tunnel) formed by the TPR2, palm and fingers subdomains together with residues from the exonuclease domain, and whose involvement in strand displacement will be discussed later (see Figure 3).

Additionally, the phosphate moiety of the priming nucleotide interacts with the invariant residue Y500 in motif KxY, as predicted (Blasco et al., 1995). Catalytic metals occupy de polymerization active site and are coordinated by the aspartate residues D249 [from conserved motif A, (Blasco et al., 1993)] and D458 [belonging to eukaryotic motif C (Bernad et al., 1990a, Bernad et al., 1990b)], and the carbonyl group of V250 (a non-conserved residue from motif A) of the palm subdomain [(Berman et al., 2007), see Figure 4].

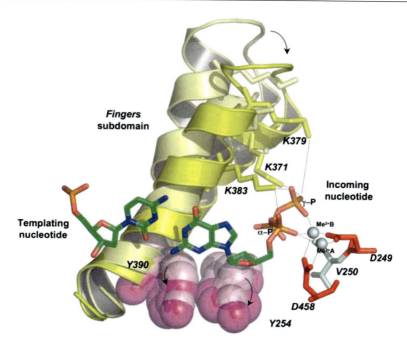

Figure 4. Concerted movement of the φ29 DNA polymerase fingers subdomain and residues Y254 and Y390 during incoming dNTP binding. The figure has been made by superimposing the binary and ternary complexes of φ29 DNA polymerase. Fingers subdomain is coloured in light and dark yellow in its open (binary complex) and closed (ternary complex) conformations, respectively. Both, Y254 and Y390 are represented as semitransparent pink and magenta spheres in the binary and ternary complexes, respectively. Metal binding residues D249 and D458 are depicted as red sticks. Metal ions A and B are represented as grey spheres. dNTP ligands K371, K379 and K383, placed at the fingers subdomains are depicted as yellow sticks. Black dashed lines represent the interactions between the dNTP phosphates and the protein. Adapted from (Berman et al., 2007).

Binding of the Incoming dNTP

The incoming dNTP gains access to the nucleotide insertion site of φ29 DNA polymerase by diffusing through a large pore enclosed by residues from the exonuclease domain as well as from the palm, fingers, and thumb subdomains (Kamtekar et al., 2004, Berman et al., 2007) and that leads to the polymerization active site. The insertion site is initially occupied by the aromatic ring of the two conserved residues Y390 (from motif B of the fingers subdomain) and Y254 (from motif A at the palm subdomain; see Figure 4). Binding of the nucleotide triggers a movement of the fingers subdomain towards the polymerization active site, rotating 14° from an opened to a closed state. This subdomain contains electropositively charged residues, conserved in most family B DNA polymerases, which bind the phosphates of the dNTP. Thus, γ-phosphate interacts with K383 from motif B (Saturno et al., 1997), whereas α-phosphate is bound by K383, K371 and K379 [from motif pre-B, (Truniger et al., 2002b, Truniger et al., 2004b)]. In addition, γ- and α-phosphates of the incoming nucleotide also participate in the coordination of the catalytic ions, the dNTP acting as a crosslinker between the fingers residues and the metal ions chelated to the conserved carboxylates mentioned above [(Berman et al., 2007), see Figure 4]. Closing of the fingers moves Y390 and Y254 out of the nucleotide insertion site into their position in the nascent base pair binding pocket, allowing the base moiety of the incoming nucleotide to form a Watson-Crick base pair with the templating nucleotide, whereas the deoxyribose ring stacks on the phenolic group of Y254. This residue has been involved in selecting for dNTPs by steric clashing between the aromatic group and the 2′-OH group of ribonucleotides (Bonnin et al., 1999). In addition, substitutions at ϕ29 DNA polymerase residue Y390 of the highly conserved B motif of family B DNA polymerases (Blanco et al., 1991) allowed us to propose a principal and dual role for this residue as one of the key determinants that dictate the nucleotide insertion and extension preferences during translesion synthesis past 8oxodG by family B replicases (de Vega and Salas, 2007).

The Mechanism of Translocation

The changes occurring upon dNTP binding have provided the structural basis for the mechanism of translocation. Thus, once the catalysis of the phosphodiester bond formation between the γ-phosphate of the incoming dNTP and the OH- group of the priming nucleotide takes place, the originated pyrophosphate leaves the DNA polymerase, breaking the electrostatic crosslink that kept the fingers subdomain in the closed state. Concomitantly to the fingers opening, residues Y254 and Y390 move back into the nucleotide insertion site, leading to the translocation of the nascent base pair out of the nascent base pair binding pocket one position, as now the nucleotide insertion site is sterically inaccessible (Berman et al., 2007). This translocation allows the 3′ OH-group of the newly added nucleotide to be in a competent position to attack nucleophilically the γ-phosphate of the incoming nucleotide during the next nucleotide insertion event (Berman et al., 2007).

The Structural Bases for Processivity and Strand Displacement Capacities of ϕ29 DNA Polymerase

Replicative DNA polymerases associate to helicases that unwind de dsDNA as replication fork progresses, and to processivity factors that stabilize the binding of the replicase to the DNA, guaranteeing an efficient DNA replication (Kornberg and Baker, 1992, Watson et al., 2004). By the contrary, ϕ29 DNA polymerase is endowed with two intrinsic and distinctive properties, processivity and strand displacement capacities, that allow it to accomplish replication of the entire ϕ29 double-stranded linear genome (19285 bp) in the absence of processivity factors and DNA helicases (Blanco et al., 1989). The crystallographic resolution of ϕ29 DNA polymerase, and biochemical analyses of a deletion mutant have indicated a functional role for the specific insertion TPR2 in these abilities (Kamtekar et al., 2004, Rodríguez et al., 2005, Berman et al., 2007). As mentioned above, this insertion forms a doughnut-shaped structure together with the thumb, palm and TPR1 subdomains (upstream duplex DNA tunnel) encircling the product DNA at the polymerization domain during replication (Berman et al., 2007). A ϕ29 DNA polymerase deletion mutant lacking most of the TPR2 insertion displayed poor ability to bind DNA that led to a dramatic loss of processivity (Rodríguez et al., 2005). The results supported the hypothesis that the TPR2 insertion is a processivity-enhancing subdomain, and that the upstream DNA tunnel constitutes an internal clamp to provide the enzyme with the maximal DNA-binding stability required to replicate the entire genome from a unique DNA polymerase-binding event (Kamtekar et al., 2004, Rodríguez et al., 2005).

In addition, φ29 DNA polymerase binary and ternary complexes showed that the template strand passes through the narrow downstream template tunnel formed by TPR2, palm and fingers subdomains together with residues from the exonuclease domain, to gain access to the polymerization active site. Thus, prior to entering the DNA polymerase, unwinding of the template and non-template strands has to take place. The fact that removal of the TPR2 insertion also abolished the capacity of φ29 DNA polymerase to couple polymerization to strand displacement (Rodríguez et al., 2005), led us to propose the TPR2 insertion acting as a "molecular wedge" to separate the parental DNA strands, mimicking the single-stranded encircling action mode of helicases (Figure 5).

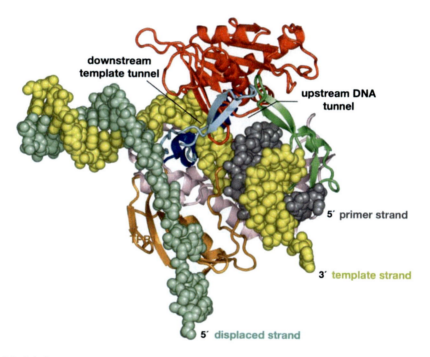

Figure 5. Model for coupling processive DNA polymerization to strand displacement by φ29 DNA polymerase. The product dsDNA is completely wrapped by the toroidal structure formed by TPR2, thumb, palm and fingers subdomains to prevent its dissociation from the polymerase, conferring a remarkable processivity. In addition, the encirclement of the template strand by the downstream template tunnel formed by TPR2, fingers and palm subdomains together with residues of the exonuclease domain, confers to TPR2 the ability to act as a structural barrier, forcing the DNA strands of the parental DNA to diverge. φ29 DNA polymerase subdomains are coloured as indicated in Figure 1 (reproduced from Rodríguez, I, Lázaro, JM, Blanco, L, Kamtekar, S, Berman, AJ, Wang, J, Steitz, TA, Salas, M, de Vega, M (2005) A specific subdomain in φ29 DNA polymerase confers both processivity and strand-displacement capacity. *Proc. Natl. Acad. Sci. USA* 102: 6407-6412. Copyright (2005) National Academy of Sciences of USA).

Whether the TPR2 insertion merely represents a steric hindrance to force the unwinding of dsDNA, or, on the contrary, plays an active role in such a helicase-like activity, involving specific residues, remains to be elucidated.

Interaction with the Terminal Protein

The end-replication problem described above has been solved by several prokaryotic and eukaryotic viruses, as well as linear plasmids from bacteria, fungi and higher plants, and even *Streptomyces spp.* by using a protein (called Terminal Protein; TP) to prime DNA synthesis from the end of their linear genomes (Salas, 1991, 1999, Salas and de Vega, 2006, Salas and de Vega, 2008). In these cases, an amino acid residue of the primer TP provides the priming OH group, becoming covalently linked to the 5′-end of the DNA (parental TP). The extensive in vitro studies performed with bacteriophage φ29, have laid the foundations for this so-called protein-primed replication mechanism. The complex formed between the replicative DNA polymerase and a free TP molecule interacts with the replication origins at both ends of the genome by the specific recognition of the parental TP and DNA sequences. A chimerical φ29 DNA polymerase containing the TPR1 subdomain of the φ29-related phage GA-1 DNA polymerase, could perform initiation reaction using GA-1 TP as primer, but only in the presence of φ29 TP-DNA as template, indicating that parental TP recognition is mainly accomplished by the DNA polymerase (Pérez-Arnaiz et al., 2007). In this respect, mutations at φ29 DNA polymerase residues K110, K112 and R113, placed at an external loop, specifically affected the recognition of the parental TP (Rodríguez et al., 2004). Once the origin has been recognized, the DNA polymerase catalyses the incorporation of dAMP onto the priming OH group of the TP residue S232, in a reaction directed by the 3′ penultimate dTMP in the template strand (initiation reaction). The φ29 DNA polymerase/primer TP heterodimer does not dissociate after the initiation step. There is a transition stage in which the DNA polymerase synthesizes a 5-nt-long DNA molecule while complexed with the primer TP, undergoes some structural change during incorporation of nucleotides 6–9 (transition) and dissociates from the primer TP when nucleotide 10 is incorporated into the nascent DNA chain (elongation mode) (Méndez et al., 1997). Finally, the same DNA polymerase catalyses chain elongation via strand displacement to fulfil TP-DNA replication (Blanco et al., 1989).

As commented, the first step in the protein-priming mechanism consists in the formation of a stable DNA polymerase/TP heterodimer (Blanco et al., 1987). Based on site-directed mutagenesis and on proteolytic analyses of φ29 DNA polymerase, it was concluded that protein-primed initiation and DNA polymerization share the same active site and both occupy a common binding site that is sequentially used by TP and DNA (Blanco and Salas, 1995, 1996, de Vega et al., 1998a, Truniger et al., 2000). In addition, the expression and purification of the C-terminal portion of φ29 DNA polymerase as an isolated domain rendered an active protein unable to interact stably with the TP (Truniger et al., 1998). These results, together with site-directed mutagenesis at residues specifically conserved in the exonuclease domain of the protein-primed subgroup of family B DNA polymerases, allowed us to conclude that both the C-terminal and N-terminal domains of φ29 DNA polymerase contribute in the interaction with the TP (de Vega et al., 1998a, Eisenbrandt et al., 2002, Rodríguez et al., 2003). Similar mutagenesis studies revealed the involvement in TP interaction of residues from the palm subdomain, as those belonging to the YxG(G/A) (Truniger et al., 1999) and Tx$_2$G/AR motifs (Méndez et al., 1994). On the other hand, deletion of the last 13 amino acid residues, located at the thumb subdomain of φ29 DNA polymerase, rendered a mutant enzyme unable to bind TP (Truniger et al., 2004a). In addition, TPR1 residues K305, Y315 and D332 were shown to be involved in the formation of the network of interactions with the TP (Dufour et al., 2000, Dufour et al., 2003). All these studies led us to conclude that the residues that interact with the TP are distributed along most of the subdomains of the φ29 DNA polymerase.

Recent crystallographic resolution of the φ29 DNA polymerase/TP complex shows that the TP forms an extended structure that is complementary to the DNA polymerase surface [(Kamtekar et al., 2006) see Figure 6A]. TP is folded into a N-terminal domain, an intermediate domain and a priming domain comprised of a four-helix bundle structure. These results validate our proposal of multiple interactions between the TP and residues of most of the φ29 DNA polymerase subdomains (de Vega et al., 1998a). Thus, the electronegativelly charged TP priming domain interacts with positive residues of the φ29 DNA polymerase thumb and TPR2 subdomains, as well as with residue R96 from the exonuclease domain, as previously proposed (Rodríguez et al., 2004). Once bound to the DNA polymerase, the exonuclease domain and the palm, thumb, TPR1 and TPR2 subdomains (upstream duplex tunnel) of the polymerase encircle the priming domain of the TP. Both, the negative charge of the priming domain and its overall dimensions mimic DNA in its interactions with the polymerase, as it occupies the DNA-binding site in the polymerase, as initially proposed (de Vega et al., 1998a). Residue S232 of the φ29 TP, that lies in a loop at the end of the priming domain, will be placed at the polymerization active site, in an orientation suitable to prime formation of the initiation product (TP-dAMP) by the DNA polymerase (see Figure 6B). The use of the φ29 DNA polymerase upstream DNA tunnel by both, the TP priming domain and the DNA product precludes the initiation at internal sites, as an upstream 3´ template would sterically clash with the TP, restricting the beginning of DNA synthesis at the ends of the genome (Kamtekar et al., 2006).

The TP intermediate domain, containing two long α-helices fits structurally with the TPR1 subdomain of the polymerase, packing against it and burying a 575Å^2 area. This interaction is maintained by means of multiple contacts, as two salt bridges between the R158 and R169 TP residues and the E291 and E322 residues of the φ29 DNA polymerase TPR1 subdomain.

Previous development of replication systems with purified proteins and DNAs from the φ29-related bacteriophages Nf and GA-1 showed that the reactions catalyzed by each of these three systems are similar.

However, and in spite of their resemblance, DNA polymerases and TPs cannot be interchanged because of the high specificity of such protein-protein interactions (Bravo et al., 1994, González-Huici et al., 2000, Longás et al., 2006). By using chimerical proteins, generated by swapping specific domains of the TP and DNA polymerase of both φ29 and GA-1, we have demonstrated that the specificity of the DNA polymerase–TP interaction is mainly contributed by the contacts established between the TP intermediate domain and the

DNA polymerase TPR1 subdomain, although both, the TP priming and intermediate domains are required to confer stability to such interaction (Pérez-Arnaiz et al., 2007).

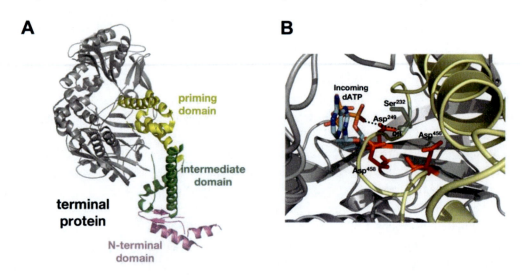

Figure 6. (A) *Ribbon representation of φ29 DNA polymerase forming a heterodimer with the TP.* φ29 DNA polymerase is coloured in grey. The different TP domains are coloured as indicated. The structure of the φ29 DNA polymerase/TP heterodimer was obtained from the coordinates PDB 2EX3 (Kamtekar et al., 2006). (B) *A detailed view of the proposed TP-deoxyadenylylation reaction catalyzed by the φ29 DNA polymerase is presented.* The figure shows the catalytic residues of the DNA polymerase (in red), the TP priming loop (yellow), TP priming Ser232 residue (grey) and incoming dATP. The dot line represents the predicted nucleophilic attack to the α phosphate of the incoming nucleotide by the hydroxyl group of Ser232 of the TP. The figure was made by structural superposition of the φ29 DNA polymerase/TP heterodimer [PDB 2EX3, (Kamtekar et al., 2006)] and the φ29 ternary complex [PDB 2PYL, (Berman et al., 2007)] to determine spatial coordinates of the incoming nucleotide.

In addition, cloning of the ϕ29 TP priming and intermediate domains independently showed that the latter assists to the former to prime the TP-dAMP formation through a conformational change in the DNA polymerase (Pérez-Arnaiz et al., 2007). The structure of the ϕ29 DNA polymerase forming a complex with the TP is very similar to that of the apo enzyme, the main conformational changes being restricted to TPR1 residues 304-315 [(Kamtekar et al., 2006); Figure 7]. Such residues form loops with a high degree of flexibility in the apo enzyme. By the contrary, the ϕ29 heterodimer structure shows that this loop moves out to allow the TP to access the polymerase active site. Altogether, these results led us to propose a model for the DNA polymerase-TP interaction (Figure 7). Thus, the TP intermediate domain would recognize specifically and interact with the DNA polymerase TPR1 subdomain. Such interaction would promote the change of the TPR1 loop from a flexible (coloured in red) to the stable moved out conformation (coloured in magenta) that now would allow the proper (prone to catalysis) placement of the TP priming domain into the DNA polymerase structure (Pérez-Arnaiz et al., 2007).

As described above, ϕ29 DNA polymerase remains bound to the TP until the tenth nucleotide is inserted (transition state) (Méndez et al., 1997). Such an interaction could preclude premature dissociation of the DNA polymerase from the very short product DNA that would compromise the efficiency of the ϕ29 genome replication. Structural resolution of the φ29 TP shows that the intermediate domain is linked to the priming domain through a flexible region (Kamtekar et al., 2006). Thus, as the DNA is synthesized, the free rotation of the priming domain would permit the synthesis of the DNA while the TP remains bound to the DNA polymerase through the interaction of the TP-intermediate domain and the DNA polymerase TPR1 subdomain (Méndez et al., 1997, Kamtekar et al., 2006).

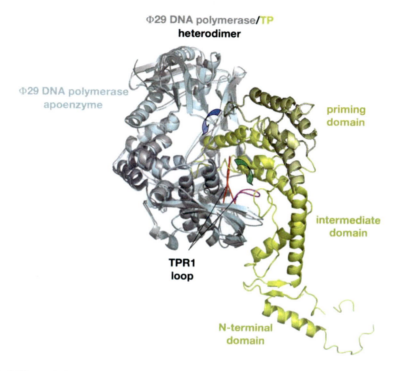

Figure 7. Modelling of the conformational changes occurring during the specific recognition of the DNA polymerase by the TP. φ29 DNA polymerase structures corresponding to the apoenzyme (in light blue) and the heterodimer (in grey) were overlapped. The TPR1 loop is coloured in red in the apoenzyme [PDB code 1XHX (Kamtekar et al., 2004)] and in magenta in the heterodimer [PDB code 2EX3; (Kamtekar et al., 2006)]. The proposed conformational change in the TPR1 loop after binding of the TP intermediate domain to the DNA polymerase TPR1 subdomain is indicated with a green arrow. TP is coloured in yellow (TP priming domain in the proposed non-productive orientation is indicated with light brown). Magenta arrow indicates the proposed change in the TP priming domain orientation after TPR1 loop movement.

The "Sliding-Back" Mechanism to Initiate TP-Primed DNA Replication

As mentioned before, φ29 DNA polymerase does not start replication at the first 3′ nucleotide of the genome but employs the second position from the 3′-end of the template for the initial base pairing and formation of the corresponding TP-dAMP complex at each DNA end. The DNA ends are recovered by a specific mechanism called "sliding-back," that is based on a 3′-terminal repetition of two T residues. This reiteration permits, prior to DNA elongation, the asymmetric translocation of the initiation product, TP-dAMP, to be paired with the first T residue (Méndez et al., 1992).

Depending on the terminal repetition, various terminal sequence recovery mechanisms have evolved: "sliding-back", by which the initiation product translocates back one position, enabling the nucleotide used as template for the initiation reaction to direct also the insertion of the second nucleotide, as it has been described to occur in ϕ29 (Méndez et al., 1992) and in the ϕ29-related bacteriophage GA-1 (Illana et al., 1996); the "stepwise sliding-back", that takes place in the *Streptococcus pneumoniae* phage Cp-1 and in the ϕ29-related phage Nf which initiate at the 3′ third nucleotide of its terminal repetition (3′-TTT) (Martín et al., 1996, Longás et al., 2008), and in the *E. coli* phage PRD1 that initiates at the fourth nucleotide (3′-CCCC) (Caldentey et al., 1993), requiring two and three consecutive sliding-back steps, respectively, to recover the DNA end information; and the "jumping-back", described to occur in adenovirus, in which the initiation product TP-CAT, initially synthesized using as template the GTA sequence at positions 4 to 6 in the template, jumps back to the sequence GTA at positions 1 to 3 (King and van der Vliet, 1994).

These mechanisms have been envisaged to increase the fidelity during the initiation reaction, as several base pairing checking steps have to occur before definitive elongation of the initiation product takes place (Méndez et al., 1992, King and van der Vliet, 1994). The fact that other TP-containing genomes, either from eukaryotic viruses, linear chromosomes and plasmids of *Streptomyces* (Salas, 1999), and virus infecting Archaea, as halovirus (Bamford et al., 2005, Bath et al., 2006), also contain some kind of sequence repetitions at their ends supports the hypothesis that the "sliding-back" mechanism, or variants of it, could be a common feature of protein-primed replication systems (Méndez et al., 1992). However, several questions regarding the initiation at internal positions of the TP-containing genomes remain to be elucidated. On the one hand, the basis for the recognition of the internal templating nucleotide. In this respect, by using chimerical TPs, constructed by swapping the priming domains of the related ϕ29 and Nf proteins, we have shown recently that this domain is the main structural determinant that dictates the internal 3′ nucleotide used as template during initiation (Longás et al., 2008).

On the other hand, the structural mechanism that accounts for this energetically unfavored mechanism, as one (in the case of bacteriophages ϕ29, Nf, GA-1, Cp1 and PRD1) or most drastically three (as in adenovirus) base pairs should be broken before the TP backwards motion step.

ϕ29 DNA Polymerase: A Potent DNA Amplification Enzyme

The two distinctive properties displayed by ϕ29 DNA polymerase, strand displacement capacity and processivity, together with its high polymerization fidelity due to the combination of a high dNMP insertion discrimination and a strong proofreading activity, led us to envisage ϕ29 DNA polymerase as an ideal tool to achieve strand displacement amplification. Thus, two aspects that were frequent limitations in the existing amplification procedures, fidelity of synthesis and length of the amplified products, would no longer be an issue. Thus, it was demonstrated that the extremely high processivity and strand displacement capacities of φ29 DNA polymerase permit the enzyme to generate concatemeric ssDNA amplicons of a unique polarity by Rolling Circle Amplification of circular DNA templates [(Blanco et al., 1989); see Figure 8A].

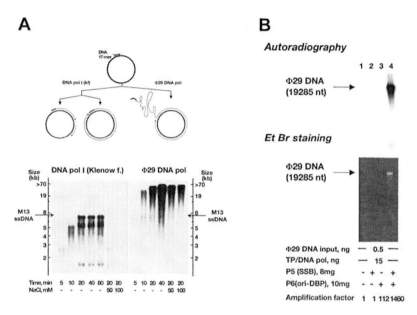

Figure 8. (A) *Rolling Circle Amplification of circular M13 ssDNA by the oligonucleotide-primed φ29 DNA polymerase.* In the upper panel, a scheme of the replicating DNA molecules is depicted. In the lower panel an autoradiogram of the replication products is presented (reproduced from Blanco, L, Bernad, A, Lázaro, JM, Martín, G, Garmendia, C, Salas, M (1989) Highly efficient DNA synthesis by the phage φ29 DNA polymerase. Symmetrical mode of DNA replication. *J. Biol. Chem.* 264: 8935-8940). (B) *Efficiency of the in vitro φ29 DNA amplification with the TP-primed φ29 DNA polymerase.* The amplified DNA was analyzed by alkaline agarose gel electrophoresis and autoradiography (upper panel) as well as ethidium bromide staining (lower panel). The amplification factor obtained in each experimental condition is shown (reproduced from Blanco, L, Lázaro, JM, de Vega, M, Bonnin, A, Salas, M (1994) Terminal protein-primed DNA amplification. *Proc. Natl. Acad. Sci. USA* 91: 12198-12202. Copyright (1994) National Academy of Sciences of USA).

In addition, an efficient amplification of full-length φ29 TP-DNA was accomplished by an *in vitro* system that, in addition to TP and φ29 DNA polymerase also required φ29 SSB (to preclude the appearance of short palindromic DNAs) and φ29 DBP (for optimal activation of the origins). Using this "viral system", φ29 DNA polymerase could amplify limited starting amounts of φ29 TP-DNA molecules (see Figure 8B). The amplification factor was over three orders of magnitude after 1 h of incubation at 30°C (Blanco et al., 1994), the amplified material being fully infective as compared with φ29 TP-DNA obtained from virions, establishing the basis for the development of heterologous DNA amplification procedures dependent on the phage φ29 replication machinery (Blanco et al., 1994).

The exceptional abilities of φ29 DNA polymerase described above have led to the development of one of the most efficient procedures for isothermal dsDNA amplification (Dean et al., 2001, Dean et al., 2002), developed by Amersham Biosciences/Molecular Staging. φ29 DNA polymerase is combined with random hexamer primers to achieve isothermal and faithful 10^4 to 10^6-fold amplification via strand displacement of either circular [Templiphi™ (www.templiphi.com)] or linear [Genomiphi™ (www.genomiphi.com)] genomes, yielding high quality amplification products that can be either digested or sequenced directly without any further purification steps.

Acknowledgements

This work was supported by research grant BFU 2008-00215 from Ministerio de Educación y Ciencia, grant S-0505/MAT-0283 from Comunidad Autónoma de Madrid and by an Institutional grant from Fundación Ramón Areces to the Centro de Biología Molecular "Severo Ochoa".

References

Bamford, DH, Ravantti, JJ, Ronnholm, G, Laurinavicius, S, Kukkaro, P, Dyall-Smith, M, Somerharju, P, Kalkkinen, N, Bamford, JK (2005). Constituents of SH1, a novel lipid-containing virus infecting the halophilic euryarchaeon *Haloarcula hispanica*. *J. Virol.* 79: 9097-9107.

Bath, C, Cukalac, T, Porter, K, Dyall-Smith, ML (2006). His1 and His2 are distantly related, spindle-shaped haloviruses belonging to the novel virus group, *Salterprovirus*. *Virology* 350: 228-239.

Beese, LS, Steitz, TA (1991). Structural basis for the 3'-5' exonuclease activity of *Escherichia coli* DNA polymerase I: a two metal ion mechanism. *EMBO J.* 10: 25-33.

Beese, LS, Derbyshire, V, Steitz, TA (1993). Structure of DNA polymerase I Klenow fragment bound to duplex DNA. *Science* 260: 352-355.

Berman, AJ, Kamtekar, S, Goodman, JL, Lázaro, JM, de Vega, M, Blanco, L, Salas, M, Steitz, TA (2007). Structures of phi29 DNA polymerase complexed with substrate: the mechanism of translocation in B-family polymerases. *EMBO J.* 26: 3494-3505.

Bernad, A, Zaballos, A, Salas, M, Blanco, L (1987). Structural and functional relationships between prokaryotic and eukaryotic DNA polymerases. *EMBO J.* 6: 4219-4225.

Bernad, A, Blanco, L, Lázaro, JM, Martín, G, Salas, M (1989). A conserved 3'-5' exonuclease active site in prokaryotic and eukaryotic DNA polymerases. *Cell* 59: 219-228.

Bernad, A, Blanco, L, Salas, M (1990a). Site-directed mutagenesis of the YCDTDS amino acid motif of the φ29 DNA polymerase. *Gene* 94: 45-51.

Bernad, A, Lázaro, JM, Salas, M, Blanco, L (1990b). The highly conserved amino acid sequence motif Tyr-Gly-Asp-Thr-Asp-Ser in alpha-like DNA polymerases is required by phage φ29 DNA polymerase for protein-primed initiation and polymerization. *Proc. Natl. Acad. Sci. USA* 87: 4610-4614.

Blanco, L, Salas, M (1984). Characterization and purification of a phage φ29-encoded DNA polymerase required for the initiation of replication. *Proc. Natl. Acad. Sci. USA* 81: 5325-5329.

Blanco, L, Salas, M (1985a). Characterization of a 3'-5' exonuclease activity in the phage ϕ29-encoded DNA polymerase. *Nucleic Acids Res.* 13: 1239-1249.

Blanco, L, Salas, M (1985b). Replication of phage ϕ29 DNA with purified terminal protein and DNA polymerase: synthesis of full-length ϕ29 DNA. *Proc. Natl. Acad. Sci. USA* 82: 6404-6408.

Blanco, L, Salas, M (1986). Effect of aphidicolin and nucleotide analogs on the phage ϕ29 DNA polymerase. *Virology* 153: 179-187.

Blanco, L, Prieto, I, Gutiérrez, J, Bernad, A, Lázaro, JM, Hermoso, JM, Salas, M (1987). Effect of NH4+ ions on ϕ29 DNA-protein p3 replication: formation of a complex between the terminal protein and the DNA polymerase. *J. Virol.* 61: 3983-3991.

Blanco, L, Bernad, A, Lázaro, JM, Martín, G, Garmendia, C, Salas, M (1989). Highly efficient DNA synthesis by the phage ϕ29 DNA polymerase. Symmetrical mode of DNA replication. *J. Biol. Chem.* 264: 8935-8940.

Blanco, L, Bernad, A, Blasco, MA, Salas, M (1991). A general structure for DNA-dependent DNA polymerases. *Gene* 100: 27-38.

Blanco, L, Lázaro, JM, de Vega, M, Bonnin, A, Salas, M (1994). Terminal protein-primed DNA amplification. *Proc. Natl. Acad. Sci. USA* 91: 12198-12202.

Blanco, L, Salas, M (1995). Mutational analysis of bacteriophage ϕ29 DNA polymerase. *Methods Enzymol* 262: 283-294.

Blanco, L, Salas, M (1996). Relating structure to function in ϕ29 DNA polymerase. *J. Biol. Chem.* 271: 8509-8512.

Blasco, MA, Bernad, A, Blanco, L, Salas, M (1991). Characterization and mapping of the pyrophosphorolytic activity of the phage ϕ29 DNA polymerase. Involvement of amino acid motifs highly conserved in alpha-like DNA polymerases. *J. Biol. Chem.* 266: 7904-7909.

Blasco, MA, Lázaro, JM, Blanco, L, Salas, M (1993). ϕ29 DNA polymerase active site. Residue Asp249 of conserved amino acid motif "Dx2SLYP" is critical for synthetic activities. *J. Biol. Chem.* 268: 24106-24113.

Blasco, MA, Méndez, J, Lázaro, JM, Blanco, L, Salas, M (1995). Primer terminus stabilization at the ϕ29 DNA polymerase active site. Mutational analysis of conserved motif KXY. *J. Biol. Chem.* 270: 2735-2740.

Bonnin, A, Lázaro, JM, Blanco, L, Salas, M (1999). A single tyrosine prevents insertion of ribonucleotides in the eukaryotic-type ϕ29 DNA polymerase. *J. Mol. Biol.* 290: 241-251.

Bravo, A, Hermoso, JM, Salas, M (1994). In vivo functional relationships among terminal proteins of *Bacillus subtilis* φ29-related phages. *Gene* 148: 107-112.

Caldentey, J, Blanco, L, Bamford, DH, Salas, M (1993). *In vitro* replication of bacteriophage PRD1 DNA. Characterization of the protein-primed initiation site. *Nucleic Acids Res* 21: 3725-3730.

de Vega, M, Lázaro, JM, Salas, M, Blanco, L (1996). Primer-terminus stabilization at the 3'-5' exonuclease active site of ϕ29 DNA polymerase. Involvement of two amino acid residues highly conserved in proofreading DNA polymerases. *EMBO J.* 15: 1182-1192.

de Vega, M, Ilyina, T, Lázaro, JM, Salas, M, Blanco, L (1997). An invariant lysine residue is involved in catalysis at the 3'-5' exonuclease active site of eukaryotic-type DNA polymerases. *J. Mol. Biol.* 270: 65-78.

de Vega, M, Blanco, L, Salas, M (1998a). φ29 DNA polymerase residue Ser122, a single-stranded DNA ligand for 3'-5' exonucleolysis, is required to interact with the terminal protein. *J. Biol. Chem.* 273: 28966-28977.

de Vega, M, Lázaro, JM, Salas, M, Blanco, L (1998b). Mutational analysis of φ29 DNA polymerase residues acting as ssDNA ligands for 3'-5' exonucleolysis. *J. Mol. Biol.* 279: 807-822.

de Vega, M, Lázaro, JM, Salas, M (2000). Phage φ29 DNA polymerase residues involved in the proper stabilisation of the primer-terminus at the 3'-5' exonuclease active site. *J. Mol. Biol.* 304: 1-9.

de Vega, M, Salas, M (2007). A highly conserved Tyrosine residue of family B DNA polymerases contributes to dictate translesion synthesis past 8-oxo-7,8-dihydro-2'-deoxyguanosine. *Nucleic Acids Res* 35: 5096-5107.

Dean, FB, Nelson, JR, Giesler, TL, Lasken, RS (2001). Rapid amplification of plasmid and phage DNA using Phi 29 DNA polymerase and multiply-primed rolling circle amplification. *Genome Res.* 11: 1095-1099.

Dean, FB, Hosono, S, Fang, L, Wu, X, Faruqi, AF, Bray-Ward, P, Sun, Z, Zong, Q, Du, Y, Du, J, Driscoll, M, Song, W, Kingsmore, SF, Egholm, M, Lasken, RS (2002). Comprehensive human genome amplification using multiple displacement amplification. *Proc. Natl. Acad. Sci. USA* 99: 5261-5266.

Doublié, S, Tabor, S, Long, AM, Richardson, CC, Ellenberger, T (1998). Crystal structure of a bacteriophage T7 DNA replication complex at 2.2 Å resolution. *Nature* 391: 251-258.

Dufour, E, Méndez, J, Lázaro, JM, de Vega, M, Blanco, L, Salas, M (2000). An aspartic acid residue in TPR-1, a specific region of protein-priming DNA polymerases, is required for the functional interaction with primer terminal protein. *J. Mol. Biol.* 304: 289-300.

Dufour, E, Rodríguez, I, Lázaro, JM, de Vega, M, Salas, M (2003). A conserved insertion in protein-primed DNA polymerases is involved in primer terminus stabilisation. *J. Mol. Biol.* 331: 781-794.

Eisenbrandt, R, Lázaro, JM, Salas, M, de Vega, M (2002). φ29 DNA polymerase residues Tyr59, His61 and Phe69 of the highly conserved ExoII motif are essential for interaction with the terminal protein. *Nucleic Acids Res.* 30: 1379-1386.

Esteban, JA, Salas, M, Blanco, L (1993). Fidelity of φ29 DNA polymerase. Comparison between protein-primed initiation and DNA polymerization. *J. Biol. Chem.* 268: 2719-2726.

Esteban, JA, Soengas, MS, Salas, M, Blanco, L (1994). 3'-5' exonuclease active site of φ29 DNA polymerase. Evidence favoring a metal ion-assisted reaction mechanism. *J. Biol. Chem.* 269: 31946-31954.

Freemont, PS, Friedman, JM, Beese, LS, Sanderson, MR, Steitz, TA (1988). Cocrystal structure of an editing complex of Klenow fragment with DNA. *Proc. Natl. Acad. Sci. USA* 85: 8924-8928.

Garmendia, C, Bernad, A, Esteban, JA, Blanco, L, Salas, M (1992). The bacteriophage φ29 DNA polymerase, a proofreading enzyme. *J. Biol. Chem.* 267: 2594-2599.

González-Huici, V, Lázaro, JM, Salas, M, Hermoso, JM (2000). Specific recognition of parental terminal protein by DNA polymerase for initiation of protein-primed DNA replication. *J. Biol. Chem.* 275: 14678-14683.

Hashimoto, H, Nishioka, M, Fujiwara, S, Takagi, M, Imanaka, T, Inoue, T, Kai, Y (2001). Crystal structure of DNA polymerase from hyperthermophilic archaeon *Pyrococcus kodakaraensis* KOD1. *J. Mol. Biol.* 306: 469-477.

Hopfner, KP, Eichinger, A, Engh, RA, Laue, F, Ankenbauer, W, Huber, R, Angerer, B (1999). Crystal structure of a thermostable type B DNA polymerase from *Thermococcus gorgonarius*. *Proc. Natl. Acad. Sci. USA* 96: 3600-3605.

Illana, B, Blanco, L, Salas, M (1996). Functional characterization of the genes coding for the terminal protein and DNA polymerase from bacteriophage GA-1. Evidence for a sliding-back mechanism during protein-primed GA-1 DNA replication. *J. Mol. Biol.* 264: 453-464.

Joyce, CM, Steitz, TA (1994). Function and structure relationships in DNA polymerases. *Annu. Rev. Biochem.* 63: 777-822.

Kamtekar, S, Berman, AJ, Wang, J, Lázaro, JM, de Vega, M, Blanco, L, Salas, M, Steitz, TA (2004). Insights into strand displacement and processivity from the crystal structure of the protein-primed DNA polymerase of bacteriophage ϕ29. *Mol. Cell* 16: 609-618.

Kamtekar, S, Berman, AJ, Wang, J, Lázaro, JM, de Vega, M, Blanco, L, Salas, M, Steitz, TA (2006). The phi29 DNA polymerase:protein-primer structure suggests a model for the initiation to elongation transition. *EMBO J.* 25: 1335-1343.

King, AJ, van der Vliet, PC (1994). A precursor terminal protein-trinucleotide intermediate during initiation of adenovirus DNA replication: regeneration of molecular ends in vitro by a jumping back mechanism. *EMBO J.* 13: 5786-5792.

Kornberg, A, Baker, T. (1992) *DNA Replication*. W.H. Freeman, New York.

Longás, E, de Vega, M, Lázaro, JM, Salas, M (2006). Functional characterization of highly processive protein-primed DNA polymerases from phages Nf and GA-1, endowed with a potent strand displacement capacity. *Nucleic Acids Res.* 34: 6051-6063.

Longás, E, Villar, L, de Vega, M, Salas, M (2008). Involvement of the terminal protein-priming domain in specifying the internal template nucleotide used to initiate DNA replication. *Proc. Natl. Acad. Sci. U S A.* 105: 18290-18295.

Martín, AC, Blanco, L, García, P, Salas, M, Méndez, J (1996). *In vitro* protein-primed initiation of pneumococcal phage Cp-1 DNA replication occurs at the third 3' nucleotide of the linear template: a stepwise sliding-back mechanism. *J. Mol. Biol.* 260: 369-377.

Méndez, J, Blanco, L, Esteban, JA, Bernad, A, Salas, M (1992). Initiation of ϕ29 DNA replication occurs at the second 3' nucleotide of the linear template: a sliding-back mechanism for protein-primed DNA replication. *Proc. Natl. Acad. Sci. USA* 89: 9579-9583.

Méndez, J, Blanco, L, Lázaro, JM, Salas, M (1994). Primer-terminus stabilization at the ϕ29 DNA polymerase active site. Mutational analysis of conserved motif TX2GR. *J. Biol. Chem.* 269: 30030-30038.

Méndez, J, Blanco, L, Salas, M (1997). Protein-primed DNA replication: a transition between two modes of priming by a unique DNA polymerase. *EMBO J.* 16: 2519-2527.

Ollis, DL, Brick, P, Hamlin, R, Xuong, NG, Steitz, TA (1985). Structure of large fragment of *Escherichia coli* DNA polymerase I complexed with dTMP. *Nature* 313: 762-766.

Pérez-Arnaiz, P, Lázaro, JM, Salas, M, de Vega, M (2006). Involvement of φ29 DNA polymerase thumb subdomain in the proper coordination of synthesis and degradation during DNA replication. *Nucleic Acids Res.* 34: 3107-3115.

Pérez-Arnaiz, P, Longás, E, Villar, L, Lázaro, JM, Salas, M, de Vega, M (2007). Involvement of phage φ29 DNA polymerase and terminal protein subdomains in conferring specificity during initiation of protein-primed DNA replication. *Nucleic Acids Res.* 35: 7061-7073.

Rodriguez, AC, Park, HW, Mao, C, Beese, LS (2000). Crystal structure of a pol alpha family DNA polymerase from the hyperthermophilic archaeon *Thermococcus* sp. 9 degrees N-7. *J. Mol. Biol.* 299: 447-462.

Rodríguez, I, Lázaro, JM, Salas, M, de Vega, M (2003). φ29 DNA polymerase residue Phe128 of the highly conserved (S/T)Lx2h motif is required for a stable and functional interaction with the terminal protein. *J. Mol. Biol.* 325: 85-97.

Rodríguez, I, Lázaro, JM, Salas, M, de Vega, M (2004). φ29 DNA polymerase-terminal protein interaction. Involvement of residues specifically conserved among protein-primed DNA polymerases. *J. Mol. Biol.* 337: 829-841.

Rodríguez, I, Lázaro, JM, Blanco, L, Kamtekar, S, Berman, AJ, Wang, J, Steitz, TA, Salas, M, de Vega, M (2005). A specific subdomain in φ29 DNA polymerase confers both processivity and strand-displacement capacity. *Proc. Natl. Acad. Sci. USA* 102: 6407-6412.

Salas, M (1991). Protein-priming of DNA replication. *Annu. Rev. Biochem.* 60: 39-71.

Salas, M, Miller, J, Leis, J, DePamphilis, M. (1996) *Mechanisms for Priming DNA Synthesis.* Cold Spring Harbor Laboratory Press, New York.

Salas, M (1999). Mechanisms of initiation of linear DNA replication in prokaryotes. *Genet Eng (N Y)* 21: 159-171.

Salas, M, de Vega, M. (2006) *Bacteriophage protein-primed DNA replication.* Hefferon, KL (ed.), Ithaca.

Salas, M, de Vega, M. (2008) Replication of bacterial viruses. In Mahy, B.W.J. and Regenmortel, M.H.V.v. (eds.), *Encyclopedia of Virology.* Elsevier, Oxford, Vol. 4, pp. 339-406.

Saturno, J, Lázaro, JM, Esteban, FJ, Blanco, L, Salas, M (1997). φ29 DNA polymerase residue Lys383, invariant at motif B of DNA-dependent polymerases, is involved in dNTP binding. *J. Mol. Biol.* 269: 313-325.

Soengas, MS, Esteban, JA, Lázaro, JM, Bernad, A, Blasco, MA, Salas, M, Blanco, L (1992). Site-directed mutagenesis at the Exo III motif of φ29 DNA polymerase; overlapping structural domains for the 3'-5' exonuclease and strand-displacement activities. *EMBO J.* 11: 4227-4237.

Truniger, V, Lázaro, JM, Salas, M, Blanco, L (1998). φ29 DNA polymerase requires the N-terminal domain to bind terminal protein and DNA primer substrates. *J. Mol. Biol.* 278: 741-755.

Truniger, V, Blanco, L, Salas, M (1999). Role of the "YxGG/A" motif of φ29 DNA polymerase in protein-primed replication. *J. Mol. Biol.* 286: 57-69.

Truniger, V, Blanco, L, Salas, M (2000). Analysis of φ29 DNA polymerase by partial proteolysis: binding of terminal protein in the double-stranded DNA channel. *J. Mol. Biol.* 295: 441-453.

Truniger, V, Lázaro, JM, Blanco, L, Salas, M (2002a). A highly conserved lysine residue in φ29 DNA polymerase is important for correct binding of the templating nucleotide during initiation of phi29 DNA replication. *J. Mol. Biol.* 318: 83-96.

Truniger, V, Lázaro, JM, Esteban, FJ, Blanco, L, Salas, M (2002b). A positively charged residue of φ29 DNA polymerase, highly conserved in DNA polymerases from families A and B, is involved in binding the incoming nucleotide. *Nucleic Acids Res* 30: 1483-1492.

Truniger, V, Lázaro, JM, de Vega, M, Blanco, L, Salas, M (2003). φ29 DNA polymerase residue Leu384, highly conserved in motif B of eukaryotic type DNA replicases, is involved in nucleotide insertion fidelity. *J. Biol. Chem.* 278: 33482-33491.

Truniger, V, Lázaro, JM, Salas, M (2004a). Function of the C-terminus of φ29 DNA polymerase in DNA and terminal protein binding. *Nucleic Acids Res.* 32: 361-370.

Truniger, V, Lázaro, JM, Salas, M (2004b). Two positively charged residues of φ29 DNA polymerase, conserved in protein-primed DNA polymerases, are involved in stabilisation of the incoming nucleotide. *J. Mol. Biol.* 335: 481-494.

Verdun, RE, Karlseder, J (2007). Replication and protection of telomeres. *Nature* 447: 924-931.

Wang, J, Yu, P, Lin, TC, Konigsberg, WH, Steitz, TA (1996). Crystal structures of an NH2-terminal fragment of T4 DNA polymerase and its complexes with single-stranded DNA and with divalent metal ions. *Biochemistry* 35: 8110-8119.

Wang, J, Sattar, AK, Wang, CC, Karam, JD, Konigsberg, WH, Steitz, TA (1997). Crystal structure of a pol α family replication DNA polymerase from bacteriophage RB69. *Cell* 89: 1087-1099.

Watson, J, Baker, T, Bell, S, Gann, A, Levine, M, Losick, R. (2004) *Molecular Biology of the Gene*. Cold Spring Harbor Lab. Press, Plainview, New York.

Zhao, Y, Jeruzalmi, D, Moarefi, I, Leighton, L, Lasken, R, Kuriyan, J (1999). Crystal structure of an archaebacterial DNA polymerase. *Structure* 7: 1189-1199.

In: Bacterial DNA, DNA Polymerase and DNA Helicases
Editor: Walter D. Knudsen and Sam S. Bruns

ISBN 978-1-60741-094-2
© 2009 Nova Science Publishers, Inc.

Chapter XII

Bacterial Replicative DNA Helicases

Subhasis B. Biswas[1], and Jessica J. Clark[1], Deepa S. Kurpad[2] and Esther E. Biswas[2]

[1]Department of Molecular Biology, Graduate School of Biomedical Sciences, University of Medicine and Dentistry of New Jersey, Stratford, NJ, USA
[2]Program in Biotechnology, Department of Bioscience Technologies
Thomas Jefferson University, Philadelphia, PA, USA

Abstract

Replicative DNA helicases are energy-transducing enzymes, which act as the engine of the cellular replication apparatus and unwind duplex DNA in the replication fork. These enzymes are true multifunctional enzymes that are responsible for organization and execution of DNA replication from initiation to termination. The most well studied of them all is DnaB helicase of *Escherichia coli*, which is the prototype of this class of DNA helicases. Early genetic studies of the *dnaB* gene defined important roles of DnaB protein in the elongation stage of DNA replication. DnaB protein acts as the organizer of the replisome by engaging in multiple protein-protein and protein-DNA interactions during the initiation, elongation, and termination stages of DNA replication. DnaB protein and its orthologs have homo-hexameric structure in solution and in the ssDNA-bound state. This homo-hexameric subunit structure allows for the formation of a ring-structure; ssDNA passes through the ring's central hole. Formation of this unique protein-ssDNA complex is a key element in this enzyme's ability to carry out its unique functions. Due to its stable ring structure, *E. coli* DnaB helicase requires the assistance of DnaC protein in order to form the initial DnaB•ssDNA complex. It forms a DnaB•DnaC•ATP complex that attenuates its nucleotidase activity but helps in binding ssDNA. DnaC protein is released upon slow ATP hydrolysis leading to organization of the replication proteins in the fork and replisome formation. DnaB protein engineers the assembly of the replisome through its specialized interaction with DnaG primase and DNA polymerase III holoenzyme. DnaB•ssDNA complex formation stimulates its ATPase activity of DnaB hexamer several fold. The energy of ATP hydrolysis powers translocation of DnaB protein on ssDNA in a 5′→3′ direction and helps unwind duplex

DNA during translocation. The rapid and processive DNA unwinding also requires direct participation of a topoisomerase to remove the DnaB-generated super-twists without which the DNA unwinding and the replication fork would likely stall. In this review article, we summarize the structural and functional properties of the *E. coli* replicative DNA helicase and compare it to analogous replicative helicases from other gram-negative and gram-positive bacteria.

Introduction

Chromosomal DNA replication is probably the most complex event in all living organisms and this process requires numerous proteins and enzymes, all of which must act in concert [1-4]. DNA helicases are essential in this process due to the fact that the initiation, priming, and elongation stages of replication use ssDNA as the template [5, 6]. Thus, DNA helicases play a significant role in the proper execution of chromosomal DNA replication [1, 7]. Genetic studies initially established the indispensable role of the dnaB gene-product in DNA replication of the *E. coli* genome [2, 3, 8-10]. Several DNA helicases have been identified in prokaryotes, including fourteen from *E. coli* [11-13]. Howevr, DnaB protein has been proven to be the replicative DNA helicase in *E. coli* using a variety of *in vitro* DNA replication systems, while other DNA helicases function in a other cellular processes such as transcription, recombination, and repair. DnaB homologs have also been shown to be involved in the replication of bacteriophages T4, T7, and P1 as well as eukaryotic SV40 viral genome [14-16]. In addition to *E. coli*, *in vitro* studies have shown that DnaB protein also acts as a DNA helicase in the replication of bacteriophage λ and plasmids including those containing *E. coli* origin of replication, oriC [1, 5, 6, 17, 18].

Earlier studies [1, 19-23] led to the development of a putative model of DnaB protein function in *E. coli* and λ genome replication and elucidated its roles in various stages of DNA replication [8, 18, 21, 24, 25]. Bound to the lagging strand of the replication fork, it interacts with a variety of replication proteins including DnaC protein, DnaG primase, holoenzyme (DNA polymerase III holoenzyme), and λP protein [26-30]. Due to its ability to physically interact with other replication proteins, DnaB protein plays a pivotal role in the assembly of the replisome and its subsequent progression of the replication fork [24, 31-35]. With the aid of DnaC protein, it forms a DnaB$_6$•DnaC$_6$ complex, with suppressed ATPase and helicase activities, which in turn delivers it to the origin of DNA replication activated by an initiator protein. The DNA replication initiator of *E. coli*, DnaA protein, forms a complex with the *E. coli* origin of replication, *oriC*, and opens a very small region of duplex DNA. It then allows DnaB$_6$•DnaC$_6$ complex to introduce the DnaB helicase to the uninitiated replication fork [36, 37]. Mechanism of this process remains unclear. Consequently, DnaB also plays an important role in the initiation of DNA replication [24, 31-35, 38].

This review focuses on the structure and function of the *E.coli* DnaB helicase, which is a true prototype of all replicative DNA helicases including both prokaryotic and eukaryotic organisms as well as its roles in the replication of bacterial and phage genomes.

I. Historical Context of Major Discoveries Pertaining to *E. coli* DnaB Helicase

Lest we forget the contributions of the scientists whose seminal studies opened our eyes to this class of proteins, we will first discuss these early discoveries of Francois Jacob. Studies by Francois Jacob and colleagues made the major initial contributions to unraveling the genetic apparatus that carries out DNA replication in *E. coli* [2, 3]. These studies identified and elucidated the roles of many of the *E. coli* DNA replication genes. They correctly identified the genes that are required for DNA synthesis, such as *dnaA*, *dnaB*, *dnaC*, and delineated their temporal order of operation. Using temperature-sensitive mutants, the *dnaA* gene-product was identified as functioning in the initiation of chromosomal DNA replication, whereas the *dnaB* gene-product was identified as being involved in the elongation phase. These findings established dnaB gene-product or DnaB protein as an important component of elongation phase of chromosomal DNA replication.

These early genetic studies paved the way for the elucidation of the mechanism of action of the encoded proteins through detailed biochemical studies in several laboratories, particularly the laboratories of Hurwitz and Kornberg [39]. Wickner et al. first established that DnaB protein is a DNA-dependent ATPase/ribonucleotidase and that it interacts with the *dnaC* gene-product *in vitro* [40, 41]. Reha-Krantz and Hurwitz also demonstrated that purified recombinant DnaB protein is a 52-kDa polypeptide that forms a hexamer [42, 43]. Following these discoveries by Reha-Krantz & Hurwitz, Arai and Kornberg made further advancements through detailed enzymological studies and developing an in vitro replication system employing ϕX174 DNA as the substrate [44]. These studies led to a better understanding of the DnaB protein's role in formation of the primosome and replisome, and its unique ability to interact with other replication proteins, including DNA primase (*dnaG* gene product). Kornberg et al. recognized many of the DnaB protein's multifaceted functions in DNA synthesis and called it a "mobile replication promoter", which appears to be one of its most important roles. Although, at the time, it was understood that enzymological unwinding of the double helix was required for semi-conservative DNA replication, Rep protein was thought to be the replicative DNA helicase in *E. coli*. However, experimental evidence from *in vitro* replication systems did not support this hypothesis [1]. In 1986, LeBowitz and McMacken demonstrated that DnaB protein possessed DNA unwinding activity, ultimately leading to the consensus that DnaB protein functions as the replicative helicase for *E. coli* chromosomal and bacteriophage λ DNA replication [6].

II. Structure of DnaB Helicase

Orthologs of DnaB protein are present in most prokaryotic systems, including gram-positive, gram-negative and thermophilic bacteria. In addition, many bacteriophages such as P1, T4 and T7 contain similar DNA helicases, which form hexameric structure in the DNA replication fork, which is a prerequisite for the activation of their helicase activity. Bacteriophage λ, which does not code for its own DNA helicase, appears to initiate replication of its DNA by utilization of the host DnaB helicase.

DNA helicases can be categorized into six super-families based on the conserved sequence motifs [45]. These enzymes can also be classified based on mechanistic (specificity for RNA or DNA), functional (cellular role) or structural properties (hexameric versus monomeric). *E. coli* DnaB is the prototypical hexameric DNA helicase, and is probably the most well-studied. Some replicative DNA helicases appear to function as monomers or dimmers, for example the T4 Dda helicase and *E. coli* rep helicase, respectively [46]. The T7gp4 DNA helicase is a monomer that forms a hexamer upon binding to ssDNA and dTTP and the *E. coli* DnaB helicase (DnaB$_{EC}$) is a very stable hexamer under normal physiological conditions [7, 19, 47, 48]. Like all helicase enzymes, DnaB$_{EC}$ contains the characteristic "core" domains, which include conserved residues involved in nucleotide binding and hydrolysis (Walker A and B motifs), an arginine (Zinc) finger-like DNA binding domain, and a RecA like fold [45].

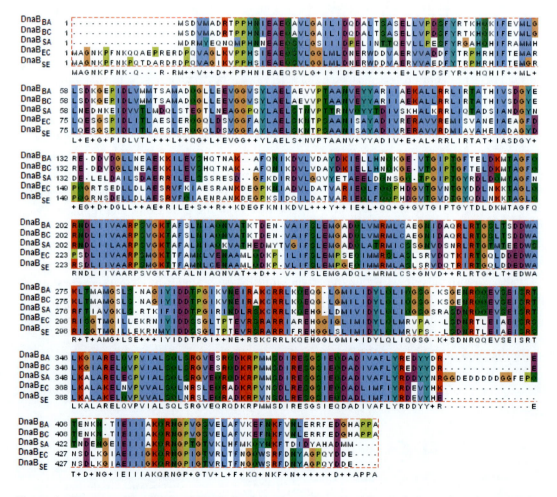

Figure 1. Sequence Homology of DnaB Proteins from Gram-Positive and Gram-Negative Bacteria: DnaB proteins from *B. anthracis* (DnaB$_{BA}$), *B. cereus* (DnaB$_{BC}$), *S. aureus* (DnaB$_{SA}$), *E. coli* (DnaB$_{EC}$) and *Salmonella enteric* (DnaB$_{SE}$) were aligned. The color codings are as follows: red, basic; blue, hydrophobic; green, hydrophilic; orange, neutral; pink, acidic; and light green, proline. Consensus sequence is presented in black and white.

An alignment of DnaB proteins from several gram-positive and gram-negative bacteria is shown in Figure 1. A comparison of the sequences of DnaB from *E. coli, S. areus, S. enterica, B. anthracis* and *B. cereus* indicate that amongst bacterial DnaB proteins, a high degree of homology is retained. The overall homology (identity and similarity) between these five DnaB proteins was ≥60%. When compared to DnaB$_{EC}$, the sequence alignment shows that DnaB$_{BA}$ lacks 17 N-terminal amino acid residues that are present in *E. coli*. Additionally, the *S. aureus* DnaB$_{SA}$ contains a 14 amino acid stretch spanning N-terminal residues 407-421, which is not observed in the other four DnaB proteins. In contrast, the C-terminal half of the DnaB molecule is highly conserved and this region is also the location of major enzymatic activity centers that include ATPase and DNA binding.

III. Structural and Functional Domains of DnaB Helicase

Electron microscopic studies of the hexameric DnaB helicase are suggestive of a doughnut-shaped, rosette structure with an internal diameter of ~40 Å [28, 31]. Recent studies suggest that the DnaB helicase and T7gp4 helicase hexamers are trimers made up of apparent dimeric units [19, 20, 28]. The hexameric structure is a common feature of primary replicative DNA helicases. Many replicative DNA helicases such as the DnaB helicase, T4 gene41 helicase, SV40 T-antigen helicase, and eukaryotic MCM helicase are hexameric in solution [49, 50]. Yet, there are other DNA helicases that do not form hexamers. For example, the T4 Dda helicase and *E. coli* Rep helicase are monomeric and dimeric respectively. The T7gp4 exists as a monomer, which forms a hexamer upon binding to ssDNA. The T7gp4 DNA helicase also has DNA primase activity in the same polypeptide. Replicative DNA helicases from gram-positive bacteria such as *B. subtilis* are monomeric; their oligomeric structures upon binding DNA have not yet been determined [51-53]. Thus, DNA helicases are structurally and functionally diverse. Similar diversity in the mechanism of action also appears likely.

Table 1. Functional Domains of *E. coli* DnaB Helicase

Domain α Residue 1-156	Domain β Residue 157-302	Domain γ Residue 303-471
Energy transduction	ATP binding site	ssDNA binding site
Protein-protein interaction site for hexamer formation	ATPase active site	ATPase regulatory site
DnaG primase and DnaC binding sites		Protein-protein interaction site for hexamer formation

The *E. coli* DnaB helicase is comprised of 471 amino acid residues. DnaB helicase has multiple structural and functional domains, initially defined through biochemical studies of truncated DnaB polypeptides. Arai et al. identified a small flexible sequence at amino acid residue 156 that is known to be susceptible to trypsin proteolytic cleavage and defined the boundary between the N- and C-terminal domains of DnaB helicase [54]. Further mechanistic and structural studies led to the definition of the N- and C-terminal domains in terms of their various functions, as shown in Figure 2 [19, 20]. These domains are defined as follows: N-terminal domain α: residues 1-156, C-terminal domain β: residues 157-302, and C-terminal domain γ: residues 303-471. Among these domains, domain γ is the most conserved (~60%) when compared to other DnaB helicases (Figure 2).

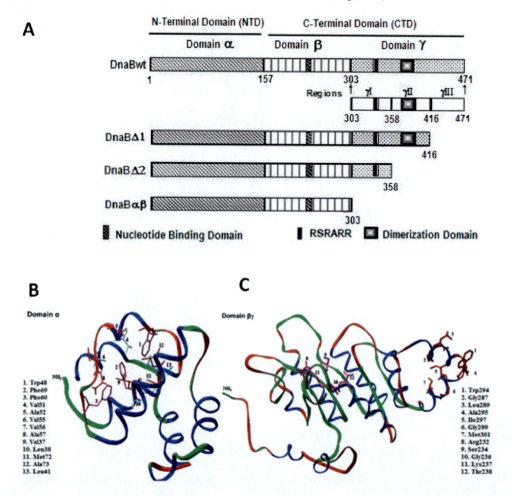

Figure 2. Structural and Functional domains of *E. coli* DnaB Protein [83]. (A) A graphic presentation of the functional domains (α, β, and γ) of *E. coli* DnaB protein and their locations in DnaB protein sequence. (B) Structure of the α domain. The residues that form the "hydrophobic core" involving the indole side chain of Trp48 are highlighted in red. The Trp48, Phe60, and phe69, all highlighted in magenta color, appears to stack to each other, which may reduce fluorescence quantum yield of Trp48. (C) Structure of the βγ domains by homology modeling. Homology modeling of domains βγ using SYBYL (Tripos Inc., St Louis, MO.) using the atomic coordinates from the X-ray structure of the helicase domain of gene 4 DNA helicase/primase of bacteriophage T7. Residues that contribute to a partial hydrophobic environment near Trp256 are highlighted in red. The residues in the Walker A motif, Ala231, Ser234, Gly236, Lys237, and Thr238 are highlighted in pink color. This is the putative ATP binding pocket of the DnaB helicase.

A deletion analysis of DnaB helicase demonstrated that the N-terminal domain α is essential for DNA helicase activity. Interestingly, domain α is devoid of any enzymatic activity. It is likely that domain α plays a key role in transducing the energy derived from ATP hydrolysis into unwinding of duplex DNA; it may also serve as a movable mechanical arm and drive DNA unwinding.

The C-terminal domains β and γ form the mechanistic core of the enzyme and they function in ATP binding, ATP hydrolysis, hexamer formation, as well as ssDNA binding [19,55]. Domain β is the central catalytic site for ATPase activity and contains all of the structural features required for ATP hydrolysis [19, 55]. The ATP binding is mediated by a "type I ATP/GTP binding motif" [56]. Cloned, expressed, and purified domain β retains the ability to hydrolyze ATP in ssDNA independent manner, but at a reduced rate. The presence of domain γ, coupled with domain β, gives the polypeptide the full DNA-dependent ATPase activity. These findings support the hypothesis that the DNA binding domain of DnaB helicase is located within domain γ.

Further analyses of highly conserved domain γ using truncated recombinant polypeptides and site directed mutagenesis as shown in Figure 2, divide domain γ into three functional regions: γI (residues 303-358), γII (residues 359-416) and γIII (residues 417-471). Region γI contains an RSRARR amino acid sequence that may mediate the ssDNA stimulation of ATPase activity, and contribute to ssDNA binding [55]. Region γII is involved in the ssDNA stimulation of the ATPase activity based on deletion studies. Region γII is also essential for ssDNA binding. Region γIII is involved in stimulating the basal ATPase activity in the presence or absence of DNA. Both regions γII and γIII are required for a stable dimer formation and stimulation of the ATPase activity.

IV. Three-Dimensional Structures of Bacterial Replicative DNA Helicases

(i). Electron Microscopic Studies: Early electron microscopic studies were carried out to determine the structure of DnaB helicase hexamer. These studies (Biswas, S.B., Williams, R.C. and Kornberg, A., 1982, unpublished results) indicated that the DnaB protein has a hexagonal subunit arrangement with at least two conformations, with probable six-fold and three-fold axes of symmetry. The electron micrograph (Figure 3) suggests that the hexameric ring is not rigid; rather it is highly flexible. The structures ranged from imperfect circles to triangular geometry with smooth and jagged edges.

The DnaB hexamer undergoes or at least has the ability to undergo conformational transitions particularly upon interaction with various ligands like ATP, ADP, ssDNA, and dsDNA. Similar results have been obtained in other studies involving DnaB helicase and bacteriophage T7gp4 DNA helicase [57-60]. The toroidal structure with a central hole that is needed for ssDNA binding for hexameric DNA helicases is a recurring theme [59, 61, 62]. Similar to the electron micrograph in Figure 3, other studies also show that the hexamer can have six-fold or three-fold symmetry, and suggest large domain movements [57, 58, 63].

Cryo-electron microscopy of the DnaB hexamer showed that in complex formation with DnaC arrangements of the complex were dependent on the interactions between DnaB•DnaC dimers [63]. DnaC monomers appeared arranged as three dumb-bell shaped dimers that interlocked with one of the faces of the helicase.

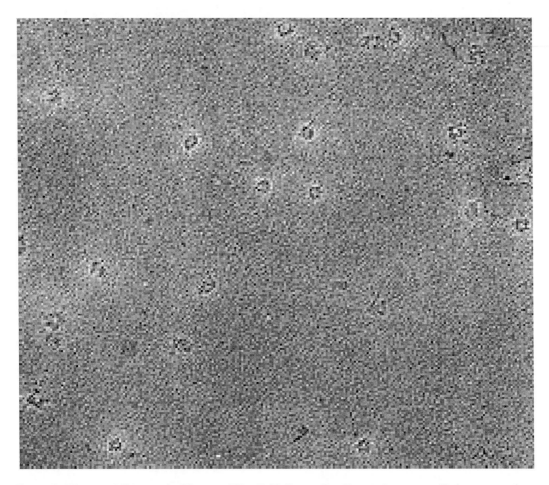

Figure 3. Electron Microscopic Picture of DnaB Helicase. DnaB protein was applied on a specimen film, air dried, negatively stained with uranyl acetate and palladium shadowed and a JEOL 100B electron microscope at the University of California, Berkley was used with a minimal exposure technique to obtain the photographic record. (Subhasis B. Biswas, Robley C. Williams, and Arthur Kornberg, 1982, unpublished results).

(ii) X-Ray Crystallographic Studies: Despite many attempts, the crystal structure of DnaB helicase has not been resolved successfully. Although crystals of DnaB helicase were generated, however, no X-ray crystal structure could be derived as these crystals were always amorphous. Therefore, attempts were made to crystallize DnaB proteins from bacteria that grow at high-temperatures e.g., *Thermus aquaticus* and *Bacillus stearothermophilus*. The crystal structure of the full-length *Thermus aquaticus* DnaB (DnaB$_{TA}$) monomer was resolved at 2.9 Å resolutions [64]. As anticipated, many important structural features of the protein were revealed at the molecular level. The N-terminal domain (NTD) or domain α of DnaB$_{TA}$, has a α-helical structure containing a large spherical bundle of helices that are terminated by an extended helical hairpin. The C-terminal domain (CTD) or domain βγ has prominently parallel β-sheets which are flanked by α-helices. The flexible linker is made up of 34 residues

(151-185) and contains an extended structure that has one α-helix flanked by two loops. These loops connect the helix to the domain α and domain βγ [64].

DnaB helicase could be easily cleaved into N-terminal and C-terminal fragments, 12kDa (P12) and 33kDa, respectively, upon digestion with trypsin [64, 65]. Trypsin cleavage takes place in the flexible linker region [20, 54]. The coordinates of domain α of DnaB$_{TA}$ and that of DnaB helicase revealed similarity in structure. The structure of domain α of DnaB$_{TA}$ formed a large part of the primase interaction surface. Helical hairpin sub-domains formed the interface of each of the dimer and packed together to form a four-helix bundle. The short chain hydrophobic residues were observed dominating the interface. The formation of the domain α dimer stabilized the hairpin in the context of the hexamer assembly, possibly in the three-fold symmetry axes. Recent structures of the C-terminal helicase-binding domain (HBD) of the DnaG primase indicated that the folding in this domain was related to that of the DnaB domain α fragment [55, 66-73]. Comparing the structure of domain α of intact DnaB$_{TA}$ and the NMR-elucidated structure of *E. coli* DnaG established that the HBD at the C-terminal of DnaG primase and domain α of DnaB$_{TA}$ had related folds [68]. The HBD was the only domain of DnaG primase, which was essential for the interaction, and stimulation of the helicase [74]. The five helicase motifs of DnaB$_{TA}$ are located in its domain βγ out of which four motifs (H1, H1a, H2 and H3) form the core of the β-sheet and provide the residues that line the nucleotide pocket. The 5th motif, H5 forms a helix and the N-terminus of β-sheet and their connecting loop comprises of residues that are involved in DNA binding [75]. The domain βγ of Taq DnaB was shown to resemble the helicase domain of T7gp4 and that of RecA core. In DnaB$_{TA}$ protein, inspection of the domain βγ folds showed that the location of the basic sequence RARARR, a putative ssDNA-binding signature, was on a helix, which lay on the opposite face of the β-sheet from the nucleotide-binding pocket. It was observed that the residues formed a leucine zipper (an all α-helical structure) and a helix-loop-strand motif in the core of the RecA fold of domain βγ [67].

V. Functions of DnaB *Helicase*: One Protein, Many Roles

DnaB protein is often referred to as DnaB helicase in scientific literature. However, DNA helicase is only one of this protein's many important functions and may be only one of the important functions in chromosomal DNA replication. Therefore, DnaB protein is the most appropriate terminology for this multifaceted enzyme.

(i) DnaB Protein is Powered by a DNA-Dependent Ribonucleotide Triphosphatase Activity: The *E. coli* DnaB protein exhibits multi-functional roles in chromosomal DNA replication, including its unique role in the replisome as the mobile replication promoter and its DNA unwinding activity [1, 6]. To execute these activities in the replication of *E. coli* chromosome, the DnaB protein requires a significant amount of energy. The required energy is provided by its potent ribonucleotide triphosphatase activity, which is allosterically regulated by DNA. Hurwitz and coworkers first reported the ribonucleotide triphosphatase activity of DnaB in 1978 [42]. The enzyme hydrolyzes all four ribonucleotides to their corresponding diphosphates at comparable rates, but the rate is significantly attenuated for

deoxy-ribonucleotides. The rate of ATP hydrolysis is considerably (six to twelve-fold) stimulated by single stranded bacteriophage φX174 or Fd DNA and to a lesser extent (two to three fold) by double stranded DNA, which established that DnaB helicase protein is an ssDNA-dependent ribonucleotidase.

(ii) Dual roles of Ribonucleotide Triphosphates: In 1981, Arai and Kornberg confirmed the ssDNA-dependent ribonucleotidase activity of DnaB helicase protein [76, 77]. However, it was observed that uridine triphosphates were hydrolyzed at a rate much lower than that observed with other ribonucleotides. It was also revealed that hydrolyzable rNTPs as well as the non-hydrolyzable analogs such as ATPγS or AMPPNP significantly stimulated ssDNA binding by DnaB protein. In addition, the non-hydrolyzable analogs were more effective stimulators. This fact demonstrated that nucleotide binding, but not hydrolysis enables DnaB protein to bind ssDNA (Figure 4).

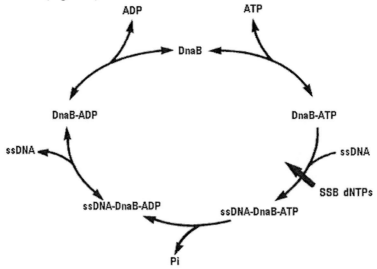

Figure 4. ATP Hydrolysis Cycle of DnaB Protein with ssDNA as Positive Effector [76].

This was the first demonstration that ribonucleotide triphosphates play dual roles in DnaB protein action: rNTP hydrolysis provides the energy for its various functions, and rNTPs are positive allosteric effectors of the DnaB•ssDNA interactions, whereas, the corresponding diphosphates (rNDPs) are negative allosteric regulators [76, 77]. Together, these findings painted a picture of regulation of association and dissociation of DnaB•ssDNA complex by rNTPs and their hydrolysis products (rNDPs), which is essential in understanding the mechanism of DNA helicase today. In summary, rNTP hydrolysis provides the energy for all enzymatic functions of DnaB protein and rNTPs act as positive allosteric effectors, whereas, the corresponding diphosphates act as negative allosteric regulators.

(iii) Asymmetry in the rNTP Binding Sites in DnaB Helicase and its Implications: In the DnaB hexamer homo-hexamer, all monomers are potentially identical [43]. Arai and Kornberg used equilibrium gel filtration analysis to determine binding of rNTPs and analogs to DnaB protein. They determined a 1:1 binding of ATP per DnaB monomer [77]. To investigate this further, a fluorescent analog of ATP, tri-nitrophenyl-ATP (TNP-ATP), which undergoes fluorescent enhancement upon binding to DnaB, was used to study DnaB-ATP

interactions [5, 17]. Scatchard analysis of TNP-ATP binding by DnaB indicated only three molecules of TNP-ATP bound per hexamer. In addition, analysis of binding of the corresponding diphosphates, TNP-ADP, revealed the same stoichiometry.

These studies suggest that all ATP (or rNTP) binding sites in the DnaB hexamer are not identical. Given that in the apo-enzyme all ATP binding sites are undoubtedly equivalent with a six-fold axis of symmetry, conformational change(s) upon the binding of first nucleotide (ATP or ADP) molecule likely leads to a hexamer with a three-fold (or a pseudo-three-fold) axis of symmetry. Incidentally, insulin also forms a homo-hexamer, which is actually a trimer of insulin dimers. Zn2+-insulin also forms a toroidal structure like that of DnaB protein, as described above. As the DnaB hexamer has a structure similar to that observed with insulin homo-hexamer by Dorothy Hodgkin [78, 79], then it is possible that each dimer consists of only one functional ATP or rNTP binding site and the second remains silent. This phenomenon of three ATP binding sites per hexamer was later observed also in other hexameric DNA helicases. It is important to note that there is a possibility that this ATP binding characteristics, as described above, may alter during DNA unwinding and/or movement on ssDNA.

(iv) Mechanisms and Intricacies of DNA Binding by DnaB Helicase: We have discussed above, specific ATP or rNTP binding properties of the DnaB hexamer and the possible implications in its mechanism of action. Arai and Kornberg established that ribonucleotide binding is a prerequisite for ssDNA binding to DnaB helicase [77]. Thus, three TNP-ATP molecules binding per hexamer likely exerts specific influence on the ssDNA binding to the DnaB hexamer. Analysis of DNA binding by DnaB helicase and ^{32}P-labeled oligo(dT)$_{25}$ and a Scatchard analysis of the binding data demonstrated that there are likely three ssDNA (or oligonucleotides) binding sites in the DnaB helicase hexamer, in parallel to three TNP-ATP binding sites in DnaB helicase hexamer [17]. It is possible that these phenomena are unrelated, but together these two phenomena are unlikely to be fortuitous. If DnaB hexamer were functionally a trimer of dimers, then each functional dimeric unit would probably bind one rNTP and one ssDNA molecule. A number of studies have been carried out on the ssDNA binding and dissociation of the DnaB-ssDNA complex [55, 76, 77, 80-82]. The dissociation constant of DnaB-ssDNA complex, determined from kinetic or binding studies, indicate a value of ~1.5-5 x 10^{-8} M in the presence of ATPγS [55, 77, 80]. A sensitive fluorescence anisotropy based analysis was carried out to measure equilibrium binding constants in solution with a number of different ribonucleotides. In the presence of ATPγS, K_d was 5.1 ± 0.3 x 10^{-8} M, which increased to 2.1 ± 0.3 x 10^{-7} M in the presence of ADP or other hydrolyzable ribonucleotides. Most reports indicate that ssDNA binding by DnaB helicase in the absence of a ribonucleotide is close to negligible. In addition, ssDNA binding is also influenced by ionic strength and the binding affinity decreases with increasing salt concentration. The binding affinity changes only slightly with temperature (25-42 °C). Taken together, the data would suggest that the DnaB-ssDNA interaction is primarily ionic with a small hydrophobic component [80].

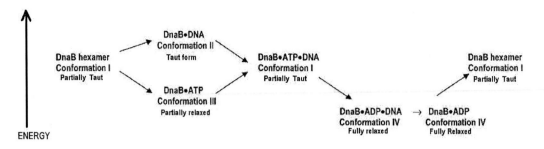

Scheme I. Conformations and Energetics of E. coli DnaB Complexes Involved in DNA [83].

DnaB protein has three tryptophan residues: Trp48 in domain α, Trp294 in domain β and Trp456 in domain γ [83]. Analysis of the fluorescence properties of these three Trp residues, using fluorescence quenching and mutagenesis, indicated that the secondary and tertiary conformation of DnaB protein changes with nucleotide and ssDNA binding. Based on these intrinsic Trp fluorescence studies, a model of conformational changes during ligand binding was proposed. DnaB protein in the apo-enzyme state has a relatively taut conformation (Conformation I), which becomes more taut (Conformation II) upon ssDNA binding (Scheme I). However, upon ATP binding, it undergoes partial relaxation (Conformation III), which allows it to bind to ssDNA. The formation of DnaB•ATP•ssDNA complex returns DnaB to a partial taut conformation (Conformation I). Upon hydrolysis of ATP, DnaB•ADP•ssDNA complex is formed and DnaB undergoes significant relaxation, primarily in domain α (Conformation IV). With ADP bound form, it remains in the fully relaxed state and returns to taut conformation (Conformation I) upon ADP release.

(v) *Implications of Nucleotide and DNA Binding:* The structural and biochemical data available to date provide a reasonable picture of the mechanism of DnaB action and allow us to construct a model of DnaB protein as the mobile replication promoter [8, 24]. The DnaB hexamer has a toroidal structure, as a result, once its binds the lagging strand (or ssDNA) in the replication fork it does not appear to dissociate from the template, and likely contributes to the processivity of the replisome. We can assume that each dimer in the hexamer is capable of binding ssDNA and a molecule of ATP or rNTP. Upon ATP binding, the dimer's affinity for ssDNA increases, leading to binding and formation of DnaB•ATP•ssDNA complex, which in turn increases it's V_{MAX} for ATP hydrolysis. Once ATP is hydrolyzed to ADP forming DnaB•ADP•ssDNA, its affinity for ssDNA decreases. Upon dissociation of ADP, DnaB•ssDNA becomes unstable and ssDNA is released from the dimer.

(vi) *Possible Roles of DnaC Protein in ssDNA Binding by DnaB Helicase:* DnaC protein of *E.coli* transfers DnaB helicase on to a nascent chromosomal replication fork at the origin of DNA replication. There are two possible pathways that may be involved in this process.

(a) *ssDNA binding by DnaC Protein:* The DnaC protein can bind ssDNA very weakly [80, 84-86]. This ssDNA binding by DnaC protein is strictly modulated by nucleotides such as ATP and ADP. A moderate affinity and saturable equilibrium DNA binding is observed only in the presence of ATP. Interestingly, the interaction of ATP and DnaC is so highly specific that the nonhydrolyzable ATP analog, ATPγS, did not replace ATP as a cofactor for DNA binding. Most importantly, the DNA binding affinity of DnaC decreases ~79-fold with change of cofactor from ATP to ADP. DnaC•ATP binds ssDNA and concomitant formation of DnaC•ADP results in an immediate dissociation from the DNA. The intracellular concentration of ATP is reportedly 12-fold higher than ADP. Therefore, DnaC proteins should be in the predominantly ATP bound state and DnaC•ATP should bind ssDNA in the open complex of *oriC* with DnaA protein. However, the formation of ADP generated by ATPases such as DnaA, DnaB and other pre-primosomal proteins may increase the ADP concentration in the immediate vicinity of *oriC*, which may result in the displacement of ATP from the ATP•DnaC•ssDNA complex by ADP. Davey et al. reported a very weak ATP hydrolysis by DnaC in the presence of DNA, which may also lead to the formation of ADP•DnaC•ssDNA complex [87]. Thus, it seems likely that DnaC should dissociate rapidly from the resultant ADP•DnaC•ssDNA complex. The higher concentration of ADP, due to continuous ATP hydrolysis, in the vicinity of the replication fork would prevent re-association of DnaC to the unwound ssDNA during the priming and elongation stages of *E. coli* DNA replication [37, 86, 88, 89]. We propose that ATP/ADP concentrations in the immediate vicinity of *oriC* and the replication fork modulate the binding and the dissociation of DnaC and DNA and the global cellular concentrations of ATP and ADP play a rather minor role.

(b) DnaB•DnaC Complex Formation: DnaB helicase appears to interact with a number of other initiator proteins, including DnaC, DnaG, DnaA, and λP protein [32, 88, 90-94]. Each of these interactions has a very specific role in the DNA replication. Kobori and Kornberg first demonstrated the formation of an isolatable complex of DnaC protein with DnaB helicase [26]. The stoichiometry of the complex is six DnaC monomers to one DnaB hexamer leading to a DnaB$_6$•DnaC$_6$ complex at *oriC*. Affinity sensor analysis of DNA binding to DnaC indicated that approximately six monomers of DnaC bound to ssDNA. The open complex at *oriC*, thus, could be bound by multiple monomers or dimers of DnaC proteins. Various modes of DnaB$_6$•DnaC$_6$ complex formation in the presence of various complexes of DnaC with different combinations of ATP, ADP, and DNA were analyzed and reported [80, 85].

In vitro, in the complete absence of nucleotides and ssDNA, DnaC and DnaB form the DnaB$_6$•DnaC$_6$ complex with a dissociation constant of 3.5 x 10^{-7} M [80]. However, this situation is not possible at or near an active replication fork *in vivo*. It is likely that the formation of the DnaB$_6$•DnaC$_6$ complex will be modulated by the cofactors such as nucleotides and DNA *in vivo*. ATP, ADP, ATPγS, and oligo(dT)$_{25}$ inhibited the DnaB$_6$•DnaC$_6$ complex formation to a varying degree. It was inhibited approximately three-fold with ADP and five-fold with ATP. A four-fold inhibition was observed with oligo(dT)$_{25}$ alone. In the presence of oligo(dT)$_{25}$ and ATP, the dissociation constant was 2.6 x 10^{-7} M and it increased to 39 x 10^{-7} M in the presence of oligo(dT)$_{25}$ and ADP. The change in the equilibrium constant was ~15-fold, which would likely promote a rapid and facile dissociation of the complex. The DnaB helicase forms a stable hexameric ring around the DNA. Consequently, DnaB would remain bound to the ssDNA following dissociation of the DnaB$_6$•DnaC$_6$ complex. . Consequently, the net result of hydrolysis of ATP to ADP in the [ATP•DnaB$_6$•DnaC$_6$•DNA] complex in *oriC* open complex would likely be its conversion to DnaB$_6$• DNA complex, upon dissociation of DnaC and ADP.

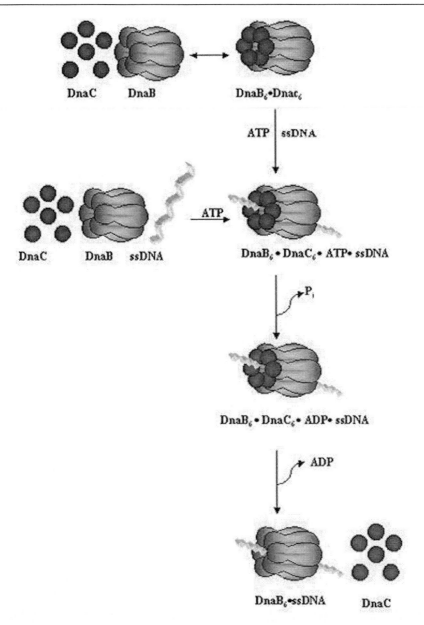

Figure 5. Proposed Models Depicting Putative Pathways of DnaC Mediated DnaB Helicase Binding to ssDNA. The steps in this model depicting DnaC protein's role are as follows: (A) DnaC forms an apparent hexamer around the ssDNA to form [ATP•DnaC$_6$•ssDNA]. (B) DnaB hexamer slowly associates with the hexameric DnaC already bound to the DNA, driven by the affinity of the two proteins for each other in the presence of ATP. During the completion of the DnaC•DnaB association, ssDNA in the [ATP•DnaC$_6$•ssDNA] complex slides through the subunit interface of the DnaB subunits. (C) The energy that is required to pry open the DnaB•DnaB interface, and to slide ssDNA into the hexamer, is actually provided by the free energy of the DnaB-DnaC association leading to the formation of [ATP•DnaC$_6$•ssDNA•DnaB$_6$] complex. (D) Once the [ATP•DnaC$_6$ssDNA•DnaB$_6$] complex is formed and ATP is hydrolyzed to ADP; the complex dissociates leaving the DnaB hexamer as a hexameric ring around the ssDNA and triggering the ssDNA-dependent ATPase of DnaB protein and its DNA helicase function. The complex, DnaB6•ssDNA is stable because of the lack of energy that is required to pry open DnaB•DnaB interface.

(c) DnaC Protein Assisted ssDNA Binding by DnaB Helicase: The DnaC protein binds to DnaB$_6$ hexamer in the N-terminal domain α as a trimer of three dimers [19, 95]. Recent conformational analysis of DnaB protein in complex with various substrates indicated that the α domain appear to undergo major conformational changes during its catalytic cycle [83]. Electron microscopic studies clearly demonstrate the organization of the DnaB$_6$•DnaC$_6$ complex as a ring of six DnaC attached to the N-terminal α domain face of the DnaB$_6$ ring [95]. In addition to inhibition of the ATPase activity of DnaB, DnaC may also block any conformational changes during the catalytic cycle of DnaB and thereby modulating the helicase function of DnaB. We have proposed two possible avenues for the formation of [ATP•DnaB$_6$•DnaC$_6$•ssDNA] complex. One possible scenario is the formation of the DnaB$_6$•DnaC$_6$ complex is followed by DNA binding in the presence of ATP and the other is through the formation of [ATP•DnaC$_n$•ssDNA] complex and followed by assembly of DnaB$_6$ hexamer to the complex. Once the [ATP•DnaB$_6$•DnaC$_6$•ssDNA] complex is formed and ATP is hydrolyzed to ADP, the complex dissociates leaving the DnaB hexamer as a hexameric ring around the ssDNA triggering the DNA dependent ATPase of DnaB and the DNA helicase function.

(vii) DnaB Helicase as the Organizer of the Replication Fork: Chromosomal DNA replication is initiated in *E. coli* at a single replication origin, *ori*C. DnaA protein activates the origin of DNA replication. Its major function in *E. coli* is to locate the 245 bp *ori*C sequences in *E. coli*, bind to four AT rich sequences and open three 13-mer sequences downstream for the primosome to assemble. Bramhill and Kornberg [36, 37] demonstrated that 20 to 40 monomers of DnaA proteins assemble in the *ori*C sequence prior to its activation. Although DnaA is required for the assembly of the preprimosome, the DnaC protein is essential in *E. coli*. Therefore, DnaA alone cannot create the preprimosome by recruiting DnaB. As DnaC forms a complex with DnaB, the general notion is that DnaA transfers DnaB$_6$•DnaC$_6$ complex (BC complex) to *ori*C [96-98]. However, DnaC itself binds ssDNA in an ATP dependent manner and appears to form a hexamer on the ssDNA template [99]. It is possible that ssDNA bound DnaC then attracts and transfers DnaB to ssDNA. It is more than likely that *ori*C binds DnaA with DnaB, DnaC or both and helps form the DnaA•DnaB•DnaC (ABC) complex at *ori*C creating the replication bubble with forks.

Earlier studies demonstrated that DnaC binds ATP and in the BC complex, it forms the dead-end ternary complex, DnaB$_6$•DnaC$_6$•ATP, by inhibiting the ATPase activity of DnaB shares in ATP binding [5]. Photo-crosslinking studies indicate that DnaC could be photocrosslinked to ATP in the ternary complex. Incidentally, in the DnaB•DnaC complex, ATP appears to be shared between the two proteins [5].

Consequently, by binding ATP, DnaC blocks many functions of DnaB protein. These functions are not activated until DnaC is dissociated from the ABC complex. Recent studies demonstrate that DnaC hydrolyzes the bound ATP molecule forming DnaB$_6$•DnaC$_6$•ADP complex, which likely leads to the dissociation of the complex and departure of DnaC from the ABC complex at oriC [100]. Dissociation of DnaC leads to the initiation of the assembly of the replisome through the actions of DnaB, which is firmly bound to oriC. Initiations of DNA replication in bacteriophage λ and plasmid pRK2 appear to follow similar strategies [13, 101-105].

(viii) DnaB Helicase Act As A Switch for Replication Initiation in Competing DNA Replication Systems: The initiation of DNA replication in bacteriophage λ, the process mirrors that of its *E. coli* host (Figure 6) [101, 102].

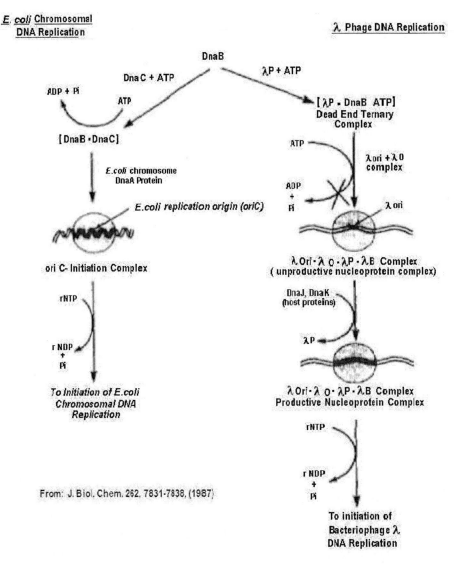

Figure 6. Regulation of DnaB Protein's Functions by DnaC and λP Proteins. A scheme proposing possible modes of regulation of DnaB protein by DnaC and λP proteins in the initial events of initiation of replication of *E. coli* and λ bacteriophage chromosomes. rNTP and rNDP, ribonucleotide tri- and diphosphates, respectively [5].

After infecting *E. coli*, bacteriophage λ expresses two of its own replication proteins: λO and λP. λO protein is the functional equivalent of the *E. coli* DnaA protein. Its role is to bind to the DNA replication origin of λ, λori, and form the λO•λori complex similar to the DnaA•oriC complex. On the other hand, λP protein is the functional equivalent of the DnaC protein as described above. It competes with host DnaC protein in capturing DnaB protein.

Similar to DnaC protein, λP forms DnaB•λP•ATP ternary complex [5]. The ATPase activity of DnaB is completely inhibited in this complex. This ternary complex is recognized and joins the λO•λori complex [101, 102]. Therefore, like the ABC complex of *E. coli* (oriC•DnaA•DnaB•DnaC), the λori•λO•DnaB•λP complex is formed. In *E. coli,* departure of DnaC is necessary for the initiation of DNA replication. This is followed by the joining of DnaG primase to the complex creates a functional open complex and priming system at the origin of replication. In the case of phage λ, departure of λP and joining of DnaG primase to the λ-complex creates the system that primes DNA replication of λ, as shown in Figure 6 [5].

Interestingly, a similar initiation mechanism for pRK2 broad host range plasmid DNA replication has been reported [13, 104, 105]. The pRK2 plasmid in *E. coli* normally requires DnaA protein activation of its origin or replication. However, pRK2 also encodes an initiator protein of its own, Trf-A, which exists in two forms Trf-A33 and Trf-A44, one of which is necessary for origin activation and DnaB recruitment. However, TrfA-44 can independently interact with the DnaB protein of *Pseudomonas putida* and *Pseudomonas aeruginosa* and create a pre-priming open complex, bypassing the DnaA involvement. Therefore, the TrfA-44 initiation protein of pRK2 can act like DnaC protein of *E. coli* and recruit DnaB protein to the RK2 replication origin by a DnaA-independent process.

Thus, the multifunctional DnaB protein appears to play a major "switch" role and decides which activated origin (initiator protein•origin complex) will undergo replication initiation. It is very likely that the free energies (ΔG^o) of interaction between DnaB and its partners like DnaC, λP, Trf-A etc. play a significant role in directing the DnaB protein to their cognate activated origin. Perhaps the replication systems in prokaryotes have evolved to this mechanism. The DnaB protein is responsible for the assembly of the primosome. Thus, once DnaB protein is recruited followed by the recruitment of DnaG primase to the pre-priming complex, priming of the leading and lagging strands is initiated. In some ways, DnaB also mediates the next important event in the replication fork, which is the joining of DNA polymerase III holoenzyme (holoenzyme) to the priming complex leading to the formation of what Arthur Kornberg called "Replisome" [106]. The DNA polymerase III holoenzyme is recruited and stabilized in the replisome by the direct physical interaction of DnaB with the τ-subunit of holoenzyme [29]. With the addition of holoenzyme and formation of the replisome, the replication fork is then ready to replicate the leading and lagging strands. Therefore, recruitment of DnaB protein to an initiated origin is the most pivotal event in the initiation of DNA replication.

One of the most important partners of DnaB protein in the *E. coli* replication fork is the DnaG primase [93, 94, 107-111]. DnaG primase and DnaB protein appear to interact both physically and functionally. Both of these interactions are well documented and have been studied *in vitro*. Arai and Kornberg first demonstrated the interaction of DnaB and DnaG proteins that lead to stimulation of synthesis of primers *in vitro*. This interaction is of great significance for Okazaki fragment synthesis in the lagging strand. Recent studies indicate that such interaction and stimulation of primer synthesis is equally true in other organisms like gram-positive *Bacillus anthracis* (E. Biswas and S. Biswas, unpublished results). The C-terminal end of DnaG protein is involved in the interaction with DnaB protein. Incidentally, unlike the N-terminus of DnaG protein, the C-terminus of DnaG is poorly conserved pointing to the possibility that DnaG•DnaB interaction is species specific (Figure 1). Further studies indicated that the interaction of cognate DnaB and DnaG proteins leads to enhancement of primer synthesis; however, interaction of DnaB and DnaG proteins from heterologous organisms may lead to interaction, but negative regulation or inhibition of the primer synthesis. Consequently, the interactions between these two enzymes and the ensuing conformational change(s) that control primer synthesis are complex.

In general, the DnaG-DnaB interaction and intra DnaG•DnaG interaction in the DnaG trimer are both energetically weak. Such weak interactions are important and may play key roles in the priming of the lagging strand. Weak protein-protein interaction with DnaB protein allows multiple DnaG molecules to associate with DnaB. This association positions the DnaG molecules in the appropriate site of the replication fork and lagging strand. In addition, the weak interaction between DnaG and DnaB proteins allow facile dissociation of a primase monomer from the complex, whenever needed. Thus, the DnaG•DnaB complex is a dynamic assembly in the replisome (Figure 7) [111].

It is probable that the ssDNA-binding site in DnaG aligns with the ssDNA-binding site of the DnaB dimeric unit in the DnaB•DnaG complex, which enables rapid movement of the ssDNA and the replication apparatus. One hypothesis is that DnaG does not synthesize primer while in the DnaB•DnaG complex in the replisome, but DnaG breaks away from the complex as a monomer when it detects and binds to a correct priming sequence on the ssDNA template (Figure 7). In this model, at high primase concentration, most of the "correct" priming sites get primed and at low concentration, only few of the "correct" sites are primed. Consequently, the number of primers synthesized on the lagging strand is directly proportional to the concentration of primase in the reaction. Thus, the average length of the Okazaki fragments is proportional to the average gap between neighboring primers, which is inversely proportional to the concentration of the primase in the reaction system. Therefore, at low primase concentrations, Okazaki fragments will be longer, but at high primase concentration, it will be shorter. As described above, the DnaG•DnaB complex is likely a dynamic assembly, which makes the primer synthesis in prokaryotic replisome extremely efficient. The mechanism is such that frequent priming of the lagging strand takes place without impeding the rapid replication fork progression. If primase follows this mechanism of lagging strand priming, the question is whether the lagging strand polymerase also breaks off from the replisome assembly following primase, synthesizes an Okazaki fragment and recycles. However, with a lagging strand loop it is not necessary for polymerase to dissociate from the lagging strand.

The mechanism proposed in the model in Figure 7 could be delineated is as follows. The DnaB hexamer and three DnaG primase monomers form a complex on the lagging strand. Once an appropriate DNA sequence, passing through the toroidal assembly of DnaG•DnaB complex, with presumably slightly higher affinity, it comes in contact with a primase subunit, that primase subunit binds to the ssDNA and breaks off from the complex as a monomer leaving a gap in the primase trimeric ring. This monomeric primase then synthesizes the primer followed by dissociation from the ssDNA. In the mean time, a new primase monomer joins the DnaB•DnaG assembly to fill the gap in the trimeric primase ring and restoring the 3:6 complexes that starts a new cycle of priming. In this model, primase continually recycles and the priming does not impede or stall the movement of the replisome powered by DnaB protein during the actual primer synthesis.

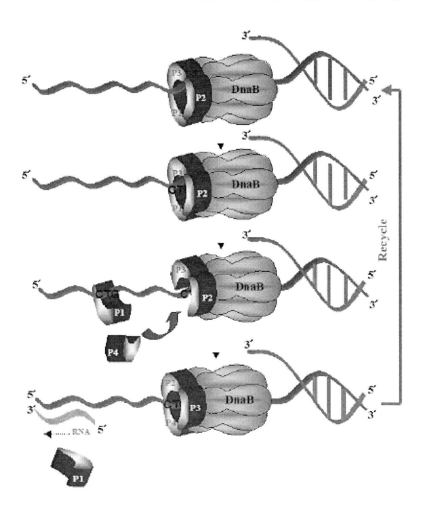

Figure 7. Primase Recycling Model of Distributive RNA Priming of the E. coli Lagging Strand. According to this model, priming in the lagging strand takes as follows: (i), DnaB hexamer (already bound to the lagging strand) and three DnaG primase monomers form a complex on the lagging strand. (ii) One of the primase monomer binds to a preferred triplet sequence, such as CTG, breaks of from the complex and initiates primer synthesis. (iii) A free primase monomer joins the vacant primase monomer position in the DnaB•DnaG complex. (iv) Primase molecule completes primer synthesis and dissociates from the primed lagging strand. A new cycle of priming begins.

(ix) DnaB Helicase as the Engine of the Replisome: DnaB has been called an engineer because it restructures the ssDNA in such a manner that DnaG primase can synthesize the RNA primers and holoenzyme can extend the primers and synthesize DNA in the leading and lagging strands [106, 112]. It has also described as the mobile replication promoter because of its powerful DNA dependent ATPase activity. It was clearly understood that DnaB, with its ATP hydrolysis power, could move the replisome, hence mobile replication promoter [8, 24]. Thus, we can translate DnaB protein's role as a mobile replication promoter to the role of DnaB as an engine of the replisome. DnaB may act simultaneously as the engine and the engineer of the replication fork. DnaB helicase has potent ATPase activity even in the absence of DNA, which can be referred to as an idle ATPase activity. As a result, it is quite

difficult to assess the extent of ATP hydrolysis required for its translocation on DNA. However, other DnaB proteins e.g., DnaB protein of *Bacillus anthracis* (DnaB$_{BA}$) displays minimal idle ATPase activity [136]. DnaB$_{BA}$ is significantly stimulated by long ssDNA such as M13mp19 ssDNA, but not as much by small oligonucleotides, even though DnaB$_{BA}$ is capable of binding short (≥15 bp) oligonucleotides and it does not require ATP for DNA binding [136]. Thus, DnaB$_{BA}$ binds DNA without ATP and does not hydrolyze ATP significantly when bound to a short oligonucleotide without much room to move, but the rate of hydrolysis increases significantly when bound to long ssDNA with room for translocation. Therefore, ATP hydrolysis with DnaB$_{BA}$ is directly connected to its translocation on long ssDNA or its function as an engine.

(x) DnaB Helicase as Driver of the Replisome from the Beginning to the End: The *E. coli* has a single origin of replication where DNA replication is initiated and at the opposite end of the circular chromosome, DNA replication ends. DNA replication is specifically terminated at specialized sequences called replication termini (Ter) [113-118]. During initiation of DNA replication at oriC, two replication forks are initiated and these two forks migrate at opposite directions presumably at the same speed, ending at the Ter sequence. Several laboratories have shown that the replication termination proteins of *E. coli* and *B. subtilis* cause unidirectional arrest of the replicative helicase DnaB upon binding to the Ter sequences [119-122]. The replication terminator protein (RTP) of *B. subtilis* and that of *E. coli*, called Tus have very different structures, however, both proteins interact in vitro with their cognate binding sites to arrest DnaB helicase [122, 123].

The Tus-Ter complex arrests only some DNA helicases such as DnaB but not others such as PcrA, helicase I, UvrD helicase in vitro [124]. It is now clear that Tus physically interacts with DnaB. Mutations on the blocking face of Tus reduce these interactions *in vitro* and these mutants are defective in arresting DnaB *in vitro* and the replication forks *in vivo*. Perhaps, both Tus-DnaB protein-protein interaction and Tus-Ter binding are necessary for polar fork arrest. It is also possible that polarity is generated strictly by DNA-protein interaction caused by helicase-mediated remodeling of Ter-Tus complex. Two opposing forks approaching the non-blocking end are postulated to melt the DNA and dislodge Tus from the Ter site.

It has been reported recently that the Tus-*Ter* complex promoted polar arrest of energy-dependent translocation of DnaB on the DNA duplex but does not cause net DNA unwinding or formation of a transient denaturation bubble at the *Ter* site [125]. In summary, the functions of DnaB protein starts at the initiation of chromosomal DNA replication at the origin, *oriC* and it continues through priming, elongation, and ends at the termination of chromosomal DNA replication at *ter* sequence by the Tus protein bound to the *ter* site.

VI. Mechanism of DNA Unwinding

Although the crystal structures of several hexameric DNA helicases, including DnaB protein from *Bacillus stearothermophilus* and *Thermus aquaticus* have been solved, [67, 109, 126]. The mechanism of duplex DNA unwinding by any one of these hexameric DNA helicases remains in the dark. Many hypothetical models have been proposed but most of these models require experimental validation. Our previous studies led to a simple mechanism of DNA unwinding by *E. coli* DnaB helicase. A complete account of this model is given below.

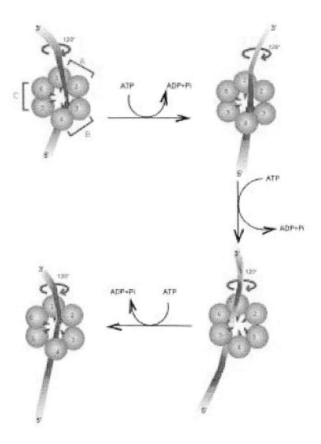

Figure 8. A Hypothetical Model of DNA Unwinding by DnaB Helicase in the Lagging Strand of the Replication Fork [55]. The proposed model is based on our finding that there are three ssDNA binding sites and three nucleotide-binding sites per DnaB hexamer and suggests the possible roles of the three binding sites in DNA unwinding. The view of the hexamer-DNA complex is from the top or 3' →5' directions. The DNA strand shown here is the lagging strand of the replication fork and is presented as a ribbon and colored red on one side and blue on the other. The monomers are labeled consecutively in a clockwise manner from 1 to 6, and this order is maintained throughout this diagram. A, B, and C refers to putative dimeric units of monomers 1 + 2, 3 + 4, and 5 + 6, respectively, each containing a dimerization site and a DNA binding motif. Due to a 120° angle between the neighboring DNA binding sites, movement of ssDNA from one binding site to the other must involve an exact 120° rotation of the plane of the ssDNA molecule; hydrolysis of ATP molecule(s) provides the energy for the rotation of ssDNA by 120°; the DnaB hexamer in each complete cycle should hydrolyze at least three ATP molecules and should rotate the plane of the ssDNA by 3 X 120° or 360°.

A Proposed Model of DNA Unwinding by the DnaB Helicase: Studies carried out in several laboratories strongly support the notion that DnaB-family helicases are homo-hexameric in nature [20, 43, 50, 54, 127-130]. These studies and those presented here have delineated the following parameters of the DnaB helicase: (i) the hexamer is comprised of three apparent dimeric units; (ii) nucleotide binding studies have established that three of the six potential nucleotide-binding sites participate actively in DnaB hexamer, possibly due to reduced nucleotide binding affinity of the subunits adjacent to the nucleotide-bound subunit in the hexamer [17]; and (iii) the polarity of movement is 5'→3'.

The observed nucleotide and ssDNA binding are consistent with the following spatial geometry of the hexamer. If the six subunits in the DnaB hexamer are numbered 1-6, then the occupied subunits would be either 1-3-5 or 2-4-6 (figure 8). Three apparent dimeric units, as shown here, are 1-2 (A), 3-4 (B), and 5-6 (C), each of these units contain one ssDNA binding site and one active ATP hydrolysis site.

A hypothetical model was proposed based on the aforementioned observations and presented in Figure 8. The important steps of DNA unwinding in this model are as follows: (i) apparent dimers in the DnaB hexamer create three independent DNA binding sites, which are oriented at 120° angles to each other; (ii) due to a 120° angle between the neighboring DNA binding sites, movement of ssDNA from one to the other binding site must involve exactly a 120° rotation of the plane of the ssDNA molecule; (iii) hydrolysis of ATP and movement of ssDNA from one dimer to the next would result in the rotation of the plane of ssDNA by 120°; (iv) the DnaB hexamer in each complete cycle should hydrolyze several ATP molecules and rotate the plane of the ssDNA by 3 x 120° or 360°; and (v) to unwind DNA, ssDNA should move in a clockwise manner. Apparently, the DnaB hexamer would be prohibited from moving DNA in an anticlockwise manner, as it will lead to further twisting of DNA and the resultant strain would hinder further movement in that direction. This cycle will continue as long as ATP (or rNTPs) and double stranded DNA are available.

The mechanism of unwinding proposed here requires binding of DnaB helicase only to the lagging strand of the fork. In the replication fork with a closed circular DNA template, uninterrupted helicase action would require the function of a DNA topoisomerase upstream of the helicase.

The DnaB helicase is a potent ssDNA dependent ATPase and consequently, it is capable of attaining a very rapid rate of DNA unwinding. The rate of DNA unwinding will also be dependent on other replication proteins such as DNA primase, DNA polymerase III holoenzyme, topoisomerase, and ssDNA binding protein. The model presented here can be applied equally well to other hexameric DNA helicases. Recent data from various laboratories suggest that the T7gp4 DNA helicase and T4 gene 41 helicase act in an identical manner to that of DnaB helicase [129, 131-133]. Indeed, T7gp4 DNA helicase has a very similar hexameric structure, nucleotide (dTTP) binding properties, and perhaps DNA binding properties as judged by electron microscopic studies [59, 134]. Notarnicola et al., have recently demonstrated that mutants of T7gp4 DNA helicase defective in hexamer formation are also defective in nucleotide hydrolysis, ssDNA binding, and DNA unwinding that is in excellent agreement with results presented here [129]. This data perhaps indicate that the ssDNA binding also requires two adjacent subunits supporting three ssDNA binding sites per hexamer.

Notarnicola et al., have also shown that the T7gp4 helicase can also form dimers and proposed two separate and distinct protein-protein interfaces [129]. This proposed model is consistent with and brings together our current knowledge regarding the ATP hydrolysis, DNA binding, and structure of DnaB protein as they relate to DNA unwinding in the double helix.

VII. Bacterial DnaB Helicases from Gram-Positive and Gram-Negative Bacteria

The replicative DNA helicases (DnaB Orthologs) have been studied in other prokaryotes such as: *Helicobacter pylori, Bacillus anthracis, Bacillus Stearothermophilus, Bacillus subtilis, and Staphylococcus aureus* with high degree of sequence homology (Figure 1). However, despite the sequence homology, these DnaB helicases differ, in some cases significantly, in their enzymatic activities and mechanism. It is likely that these enzymatic and mechanistic differences evolved to suit or take advantage of their unique growth conditions.

(i) DnaB Helicase of Helicobacter Pylori [127]: *Helicobacter Pylori* is a gram-negative bacterial pathogen and its DnaB helicase (DnaB$_{HP}$) is hexameric [130]. It is anticipated that the DNA replication process of *H. pylori* occurs in a similar manner to *E. coli*. However, the replicative enzymes involved in this process have their own unique characteristics. For instance, *H. pylori* does not appear to have a DnaC homolog, which is essential for DnaB loading onto the DNA replication origin (*oriC*) [127]. DnaB$_{HP}$ apparently does not need a DnaC protein for its function and it can autonomously load on to *oriC* [130]. It appears to interact with *E. coli* DnaC as it forms a complex and co-elutes with DnaB$_{HP}$ in gel filtration analysis. The results of this study suggested that these proteins may physically interact with each other [127].

The N-terminal region consists of amino acid residues 1-120, whereas the C-Terminal domain can be found within the residues 175 and 488 [130]. A unique feature of DnaB$_{HP}$ is that the extreme C-Terminal domain possesses a supplementary 34-amino acid residue insertion region [127, 130] and can be found within residues 400-403 [130]. The additional 34 amino acid residues at the C-Terminal region is believed to play an integral role in the structural integrity and function of DnaB$_{HP}$ [130]. Studies have demonstrated that the N-Terminal domain is not essential for *in vitro* helicase activity, but required for conformational changes of the hexamer [130]. Studies have also shown that the C-terminal domain and the linker region are vital for DNA helicase activity [130]. The linker region is also important for intermolecular interaction and structural flexibility required to stabilize hexameric conformations that are essential to the helicase activity [130]. There is diminutive homology at the amino acid level of the N-terminal domain among the different DnaB helicases, but there is more prominent homology at the C-terminal helicase domain [130, 135]. Studies have shown that DnaB$_{HP}$ shows an overall 32% identity and 57% homology with DnaB$_{EC}$ [127, 130].

(ii) DnaB helicase of Bacillus anthracis [136]: The *B. anthracis* orthologs of DnaB$_{EC}$ with 27% identity, 23% similarity, 61% overall homology. However, DnaB$_{BA}$ lacks 17 N-terminal amino acid residues that are present in *DnaB$_{EC}$* and it has a lower degree of homology in the N-terminus and an appreciably higher degree of homology at the C-terminus.

The genome of the gram-positive pathogen has been completely sequenced and two genes with homology to *E. coli* DnaB gene have been identified: BAS0880 and BAS5321. These genes have been analyzed by *in vivo* gene disruption. To examine the roles of each gene in chromosomal DNA replication, an allele exchange experiment was performed. The results of this experiment illustrated that BAS5321 was the essential gene. The BAS5321 ORF is 1359 base pairs and codes for the DnaB$_{BA}$, a soluble 50-kDa polypeptide comprising of 453 amino acid residues [136].

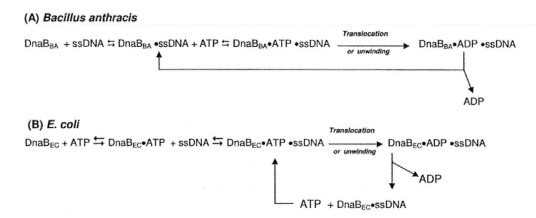

Scheme II. DNA Binding Cycles of DnaB Helicases. (A) DNA binding cycle for DnaB$_{EC}$ where nucleotide binding is a prerequisite for ssDNA could be bound. (B) DNA binding cycle for DnaB$_{BA}$ where DNA binding is independent of nucleotide binding status of DnaB$_{BA}$ protein [136].

An analysis of DnaB$_{BA}$ ATPase activity revealed a very weak ssDNA independent ATPase activity. The ATPase activity was stimulated by ssDNA cofactors, a property common to DNA helicases. Kinetic results demonstrated that the stimulation of ATPase activity of DnaB$_{BA}$ was dependent on the length of ssDNA cofactors and not just DNA binding, as in *E. coli* DnaB. The maximum ATPase activity was observed with long ssDNA templates. This feature appears to be a unique property of the *B. anthracis* enzyme. Even though, DnaB$_{BA}$ has significantly lower ATPase activity, but considerably higher DNA helicase activity than DnaB$_{EC}$ of *E. coli*. The specific helicase activity of DnaB$_{BA}$ was approximately 50-fold higher than DnaB$_{EC}$ and further analysis of dATPase activity showed that DnaB$_{BA}$ was able to hydrolyze ATP or dATP with equal efficiency unlike DnaB$_{EC}$. The DNA helicase activity of DnaB$_{BA}$ is strictly ATP or dATP dependent. When compared, dATP functioned as a better cofactor than ATP. On the contrary, non-adenine nucleotides like GTP, CTP, and UTP were unable to replace ATP/dATP in the DNA helicase activity. Perhaps these nucleotides may not bind to DnaB$_{BA}$ or it cannot hydrolyze them. In contrast to DnaB$_{EC}$ and other DnaB helicases, DnaB$_{BA}$ can bind to ssDNA without the presence of nucleotides (Scheme II). This is significant mechanistic difference with other DnaB helicases. Once the DnaB$_{BA}$•ssDNA complex is formed, the complex binds to ATP. Following ATP binding, translocation is initiated and ATP is hydrolyzed to provide energy for translocation as shown in the equation below [136].

Similar to DnaB$_{HP}$, DnaB$_{BA}$ lacks the DnaC protein and is capable of binding ssDNA autonomously. During DnaB$_{BA}$'s accelerated replication process, the polarity of migration occurs in the usual 5'→3' direction. *In vitro* studies also demonstrate that DnaB$_{BA}$ interacts with *B. anthracis* DnaG primase (DnaG$_{BA}$) and stimulates the RNA primer synthesis while heterologous *E. coli* DnaB helicase inhibits primer synthesis [137].

(iii) DnaB helicase of Bacillus stearothermophilus [138, 139]: *B. stearothermophilus* is a gram-positive, high-temperature bacterium. Analysis of its hexameric replicative helicase, DnaB$_{BS}$ is a polypeptide of 454 amino acids with regions of homology and is 45% identical and 69% similar to *E. coli* DnaB [138, 139]. By contrast, DnaB$_{BS}$ is 82% identical and 92% similar to *B. subtilis*. The 33kDa C-Terminal domain (50% identity and 73% similarity to *E. coli*) shows a greater degree of conservation than the N-Terminal domain (36% identical and 61% similar). Conserved helicase motifs in the C-Terminus may account for the high degree of sequence conservation in this domain.

Site-directed mutagenesis studies on the conserved amino acid residues within motifs H1, H1a, H2 and H3 of the DnaB$_{BS}$ helicase identified the residues that coordinate the Mg^{2+} ions essential for protein activity. Interestingly, these inactive proteins were still able to form stable DnaB$_{BS}$ hexamers. This supports the notion that Mg^{2+} ion is not necessary for stable hexamer formation. Further studies target a glutamic acid (E241) in DnaB$_{BS}$. The mutation of E241 led to a dramatic reduction in ATPase activity; there was no detectable helicase activity and minor alteration in DNA binding affinity. DnaB$_{BS}$ helicase possesses a highly conserved glutamine residue (Q362 and Q254, respectively), which is believed to participate in coupling conformational changes to ATP binding/hydrolysis [139, 140]. A mutation of the region affects the ability of DnaB$_{BS}$ to bind both ssDNA and dsDNA in response to nucleotide binding. It has been observed that in the absence of nucleotide the wild-type DnaB$_{BS}$ and the mutant Q362A proteins bind ssDNA or dsDNA weakly. However, in the presence of a nucleotide, the protein's affinity increases significantly. Studies have also illustrated that the primase-induced stimulation of DnaB activity is mediated by the same conformational coupling pathway as the nucleotide-induced modulation of DNA binding.

(iv) Staphylococcus aureus and Bacillus subtilis DnaB Helicases [141-145]: When compared to gram-positive *B. subtilis*, the genomic sequences of *S. aureus* replicative proteins appear similar [141]. The nomenclature and functional activities of these proteins share similar characteristics as well. However, when compared to the replicative proteins of *E. coli* and most other microbial system, notable differences are observed. Unlike *E. coli* DnaB helicase, the DnaB helicases in *S. aureus* and *B. subtilis* are also referred to as DnaC. The *S. aureus* DnaB helicase (DnaB$_{SA}$) shares a 44% homology with *E. coli* DnaB helicase and 58% homology with *B. subtilis* DnaB helicase DnaB$_{BS}$. Furthermore, the orthologs of *E. coli* DnaC protein found in *S. aureus* and *B. subtilis* share the responsibility of loading the DnaB$_{SA}$ helicase to the target regions of DNA [141-145].

The cellular functions of DnaB$_{SA}$ have been probed through studies of positive mutants in DNA replication. These studies demonstrated that DnaB$_{SA}$ functions in both the initiation and elongation stages. Interestingly it appears that DnaB$_{SA}$ may support other functions in *S. aureus*. Studies of UV sensitive mutants support that DnaB$_{SA}$ plays an important role in DNA recombination and/or repair.

Conclusions

Early genetic studies correctly identified the *dnaB* gene as important in the initiation and elongation stages of *E. coli* chromosomal DNA replication. Biochemical studies of DnaB protein slowly revealed its multifaceted cellular functions, one of which is the DNA unwinding.

DnaA initiator protein activates *ori*C, but DnaB protein assembles the replisome by specific protein-protein interactions with DnaG primase and pol III holoenzyme. Thus, DnaB acts as a scaffold for the replisome to form during initiation. With its strong interaction with ssDNA, it provides a stable attachment of the entire replisome to the forked DNA during the elongation phase. Using ATPase activity, it provides power that translocates the entire replisome at an exceptional speed. It is also clear that DnaB protein modulates the rate of synthesis of primers by DnaG primase as well as Okazaki fragment length and synthesis. In addition, the rate of overall DNA synthesis by the holoenzyme is also modulated by DnaB protein. At the end, when DNA synthesis of both parental strands are complete at the termination site, *ter*, it ends the process of replication through its interaction with the Tus-ter complex.

Thus, the role of DnaB protein in *E. coli* DNA replication starts at the initiation and continues through elongation to the termination. DnaB protein acts as the organizer of the DNA replication in addition to carrying out its enzymatic roles. DnaB protein is likely the single most important protein/enzyme in the DNA replication in *E. coli*. Thus, DnaB protein could be a valuable target for developing a new generation of antibiotics [146].

Orthologs of *E. coli* DnaB protein have been cloned, purified, and studied from a number of prokaryotic organisms with unique mechanistic features, which evolved as a result of each organism's unique growth conditions. These studies help to understand the global and diverse mechanism of DnaB protein. Much remains to be understood regarding this protein family which plays a critical first step in DNA replication in addition to its role as the mobile replication promoter.

Acknowledgements

Authors gratefully acknowledge support of this work by grants from the National Institute of Allergy & Infectious Diseases, National Institutes of Health (STTR AI064974, PI:Donald Moir, Microbiotix Inc., Worcester, MA), and the UMDNJ Foundation. Authors wish to thank Leelabati Biswas and Julia Crawford for critical editorial review of the manuscript and former members of Biswas laboratory for their many important contributions.

References

[1] Kornberg, A., and Baker, T. A. (1992) DNA Replication, W.H. Freeman and Co., New York, NY.

[2] Hirota, Y., Mordoh, J., and Jacob, F. (1970) On the process of cellular division in Escherichia coli. 3. Thermosensitive mutants of Escherichia coli altered in the process of DNA initiation, *J. Mol. Biol. 53*, 369-387.

[3] Mordoh, J., Hirota, Y., and Jacob, F. (1970) On the process of cellular division in Escherichia coli. V. Incorporation of deoxynucleoside triphosphates by DNA

thermosensitive mutants of Escherichia coli also lacking DNA polymerase activity, *Proc. Nat'l. Acad. Sci. USA 67*, 773-778.

[4] Sumida-Yasumoto, C., Yudelevich, A., and Hurwitz, J. (1976) DNA synthesis in vitro dependent upon phiX174 replicative form I DNA, *Proc. Nat'l. Acad. Sci. USA 73*, 1887-1891.

[5] Biswas, S. B., and Biswas, E. E. (1987) Regulation of dnaB function in DNA replication in Escherichia coli by dnaC and lambda P gene products, *J. Biol.Chem. 262*, 7831-7838.

[6] LeBowitz, J. H., and McMacken, R. (1986) The Escherichia coli dnaB replication protein is a DNA helicase, *J. Biol.Chem. 261*, 4738-4748.

[7] Patel, S. S., and Picha, K. M. (2000) Structure and function of hexameric helicases, *Annu. Rev. Biochem. 69*, 651-697.

[8] McMacken, R., Ueda, K., and Kornberg, A. (1977) Migration of Escherichia coli dnaB protein on the template DNA strand as a mechanism in initiating DNA replication, *Proc. Nat'l. Acad. Sci. USA 74*, 4190-4194.

[9] Schuster, H., Schlicht, M., Lanka, E., Mikolajczyk, M., and Edelbluth, C. (1977) DNA synthesis in an Escherichia coli dnaB dnaC mutant, *Mol. Gen. Genet. 151*, 11-16.

[10] Ueda, K., McMacken, R., and Kornberg, A. (1978) dnaB protein of Escherichia coli. Purification and role in the replication of phiX174 DNA, *J. Biol.Chem. 253*, 261-269.

[11] Arai, N., Arai, K., and Kornberg, A. (1981) Complexes of Rep protein with ATP and DNA as a basis for helicase action, *J Biol Chem 256*, 5287-5293.

[12] Ratnakar, P. V., Mohanty, B. K., Lobert, M., and Bastia, D. (1996) The replication initiator protein pi of the plasmid R6K specifically interacts with the host-encoded helicase DnaB, *Proc. Nat'l. Acad. Sci. USA 93*, 5522-5526.

[13] Konieczny, I., and Helinski, D. R. (1997) Helicase delivery and activation by DnaA and TrfA proteins during the initiation of replication of the broad host range plasmid RK2, *J. Biol.Chem. 272*, 33312-33318.

[14] Dodson, M., Dean, F. B., Bullock, P., Echols, H., and Hurwitz, J. (1987) Unwinding of duplex DNA from the SV40 origin of replication by T antigen, *Science 238*, 964-967.

[15] Venkatesan, M., Silver, L. L., and Nossal, N. G. (1982) Bacteriophage T4 gene 41 protein, required for the synthesis of RNA primers, is also a DNA helicase, *J. Biol.Chem.257*, 12426-12434.

[16] Matson, S. W., Tabor, S., and Richardson, C. C. (1983) The gene 4 protein of bacteriophage T7. Characterization of helicase activity, *J. Biol.Chem. 258*, 14017-14024.

[17] Biswas, E. E., Biswas, S. B., and Bishop, J. E. (1986) The dnaB protein of Escherichia coli: mechanism of nucleotide binding, hydrolysis, and modulation by dnaC protein, *Biochemistry 25*, 7368-7374.

[18] Mallory, J. B., Alfano, C., and McMacken, R. (1990) Host virus interactions in the initiation of bacteriophage lambda DNA replication. Recruitment of Escherichia coli DnaB helicase by lambda P replication protein, *J. Biol.Chem. 265*, 13297-13307.

[19] Biswas, E. E., and Biswas, S. B. (1999) Mechanism of DnaB helicase of Escherichia coli: structural domains involved in ATP hydrolysis, DNA binding, and oligomerization, *Biochemistry 38*, 10919-10928.

[20] Biswas, S. B., Chen, P. H., and Biswas, E. E. (1994) Structure and function of Escherichia coli DnaB protein: role of the N-terminal domain in helicase activity, *Biochemistry 33*, 11307-11314.
[21] Arai, K., Yasuda, S., and Kornberg, A. (1981) Mechanism of dnaB protein action. I. Crystallization and properties of dnaB protein, an essential replication protein in Escherichia coli, *J. Biol. Chem. 256*, 5247-5252.
[22] Arai, K., and Kornberg, A. (1981) Mechanism of dnaB protein action. II. ATP hydrolysis by dnaB protein dependent on single- or double-stranded DNA, *J. Biol.Chem. 256*, 5253-5259.
[23] Dodson, M., Echols, H., Wickner, S., Alfano, C., Mensa-Wilmot, K., Gomes, B., LeBowitz, J., Roberts, J. D., and McMacken, R. (1986) Specialized nucleoprotein structures at the origin of replication of bacteriophage lambda: localized unwinding of duplex DNA by a six-protein reaction, *Proc. Nat'l. Acad. Sci. USA 83*, 7638-7642.
[24] Arai, K., and Kornberg, A. (1979) A general priming system employing only dnaB protein and primase for DNA replication, *Proc. Nat'l. Acad. Sci. USA 76*, 4308-4312.
[25] Wickner, S., and Hurwitz, J. (1975) Association of phiX174 DNA-dependent ATPase activity with an Escherichia coli protein, replication factor Y, required for in vitro synthesis of phiX174 DNA, *Proc. Nat'l. Acad. Sci. USA 72*, 3342-3346.
[26] Kobori, J. A., and Kornberg, A. (1982) The Escherichia coli dnaC gene product. III. Properties of the dnaB-dnaC protein complex, *J. Biol.Chem. 257*, 13770-13775.
[27] Seitz, H., Weigel, C., and Messer, W. (2000) The interaction domains of the DnaA and DnaB replication proteins of Escherichia coli, *Mol. Microbiol .37*, 1270-1279.
[28] San Martin, M. C., Stamford, N. P., Dammerova, N., Dixon, N. E., and Carazo, J. M. (1995) A structural model for the Escherichia coli DnaB helicase based on electron microscopy data, *J. Struct. Biol. 114*, 167-176.
[29] Kim, S., Dallmann, H. G., McHenry, C. S., and Marians, K. J. (1996) Coupling of a replicative polymerase and helicase: a tau-DnaB interaction mediates rapid replication fork movement, *Cell 84*, 643-650.
[30] Dodson, M., Roberts, J., McMacken, R., and Echols, H. (1985) Specialized nucleoprotein structures at the origin of replication of bacteriophage lambda: complexes with lambda O protein and with lambda O, lambda P, and Escherichia coli DnaB proteins, *Proc. Nat'l. Acad. Sci. USA 82*, 4678-4682.
[31] Arai, K., Low, R., Kobori, J., Shlomai, J., and Kornberg, A. (1981) Mechanism of dnaB protein action. V. Association of dnaB protein, protein n', and other repriming proteins in the primosome of DNA replication, *J. Biol.Chem. 256*, 5273-5280.
[32] Datta, H. J., Khatri, G. S., and Bastia, D. (1999) Mechanism of recruitment of DnaB helicase to the replication origin of the plasmid pSC101, *Proc. Nat'l. Acad. Sci. USA 96*, 73-78.
[33] Gao, D., and McHenry, C. S. (2001) tau binds and organizes Escherichia coli replication proteins through distinct domains. Domain IV, located within the unique C terminus of tau, binds the replication fork, helicase, DnaB, *J. Biol. Chem. 276*, 4441-4446.

[34] Masai, H., Nomura, N., and Arai, K. (1990) The ABC-primosome. A novel priming system employing dnaA, dnaB, dnaC, and primase on a hairpin containing a dnaA box sequence, *J. Biol.Chem. 265*, 15134-15144.

[35] Wickner, S., and Hurwitz, J. (1975) Interaction of Escherichia coli dnaB and dnaC(D) gene products in vitro, *Proc. Nat'l. Acad. Sci. USA 72*, 921-925.

[36] Bramhill, D., and Kornberg, A. (1988) Duplex opening by dnaA protein at novel sequences in initiation of replication at the origin of the E. coli chromosome, *Cell 52*, 743-755.

[37] Bramhill, D., and Kornberg, A. (1988) A model for initiation at origins of DNA replication, *Cell 54*, 915-918.

[38] Sutton, M. D., Carr, K. M., Vicente, M., and Kaguni, J. M. (1998) Escherichia coli DnaA protein. The N-terminal domain and loading of DnaB helicase at the E. coli chromosomal origin, *J. Biol.Chem. 273*, 34255-34262.

[39] Hurwitz, J., Dean, F. B., Kwong, A. D., and Lee, S. H. (1990) The in vitro replication of DNA containing the SV40 origin, *J. Biol.Chem. 265*, 18043-18046.

[40] Wickner, S., and Hurwitz, J. (1975) Interaction of Escherichia coli dnaB and dnaC(D) gene products in vitro, *Proc. Nat'l. Acad. Sci. USA 72*, 921-925.

[41] Wickner, S., Wright, M., and Hurwitz, J. (1974) Association of DNA-dependent and -independent ribonucleoside triphosphatase activities with dnaB gene product of Escherichia coli, *Proc. Nat'l. Acad. Sci. USA 71*, 783-787.

[42] Reha-Krantz, L. J., and Hurwitz, J. (1978) The dnaB gene product of Escherichia coli. II. Single stranded DNA-dependent ribonucleoside triphosphatase activity, *J. Biol.Chem. 253*, 4051-4057.

[43] Reha-Krantz, L. J., and Hurwitz, J. (1978) The dnaB gene product of Escherichia coli. I. Purification, homogeneity, and physical properties, *J. Biol.Chem. 253*, 4043-4050.

[44] Arai, K., Arai, N., Shlomai, J., Kobori, J., Polder, L., Low, R., Hubscher, U., Bertsch, L., and Kornberg, A. (1981) Enzyme studies of phi X174 DNA replication, *Prog. Nucleic Acid Res. Mol. Biol. 26*, 9-32.

[45] Singleton, M. R., Dillingham, M. S., and Wigley, D. B. (2007) Structure and mechanism of helicases and nucleic acid translocases, *Annu. Rev. Biochem. 76*, 23-50.

[46] Arai, K., Arai, N., Shlomai, J., and Kornberg, A. (1980) Replication of duplex DNA of phage phi X174 reconstituted with purified enzymes, *Proc. Nat'l. Acad. Sci. USA 77*, 3322-3326.

[47] Patel, S. S., and Hingorani, M. M. (1993) Oligomeric structure of bacteriophage T7 DNA primase/helicase proteins, *J. Biol.Chem. 268*, 10668-10675.

[48] Patel, S. S., Rosenberg, A. H., Studier, F. W., and Johnson, K. A. (1992) Large scale purification and biochemical characterization of T7 primase/helicase proteins. Evidence for homodimer and heterodimer formation, *J. Biol.Chem. 267*, 15013-15021.

[49] Takata, M., Guo, L., Katayama, T., Hase, M., Seyama, Y., Miki, T., and Sekimizu, K. (2000) Mutant DnaA proteins defective in duplex opening of oriC, the origin of chromosomal DNA replication in Escherichia coli, *Mol. Microbiol. 35*, 454-462.

[50] Dong, F., Gogol, E. P., and von Hippel, P. H. (1995) The phage T4-coded DNA replication helicase (gp41) forms a hexamer upon activation by nucleoside triphosphate, *J. Biol.Chem. 270*, 7462-7473.

[51] Marsin, S., McGovern, S., Ehrlich, S. D., Bruand, C., and Polard, P. (2001) Early steps of Bacillus subtilis primosome assembly, *J. Biol.Chem. 276*, 45818-45825.

[52] Bruand, C., Farache, M., McGovern, S., Ehrlich, S. D., and Polard, P. (2001) DnaB, DnaD and DnaI proteins are components of the Bacillus subtilis replication restart primosome, *Mol. Microbiol. 42*, 245-255.

[53] Sakamoto, Y., Nakai, S., Moriya, S., Yoshikawa, H., and Ogasawara, N. (1995) The Bacillus subtilis dnaC gene encodes a protein homologous to the DnaB helicase of Escherichia coli, *Microbiology 141 (Pt 3)*, 641-644.

[54] Nakayama, N., Arai, N., Kaziro, Y., and Arai, K. (1984) Structural and functional studies of the dnaB protein using limited proteolysis. Characterization of domains for DNA-dependent ATP hydrolysis and for protein association in the primosome, *J. Biol.Chem. 259*, 88-96.

[55] Biswas, E. E., and Biswas, S. B. (1999) Mechanism of DNA binding by the DnaB helicase of Escherichia coli: analysis of the roles of domain gamma in DNA binding, *Biochemistry 38*, 10929-10939.

[56] Walker, J. E., Saraste, M., Runswick, M. J., and Gay, N. J. (1982) Distantly related sequences in the alpha- and beta-subunits of ATP synthase, myosin, kinases and other ATP-requiring enzymes and a common nucleotide binding fold, *EMBO J. 1*, 945-951.

[57] Yang, S., Yu, X., VanLoock, M. S., Jezewska, M. J., Bujalowski, W., and Egelman, E. H. (2002) Flexibility of the rings: structural asymmetry in the DnaB hexameric helicase, *J. Mol. Biol. 321*, 839-849.

[58] Yu, X., Jezewska, M. J., Bujalowski, W., and Egelman, E. H. (1996) The hexameric E. coli DnaB helicase can exist in different Quaternary states, *J. Mol. Biol. 259*, 7-14.

[59] Egelman, E. H., Yu, X., Wild, R., Hingorani, M. M., and Patel, S. S. (1995) Bacteriophage T7 helicase/primase proteins form rings around single-stranded DNA that suggest a general structure for hexameric helicases, *Proc. Nat'l. Acad. Sci. USA 92*, 3869-3873.

[60] Crampton, D. J., Ohi, M., Qimron, U., Walz, T., and Richardson, C. C. (2006) Oligomeric states of bacteriophage T7 gene 4 primase/helicase, *J. Mol. Biol. 360*, 667-677.

[61] VanLoock, M. S., Alexandrov, A., Yu, X., Cozzarelli, N. R., and Egelman, E. H. (2002) SV40 large T antigen hexamer structure: domain organization and DNA-induced conformational changes, *Current Biology 12*, 472-476.

[62] Fouts, E. T., Yu, X., Egelman, E. H., and Botchan, M. R. (1999) Biochemical and electron microscopic image analysis of the hexameric E1 helicase, *J. Biol.Chem. 274*, 4447-4458.

[63] San Martin, C., Radermacher, M., Wolpensinger, B., Engel, A., Miles, C. S., Dixon, N. E., and Carazo, J. M. (1998) Three-dimensional reconstructions from cryoelectron microscopy images reveal an intimate complex between helicase DnaB and its loading partner DnaC, *Structure 6*, 501-509.

[64] Bailey, S., Eliason, W. K., and Steitz, T. A. (2007) The crystal structure of the Thermus aquaticus DnaB helicase monomer, *Nucleic Acids Res. 35*, 4728-4736.

[65] Thirlway, J., and Soultanas, P. (2006) In the Bacillus stearothermophilus DnaB-DnaG complex, the activities of the two proteins are modulated by distinct but overlapping networks of residues, *J. Bacteriol. 188*, 1534-1539.

[66] Oakley, A. J., Loscha, K. V., Schaeffer, P. M., Liepinsh, E., Pintacuda, G., Wilce, M. C., Otting, G., and Dixon, N. E. (2005) Crystal and solution structures of the helicase-binding domain of Escherichia coli primase, *J. Biol.Chem. 280*, 11495-11504.

[67] Bailey, S., Eliason, W. K., and Steitz, T. A. (2007) The crystal structure of the Thermus aquaticus DnaB helicase monomer, *Nucleic Acids Res. 35*, 4728-4736.

[68] Syson, K., Thirlway, J., Hounslow, A. M., Soultanas, P., and Waltho, J. P. (2005) Solution structure of the helicase-interaction domain of the primase DnaG: a model for helicase activation, *Structure 13*, 609-616.

[69] Jezewska, M. J., Rajendran, S., and Bujalowski, W. (1998) Complex of Escherichia coli primary replicative helicase DnaB protein with a replication fork: recognition and structure, *Biochemistry 37*, 3116-3136.

[70] Soultanas, P. (2005) The bacterial helicase-primase interaction: a common structural/functional module, *Structure 13*, 839-844.

[71] Mesa, P., Alonso, J. C., and Ayora, S. (2006) Bacillus subtilis bacteriophage SPP1 G40P helicase lacking the n-terminal domain unwinds DNA bidirectionally, *J. Mol. Biol. 357*, 1077-1088.

[72] Enemark, E. J., and Joshua-Tor, L. (2006) Mechanism of DNA translocation in a replicative hexameric helicase, *Nature 442*, 270-275.

[73] Leipe, D. D., Aravind, L., Grishin, N. V., and Koonin, E. V. (2000) The bacterial replicative helicase DnaB evolved from a RecA duplication, *Genome Res. 10*, 5-16.

[74] Story, R. M., Weber, I. T., and Steitz, T. A. (1992) The structure of the E. coli recA protein monomer and polymer, *Nature 355*, 318-325.

[75] Toth, E. A., Li, Y., Sawaya, M. R., Cheng, Y., and Ellenberger, T. (2003) The crystal structure of the bifunctional primase-helicase of bacteriophage T7, *Mol. cell 12*, 1113-1123.

[76] Arai, K., and Kornberg, A. (1981) Mechanism of dnaB protein action. II. ATP hydrolysis by dnaB protein dependent on single- or double-stranded DNA, *J. Biol.Chem.256*, 5253-5259.

[77] Arai, K., and Kornberg, A. (1981) Mechanism of dnaB protein action. III. Allosteric role of ATP in the alteration of DNA structure by dnaB protein in priming replication, *J. Biol.Chem. 256*, 5260-5266.

[78] Blundell, T. L., Cutfield, J. F., Cutfield, S. M., Dodson, E. J., Dodson, G. G., Hodgkin, D. C., Mercola, D. A., and Vijayan, M. (1971) Atomic positions in rhombohedral 2-zinc insulin crystals, *Nature 231*, 506-511.

[79] Blundell, T. L., Cutfield, J. F., Dodson, E. J., Dodson, G. G., Hodgkin, D. C., and Mercola, D. A. (1972) The crystal structure of rhombohedral 2 zinc insulin, *Cold Spring Harb. Symp. Quant. Biol. 36*, 233-241.

[80] Biswas, S. B., and Biswas-Fiss, E. E. (2006) Quantitative analysis of binding of single-stranded DNA by Escherichia coli DnaB helicase and the DnaB x DnaC complex, *Biochemistry 45*, 11505-11513.

[81] Bujalowski, W., Klonowska, M. M., and Jezewska, M. J. (1994) Oligomeric structure of Escherichia coli primary replicative helicase DnaB protein, *J. Biol.Chem. 269*, 31350-31358.

[82] Jezewska, M. J., Rajendran, S., Bujalowska, D., and Bujalowski, W. (1998) Does single-stranded DNA pass through the inner channel of the protein hexamer in the complex with the Escherichia coli DnaB Helicase? Fluorescence energy transfer studies, *J. Biol.Chem. 273*, 10515-10529.

[83] Flowers, S., Biswas, E. E., and Biswas, S. B. (2003) Conformational dynamics of DnaB helicase upon DNA and nucleotide binding: analysis by intrinsic tryptophan fluorescence quenching, *Biochemistry 42*, 1910-1921.

[84] Biswas, S. B., Flowers, S., and Biswas-Fiss, E. E. (2004) Quantitative analysis of nucleotide modulation of DNA binding by DnaC protein of Escherichia coli, *Biochem. J. 379*, 553-562.

[85] Galletto, R., Jezewska, M. J., and Bujalowski, W. (2003) Interactions of the Escherichia coli DnaB helicase hexamer with the replication factor the DnaC protein. Effect of nucleotide cofactors and the ssDNA on protein-protein interactions and the topology of the complex, *J. Mol. Biol. 329*, 441-465.

[86] Learn, B. A., Um, S. J., Huang, L., and McMacken, R. (1997) Cryptic single-stranded-DNA binding activities of the phage lambda P and Escherichia coli DnaC replication initiation proteins facilitate the transfer of E. coli DnaB helicase onto DNA, *Proc. Nat'l. Acad. Sci. USA 94*, 1154-1159.

[87] Davey, M. J., Fang, L., McInerney, P., Georgescu, R. E., and O'Donnell, M. (2002) The DnaC helicase loader is a dual ATP/ADP switch protein, *The EMBO J. 21*, 3148-3159.

[88] Biswas, S. B., and Biswas, E. E. (1987) Regulation of dnaB function in DNA replication in Escherichia coli by dnaC and lambda P gene products, *J. Biol.Chem. 262*, 7831-7838.

[89] Funnell, B. E., Baker, T. A., and Kornberg, A. (1987) In vitro assembly of a prepriming complex at the origin of the Escherichia coli chromosome, *J. Biol.Chem. 262*, 10327-10334.

[90] Marszalek, J., Zhang, W., Hupp, T. R., Margulies, C., Carr, K. M., Cherry, S., and Kaguni, J. M. (1996) Domains of DnaA protein involved in interaction with DnaB protein, and in unwinding the Escherichia coli chromosomal origin, *J. Biol.Chem. 271*, 18535-18542.

[91] Alfano, C., and McMacken, R. (1989) Ordered assembly of nucleoprotein structures at the bacteriophage lambda replication origin during the initiation of DNA replication, *J. Biol.Chem. 264*, 10699-10708.

[92] Liang, C., Weinreich, M., and Stillman, B. (1995) ORC and Cdc6p interact and determine the frequency of initiation of DNA replication in the genome, *Cell 81*, 667-676.

[93] Lu, Y. B., Ratnakar, P. V., Mohanty, B. K., and Bastia, D. (1996) Direct physical interaction between DnaG primase and DnaB helicase of Escherichia coli is necessary for optimal synthesis of primer RNA, *Proc. Nat'l. Acad. Sci. USA 93*, 12902-12907.

[94] Khopde, S., Biswas, E., and Biswas, S. (2002) Affinity and sequence specificity of DNA binding and site selection for primer synthesis by Escherichia coli primase, *Biochemistry 41*, 14820-14830.

[95] Barcena, M., Ruiz, T., Donate, L. E., Brown, S. E., Dixon, N. E., Radermacher, M., and Carazo, J. M. (2001) The DnaB.DnaC complex: a structure based on dimers assembled around an occluded channel, *EMBO J. 20*, 1462-1468.

[96] Fuller, R. S., Kaguni, J. M., and Kornberg, A. (1981) Enzymatic replication of the origin of the Escherichia coli chromosome, *Proc. Nat'l. Acad. Sci. USA 78*, 7370-7374.

[97] Fuller, R. S., and Kornberg, A. (1983) Purified dnaA protein in initiation of replication at the Escherichia coli chromosomal origin of replication, *Proc. Nat'l. Acad. Sci. USA 80*, 5817-5821.

[98] Fuller, R. S., Funnell, B. E., and Kornberg, A. (1984) The dnaA protein complex with the E. coli chromosomal replication origin (oriC) and other DNA sites, *Cell 38*, 889-900.

[99] Biswas, S. B., Flowers, S., and Biswas-Fiss, E. E. (2004) Quantitative analysis of nucleotide modulation of DNA binding by DnaC protein of Escherichia coli, *Biochem. J. 379*, 553-562.

[100] Davey, M. J., Fang, L., McInerney, P., Georgescu, R. E., and O'Donnell, M. (2002) The DnaC helicase loader is a dual ATP/ADP switch protein, *EMBO J.21*, 3148-3159.

[101] Mallory, J. B., Alfano, C., and McMacken, R. (1990) Host virus interactions in the initiation of bacteriophage lambda DNA replication. Recruitment of Escherichia coli DnaB helicase by lambda P replication protein, *J. Biol.Chem. 265*, 13297-13307.

[102] Alfano, C., and McMacken, R. (1989) Ordered assembly of nucleoprotein structures at the bacteriophage lambda replication origin during the initiation of DNA replication, *J. Biol.Chem. 264*, 10699-10708.

[103] Pinkney, M., Diaz, R., Lanka, E., and Thomas, C. M. (1988) Replication of mini RK2 plasmid in extracts of Escherichia coli requires plasmid-encoded protein TrfA and host-encoded proteins DnaA, B, G DNA gyrase and DNA polymerase III, *J. Mol. Biol. 203*, 927-938.

[104] Doran, K. S., Helinski, D. R., and Konieczny, I. (1999) A critical DnaA box directs the cooperative binding of the Escherichia coli DnaA protein to the plasmid RK2 replication origin, *J. Biol.Chem. 274*, 17918-17923.

[105] Gaylo, P. J., Turjman, N., and Bastia, D. (1987) DnaA protein is required for replication of the minimal replicon of the broad-host-range plasmid RK2 in Escherichia coli, *J. Bacteriol. 169*, 4703-4709.

[106] Kornberg, A. (1984) Enzyme studies of replication of the Escherichia coli chromosome, *Adv. Exp. Med. Biol. 179*, 3-16.

[107] Thirlway, J., and Soultanas, P. (2006) In the Bacillus stearothermophilus DnaB-DnaG complex, the activities of the two proteins are modulated by distinct but overlapping networks of residues, *J. Bacteriol. 188*, 1534-1539.

[108] Su, X. C., Schaeffer, P. M., Loscha, K. V., Gan, P. H., Dixon, N. E., and Otting, G. (2006) Monomeric solution structure of the helicase-binding domain of Escherichia coli DnaG primase, *FEBS J. 273*, 4997-5009.

[109] Bailey, S., Eliason, W. K., and Steitz, T. A. (2007) Structure of hexameric DnaB helicase and its complex with a domain of DnaG primase, *Science 318*, 459-463.

[110] Bhattacharyya, S., and Griep, M. A. (2000) DnaB helicase affects the initiation specificity of Escherichia coli primase on single-stranded DNA templates, *Biochemistry 39*, 745-752.

[111] Mitkova, A. V., Khopde, S. M., and Biswas, S. B. (2003) Mechanism and stoichiometry of interaction of DnaG primase with DnaB helicase of Escherichia coli in RNA primer synthesis, *J. Biol.Chem. 278*, 52253-52261.

[112] Arai, K., Low, R. L., and Kornberg, A. (1981) Movement and site selection for priming by the primosome in phage phi X174 DNA replication, *Proc. Nat'l. Acad. Sci. USA 78*, 707-711.

[113] Mulugu, S., Potnis, A., Shamsuzzaman, Taylor, J., Alexander, K., and Bastia, D. (2001) Mechanism of termination of DNA replication of Escherichia coli involves helicase-contrahelicase interaction, *Proc. Nat'l. Acad. Sci. USA 98*, 9569-9574.

[114] Hidaka, M., Kobayashi, T., and Horiuchi, T. (1991) A newly identified DNA replication terminus site, TerE, on the Escherichia coli chromosome, *J. Bacteriol. 173*, 391-393.

[115] MacAllister, T., Khatri, G. S., and Bastia, D. (1990) Sequence-specific and polarized replication termination in vitro: complementation of extracts of tus- Escherichia coli by purified Ter protein and analysis of termination intermediates, *Proc. Nat'l. Acad. Sci. USA 87*, 2828-2832.

[116] Hidaka, M., Kobayashi, T., Takenaka, S., Takeya, H., and Horiuchi, T. (1989) Purification of a DNA replication terminus (ter) site-binding protein in Escherichia coli and identification of the structural gene, *J. Biol.Chem. 264*, 21031-21037.

[117] Khatri, G. S., MacAllister, T., Sista, P. R., and Bastia, D. (1989) The replication terminator protein of E. coli is a DNA sequence-specific contra-helicase, *Cell 59*, 667-674.

[118] Kobayashi, T., Hidaka, M., and Horiuchi, T. (1989) Evidence of a ter specific binding protein essential for the termination reaction of DNA replication in Escherichia coli, *EMBO J.8*, 2435-2441.

[119] Bussiere, D. E., and Bastia, D. (1999) Termination of DNA replication of bacterial and plasmid chromosomes, *Mol. Microbiol. 31*, 1611-1618.

[120] Duggin, I. G., Andersen, P. A., Smith, M. T., Wilce, J. A., King, G. F., and Wake, R. G. (1999) Site-directed mutants of RTP of Bacillus subtilis and the mechanism of replication fork arrest, *J. Mol. Biol. 286*, 1325-1335.

[121] Wake, R. G. (1997) Replication fork arrest and termination of chromosome replication in Bacillus subtilis, *FEMS Microbiol. Lett. 153*, 247-254.

[122] Sahoo, T., Mohanty, B. K., Lobert, M., Manna, A. C., and Bastia, D. (1995) The contrahelicase activities of the replication terminator proteins of Escherichia coli and Bacillus subtilis are helicase-specific and impede both helicase translocation and authentic DNA unwinding, *J. Biol. Chem. 270*, 29138-29144.

[123] Gautam, A., Mulugu, S., Alexander, K., and Bastia, D. (2001) A single domain of the replication termination protein of Bacillus subtilis is involved in arresting both DnaB helicase and RNA polymerase, *J. Biol. Chem. 276*, 23471-23479.

[124] Neylon, C., Kralicek, A. V., Hill, T. M., and Dixon, N. E. (2005) Replication termination in Escherichia coli: structure and antihelicase activity of the Tus-Ter complex, *Microbiolo. Mol. Biol. Rev. 69*, 501-526.

[125] Bastia, D., Zzaman, S., Krings, G., Saxena, M., Peng, X., and Greenberg, M. M. (2008) Replication terminatoion mechanism as revealed by Tus-mediated polar arrest of a sliding helicase, *Proc. Nat'l. Acad. Sci. USA*, 0000-0000.

[126] Kaplan, D. L., and Steitz, T. A. (1999) DnaB from Thermus aquaticus unwinds forked duplex DNA with an asymmetric tail length dependence, *J. Biol. Chem. 274*, 6889-6897.

[127] Nitharwal, R. G., Paul, S., Dar, A., Choudhury, N. R., Soni, R. K., Prusty, D., Sinha, S., Kashav, T., Mukhopadhyay, G., Chaudhuri, T. K., Gourinath, S., and Dhar, S. K. (2007) The domain structure of Helicobacter pylori DnaB helicase: the N-terminal domain can be dispensable for helicase activity whereas the extreme C-terminal region is essential for its function, *Nucleic Acids Res. 35*, 2861-2874.

[128] Arai, K., Yasuda, S., and Kornberg, A. (1981) Mechanism of dnaB protein action. I. Crystallization and properties of dnaB protein, an essential replication protein in Escherichia coli, *J. Biol. Chem. 256*, 5247-5252.

[129] Notarnicola, S. M., Park, K., Griffith, J. D., and Richardson, C. C. (1995) A domain of the gene 4 helicase/primase of bacteriophage T7 required for the formation of an active hexamer, *J. Biol.Chem. 270*, 20215-20224.

[130] Soni, R. K., Mehra, P., Choudhury, N. R., Mukhopadhyay, G., and Dhar, S. K. (2003) Functional characterization of Helicobacter pylori DnaB helicase, *Nucleic Acids Res. 31*, 6828-6840.

[131] Young, M. C., Schultz, D. E., Ring, D., and von Hippel, P. H. (1994) Kinetic parameters of the translocation of bacteriophage T4 gene 41 protein helicase on single-stranded DNA, *J. Mol. Biol. 235*, 1447-1458.

[132] Richardson, R. W., and Nossal, N. G. (1989) Characterization of the bacteriophage T4 gene 41 DNA helicase, *J. Biol.Chem. 264*, 4725-4731.

[133] Patel, S. S., and Picha, K. M. (2000) Structure and function of hexameric helicases., *Annu. Rev. Biochem. 69*, 651-697.

[134] Yu, X., Hingorani, M. M., Patel, S. S., and Egelman, E. H. (1996) DNA is bound within the central hole to one or two of the six subunits of the T7 DNA helicase, *Nature (Struct. Biol.) 3*, 740-743.

[135] Kaito, C., Kurokawa, K., Hossain, M. S., Akimitsu, N., and Sekimizu, K. (2002) Isolation and characterization of temperature-sensitive mutants of the Staphylococcus aureus dnaC gene, *FEMS Microbiol. Let. 210*, 157-164.

[136] Biswas E. E., Barnes, M., Moir D. T. and Biswas S. B. (2009) An essential DnaB helicase of Bacillus anthracis: identification, characterization, and mechanism of action. *J. Bacteriol.* **191**, 249-260.

[137] Biswas S. B., Wydra, E. and Biswas E. E. (2009) Mechanisms of DNA binding and regulation of Bacillus anthracis DNA primase. *Biochemistry, 48*, 0000-0000.

[138] Bird, L. E., and Wigley, D. B. (1999) The Bacillus stearothermophilus replicative helicase: cloning, overexpression and activity, *Biochim. Biophys. Acta 1444*, 424-428.

[139] Soultanas, P., and Wigley, D. B. (2002) Site-directed mutagenesis reveals roles for conserved amino acid residues in the hexameric DNA helicase DnaB from Bacillus stearothermophilus, *Nucleic Acids Res. 30*, 4051-4060.
[140] Velten, M., McGovern, S., MArsin, S., Ehrlich, S., Noirot, P., and Polard, P. (2003) A Two-Protein Strategy for the Functional Loading of a Cellular Replicative DNA Helicase, *Mol. Cell 11*, 1009-1020.
[141] Li, Y., Kurokawa, K., Reutimann, L., Mizumura, H., Matsuo, M., and Sekimizu, K. (2007) DnaB and DnaI temperature-sensitive mutants of Staphylococcus aureus: evidence for involvement of DnaB and DnaI in synchrony regulation of chromosome replication, *Microbiology 153*, 3370-3379.
[142] Bruand, C., and Ehrlich, S. D. (1995) The Bacillus subtilis dnaI gene is part of the dnaB operon, *Microbiology (Reading, England) 141 (Pt 5)*, 1199-1200.
[143] Bruand, C., Sorokin, A., Serror, P., and Ehrlich, S. D. (1995) Nucleotide sequence of the Bacillus subtilis dnaD gene, *Microbiology 141 (Pt 2)*, 321-322.
[144] Moriya, S., Imai, Y., Hassan, A. K., and Ogasawara, N. (1999) Regulation of initiation of Bacillus subtilis chromosome replication, *Plasmid 41*, 17-29.
[145] Ogasawara, N., Moriya, S., Mazza, P. G., and Yoshikawa, H. (1986) Nucleotide sequence and organization of dnaB gene and neighbouring genes on the Bacillus subtilis chromosome, *Nucleic Acids Res. 14*, 9989-9999.
[146] Aiello, D., Barnes, M. H., Biswas, E. E., Biswas, S. B., Gu, S., Williams, J. D., Bowlin, T. L., and Moir, D. T. (2009) Discovery, Characterization and Comparison of Inhibitors of Bacillus anthracis and Staphylococcus aureus Replicative DNA Helicases. *Bioorg. Med. Chem.*, **17**, 4466–4476.

In: Bacterial DNA, DNA Polymerase and DNA Helicases ISBN 978-1-60741-094-2
Editor: Walter D. Knudsen and Sam S. Bruns © 2009 Nova Science Publishers, Inc.

Chapter XIII

Effect of Substrate Traps on Hepatitis C Virus NS3 Helicase Catalyzed DNA Unwinding: Evidence for Enzyme Catalyzed Strand Exchange

Ryan S. Rypma, Angela M. I. Lam and David N. Frick[*]

Department of Biochemistry and Molecular Biology
New York Medical College

Abstract

The helicase encoded by the hepatitis C virus (HCV) is shown in this chapter to catalyze homologous DNA strand exchange. Single-stranded DNA oligonucleotides complementary to either the short or long strand of a partially duplex DNA substrate affected the rate and extent of unwinding catalyzed by either the full-length HCV NS3 protein fused to a portion of HCV NS4A, or a truncated NS3 protein lacking the protease domain. The oligonucleotides did not, however, sequester HCV helicase and prevent it from separating the original duplex after a single binding cycle. Furthermore, when DNA oligonucleotides were pre-incubated with HCV helicase, they did not prevent subsequent duplex separation, indicating that the enzyme catalyzes strand exchange. The protease portion of NS3 was not needed for this strand exchange. Fluorescent DNA substrates were further used to directly monitor both this ssDNA assisted unwinding and homologous strand exchange, and the effect of a protein trap (poly(U) RNA) on HCV helicase-catalyzed strand exchange was examined. The results demonstrate that HCV helicase can simultaneously bind at least three DNA strands and imply that HCV NS3 helicase could play an important role not only in viral RNA unwinding, but also in the folding of the HCV genome.

---

[*] To whom correspondence should be addressed: Department of Biochemistry and Molecular Biology, New York Medical College, Valhalla, NY 10595. Tel.:914-594-4190; Fax: 914-594-4058; E-mail: David_Frick@NYMC.edu

Abbreviations

BHQ2	Black hole quencher 2;
Cy3	cyanine 3;
FRET	fluorescence resonance energy transfer;
HCV	hepatitis C virus;
Ls	long strand;
LsT	long strand trap;
ssDNA	single stranded DNA;
Ss	short strand;
SsT	short strand trap;
Poly(U)	poly(U)ridylic acid.

Introduction

The propensity of complementary DNA and RNA chains to anneal rapidly makes the measurement of helicase activity challenging. Normally, helicase unwinding assays are performed at very low DNA (or RNA) concentrations with excess helicase proteins. The low nucleic acid concentrations and excess protein make annealing of the newly single-stranded DNA (ssDNA)[1] chains less likely. Although such assays can probe basic properties of helicases, determined rates only measure the rate of the slowest step in the unwinding process and not the actual rate of helicase-catalyzed strand separation.

To better understand the action of a helicase in unwinding assays, a trap that sequesters protein molecules not initially bound to the duplex substrate can be added at the start of the reaction. Such "pseudo-first order" reaction conditions have been used as the basis for most current models explaining the helicase-catalyzed unwinding reactions [1, 2]. The trap can be a negatively charged protein like heparin or nucleic acids. An ssDNA oligonucleotide that is complementary to one strand of the helicase substrate is also sometimes included in such assays to prevent the dissociated strands from annealing before they are analyzed. In theory, the presence of the DNA trap should not affect the rate of unwinding by the proteins originally bound to the substrate DNA; it should only prevent other enzyme molecules from binding additional substrate molecules or rebinding partially unwound substrate to which they were once bound. This study examines the effects of substrate traps on reactions catalyzed by a helicase encoded by the hepatitis C virus (HCV).

Hepatitis C (HepC) is a disease that affects about 170 million people worldwide [3] and is frequently called a "silent" killer because it causes few symptoms while the pathogen slowly destroys the liver. After a few decades of infection, during which time patients might unknowingly transmit the blood-borne (+)RNA virus to others, many HepC patients develop fibrosis, cirrhosis, or liver cancer. At this late stage of the disease, a liver transplant is the only option for survival, and as a result, HCV infection is presently the most common cause for liver transplantation in America. Current HCV therapies combine pegylated interferon alpha and ribavirin, but they are costly, produce debilitating side effects, and are not fully effective against many common HCV strains.

HCV vaccines and treatments have been delayed because the virus is extraordinarily difficult to work with in the laboratory. HCV was first identified almost two decades after either hepatitis A or B viruses, and it was only 2005 when HCV could be reliably cultivated in cell culture [4]. The HCV genome is a single long open reading frame that encodes an ~3,000 amino acid long protein that is cleaved into mature structural (core, E1, E2) and non-structural proteins (p7, NS2, NS3, NS4A, NS4B, NS5A, and NS5B). The non-structural proteins are responsible for replication and packaging of the viral genome into capsids formed by structural proteins. Four enzymatic activities reside in the HCV non-structural proteins. NS5B is an RNA–dependent RNA polymerase that synthesizes new viral RNA, and there are two proteases that cleave the polyprotein, the NS2/NS3 autocatalytic protease and the NS3-NS4A serine protease. NS3 is as a helicase capable of unwinding both DNA and RNA. Several compounds that influence the activity of the NS3 protease and the NS5B RNA-dependent RNA polymerase are currently in clinical trials, but after almost 15 years in development none are yet on the market.

The HCV helicase most likely assists RNA replication by tracking along RNA and resolving double stranded intermediates that form either as secondary structures in a single strand or between (+) sense and (-) sense RNA molecules [5]. There are also other possible roles for HCV helicase, such as in assisting translation, protein processing, or packaging RNA into virions. Regardless of its precise role, knocking out helicase with genetics [5], antibodies [6], RNA aptamers [7], or small peptides [8] prevents HCV from replicating in cells.

The mature NS3 protein comprises 5 domains: the N-terminal 2 domains form the serine protease along with the NS4A cofactor, and the C-terminal 3 domains form the helicase. The helicase portion of NS3 can be separated from the protease portion by truncating at a peptide linker, but the resulting protein (called NS3h) is somewhat less active than the NS3-NS4A complex [9]. Removing the protease allows one to express much higher levels of NS3h as a more soluble protein in *E. coli*. At the heart of HCV helicase is a "Walker-type" ATPase fold [10]. ATP most likely binds near Lys210 in the "Walker A-site" with Asp290 in the "Walker B-site" binding a bridging divalent metal cation [11]. The Walker site and a conserved catalytic base (Glu291 in NS3) [11] are packed in a cleft formed by two motor domains, each of which structurally resembles the *E. coli* RecA protein [12]. In HCV helicase, the two RecA-like motor domains correspond to NS3h domains 1 and 2. Unlike most related proteins, HCV helicase can separate RNA [13] and DNA [14] duplexes. One strand of nucleic acid binds between the motor domains and domain 3 [15, 16]. Most current models explaining HCV helicase action speculate that ATP binding and/or its hydrolysis regulates domain 2 movements which in turn, leads to movement of the protein along RNA. Our previous work has shown that Arg393 on domain 2 clamps onto RNA while a beta-loop extending from domain 2 splits the helix [17] and the protein is propelled *via* electrostatic interactions with the nucleic acid [18, 19].

Many studies have used ssDNA traps to calculate the rate at which the hepatitis C virus (HCV) helicase unwinds DNA [16, 20-23] and RNA duplexes [24]. In those reports, ssDNA traps terminated the reaction after a single cycle lasting only a few seconds. In contrast, several studies from our lab have shown that the addition of ssDNA oligonucleotides up to 1,000 times in excess of the helicase:DNA complex do not terminate unwinding after a single enzyme-DNA binding event. Rather, the HCV helicase catalyzed reactions continue to progress long after original enzyme-DNA complexes should have dissociated [9, 17, 18].

In this chapter, we have investigated this peculiar phenomenon and have found that the sequence of the trap DNA affects both the rate and extent of HCV helicase-catalyzed DNA unwinding, suggesting that the ssDNA is not simply sequestering excess enzyme, but participates in the reaction. We also show that HCV helicase actively exchanges one strand of a double helix for a third strand. Strand exchange occurs even in the presence of a true enzyme trap that does not participate in the reaction. Most interestingly, when reactions are initiated with most enzyme molecules bound to ssDNA (not the duplex substrate), strand exchange still proceeds. The data suggest that, like similar superfamily 2 (SF2) helicases involved in recombination, HCV helicase can bind three strands of nucleic acid and exchange one strand of the duplex for another.

Experimental Procedures

Materials– DNA oligonucleotides were purchased from Integrated DNA Technologies (Coralville, IA), and their concentrations were determined from provided extinction coefficients. All but one of the recombinant proteins utilized in this study have been described previously [9, 25]. Note that the truncated NS3 protein containing the helicase domain flanked by a C-terminal polyhistidine tag, designated here as "NS3h," is the same protein that was described previously by our lab as Hel-1b [25].

The single chain full-length NS3-NS4A protein (scNS3-4A) used here was similar to that described in reference [26] except that it was derived from the HCV genotype 1b J4 strain [27]. In scNS3-4A, a Ser-Gly-Ser linker region joins the protease-activating region of NS4A to the N-terminal portion of NS3. PCR was used with pJ4L6S [27] as the template to generate the scNS3-4A expression plasmid. The sequence of the upstream PCR primer was 5'-GAT ATA CAT ATG GGT TCT GTT GTT ATT GTT GGT AGA ATT ATT TTA TCT GGT AG TGG TAG TAT CAC GGC CTA CTC CCA A-3', which encodes for an *Nde*I restriction enzyme site, NS4A residues G21-G33, a SGS linker, and P2 of the NS3 protein. The downstream primer (5'-GCG CGC GAA TTC GGT CAA GTG ACG ACC TCC AGG TCA GCC GAC ATG C-3') contained an *Eco*RI site. The resulting DNA amplicon was cloned into the multiple cloning site of pET28a (Novagen). The resulting plasmid was sequenced and designated p28scNS3-4A. The plasmid-encoded scNS3-4A also encodes an N-terminal 21-amino acid long His-tag bearing sequence (MGS SHH HHH HSS GLV PRG SHM).

NS3 proteins were purified as described previously [9, 25, 26]. With scNS3-4A, its protease activity was monitored (using the EnzoLyte 490 HCV Protease Assay Kit, Anaspec, San Jose, CA) and its helicase activity (using the FRET-based assay below) was monitored throughout the purification to ensure neither activity was lost during purification. Final protein concentrations were determined by using A_{280} with the following extinction coefficients calculated from the Trp, Tyr, and Phe content of each protein: NS3h, 42.4 mM^{-1} cm^{-1}; scNS3-4A, 68.4 mM^{-1} cm^{-1}.

Table I. Oligonucleotides and unwinding substrates

Name	Sequence
Ls	5'-TAGTACCGCCACCCTCAGAACCTTTTTTTTTTTTTT-3'
ss (LsT)	5'-GGTTCTGCGGGTGGCGGTACTA-3'
ssT	5'-TAGTACCGCCACCCTCAGAACC-3'
T18	5'-TTTTTTTTTTTTTTTTTT-3'
[5'-F]Ls	Cy3-TAGTACCGCCACCCTCAGAACCTTTTTTTTTTTTTT-3'
[3'-Q]Ss ([3'-Q]LsT)	5'-GGTTCTGCGGGTGGCGGTACTA-BHQ2
[5'-F]Ss	Cy3-GGTTCTGCGGGTGGCGGTACTA-3'
[3'-Q]SsT	5'-TAGTACCGCCACCCTCAGAACC-BHQ2
Ls:Ss	5'-GGTTCTGCGGGTGGCGGTACTA-3' 3'-TTTTTTTTTTTTTTCCAAGACTCCCACCGCCATGAT-5'
[5'-F]Ls:[3'-Q]Ss	5'-GGTTCTGCGGGTGGCGGTACTA-BHQ2 3'-TTTTTTTTTTTTTTCCAAGACTCCCACCGCCATGAT-CY3
Ls:[5'-F]Ss	Cy3-GGTTCTGCGGGTGGCGGTACTA-3' 3'-TTTTTTTTTTTTTTCCAAGACTCCCACCGCCATGAT-5'
[5'-F]Ls:Ss	5'-GGTTCTGCGGGTGGCGGTACTA-3' 3'-TTTTTTTTTTTTTTCCAAGACTCCCACCGCCATGAT-BHQ2

Gel-Based Helicase Assay– To generate substrates for helicase assays, two synthetic oligonucleotides (Table I) were annealed by heating them to 95 °C and allowing them to cool slowly to room temperature. Before annealing, the shorter strand was labeled using [γ^{32}P]ATP and polynucleotide kinase. The DNA substrate (4 nM) and 100 nM HCV helicase were incubated in reaction buffer (25 mM MOPS, 3 mM MgCl$_2$, 0.1% Tween-20) with 200 nM of SsT or LsT (Table I). Reactions were initiated by the addition of 3 mM ATP, terminated at various times with 20 mM EDTA and 0.5% SDS, and analyzed using a 10% non-denaturing polyacrylamide gel.

FRET-based Unwinding/Strand Exchange Assays– Fluorescent helicase substrates were based on those described by Boguszewska-Chachulska *et al.* [28]. Reactions were carried out using substrates of identical sequence to that used in the gel-based helicase assay that were labeled either with Cy3 or BHQ2 (Table I). Reactions were performed in 25 mM MOPS (pH 6.5), 5 mM MgCl$_2$, 25% glycerol, 0.1% Tween-20, 1 mM DTT, 3 mM ATP, substrate, and helicase. All reactions were conducted at 37 °C in a reaction volume of 100 μL. Fluorescence was measured using a Varian Cary Eclipse fluorescence spectrophotometer with excitation and emission wavelengths set for 550 and 570 nm, respectively, and slit widths set to 10 nm for both the excitation and emission. Initial velocities were calculated using linear regression with datasets pruned of values outside the linear range. Non-linear regression was performed using Prism 4 (GraphPad Software, Inc.).

Results

In unwinding assays performed under single turnover conditions, HCV helicase has been shown to rapidly unwind typical DNA substrates (20-40 bp) in only a few seconds [16, 21,

23]. However, when trying to repeat some of these studies, we were surprised to note that HCV helicase-catalyzed DNA (and RNA) unwinding reactions performed in the presence of excess ssDNA continued for several minutes [9, 17, 18]. The difference between our studies and those of other labs was that the oligonucleotide we used as a trap was complementary to one strand of the helicase substrate. Experiments with other helicases have shown that addition of a trap complementary to the substrate normally prevents helicase recycling and slows steady-state unwinding rates (for example see Figure 3 of reference [29]). However, since this was apparently not the case with HCV helicase, we initiated this study to investigate this peculiar phenomenon more carefully.

Free ssDNA strands influence HCV helicase-catalyzed unwinding reactions– Figure 1A shows a typical assay using a full-length NS3 protein with the portion of NS4A needed to activate the NS3 protease fused to the N-terminus of NS3 (scNS3-4A). ScNS3-4A unwinds RNA, DNA, and cleaves peptides at rates indistinguishable from those catalyzed by full-length NS3/NS4A complexes expressed and purified from baculovirus-infected insect cells [9, 26]. The assay shown in Figure 1A monitors the unwinding of a partially duplex DNA made of a long strand and a [^{32}P]short strand.

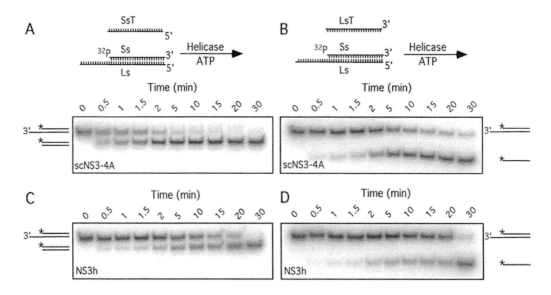

Figure 1. Time course of HCV helicase catalyzed unwinding of a [^{32}P]DNA substrate as analyzed using native polyacrylamide gels. Reactions contained either full-length NS3 with protease cofactor NS4A peptide covalently linked to the N-terminus of the native NS3 (scNS3-4A, panels *A* and *B)* or truncated NS3 lacking the protease region (NS3h, panels *C* and *D*). Either a short strand trap (SsT) complementary to the short strand of the substrate (*A* and *C*) or a long strand trap (LsT) (*B* and *D*) was used to observe differences in product formation. Reactions terminated at various times are indicated. 4 nM DNA substrate was pre-incubated with 20 nM enzyme and initiated by adding both 3 mM ATP and 200 nM oligonucleotide trap in all reactions.

When annealed, the long strand forms an ssDNA tail. This substrate was annealed, diluted to 4 nM in reaction buffer and mixed with 20 nM scNS3-4A. The reaction was initiated with 3 mM ATP and 200 nM of a DNA oligonucleotide complementary to the short strand of the substrate (short strand trap, SsT). As shown by native polyacrylamide gel analysis, the product of the reaction is double stranded DNA where the long strand was exchanged with the complement of the short strand (Figure 1A). When the reaction in Figure 1A was repeated without the SsT, no unwinding was observed, suggesting the reaction products annealed to reform the substrate before or during gel analysis.

To test the idea that the SsT simply prevents the long and short oligonucleotides from annealing after separation by the helicase, an oligonucleotide complementary to the long strand was substituted (long strand trap, LsT). The SsT and LsT are the same length and reverse complements of each other (note that LsT is simply the short strand lacking a radiolabel). As shown in Figure 1B, reactions containing LsT yielded a labeled single-stranded short oligonucleotide. Interestingly, however, reactions containing LsT proceed notably more slowly than those containing SsT (compare 1A and 1B).

It is possible that the protease portion of NS3 is responsible for the difference seen in reaction rates with either SsT or LsT. Perhaps the protease domain blocks LsT from annealing with the long strand. To examine this possibility, we repeated the assays using truncated NS3 lacking the protease domain (NS3h) from HCV genotype 1a [9], and NS3h protein from genotype 1b and NS3h from genotype 2a [25]. All recombinant HCV helicase proteins yielded similar results. Unwinding in the presence of SsT proceeded more rapidly than unwinding in the presence of LsT. Reactions with NS3h (genotype 1b(J4)), which has the same sequence as scNS3-4A but lacks the protease, are shown in Figure 1 *panels* C and D.

Since the concentrations of SsT and LsT in the reactions shown in Figure 1 exceeded the protein concentration by 10-fold, one might suspect that the reactions occurred under single-turnover conditions. In other words, the trap (SsT or LsT) should prevent the enzyme from re-cycling after it falls from the substrate to which it was initially bound. If this were the case, then reactions containing either SsT or LsT would proceed with the same rates to reach the same final amplitudes, reflective of the amount of helicase initially bound to the substrate and the processivity with which it travels along the DNA. The gels in Figure 1 reveal that single tunrnover conditions are not achieved because all the SsT and LsT reactions proceed at different rates and all reactions proceed almost to completion.

Effect of ssDNA oligonucleotides in continuous FRET-based unwinding assays– A detailed examination of the effect of traps using gel-based helicase assays is difficult because only a limited number of time points can be analyzed and because traps must always be added during electrophoresis to prevent the strands from re-annealing in the gels. Therefore, to explore this "trap-assisted" unwinding in more detail, a FRET-based unwinding assay was used to monitor unwinding. To this end, a Cy3 fluorescent probe was attached to the 5'-end of the long strand and a black hole quencher 2 (BHQ2) was attached to the 3'-end of the short strand. With this substrate, fluorescence from Cy3-labeled long strand ([5'-F]Ls) is quenched when it is annealed to the short strand labeled with BHQ2 ([3'-Q]Ss), and unwinding can be continuously monitored in a fluorescence spectrophotometer [28]. Using gel-based helicase assays, oligonucleotides containing the Cy3 and BHQ2 probes were unwound at the same rates by HCV helicase as those lacking modifications (data not shown). Also, binding of HCV helicase to [5'-F]Ls does not significantly affect Cy3 fluorescence.

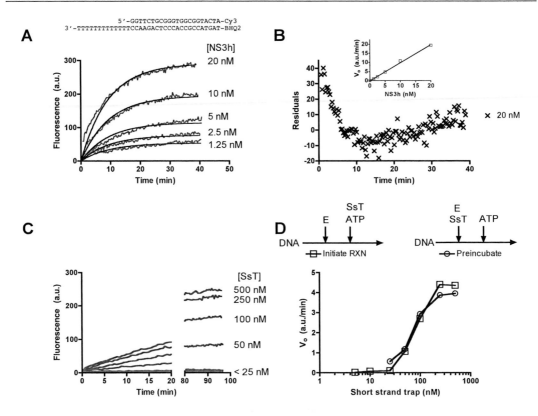

Figure 2. Effect of ssDNA on helicase unwinding as monitored using a FRET-based helicase assay. *A*, Time course for unwinding of 20 nM of the [5'-F]Ls:[3'-Q]Ss substrate (sequence above plot) in reactions initiated with 3 mM ATP and 200 nM SsT using various HCV helicase concentrations as indicated. Data are fit to a first order rate equation ($F=F_{max}(1-e^{-kt})$). *B*, Plotted are the residuals (difference between the data and model) of the 20 nM enzyme trace in *panel A*. The inset shows a linear relationship between enzyme concentration and initial velocity (V_o). *C*, Increasing SsT trap concentrations as shown on the right to stimulate rates of unwinding on 20 nM [5'-F]Ls:[3'-Q]Ss substrate by 20 nM NS3h. *D*, Various concentrations of SsT added with ATP to initiate reaction (squares) or pre-incubated with the enzyme (circles) have similar effects on initial velocity of the unwinding reaction.

Figure 2A shows the results of five typical unwinding assays using [5'-F]Ls:[3'-Q]Ss. Each was performed at different concentrations of NS3h. Reactions were initiated with ATP (3 mM) and the SsT (200 nM). In each case, fluorescence increases at a rate proportional to the amount of enzyme in solution. Gel electrophoresis of the substrate at 8 time points confirmed that the fluorescence is reflective of the percent of DNA unwound. A fully unwound substrate yields a fluorescence of about 300 arbitrary units (a.u.).

If SsT were acting to trap excess enzyme after it initially falls from the substrate, then the data in Figure 2A should fit to a first-order rate equation [30]. However, when the data are fit to a model assuming single turnover conditions with amplitudes proportional to the amount of enzyme in solution, the fit is poor (see lines in Figure 2A), and the data systematically deviate from the model as seen by plotting the residuals (Figure 2B). The initial rates of the reaction are, nevertheless, linear with enzyme concentration (Figure 2B insert).

To further test the possibility that SsT was not acting as an enzyme trap, 10 nM of helicase was incubated with 20 nM of [5'-F]Ls:[3'-Q]Ss substrate, and unwinding was initiated with ATP in the presence of different concentrations of SsT (Figure 2C). If the SsT is not an active participant in the reaction, then increasing concentrations of SsT above what is needed to prevent [3'-Q]Ss from annealing to the [5'-F]Ls (*i.e.* 20 nM) should not affect reaction rates and should decrease the final reaction amplitude (since it would prevent helicase recycling). The data (Figure 2C) reveal that SsT accelerates initial reaction rates, and even at concentrations exceeding 20-times that of the helicase, SsT does not decrease the final reaction amplitude. Furthermore, if SsT were acting as an enzyme trap, then pre-incubating SsT with the enzyme before initiating the reaction should inhibit DNA unwinding (because less helicase would be available to bind the substrate). In the reactions in Figure 2C, SsT was added with ATP to initiate the reaction. The initial velocities for the resulting unwinding are plotted in Figure 2D (squares). Virtually identical reaction rates were seen if both substrate and SsT were pre-incubated with the enzyme before adding ATP (circles, Figure 2D). The data indicate that SsT is acting as a reaction participant, not as a trap. In other words, the data suggest HCV helicase is catalyzing strand exchange rather than unwinding.

Strand exchange under multiple-turnover conditions– It is often assumed that the rate of signal change in FRET-based helicase assays reflects only the rate of helicase-catalyzed unwinding. The data above, however, suggest that the rate of signal change depends not only on the amount of helicase, but also on the amount of DNA trap (SsT). One explanation for this could be that HCV helicase is actively replacing one strand of a double helix with another strand, *i.e.* replacing the long strand with SsT. Unfortunately, the signal obtained with the [5'-F]Ls:[3'-Q]Ss substrate results from strand dissociation, not exchange. Therefore, two additional substrates were constructed that allow the direct analysis of exchange of SsT and LsT into a duplex substrate (Table I, Figure 3).

The substrates used to continually monitor strand exchange consisted of a labeled long or short strand annealed to an unlabeled strand. In one, Ls:[5'-F]Ss, the short strand was labeled with Cy3. In the other, [5'-F]Ls:Ss, the long strand was labeled with Cy3. Unwinding of either substrate alone does not result in a change in fluorescence. However, if one of the strands is annealed to a strand with an appropriately positioned quencher, fluorescence will decrease. Therefore, if strand exchange limits the unwinding rates seen with [5'-F]Ls:[3'-Q]Ss, then the same rates of fluorescence change should be observed with the substrates lacking the quencher in the presence of traps labeled with a quencher.

Figures 3A and 3C show typical reactions performed with the [5'-F]Ls:[3'-Q]Ss substrate. The two figures again highlight the differences in relative unwinding rates when two different trap sequences of equal length are used – one complementary to the short strand (SsT) (Figure 3A) and one complementary to the long strand (LsT) (Figure 3C). As shown in the previous gel-based helicase assay (Figure 1), use of the trap complementary to the short strand results in enhanced unwinding rates. Each figure plots three different enzyme concentrations to show that initial rates are dependent upon enzyme concentration and that SsT supports faster reactions than LsT.

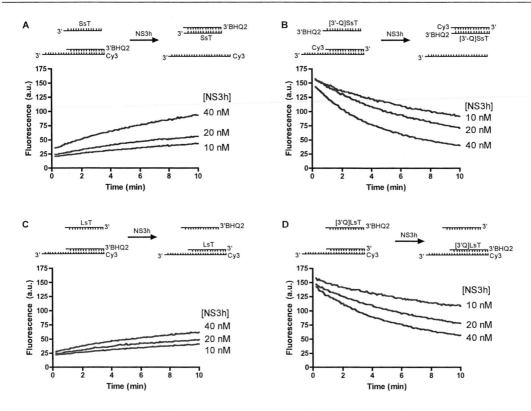

Figure 3. Comparison of HCV helicase catalyzed reactions in FRET-based unwinding and strand exchange assays. Trap-assisted unwinding assays (panels A and C) are compared with strand exchange assays (panels B and D). In each set of reactions, 10 nM substrate was incubated with enzyme concentrations as shown on each panel, and reactions were initiated with 3 mM ATP and 200 nM oligonucleotide trap. The substrate and ssDNA trap used are shown above each panel. Unwinding reactions in *A* and *B* include SsT, while LsT is included in reactions in *C* and *D*.

To perform a reaction in which fluorescence changes upon strand exchange in the presence of SsT, the Ls:[5'-F]Ss substrate was incubated with HCV helicase, and the reactions were initiated by adding [3'-Q]SsT and ATP. When the analogous strand separation reactions shown in Figure 3A are performed with the Ls:[5'-F]Ss substrate (Fig 3B), a decrease in Cy3 fluorescence is observed as the enzyme anneals [5'-F]Ss with [3'-Q]SsT. When converted to percent DNA unwound vs. time, the rates obtained in this "reverse" assay are virtually identical to those obtained in the "forward" fluorescence assay and the gel-based assay when performed under the same conditions. The [5'-F]Ls:Ss substrate was used to directly monitor the exchange of [3'-Q]LsT for the short strand. To this end, the [5'-F]Ls:Ss substrate was pre-incubated with HCV helicase and the duplex substrate, and the reaction was initiated with ATP and [3'-Q]LsT. A decrease in Cy3 fluorescence was observed that was proportional to enzyme concentration (Figure 3D). These rates were essentially the reciprocals of those seen when LsT was included in reactions using the [5'-F]Ls:[3'-Q]Ss substrate (Figure 3D). Again, the differences in fluorescence signal reduction over time in Figures 3B and 3D suggest that apparent rates are greater in the presence of SsT than they are in the presence of LsT. In fact, the percent difference using the two traps in Figures 3B and

3D is similar to the percent difference using the two traps with the [5'-F]Ls:[3'-Q]Ss substrate (Figures A and C), as measured by relative change in fluorescence at 10 minutes.

Strand exchange in the presence of a protein trap– The above data suggest that the apparent rates of DNA unwinding depend mainly on the ability of HCV helicase to exchange one strand of a duplex for another. This discovery is surprising because it implies that HCV helicase can bind at least three strands at once. We therefore explored if and how the reaction would proceed in the presence of a true enzyme trap that does not participate in the unwinding reaction.

First, we monitored the unwinding of the [5'-F]Ls:[3'-Q]Ss substrate in the presence of excess enzyme in the absence of any added oligonucleotides (Figure 4A). In the absence of any trap, the enzyme has unlimited access to other substrate molecules with which it can bind and unwind. Furthermore, because unwound duplex DNA can spontaneously anneal under assay conditions in the absence of excess complementary ssDNA traps that would prevent such re-annealing, varying amounts of enzyme can achieve varying levels of equilibrium states of substrate duplex separation. Greater amounts of enzyme in the presence of constant substrate concentrations enhances product formation, and hence, higher final amplitudes using the labeled substrate. The curves in Figure 4A are therefore composites of the spontaneous substrate rewinding (signal degrading) and enzyme unwinding activity (signal enhancing).

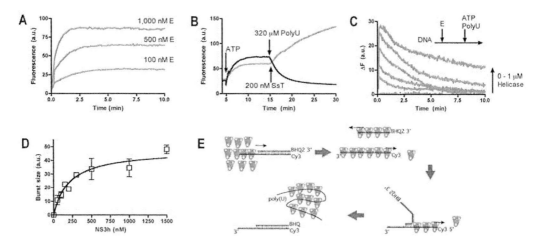

Figure 4. Effect of poly(U) RNA on HCV helicase catalyzed unwinding. *A*, Various enzyme concentrations (noted) were incubated with 10 nM [5'-F]Ls:[3'-Q]Ss substrate and reactions were initiated with ATP. *B*, Two traces of fluorescent signal generation over time are shown with ATP addition at 5 minutes for both, while 320 mM poly(U) is added to one reaction at 15 minutes (black line), and 200 nM SsT is added to the other at 15 minutes (gray line). *C*, Signal traces using increasing amounts of helicase as shown on the right. In each case, enzyme was incubated with substrate, and reactions were initiated with 3 mM ATP and 160 mM poly(U). *D*, The maximum amplitude (burst) of fluorescence increase after ATP addition plotted versus enzyme concentration. Points are the means of data from panel C, and two identical repeat experiments, with error bars depicting standard deviations. The data are fit to a one-site binding model. *E*, A model depicting how poly(U) serves as an enzyme trap once the pre-bound substrate is unwound. Decreasing signal ensues as a result of spontaneous substrate re-annealing under experimental conditions.

We next examined the effect of traps on reactions containing 10 times more helicase than DNA (100 nM helicase, 10 nM [5'-F]Ls:[3'-Q]Ss substrate). When polymers that were not complementary to either strand were added, such as poly(U) RNA (average length 2500 nt), the long and short strands spontaneously re-annealed, and a quenching of fluorescence was observed (black line, Figure 4B). On the other hand, when SsT was added at the same time to an identical reaction, fluorescence levels increased (gray line, Figure 4B).

As there is nothing inherent in the ability of poly(U) to promote or enhance spontaneous duplex DNA formation, we can conclude the poly(U) was effective as an enzyme trap to prevent or severely inhibit the enzyme's ability to bind and unwind additional substrate duplexes (Figure 4E). Conversely, addition of excess SsT served to drive equilibrium towards product formation as fluorescence signal increased. Because the addition of poly(U) to the reaction mixture can reveal spontaneous duplex reformation as shown in the previous experiment, the amplitude of a single turnover event should be measurable if poly(U) is added with ATP at initiation of the reaction. Indeed, as shown in Figure 4C, various initial bursts of fluorescence intensities are observable at different enzyme concentrations. These bursts are transient, however, as the enzyme can no longer rebind to additional substrate. What follows is a steady decline in signal intensity as the two strands re-anneal. A plot of enzyme concentration versus the initial burst amplitude observed (Figure 4D) reveals an apparent K_d of 200 ±100 nM, which represents a composite of the affinity of the enzyme for the DNA at each step as it unwinds the duplex.

Reactions were then performed using poly(U) as an enzyme trap to simulate single turnover reactions. In these reactions, 100 nM of HCV helicase was pre-incubated with 10 nM substrate, and the reactions were initiated with ATP and poly(U). As seen in Figure 4C, under these conditions the amount of DNA separated in a single turnover event was observed as a burst. Then, fluorescence decreased as the complementary strands re-associated. If SsT and LsT simply prevent re-annealing, then no additional substrate should be unwound in the presence of poly(U) and SsT (or LsT). Surprisingly, however, addition of either SsT (Figure 5A) or LsT (Figure 5B) *both* prevented the re-annealing of the substrate after the first turnover *and* allowed for additional substrate to be unwound. When oligonucleotides not complementary to either the short or long strands are added with ATP and poly(U), no effect was seen on the re-annealing of the DNA substrate, and no additional unwinding was observed (Figure 5C). As would be expected, none of the oligonucleotides affected the amount of DNA unwound in the first turnover, and the burst amplitudes were similar regardless of whether an oligonucleotide was added with the ATP and poly(U).

The ability of SsT and LsT to participate in strand exchange was calculated using equation 1

$$V_{s.e} = V_{obs} - V_{anneal} \qquad \text{Eq. 1}$$

where V_{obs} is the rate of fluorescence change in a reaction from 1 to 3 min after initiation with ATP, and V_{anneal} is the rate of change over the same period of time in the absence of added oligonucleotide. Such an analysis reveals that SsT stimulates strand exchange to a level almost twice that stimulated by LsT (Figure 5D).

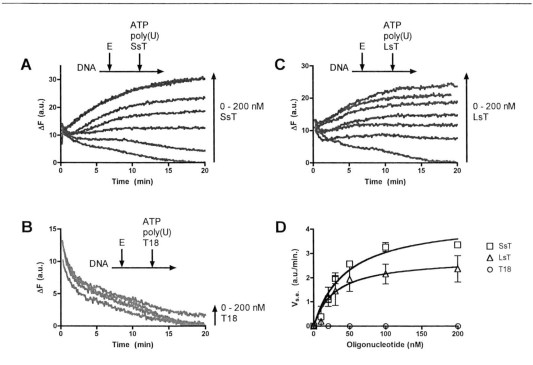

Figure 5. HCV-helicase catalyzed strand exchange in the presence of poly(U). All reactions contained 10 nM [5′-F]Ls:[3′-Q]Ss substrate which was pre-incubated with 100 nM helicase for 10 minutes. Reactions were initiated with 3 mM ATP, 160 mM poly(U) as enzyme trap, and varying concentrations of SsT (A), LsT (B), or T18 (C) as shown on the right of each panel. D, Rates of strand exchange ($V_{s.e.}$) versus oligonucleotide concentration. $V_{s.e.}$ was calculated with Equation 1 and data were fit to the Michaelis-Menten equation. Error bars depict standard deviation from three identical repeat reactions.

The single turnover experiments were then repeated, but instead of simultaneously adding oligonucleotide trap and ATP, the trap was pre-mixed with the enzyme and duplex substrate. One might expect that if SsT were pre-mixed with enzyme in the absence of ATP, the helicase would bind SsT tightly and not bind the duplex substrate when subsequently added. If enzymes bind only SsT, then they should bind the poly(U) upon reaction initiation and not unwind any substrate. Pre-mixing of either SsT (Figure 6A), LsT (Figure 6B), or T18 (Figure 6A), in fact, decreases the amount of DNA unwound in a single turnover. If burst size is plotted versus oligonucleotide concentration (Figure 6D), they all act like competitive inhibitors with similar K_i's (~20 nM). However, the fact that HCV helicase starts a reaction bound to SsT or LsT, does not prevent the helicase from exchanging the strand on which it is bound for one of the strands of [5'-Cy3]Ls:[3'-BHQ]Ss (Figure 6A, B). In fact, the strand exchange velocities are similar to those seen when SsT or LsT is added with poly(U) and ATP (compare Figures 5D and 6E). Even more remarkably, under these conditions, some strand exchange was observed when the enzyme was initially bound to an oligonucleotide not complementary to either the long or short strands (ex. T18, Figure 6C). However, such complexes were unstable and after a few minutes, the [5'-Cy3]Ls:[3'-BHQ]Ss substrate re-annealed. The presence of a control oligonucleotide (T18) has little effect on the rate of annealing of [5'-Cy3]Ls:[3'-BHQ]Ss after the enzyme has been sequestered after a single turnover (Figure 6F).

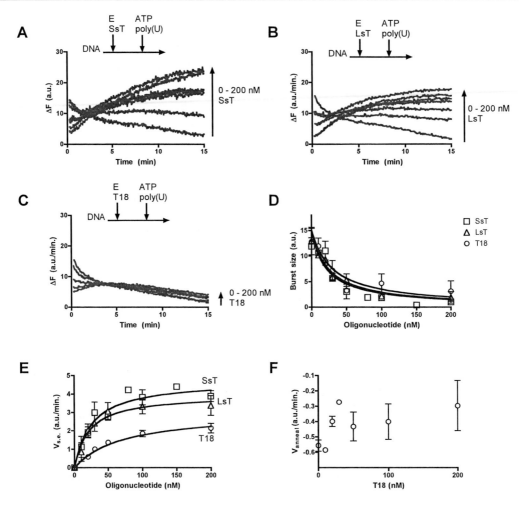

Figure 6. Helicase catalyzed unwinding in the presence of poly(U) when the enzyme is pre-incubated with ssDNA oligonucleotides. All reactions were performed using 10 nM [5'-F]Ls:[3'-Q]Ss substrate pre-incubated with 100 nM helicase and varying concentrations of either SsT (*A*), LsT (*B*), or T18 (*C*) and initiated with 3 mM ATP and 160 uM poly(U). *D*, Initial amplitude of unwinding reactions from panels *A-C* are plotted versus oligonucleotide substrate trap concentration and fit to a one-site competition model. *E*, The rate of strand exchange ($V_{s.e.}$) is plotted as a function of the free oligonucleotide. The data were fit to the Michaelis-Menten equation. *F*, Annealing rates of [5'-F]Ls:[3'-Q]Ss 9-11 minutes after reaction initiation plotted as a function of T18 concentration. Error bars in panels D-F depict standard deviation from three separate reactions.

Conclusion

The unwinding of duplex DNA and RNA substrates by HCV helicase has been extensively studied. It is clear that HCV helicase binds to one strand of a duplex and moves in a 3' to 5' direction to displace a complementary strand. Behind the helicase, the duplex can re-anneal if it does not remain bound to HCV helicase or other proteins (Figure 4E). For this reason, no unwinding can be observed unless either the amount of HCV helicase greatly exceeds the amount of DNA (or RNA) in the assay or ssDNA complementary to one strand of the substrate is added. We show here that this "trap" strand does not simply prevent substrate annealing, but is rather actively exchanged into the original duplex. In other words, HCV helicase, like RecA [31] and other related proteins, catalyzes homologous DNA strand exchange (Figure 7). HCV NS3-catalyzed strand exchange occurs not only in multiple turnover conditions, but also in the presence of a protein trap that prevents the helicase from re-binding substrates.

It should be noted that for an unwinding event to be monitored with the [5'-F]Ls:[3'-Q]Ss substrate, all that is required is the physical separation of the strands. On the other hand, with the use of the Ls:[5'-F]Ss or [5'-F]Ls:Ss substrate, not only must the strands be fully separated, but annealing to the a third strand bearing the quencher must also occur for fluorescence to change. Since, all the bases of all substrates used in Figure 3 are identical, and the only differences between the reaction pairs is the position of the Cy3 and BHQ2 probes, the data reveal that strand exchange is the rate-limiting step in the FRET-based helicase assays performed in the absence of a protein trap.

Because there have been so many HCV helicase structures published [15, 16, 32-35], it is tempting to speculate on how HCV helicase can simultaneously bind at least three nucleic acid strands in order to catalyze strand exchange. The enzyme appears to interact with both SsT (or LsT) and [5'-F]Ls:[3'-Q]Ss, because even when HCV helicase is initially bound to SsT (or LsT), it can somehow find the substrate [5'-F]Ls:[3'-Q]Ss and exchange the trap for one of the DNA strands. How the enzyme coordinates the action of three strands is still a mystery given that only one DNA binding site has been visualized on the enzyme [15], which most likely represents the site that initially binds the ssDNA region of the duplex, *i.e.*, 3'-tail of the long strand of the substrates used here. There are two basic possibilities. The first is that the enzyme is a functional oligomer and that two monomers coordinate their actions with one subunit bound to the duplex and one to the ssDNA (Figure 7). The other possibility is that, like similar SF2 helicases that function in recombination, an NS3 monomer of can bind more than two strands of DNA. Such an arrangement has been seen in a crystal structure of the related recG protein [36].

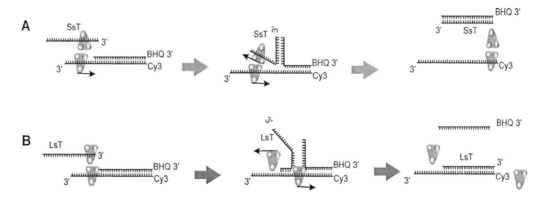

Figure 7. Model for helicase catalyzed strand exchange. A, Steps required for annealing of a strand complementary (SsT) to the short strand of the helicase substrate. A, Steps required for annealing of a strand complementary (LsT) to the long strand of the helicase substrate.

Regardless of how the strands are coordinated on the enzyme, the differences seen between the exchange of the SsT and LsT is enlightening. The data reveal that HCV helicase more readily anneals a third strand to the short strand that is released (Figure 7A). This observation supports the notion that the enzyme moves along the long strand as it unwinds the DNA, and implies that enzyme protomers could cooperate in strand exchange, with one subunit moving along the substrate and the other along the SsT. The slower exchange of the LsT may occur because of a lack of subunit cooperation or because the enzyme must fall from the long strand before LsT exchange can occur (Figure 7B).

It is important to note that strand exchange occurs at a rate several orders of magnitude slower than the maximum rate at which HCV helicase can unwind a duplex substrate. As seen in the single-turnover reactions (Figures 5 and 6), an enzyme complex bound only to [5'-F]Ls:[3'-Q]Ss unwinds bound substrate very rapidly. The time between reaction initiation and first observation in the reactions (Figures 5 and 6) was typically about 20 seconds. It was not possible to observe the accurate burst rates using our available equipment, but others have [16, 20-24], and the rapid bursts observed here are consistent with such previously published reports. We have not yet attempted to define precisely the mechanism of HCV helicase-catalyzed homologous DNA strand exchange. The protein must bind DNA strands, search for homology, form a joint molecule, migrate the branch, and release the displaced strand(s).

At present, we may also only speculate on the role HCV helicase-catalyzed strand exchange may play in viral replication. We used DNA simply for convenience and because HCV helicase unwinds DNA even better than RNA. However, we have repeated most of the assays shown here using RNA substrates and the scNS3-4A protein and seen few significant differences between RNA and DNA substrates. Many helicases related to HCV NS3 function as chaperones to fold RNA [37-40], and the HCV core protein has been recently shown to promote RNA annealing [41]. Perhaps HCV core and NS3 work together to manipulate the HCV genome in ways that are yet to be elucidated.

Acknowledgements

This work was supported by National Institutes of Health grant AI052395. We also thank Dr. Fred Jaffe and Olya Ginzburg for valuable technical assistance.

References

[1] Ali, J. A., and Lohman, T. M. (1997) Kinetic measurement of the step size of DNA unwinding by Escherichia coli UvrD helicase. *Science* 275, 377-380.

[2] Lucius, A. L., Maluf, N. K., Fischer, C. J., and Lohman, T. M. (2003) General methods for analysis of sequential "n-step" kinetic mechanisms: application to single turnover kinetics of helicase-catalyzed DNA unwinding. *Biophys. J.* 85, 2224-2239.

[3] McHutchison, J. G. (2004) Understanding hepatitis C. *Am J Manag Care* 10, S21-9.

[4] Wakita, T., Pietschmann, T., Kato, T., Date, T., Miyamoto, M., Zhao, Z., Murthy, K., Habermann, A., Krausslich, H. G., Mizokami, M., Bartenschlager, R., and Liang, T. J.

(2005) Production of infectious hepatitis C virus in tissue culture from a cloned viral genome. *Nat. Med.* 11, 791-796.
[5] Lam, A. M., and Frick, D. N. (2006) Hepatitis C virus subgenomic replicon requires an active NS3 RNA helicase. *J. Virol.* 80, 404-411.
[6] Artsaenko, O., Tessmann, K., Sack, M., Haussinger, D., and Heintges, T. (2003) Abrogation of hepatitis C virus NS3 helicase enzymatic activity by recombinant human antibodies. *J. Gen. Virol.* 84, 2323-2332.
[7] Hwang, B., Cho, J. S., Yeo, H. J., Kim, J. H., Chung, K. M., Han, K., Jang, S. K., and Lee, S. W. (2004) Isolation of specific and high-affinity RNA aptamers against NS3 helicase domain of hepatitis C virus. *RNA* 10, 1277-1290.
[8] Gozdek, A., Zhukov, I., Polkowska, A., Poznanski, J., Stankiewicz-Drogon, A., Pawlowicz, J. M., Zagorski-Ostoja, W., Borowski, P., and Boguszewska-Chachulska, A. M. (2008) NS3 peptide, a novel potent Hepatitis C virus NS3 helicase inhibitor, its mechanism of action and antiviral activity in the replicon system. *Antimicrob. Agents Chemother.* 52, 393-401.
[9] Frick, D. N., Rypma, R. S., Lam, A. M., and Gu, B. (2004) The nonstructural protein 3 protease/helicase requires an intact protease domain to unwind duplex RNA efficiently. *J. Biol. Chem.* 279, 1269-1280.
[10] Walker, J. E., Saraste, M., Runswick, M. J., and Gay, N. J. (1982) Distantly related sequences in the alpha- and beta-subunits of ATP synthase, myosin, kinases and other ATP-requiring enzymes and a common nucleotide binding fold. *EMBO J.* 1, 945-951.
[11] Frick, D. N., Banik, S., and Rypma, R. S. (2007) Role of divalent metal cations in ATP hydrolysis catalyzed by the hepatitis C virus NS3 helicase: magnesium provides a bridge for ATP to fuel unwinding. *J. Mol. Biol.* 365, 1017-1032.
[12] Story, R. M., Weber, I. T., and Steitz, T. A. (1992) The structure of the E. coli recA protein monomer and polymer. *Nature* 355, 318-325.
[13] Kim, D. W., Gwack, Y., Han, J. H., and Choe, J. (1995) C-terminal domain of the hepatitis C virus NS3 protein contains an RNA helicase activity. *Biochem. Biophys. Res. Commun.* 215, 160-166.
[14] Tai, C. L., Chi, W. K., Chen, D. S., and Hwang, L. H. (1996) The helicase activity associated with hepatitis C virus nonstructural protein 3 (NS3). *J. Virol.* 70, 8477-8484
[15] Kim, J. L., Morgenstern, K. A., Griffith, J. P., Dwyer, M. D., Thomson, J. A., Murcko, M. A., Lin, C., and Caron, P. R. (1998) Hepatitis C virus NS3 RNA helicase domain with a bound oligonucleotide: the crystal structure provides insights into the mode of unwinding. *Structure* 6, 89-100.
[16] Mackintosh, S. G., Lu, J. Z., Jordan, J. B., Harrison, M. K., Sikora, B., Sharma, S. D., Cameron, C. E., Raney, K. D., and Sakon, J. (2006) Structural and biological identification of residues on the surface of NS3 helicase required for optimal replication of the hepatitis C virus. *J. Biol. Chem.* 281, 3528-3535.
[17] Lam, A. M., Keeney, D., and Frick, D. N. (2003) Two novel conserved motifs in the hepatitis C virus NS3 protein critical for helicase action. *J. Biol. Chem.* 278, 44514-44524.

[18] Lam, A. M., Rypma, R. S., and Frick, D. N. (2004) Enhanced nucleic acid binding to ATP-bound hepatitis C virus NS3 helicase at low pH activates RNA unwinding. *Nucleic Acids Res.* 32, 4060-4070.

[19] Frick, D. N., Rypma, R. S., Lam, A. M., and Frenz, C. M. (2004) Electrostatic analysis of the hepatitis C virus NS3 helicase reveals both active and allosteric site locations. *Nucleic Acids Res.* 32, 5519-5528.

[20] Porter, D. J., Short, S. A., Hanlon, M. H., Preugschat, F., Wilson, J. E., Willard, D. H. J., and Consler, T. G. (1998) Product release is the major contributor to kcat for the hepatitis C virus helicase-catalyzed strand separation of short duplex DNA. *J. Biol. Chem.* 273, 18906-18914.

[21] Levin, M. K., Wang, Y. H., and Patel, S. S. (2004) The functional interaction of the hepatitis C virus helicase molecules is responsible for unwinding processivity. *J. Biol. Chem.* 279, 26005-26012.

[22] Levin, M. K., Gurjar, M., and Patel, S. S. (2005) A Brownian motor mechanism of translocation and strand separation by hepatitis C virus helicase. *Nat Struct Mol Biol* 12, 429-435.

[23] Tackett, A. J., Chen, Y., Cameron, C. E., and Raney, K. D. (2005) Multiple full-length NS3 molecules are required for optimal unwinding of oligonucleotide DNA in vitro. *J. Biol. Chem.* 280, 10797-10806.

[24] Serebrov, V., and Pyle, A. M. (2004) Periodic cycles of RNA unwinding and pausing by hepatitis C virus NS3 helicase. *Nature* 430, 476-480.

[25] Lam, A. M., Keeney, D., Eckert, P. Q., and Frick, D. N. (2003) Hepatitis C virus NS3 ATPases/helicases from different genotypes exhibit variations in enzymatic properties. *J. Virol.* 77, 3950-3961.

[26] Howe, A. Y., Chase, R., Taremi, S. S., Risano, C., Beyer, B., Malcolm, B., and Lau, J. Y. (1999) A novel recombinant single-chain hepatitis C virus NS3-NS4A protein with improved helicase activity. *Protein Sci* 8, 1332-1341.

[27] Yanagi, M., St Claire, M., Shapiro, M., Emerson, S. U., Purcell, R. H., and Bukh, J. (1998) Transcripts of a chimeric cDNA clone of hepatitis C virus genotype 1b are infectious in vivo. *Virology* 244, 161-172.

[28] Boguszewska-Chachulska, A. M., Krawczyk, M., Stankiewicz, A., Gozdek, A., Haenni, A. L., and Strokovskaya, L. (2004) Direct fluorometric measurement of hepatitis C virus helicase activity. *FEBS Lett.* 567, 253-258.

[29] Houston, P., and Kodadek, T. (1994) Spectrophotometric assay for enzyme-mediated unwinding of double-stranded DNA. *Proc. Natl. Acad. Sci. U S A* 91, 5471-5474.

[30] Pang, P. S., Jankowsky, E., Planet, P. J., and Pyle, A. M. (2002) The hepatitis C viral NS3 protein is a processive DNA helicase with cofactor enhanced RNA unwinding. *EMBO J.* 21, 1168-1176.

[31] Bazemore, L. R., Takahashi, M., and Radding, C. M. (1997) Kinetic analysis of pairing and strand exchange catalyzed by RecA. Detection by fluorescence energy transfer. *J. Biol. Chem.* 272, 14672-14682.

[32] Yao, N., Hesson, T., Cable, M., Hong, Z., Kwong, A. D., Le, H. V., and Weber, P. C. (1997) Structure of the hepatitis C virus RNA helicase domain. *Nat. Struct. Biol.* 4, 463-467.

[33] Cho, H. S., Ha, N. C., Kang, L. W., Chung, K. M., Back, S. H., Jang, S. K., and Oh, B. H. (1998) Crystal structure of RNA helicase from genotype 1b hepatitis C virus. A feasible mechanism of unwinding duplex RNA. *J. Biol. Chem.* 273, 15045-15052.

[34] Yao, N., Reichert, P., Taremi, S. S., Prosise, W. W., and Weber, P. C. (1999) Molecular views of viral polyprotein processing revealed by the crystal structure of the hepatitis C virus bifunctional protease-helicase. *Structure Fold Des* 7, 1353-1363.

[35] Liu, D., Wang, Y. S., Gesell, J. J., and Wyss, D. F. (2001) Solution structure and backbone dynamics of an engineered arginine-rich subdomain 2 of the hepatitis C virus NS3 RNA helicase. *J. Mol. Biol.* 314, 543-561.

[36] Singleton, M. R., Scaife, S., and Wigley, D. B. (2001) Structural analysis of DNA replication fork reversal by RecG. *Cell* 107, 79-89.

[37] Herschlag, D. (1995) RNA chaperones and the RNA folding problem. *J. Biol. Chem.* 270, 20871-20874.

[38] Lorsch, J. R. (2002) RNA chaperones exist and DEAD box proteins get a life. *Cell* 109, 797-800.

[39] Uhlmann-Schiffler, H., Jalal, C., and Stahl, H. (2006) Ddx42p--a human DEAD box protein with RNA chaperone activities. *Nucleic Acids Res.* 34, 10-22.

[40] Chamot, D., Colvin, K. R., Kujat-Choy, S. L., and Owttrim, G. W. (2005) RNA structural rearrangement via unwinding and annealing by the cyanobacterial RNA helicase, CrhR. *J. Biol. Chem.* 280, 2036-2044.

[41] Cristofari, G., Ivanyi-Nagy, R., Gabus, C., Boulant, S., Lavergne, J. P., Penin, F., and Darlix, J. L. (2004) The hepatitis C virus Core protein is a potent nucleic acid chaperone that directs dimerization of the viral (+) strand RNA in vitro. *Nucleic Acids Res.* 32, 2623-2631.

Short Communication

In: Bacterial DNA, DNA Polymerase and DNA Helicases
Editor: Walter D. Knudsen and Sam S. Bruns
ISBN 978-1-60741-094-2
© 2009 Nova Science Publishers, Inc.

Short Communication

Discovery of Orphan Helicases and Deorphanization by Genome-Wide Analyses in Two Model Organisms, *S. Cerevisiae* and *C. Elegans*

Toshihiko Eki[*,1] *and Fumio Hanaoka*[2]

[1]Division of Life Science and Biotechnology, Department of Ecological Engineering, Toyohashi University of Technology, Toyohashi, Aichi 441-8580, Japan
[2]Department of Life Science, Faculty of Science, Gakushuin University, 1-5-1 Mejiro, Toshima-ku, Tokyo 171-8588, Japan

Eukaryotic DNA helicases [1] are members of the helicase superfamily together with two other functional classes, RNA helicases [2] and chromatin remodeling ATPases [3]. Members of this superfamily play crucial roles in various nucleic acid- and chromatin-mediated cellular processes such as DNA replication, repair and recombination, pre-mRNA splicing, ribosome biogenesis, RNA interference, and chromatin remodeling. Since these reactions are essential for the maintenance, expression, and regulation of genetic information in the chromosome, dysfunctional helicase genes may lead to genetic diseases such as cancer. Indeed, genetic mutations of the human RecQ-like BLM helicase and WRN DNA helicase result in Bloom syndrome and Werner syndrome, which are characterized by the early development of various cancers and premature aging, respectively [4]. Most helicases share conserved amino acid sequence motifs; their corresponding genes are classified into five superfamilies (SF1-SF5) based on the occurrence and characteristics of conserved motifs [5].

The genome sequences of two representative eukaryotes, the budding yeast *Saccharomyces cerevisiae* and the nematode *Caenorhabditis elegans* were determined in 1996 [6] and 1998 [7], respectively. Subsequently, high-throughput genome-wide analyses have been systematically performed to clarify the cellular roles of 6000 and 19000 genes discovered in each genome, including gene expression profiling [8,9], proteome analyses

[10,11], comprehensive analyses of loss-of-function phenotypes [12,13], and genetic interaction analyses [14,15]. A huge amount of functional data from these studies has been analyzed and deposited in public databases such as the *Saccharomyces* Genome Database (SGD) [16] and the WormBase [17]. However, a few decades after the initial sequencing, a large number of genes remain functionally-unknown (i.e., orphan) in both organisms; for instance, over 1000 genes are uncharacterized even in yeast [18]. A large number of helicase-like genes have been found in the sequenced eukaryotic genomes, including many genes encoding orphan helicase-related proteins. Because of the biological importance of the helicase family, a comprehensive analysis of the functions of helicase family members in two model organisms, *S. cerevisiae* [19] and *C. elegans* [20], has been performed. In this Short Communication, we briefly summarize the results from studies of helicase-like proteins in both organisms and describe the effectiveness and limitations of a genome-wide analysis of orphan family members.

Overview of the Eukaryotic Helicase Family in S. Cerevisiae and C. Elegans

At the end of the last century, we systematically identified helicase-like genes in budding yeast by bioinformatics and examined the loss-of-function phenotypes of deletion strains of the newly identified orphan genes under nutrient-rich conditions [19]. The *S. cerevisiae* genome encodes more than 103 helicase-like proteins (note: these include two new members Yku70p and Yku80p, and exclude Bdf2p and the AAA+ members [21] with the exception of two bacterial DNA helicase RuvB (RVB)-like proteins [22] and subunits of the replicative MCM (minichromosome maintenance) complex [23], from the previously-reported 137 proteins]. In the study, we newly identified 5 essential genes (*YDL031W*[*DBP10*], *YDL084W*[*SUB2*], *YKL078W*[*DHR2*], *YLR276C*[*DBP9*], and *YMR128W*[*ECM16*]), and 16 dispensable genes including *YDL070W*[*BDF2*] and *YGL150C*[*INO80*] which were reported as essential genes [19], but later shown to be dispensable for viability by others, as well as *YBR245C*[*ISW1*], *YDR291W*[*HRQ1*], *YDR332W*[*IRC3*], *YDR334W*[*SWR1*], *YFR038W*[*IRC5*], *YGL064C*[*MRH4*], *YHR031C*[*RRM3*], *YIR002C*[*MPH1*], *YKL017C*[*HCS1*], *YLR247C*[*IRC20*], *YLR419W*, *YNL218W*[*MGS1*], *YOL095C*[*HMI1*], and *YOR304W*[*ISW2*]. In the last decade, some of these orphan genes have been characterized and their molecular functions (i.e., deorphanization) have been clarified by others, for example, *SUB2* in pre-mRNA splicing [24], *DHR2*, *ECM16* and several DEAD-box genes such as *DBP9* in ribosome biogenesis [25,26], *MRH4* and *HMI1* in the maintenance of mitochondrial DNA [27,28], *MPH1* and *MGS1* in DNA repair [29,30], and *INO80*, *ISW1*, *ISW2*, and *SWR1* in chromatin remodeling and transcription [31-33].

Recently, we have concentrated on a comprehensive study of helicase family members in a multicellular organism, the nematode *C. elegans*, using an RNA interference (RNAi) technique [20]. Comparative analysis of helicase-like proteins in yeast and *C. elegans* allowed us to identify nematode-specific proteins that are likely to play an important role in

---

[*] Tel: 81-532-44-6907; Fax: 81-532-44-6929; E-mail: eki@eco.tut.ac.jp

multicellular organism-specific functions such as morphogenesis. Identification and characterization of these higher eukaryote-specific helicases will be useful in understanding the molecular mechanisms of genetic diseases caused by mutations of human helicase-like genes. In this study, 134 *C. elegans* helicase proteins were identified in the nematode database WormBase (release WS162) and classified into 10 subfamilies (DEAD-box, DEAH-box, SKI2, UPF1, SWI2/SNF2, MCM, PIF1, MPH1, RAD3, and RECQ) on the basis of a modified classification of yeast helicase-like proteins and "other helicase-like proteins". According to consensus amino acid sequence motif alignments, the proteins of the DEAD-box, DEAH-box, and SKI2 subfamilies are DExD/H-box RNA helicases and belong to superfamily 2 (SF2) [5,34]. The SKI2 subfamily is a subgroup of the DExD/H-box protein family and members of this family show sequence similarity to yeast Ski2p involved in mRNA degradation [35]. Many proteins belonging to the three subfamilies are involved in various aspects of RNA metabolism, including transcription, RNA export, ribosome biogenesis, pre-mRNA splicing, RNA degradation, and translation [2,34,36]. UPF1 is a subfamily of superfamily 1 (SF1) helicases with sequence similarity to yeast Upf1p (Nam7p) which acts in nonsense-mediated mRNA decay [34,37]. The roles of many members of this subfamily in nucleic acids-mediated reactions remain to be determined. Members of the MPH1 subfamily are RNA helicase-like proteins with sequence similarity to the yeast DNA repair protein Mph1p [30], and in higher eukaryotes, many members of this subfamily play roles in RNA metabolism. A number of the proteins in the MCM, PIF1, RAD3, and RECQ subfamilies play roles in DNA-mediated reactions. The proteins of the MCM subfamily are members of AAA+ class of ATPases [21] and form a ring-shaped hexameric DNA helicase for DNA replication [23]. PIF1 proteins are SF1 helicases with sequence similarity to yeast DNA helicase Pif1p, which is needed for stable maintenance of the genome [38]. The members of the RAD3 subfamily share sequence similarity to DNA helicase Rad3p [39], a yeast homolog of human ERCC2 protein. RECQ subfamily members are homologous to *E. coli* RecQ DNA helicase and many members of this subfamily play a role in maintaining genome integrity [4]. SWI2/SNF2 proteins are SF2 helicases with sequence similarity to yeast Snf2p (Swi2p), and many members of this family act primarily in chromatin dynamics and/or DNA metabolism as part of larger complexes [3,40]. The total number of helicase-like proteins in *C. elegans* (134 proteins) was greater than the number of yeast helicase-like proteins (103 proteins including 21 subtelomeric helicase-like proteins [41]). Increased helicase members in the nematode are likely to be generated by gene duplications, since we detected 10 pairs of putative duplicated genes and clusters of *C. elegans*-specific *SNF2*-like genes and a novel class of mobile genetic elements called *Helitrons* [42] in the nematode genome, even though some of these duplicated genes were truncated into pseudogenes. Detection of putative orthologs in three different species (*S. cerevisiae*, *D. malanogaster*, and *H. sapiens*) corresponding to nematode proteins indicated that the helicase superfamily consists of three classes of proteins: highly conserved, evolutionarily diverged proteins, and species-specific proteins [20]. The putative orthologous pairs of yeast and nematode helicase-like proteins and unpaired proteins in each subfamily are summarized in Table 1 together with the corresponding loss-of-function phenotypes of the yeast deletion strains or RNAi-treated nematodes. Six MCM subfamily members [23] and two RVB-like proteins [22] were

Table 1. Comparison of helicase-like proteins in S. cerevisiae and C. elegans and their loss-of-function phenotypes.

Subfamily	S. cerevisiae gene	S. cerevisiae protein	Phenotype code of deletion strain[a]	Function of yeast protein	C. elegans protein	Phenotype code of RNAi-treated nematode[a]
DEAD-box subfamily						
DEAD-box	YOR046C	Dbp5p	red	Nucleo-cytoplasmic RNA transport	T07D4.4a	green
DEAD-box	YNR038W[b]	Dbp6p	red	Ribosome biogenesis (60S)	ZK686.2	orange
DEAD-box	YHR169W	Dbp8p	red	Ribosome biogenesis	H20J04.4b	orange
DEAD-box	YLR276C	Dbp9p	red	Ribosome biogenesis	C24H12.4a	orange
DEAD-box	YDL031W	Dbp10p	red	Ribosome biogenesis	Y94H6A.5a	orange
DEAD-box	YDL160C	Dhh1p	orange	Decapping and mRNA turnover	C07H6.5 (CGH-1)	red
DEAD-box	YLL008W	Drs1p	red	Ribosome biogenesis	Y71G12B.8	orange
DEAD-box	YMR290C	Has1p	red	Ribosome biogenesis	B0511.6	orange
DEAD-box	YJL033W	Hca4p	red	Ribosome biogenesis, pre-rRNA maturation (40S)	Y23H5B.6	orange
DEAD-box	YBR237W	Prp5p	red	Pre-mRNA splicing	F53H1.1	orange
DEAD-box	YBR142W	Mak5p	red	Ribosome biogenesis (60S)	F55F8.2a	orange
DEAD-box	YGL171W	Rok1p	red	Ribosome biogenesis, pre-rRNA maturation (40S)	R05D11.4	green
DEAD-box	YHR065C	Rrp3p	red	rRNA maturation (40S)	T26G10.1	orange
DEAD-box	YFL002C	Spb4p	red	Ribosome biogenesis, pre-rRNA maturation (60S)	ZK512.2	orange
DEAD-box	YDL084W	Sub2p	red	Pre-mRNA splicing, mRNA export	C26D10.2a (HEL-1)	red

Subfamily	S. cerevisiae gene	S. cerevisiae protein	Phenotype code of deletion strain[a]	Function of yeast protein	C. elegans protein	Phenotype code of RNAi-treated nematode[a]
DEAD-box	YDR021W	Fal1p		Ribosome biogenesis, pre-rRNA maturation (40S)	F33D11.10	
DEAD-box	YKR059W	Tif1p		Translation initiation	Y65B4A.6	
DEAD-box	YJL138C	Tif2p		Translation initiation	F57B9.6a (INF-1)	
DEAD-box	YDR243C	Prp28p		Pre-mRNA splicing	F01F1.7	
DEAD-box	YNL112W	Dbp2p		RNA stability, ribosome biogenesis	F58E10.3a	
DEAD-box	YOR204W[c]	Ded1p		Translation initiation	Y71H2AM.19	
DEAD-box	YPL119C	Dbp1p		Translation initiation	Y54E10A.9a (VBH-1)	
DEAD-box	YGL064C[d]	Mrh4p		Maintenance of mitochondrial DNA	F01F1.7/F53H1.1	
DEAD-box	YGL078C	Dbp3p		Ribosome biogenesis, pre-rRNA maturation (60S)	F58E10.3a	
DEAD-box	YDR194C	Mss116p		Mitochondrial gene expression	B0511.6	
DEAD-box	YKR024C	Dbp7p		Ribosome biogenesis (60S)	B0511.6	
DEAD-box					C14C11.6 (MUT-14)[e]	
DEAD-box					C46F11.4	
DEAD-box					F57B9.3	
DEAD-box					F58G11.2	
DEAD-box					H27M09.1	
DEAD-box					T06A10.1 (MEL-46)	WB

Table 1. (Continued)

Subfamily	S. cerevisiae gene	S. cerevisiae protein	Phenotype code of deletion strain[a]	Function of yeast protein	C. elegans protein	Phenotype code of RNAi-treated nematode[a]
DEAD-box					Y38A10A.6	🟩
DEAD-box					Y54G11A.3	🟩
DEAD-box					Y55F3BR.1	🟩
DEAD-box					ZC317.1	🟩
DEAD-box (glh)					B0414.6 (GLH-3)	🟩
DEAD-box (glh)					C55B7.1 (GLH-2)	🟩
DEAD-box (glh)					T12F5.3 (GLH-4)	🟩
DEAD-box (glh)					T21G5.3 (GLH-1)	🟩
DDX1-like					F20A1.9	🟩

DEAH-box subfamily

Subfamily	S. cerevisiae gene	S. cerevisiae protein	Phenotype code of deletion strain[a]	Function of yeast protein	C. elegans protein	Phenotype code of RNAi-treated nematode[a]
DEAH-box	YMR128W	Ecm16p	🟥	Ribosome biogenesis (40S)	C06E1.10 (RHA-2)	🟧
DEAH-box	YKR086W	Prp16p	🟥	Pre-mRNA splicing	K03H1.2 (MOG-1)	🟥
DEAH-box	YGL120C	Prp43p	🟥	Pre-mRNA splicing	F56D2.6a	🟧
DEAH-box	YNR011C	Prp2p	🟥	Pre-mRNA splicing	C04H5.6 (MOG-4)	🟥
DEAH-box	YER013W	Prp22p	🟩	Pre-mRNA splicing	EEED8.5 (MOG-5)	🟥
DEAH-box	YLR419W			Unknown	T07D4.3 (RHA-1)	🟩
DEAH-box	YKL078W	Dhr2p	🟥	Ribosome biogenesis (40S)	EEED8.5 (MOG-5)	

Subfamily	S. cerevisiae gene	S. cerevisiae protein	Phenotype code of deletion strain[a]	Function of yeast protein	C. elegans protein	Phenotype code of RNAi-treated nematode[a]
DEAH-box					F52B5.3	🟩
DEAH-box					T05E8.3	🟧
DEAH-box					Y108F1.5	🟩
DEAH-box					Y37E11AM.1	🟩
DEAH-box					Y67D2.6	🟩
SKI2 subfamily						
SKI2	YJL050W	Mtr4p	🟥	Ribosome biogenesis, pre-rRNA processing (60S), nuclear RNA degradation (?), mRNA transport (?)	W08D2.7	🟧
SKI2	YPL029W	Suv3p	🟩	Mitochondrial RNA degradation	C08F8.2a	🟧
SKI2	YLR398C	Ski2p	🟩	dsRNA killer propagation, cytoplasmic 3'-5' RNA degradation	F01G4.3	🟩
SKI2	YER172C	Brr2p	🟥	Pre-mRNA splicing	Y46G5A.4	🟥
SKI2	YGR271W	Slh1p	🟩	Regulation of translation	Y54E2A.6	🟩
SKI2	YGL251C	Hfm1p	🟩	Crossover control in meiosis	Y54E2A.6	
SKI2					C28H8.3	🟩
SKI2					Y46G5A.6	🟩
SKI2					Y55B1AL.3	🟩

Table 1. (Continued)

Subfamily	S. cerevisiae gene	S. cerevisiae protein	Phenotype code of deletion strain[a]	Function of yeast protein	C. elegans protein	Phenotype code of RNAi-treated nematode[a]
UPF1 subfamily						
UPF1	YHR164C	Dna2p	🟥	DNA replication, Okazaki fragment maturation	F43G6.1b (DNA-2)	🟩
UPF1	YMR080C	Nam7p	🟩	RNA stability, nonsense-mediated mRNA decay	Y48G8AL.6 (SMG-2)	🟩
UPF1	YKL017C	Hcs1p	🟥	DNA replication?	Y48G8AL.6	
UPF1	YLR430W	Sen1p	🟥	tRNA-, snRNA-, snoRNA- maturation	Y48G8AL.6	
UPF1	YER176W	Ecm32p	🟩	Translation termination	Y48G8AL.6	
UPF1					C05C10.2	🟩
UPF1					C41D11.7	🟩
UPF1					C44H9.4	🟩
UPF1					K08D10.5	🟩
UPF1					R03D7.2	🟩
UPF1					Y80D3A.2	🟧
UPF1					ZK1067.2	🟩
UPF1 (far related)					C44H9.2	

Subfamily	S. cerevisiae gene	S. cerevisiae protein	Phenotype code of deletion strain[a]	Function of yeast protein	C. elegans protein
SWI2/SNF2 subfamily					
SWI2/SNF2	YAL019W	Fun30p		DNA repair?, mitochondrial function?	M03C11.8
SWI2/SNF2	YBR245C	Isw1p		Chromatin remodeling, transcription	F37A4.8 (ISW-1)
SWI2/SNF2	YOR304W	Isw2p		Chromatin remodeling, transcription	
SWI2/SNF2	YPL082C	Mot1p		Transcription	F15D4.1 (BTF-1)
SWI2/SNF2	YDR334W	Swr1p		Chromatin remodeling, DNA repair	Y111B2A.22 (SSL-1)
SWI2/SNF2	YER164W	Chd1p		Chromatin remodeling, transcription	H06O01.2
SWI2/SNF2	YGL163C	Rad54p		DNA repair, DNA recombination	W06D4.6 (RAD-54)
SWI2/SNF2	YJR035W	Rad26p		Transcription coupled repair	F53H4.1 (CSB-1)
SWI2/SNF2	YOR290C	Snf2p		Chromatin remodeling, transcription	F01G4.1 (PSA-4)
SWI2/SNF2	YIL126W	Sth1p		G2 control, chromatin remodeling, transcription	C52B9.8
SWI2/SNF2	YBR073W	Rdh54p		DNA repair, DNA recombination	W06D4.6 (RAD-54)
SWI2/SNF2	YFR038W	Irc5p		Unkown (DNA repair?, chromatin remodeling?)	F01G4.1/C52B9.8
SWI2/SNF2	YGL150C	Ino80p		Chromatin remodeling, transcription, DNA repair	Y111B2A.22 (SSL-1)
SWI2/SNF2	YBR114W	Rad16p		DNA repair	F54E12.2

Table 1. (Continued)

Subfamily	S. cerevisiae gene	S. cerevisiae protein	Phenotype code of deletion strain[a]	Function of yeast protein	C. elegans protein	Subfamily
SWI2/SNF2	YLR032W	Rad5p		Post-replication repair	F54E12.2	
SWI2/SNF2	YOR191W	Ris1p		Chromatin structure, gene silencing	F54E12.2	
SWI2/SNF2					B0041.7 (XNP-1)	
SWI2/SNF2					C16A3.1	
SWI2/SNF2					C27B7.4	
SWI2/SNF2					F26F12.7 (LET-418)	
SWI2/SNF2					F53H4.6	
SWI2/SNF2					F54E12.2	
SWI2/SNF2					F59A7.8	
SWI2/SNF2					T04D1.4 (TAG-192)	
SWI2/SNF2					T14G8.1 (CHD-3)	
SWI2/SNF2					T23H2.3	
SWI2/SNF2					Y113G7B.14	
SWI2/SNF2					Y116A8C.13	
SWI2/SNF2 (far related)					C25F9.5	
SWI2/SNF2 (far related)					F19B2.5	
SWI2/SNF2 (far related)					M04C3.1	

Subfamily	S. cerevisiae gene	S. cerevisiae protein	Phenotype code of deletion strain[a]	Function of yeast protein	C. elegans protein	Subfamily
SWI2/SNF2 (far related)					Y43F8B.14	
SWI2/SNF2 (far related)					C25F9.4	WB
SWI2/SNF2 (far related)					M04C3.2	WB
MCM subfamily						
MCM	YBL023C	Mcm2p		DNA replication	Y17G7B.5a (MCM-2)	
MCM	YBR202W	Cdc47p		DNA replication	F32D1.10 (MCM-7)	
MCM	YEL032W	Mcm3p		DNA replication	C25D7.6 (MCM-3)	
MCM	YGL201C	Mcm6p		DNA replication	ZK632.1a (MCM-6)	
MCM	YLR274W	Cdc46p		DNA replication	R10E4.4 (MCM-5)	
MCM	YPR019W	Cdc54p		DNA replication	Y39G10AR.14 (MCM-4)	
PIF1 subfamily						
PIF1	YML061C	Pif1p		Maintenance of mitochondrial DNA and telomeres	Y18H1A.6 (PIF-1)	
PIF1	YHR031C	Rrm3p		rDNA replication, Ty1 transposition		
PIF1					C11G6.2	
PIF1					F11C3.1	
PIF1					Y116F11A.1	

Table 1. (Continued)

Subfamily	S. cerevisiae gene	S. cerevisiae protein	Phenotype code of deletion strain[a]	Function of yeast protein	C. elegans protein	Subfamily
PIF1 (Helitron)					F33H12.6	
PIF1 (Helitron)					F59H6.5	
PIF1 (Helitron)					Y16E11A.2	
PIF1 (Helitron)					Y27F2A.5	
PIF1 (Helitron)					Y46B2A.2	
PIF1 (Helitron)					ZK250.9	

MPH1 subfamily

MPH1	YIR002C	Mph1p		DNA repair	D2005.5 (DRH-3)	
MPH1					C01B10.1 (DRH-2)[f]	
MPH1					D2005.5 (DRH-3)	
MPH1					F15B10.2 (DRH-1)	
MPH1					K12H4.8 (DCR-1)	

RAD3 subfamily

RAD3	YER171W	Rad3p		DNA repair, transcription	Y50D7A.2	
RAD3	YPL008W	Chl1p		Chromosome segregation	M03C11.2	
RAD3					Y50D7A.11	

Subfamily	S. cerevisiae gene	S. cerevisiae protein	Phenotype code of deletion strain[a]	Function of yeast protein	C. elegans protein	Subfamily
RAD3					F25H2.13	
RAD3					F33H2.1 (DOG-1)	

RECQ subfamily

RECQ	YMR190C	Sgs1p		DNA repair, DNA recombination	T04A11.6 (HIM-6)	
RECQ					E03A3.2 (RCQ-5)	
RECQ					F18C5.2 (WRN-1)	
RECQ					K02F3.12	

Other helicase-related proteins

RVB	YDR190C	Rvb1p		Chromatin remodeling, transcription	C27H6.2 (RUVB-1)	
RVB	YPL235W	Rvb2p		Chromatin remodeling, transcription	T22D1.10 (RUVB-2)	
SSL2	YIL143C	Ssl2p		DNA repair, transcription	Y66D12A.15	
KU70	YMR284W	Yku70p		DNA repair, telomere maintenance	Y47D3A.4 (CKU-70)	WB
KU80	YMR106C	Yku80p		DNA repair, telomere maintenance	R07E5.8 (CKU-80)	
YLR247C	YLR247C	Irc20p		Unknown (transcriptional regulation?)	T05A12.4a	
YDR291W	YDR291W	Hrq1		Unknown	F18C5.2	
YDR332W	YDR332W	Irc3p		Unknown (mitochondrial function?)	R05D11.4	

Table 1. (Continued)

Subfamily	S. cerevisiae gene	S. cerevisiae protein	Phenotype code of deletion strain[a]	Function of yeast protein	C. elegans protein	Subfamily
HPR5	YJL092W	Hpr5p		DNA repair	Y55B1BR.3	
HMI1	YOL095C	Hmi1p		Maintenance of mitochondrial DNA	ND[g]	
Y'-Hel1	YBL111C			Unknown	ND	
Y'-Hel1	YBL113C			Unknown	ND	
Y'-Hel1	YDR545W	Yrf1-1p		Unknown	ND	
Y'-Hel1	YEL077C			Unknown	ND	
Y'-Hel1	YER190W	Yrf1-2p		Unknown	ND	
Y'-Hel1	YFL066C			Unknown	ND	
Y'-Hel1	YGR296W	Yrf1-3p		Unknown	ND	
Y'-Hel1	YHL050C			Unknown	ND	
Y'-Hel1	YHR218W			Unknown	ND	
Y'-Hel1	YHR219W			Unknown	ND	
Y'-Hel1	YIL177C			Unknown	ND	
Y'-Hel1	YJL225C			Unknown	ND	
Y'-Hel1	YLL066C			Unknown	ND	
Y'-Hel1	YLL067C			Unknown	ND	
Y'-Hel1	YLR466W	Yrf1-4p		Unknown	ND	
Y'-Hel1	YLR467W	Yrf1-5p		Unknown	ND	
Y'-Hel1	YML133C			Unknown	ND	

Subfamily	S. cerevisiae gene	S. cerevisiae protein	Phenotype code of deletion strain[a]	Function of yeast protein	C. elegans protein	Subfamily
Y'-Hel1	YNL339C	Yrf1-6p		Unknown	ND	
Y'-Hel1	YOR396W			Unknown	ND	
Y'-Hel1	YPL283C	Yrf1-7p		Unknown	ND	
Y'-Hel1	YPR204W			Unknown	ND	
Twinkle-like					F46G11.1	
DNA pol theta-like					W03A3.2 (POLQ-1)	
MOP-3-like					F20H11.2 (NSH-1)	
INTS6-like					F08B4.1b (DIC-1)	WB
Plant helicase-like					F52G3.3	
Plant helicase-like					F52G3.4	

The loss-of-function phenotypes of helicase-like proteins in S. cerevisiae and C. elegans are summarized according to the subfamily for comparison between species. This table is revised and modified from Table 2 in the reference[20]. Function-uncertain yeast genes are indicated in bold.

a Loss-of-function phenotypes of yeast gene deletion strains, phenotype codes, and functions of yeast proteins. Phenotype code: The phenotypes of RNAi-treated nematodes and yeast gene deletion strains are indicated by color coding: embryonic lethal in red, larval arrest and slow growth in orange, and wild-type (no phenotype) in green (nematode); lethal in red, slow growth in orange, viable in green, and no data in white (yeast), respectively. WB: RNAi phenotype data was obtained from the WormBase (release WS189).

b The proteins surrounded by bold lines are a putative orthologous pair based on BLASTP scores (not shown in the table). The Ku70 and Ku80 homologues in yeast and nematodes are described in the Saccharomyces Genome Database and WormBase. c Two pairs of yeast proteins (Snf2p and Sth1p, Ded1p and Dbp1p) with two C. elegans orthologs are surrounded by dashed lines. d The yeast proteins with BLAST scores lower than that of the putative orthologs or without any sequence homologies to C. elegans proteins are surrounded by a solid line and indicated separately. e C. elegans proteins without significant similarities to yeast helicase-like proteins are also indicated separately. f DRH-2 is omitted from the current database as a pseudogene. g ND: not detected. Several C. elegans proteins with E-values of over e-10 to subtelomeric helicase-like proteins were omitted because of similarities to low-complexity regions in their amino acid sequences

completely conserved and required for viability in both species. The subfamilies of DEAD-box, DEAH-box, SKI2, UPF1, and SWI2/SNF2, contained two classes of the proteins: those that are conserved in both yeast and nematode, and those that are unique in each species. For example, 21 putative orthologous pairs of the DEAD-box members were well conserved due to significant sequence homologies with their corresponding partners. In contrast, we could not detect any homologous nematode proteins corresponding to the four yeast proteins Mrh4p, Dbp3p, Dbp7p, and Mss116p, or any yeast counterparts of 15 nematode proteins (Table 1). The budding yeast contains one or two members in the PIF1, MPH1, RAD3, and RECQ subfamilies. In contrast, the family members in *C. elegans* had apparently increased, and many of these divergent proteins are conserved in humans [20]. Twenty-five yeast-specific proteins and six higher eukaryote-specific nematode proteins (DIC-1, NSH-1, POLQ-1, F46G11.1, F52G3.3, and F52G3.4) were also detected [20]. We found a high degree of conservation of loss-of-function phenotypes for putative orthologs in both organisms. The majority (20 out of 22 proteins) of putative *C. elegans* orthologs of yeast essential DEAD-box members caused embryonic lethality or growth-defective phenotypes when depleted (Table 1) [20]. Similar phenotypic conservation of putative orthologs in both species was also found in members of the DEAH-box, SKI2, MCM, RAD3 subfamilies, as well as for the RVB-like and SSL2-like proteins, suggesting that these genes play similar essential roles in *C. elegans,* as do their counterparts in yeast. In contrast, it should be noted that depletions of subfamily members that have diverged in *C. elegans* rarely induced growth-defected phenotypes. This phenomenon seems to suggest that they play dispensable roles in survival or growth under nematode test conditions, but it may also indicate a functional redundancy with paralogous proteins or diverged members in *C. elegans*. In fact, two diverged DEAD-box members (GLH-1 and GLH-4) are known to play redundant roles in germline development [43], as do two *C. elegans* Mi-2 homologues (LET-418 and CHD-3) of the SWI2/SNF2 subfamily in development [44].

Current Status of Deorphanization of Orphan Helicase Members by Genome-Wide Analyses in Yeast and Nematodes

It is well known that proteins with homologous sequences, such as paralogs, play closely related roles in the same species. Previously, we obtained hints that allowed us to speculate on the biological functions of newly discovered orphan proteins by clustering helicase-like proteins on the basis of their amino acid sequence homology. We classified yeast helicase-like proteins into 11 clusters based on their sequence similarities and assigned 21 orphan proteins to the clusters (note: use of the nomenclature "cluster" here is similar to the use of the term "subfamily" in Table 1) [19]. It was reasonable to assume that each orphan protein would play related roles to other members in the cluster, and later this assumption was proven to be correct by others. For instance, deletion of the Cluster VII orphan genes *DBP9, DBP10, SUB2,* and *MRH4* and two (*DHR2,* and *ECM16*) of the three Cluster X orphan genes affected viability or cell growth. Later these gene products were shown to play essential roles in RNA metabolism such as ribosome biogenesis and pre-mRNA splicing together with other characterized members in the corresponding cluster (Table 1). In contrast, six orphan members in Cluster I were dispensable, suggesting that they play non-essential functions in cell viability under normal culture conditions. Four out of six proteins (Isw1p, Isw2p, Swr1p, and Ino80p) were shown to be involved in chromatin remodeling and/or DNA repair like other members in the cluster. This approach also seems to be applicable to the characterization of nematode orphan genes. For example, it is well known that diverged DNA helicases of the RECQ subfamily (HIM-6 [45], RCQ-5 [46], WRN-1 [47]), and DOG-1 [48] in the RAD3 subfamily, play important roles in DNA repair. The other RAD3 member Y50D7A.2/Y50D7A.11 is a putative ortholog of human ERCC2 and RTEL-1(BCH-1) [49] is a counterpart of mouse RTEL1, which were reported to act in nucleotide excision repair and the maintenance of genome integrity, respectively. When a putative yeast ortholog is detected, we may then assign a cellular role to the corresponding nematode counterpart. However, it must be noted that we cannot unilaterally assign roles to orphan proteins based on the functional conservation of putative orthologs among species. For example, a sole yeast MPH1 subfamily protein Mph1p functions in DNA repair [30]; however, three nematode members (DRH-1, DRH-3, and DCR-1) play crucial roles in RNA interference [50,51], as described later. We should also remember that this approach is not applicable to the analysis of species-specific orphan proteins such as the *C. elegans*-specific *SNF2*-like proteins due to an absence of homologous proteins in other species.

Alternatively, a systematic analysis based on loss-of-function phenotypes is useful for finding the potential roles of newly discovered genes in a variety of organisms. In our analyses, 42 (51.2 %) out of 82 unique helicase-like genes were essential for viability in the budding yeast and 51 (39.5 %) out of 129 genes tested showed development- or growth-affected phenotypes in the nematode. The large fraction of essential genes in the helicase superfamily indicates their crucial roles in both species. The cellular functions of five essential orphan proteins were successfully determined presumably because they have roles in a few limited essential cellular processes in budding yeast. In contrast, the deorphanization of dispensable orphan proteins in yeast has not been as easy, because they could potentially be involved in many bioprocesses that may not be essential for viability under the tested conditions. We could not obtain any indication of their roles in yeast by performing conventional phenotypic analyses of the deletion strains; however, phenotypic analyses worked successfully under particular culture conditions. We observed impaired cell growth of the deletion strains of three dispensable genes (*YDR332W, YGL064C*, and *YOL095C*) at higher temperatures [19], and subsequently observed growth defects on a culture plate containing the non-fermentable carbon source glycerol (unpublished data). A similar phenotype of impaired mitochondrial respiration was also reported for these deletion strains in a systematic screen for mitochondrial proteins by Steinmetz et al. [52], obviously suggesting mitochondrial functions for these gene products. In addition, a proper yeast-based reporter assay is also useful for the characterization of orphan genes as well for screening genes involved in a particular cell process. For example, three dispensable orphan helicase-like genes (*YDR332W, YFR038W*, and *YLR247C*) were detected in a genome-wide screen for yeast deletion strains affecting levels of spontaneous Rad52p foci using a reporter plasmid containing a Rad52-YFP gene fusion. The proteins encoded by these genes were designated *IRC3, IRC5*, and *IRC20* (*Increased Recombination Centers*) and are likely to be involved in

DNA repair and/or recombination because of the intrinsic roles of Rad52p in both processes. So far, the functions of two proteins (Ydr291wp[Hrq1p] and Ylr419wp) and 21 subtelomeric helicase-like proteins remain unknown, and the cellular roles of more than 5 orphan proteins (Fun30p, Hcs1p, Irc3p, Irc5p, and Irc20p) need to be further clarified despite their putative functions.

In contrast to yeast helicase-like proteins, the functions of many helicase proteins, especially those of diverged members in higher eukaryotes, remain unclear. Recently, we screened for helicase-like genes involved in the maintenance of genome integrity based on increased sensitivity to X-ray irradiation in RNAi-treated nematodes. Unexpectedly we identified an RNA helicase-like gene *drh-3* (*D2005.5*) in the candidates, which was not implicated in DNA repair [20]; van Haaften et al. also detected *D2005.5* in an RNAi-based screen for nematode genes involved in protection against radiation [53]. Depletion of *drh-3* function caused both hypersensitivity to DNA double-strand damage and induced chromosomal abbreviations in oocytes [54].

The DRH-3 protein belongs to the MPH1 family, and a sole yeast counterpart Mph1p functions in DNA repair. These observations suggested that an RNA helicase-like DRH-3 protein might act in DNA repair. However, Duchaine et al. recently reported that DRH-3 co-purified with an essential RNAi protein, termed Dicer (DCR-1), which functions in an RNAi process in *C. elegans* [50]. Unlike yeast Mph1p or nematode DRH-3, mammalian DRH-3-like proteins such as RIG-I and MDA5 play a crucial role in sensing viral RNAs in host defense against pathogenic RNA viruses [55].

This study shows that there exists considerable difficulty in the deorphanization of function-unknown helicases and alerts us to the possibility that a putative ortholog may function differently from its counterparts in other species. In addition, this raises an interesting and fundamental question as to why (how) RNAi proteins are involved in both the maintenance of genome integrity and RNAi in multicellular organisms. Currently, DRH-3 is being studied as an interesting key factor to clarify the molecular mechanism of the RNAi pathway governing chromosome dynamics in *C. elegans*.

In our systematic analysis of helicase family, we were able to determine previously unknown roles for a novel helicase-like protein DRH-3 in nematodes. Deorphanization of orphan proteins in a protein family is reminiscent of treasure hunting. We believe we will find "hidden jewels" with novel cellular functions using properly designed RNAi-based screens of helicase family members, especially for orphan proteins that diverged in higher eukaryotes.

Acknowledgements

The authors thank their colleagues, Drs. Isao Katsura, Takeshi Ishihara, Akiko Shiratori, Mikio Arisawa, Takehiko Shibata, Yasufumi Murakami, and members of the Eki Laboratory for their support. We also apologize to colleagues whose work could not be cited in the manuscript. This work was supported in part by Grants-in-Aid for Scientific Research (C) (No. 17590057 and 20590056) from The Ministry of Education, Culture, Sports, Science and Technology (MEXT).

References

[1] Tuteja, N., and Tuteja, R. (2004). Unraveling DNA helicases. Motif, structure, mechanism and function. *Eur. J. Biochem., 271*, 1849-1863.

[2] Bleichert, F., and Baserga, S. J. (2007). The long unwinding road of RNA helicases. *Mol. Cell, 27*, 339-352.

[3] Lusser, A., and Kadonaga, J. T. (2003). Chromatin remodeling by ATP-dependent molecular machines. *Bioessays, 25*, 1192-1200.

[4] Hanada, K., and Hickson, I. D. (2007). Molecular genetics of RecQ helicase disorders. *Cell Mol. Life Sci., 64*, 2306-2322.

[5] Gorbalenya, A. E., and Koonin, E. V. (1993). Helicases: amino acid sequence comparisons and structure-function relationships. *Curr. Opin. Struct. Biol., 3*, 419-429.

[6] Goffeau, A., Barrell, B. G., Bussey, H., Davis, R. W., Dujon, B., Feldmann, H., Galibert, F., *et al.* (1996). Life with 6000 genes. *Science, 274*, 546, 563-547.

[7] The *C. elegans* Sequencing Consortium (1998). Genome sequence of the nematode *C. elegans*: a platform for investigating biology. *Science, 282*, 2012-2018.

[8] Kim, S. K., Lund, J., Kiraly, M., Duke, K., Jiang, M., Stuart, J. M., Eizinger, A., *et al.* (2001). A gene expression map for *Caenorhabditis elegans*. *Science, 293*, 2087-2092.

[9] Wodicka, L., Dong, H., Mittmann, M., Ho, M. H., and Lockhart, D. J. (1997). Genome-wide expression monitoring in *Saccharomyces cerevisiae*. *Nat. Biotechnol., 15*, 1359-1367.

[10] Krogan, N. J., Cagney, G., Yu, H., Zhong, G., Guo, X., Ignatchenko, A., Li, J., *et al.* (2006). Global landscape of protein complexes in the yeast *Saccharomyces cerevisiae*. *Nature, 440*, 637-643.

[11] Li, S., Armstrong, C. M., Bertin, N., Ge, H., Milstein, S., Boxem, M., Vidalain, P. O., *et al.* (2004). A map of the interactome network of the metazoan *C. elegans*. *Science, 303*, 540-543.

[12] Giaever, G., Chu, A. M., Ni, L., Connelly, C., Riles, L., Veronneau, S., Dow, S., *et al.* (2002). Functional profiling of the *Saccharomyces cerevisiae* genome. *Nature, 418*, 387-391.

[13] Kamath, R. S., Fraser, A. G., Dong, Y., Poulin, G., Durbin, R., Gotta, M., Kanapin, A., *et al.* (2003). Systematic functional analysis of the *Caenorhabditis elegans* genome using RNAi. *Nature, 421*, 231-237.

[14] Tong, A. H., Lesage, G., Bader, G. D., Ding, H., Xu, H., Xin, X., Young, J., *et al.* (2004). Global mapping of the yeast genetic interaction network. *Science, 303*, 808-813.

[15] Lehner, B., Crombie, C., Tischler, J., Fortunato, A., and Fraser, A. G. (2006). Systematic mapping of genetic interactions in *Caenorhabditis elegans* identifies common modifiers of diverse signaling pathways. *Nat. Genet., 38*, 896-903.

[16] Christie, K. R., Weng, S., Balakrishnan, R., Costanzo, M. C., Dolinski, K., Dwight, S. S., Engel, S. R., *et al.* (2004). *Saccharomyces* Genome Database (SGD) provides tools to identify and analyze sequences from *Saccharomyces cerevisiae* and related sequences from other organisms. *Nucleic Acids Res., 32*, D311-314.

[17] Rogers, A., Antoshechkin, I., Bieri, T., Blasiar, D., Bastiani, C., Canaran, P., Chan, J., et al. (2008). WormBase 2007. *Nucleic Acids Res., 36*, D612-617.

[18] Pena-Castillo, L., and Hughes, T. R. (2007). Why are there still over 1000 uncharacterized yeast genes? *Genetics, 176*, 7-14.

[19] Shiratori, A., Shibata, T., Arisawa, M., Hanaoka, F., Murakami, Y., and Eki, T. (1999). Systematic identification, classification, and characterization of the open reading frames which encode novel helicase-related proteins in *Saccharomyces cerevisiae* by gene disruption and Northern analysis. *Yeast, 15*, 219-253.

[20] Eki, T., Ishihara, T., Katsura, I., and Hanaoka, F. (2007). A genome-wide survey and systematic RNAi-based characterization of helicase-like genes in *Caenorhabditis elegans*. *DNA Res., 14*, 183-199.

[21] Erzberger, J. P., and Berger, J. M. (2006). Evolutionary relationships and structural mechanisms of AAA+ proteins. *Annu Rev Biophys Biomol Struct, 35*, 93-114.

[22] Jonsson, Z. O., Dhar, S. K., Narlikar, G. J., Auty, R., Wagle, N., Pellman, D., Pratt, R. E., et al. (2001). Rvb1p and Rvb2p are essential components of a chromatin remodeling complex that regulates transcription of over 5% of yeast genes. *J. Biol. Chem., 276*, 16279-16288.

[23] Ishimi, Y. (1997). A DNA helicase activity is associated with an MCM4, -6, and -7 protein complex. *J. Biol. Chem., 272*, 24508-24513.

[24] Libri, D., Graziani, N., Saguez, C., and Boulay, J. (2001). Multiple roles for the yeast SUB2/yUAP56 gene in splicing. *Genes Dev., 15*, 36-41.

[25] Colley, A., Beggs, J. D., Tollervey, D., and Lafontaine, D. L. (2000). Dhr1p, a putative DEAH-box RNA helicase, is associated with the box C+D snoRNP U3. *Mol. Cell Biol., 20*, 7238-7246.

[26] Daugeron, M. C., Kressler, D., and Linder, P. (2001). Dbp9p, a putative ATP-dependent RNA helicase involved in 60S-ribosomal-subunit biogenesis, functionally interacts with Dbp6p. *RNA, 7*, 1317-1334.

[27] Kuusk, S., Sedman, T., Joers, P., and Sedman, J. (2005). Hmi1p from *Saccharomyces cerevisiae* mitochondria is a structure-specific DNA helicase. *J. Biol. Chem., 280*, 24322-24329.

[28] Schmidt, U., Lehmann, K., and Stahl, U. (2002). A novel mitochondrial DEAD box protein (Mrh4) required for maintenance of mtDNA in *Saccharomyces cerevisiae*. *FEMS Yeast Res., 2*, 267-276.

[29] Hishida, T., Iwasaki, H., Ohno, T., Morishita, T., and Shinagawa, H. (2001). A yeast gene, *MGS1*, encoding a DNA-dependent AAA(+) ATPase is required to maintain genome stability. *Proc. Natl. Acad. Sci. U S A, 98*, 8283-8289.

[30] Scheller, J., Schurer, A., Rudolph, C., Hettwer, S., and Kramer, W. (2000). MPH1, a yeast gene encoding a DEAH protein, plays a role in protection of the genome from spontaneous and chemically induced damage. *Genetics, 155*, 1069-1081.

[31] Krogan, N. J., Keogh, M. C., Datta, N., Sawa, C., Ryan, O. W., Ding, H., Haw, R. A., et al. (2003). A Snf2 family ATPase complex required for recruitment of the histone H2A variant Htz1. *Mol. Cell, 12*, 1565-1576.

[32] Shen, X., Mizuguchi, G., Hamiche, A., and Wu, C. (2000). A chromatin remodelling complex involved in transcription and DNA processing. *Nature, 406*, 541-544.

[33] Tsukiyama, T., Palmer, J., Landel, C. C., Shiloach, J., and Wu, C. (1999). Characterization of the imitation switch subfamily of ATP-dependent chromatin-remodeling factors in *Saccharomyces cerevisiae*. *Genes Dev., 13*, 686-697.

[34] de la Cruz, J., Kressler, D., and Linder, P. (1999). Unwinding RNA in *Saccharomyces cerevisiae*: DEAD-box proteins and related families. *Trends Biochem. Sci., 24*, 192-198.

[35] Anderson, J. S., and Parker, R. P. (1998). The 3' to 5' degradation of yeast mRNAs is a general mechanism for mRNA turnover that requires the SKI2 DEVH box protein and 3' to 5' exonucleases of the exosome complex. *EMBO J., 17*, 1497-1506.

[36] Cordin, O., Banroques, J., Tanner, N. K., and Linder, P. (2006). The DEAD-box protein family of RNA helicases. *Gene, 367*, 17-37.

[37] Czaplinski, K., Weng, Y., Hagan, K. W., and Peltz, S. W. (1995). Purification and characterization of the Upf1 protein: a factor involved in translation and mRNA degradation. *RNA, 1*, 610-623.

[38] Boule, J. B., and Zakian, V. A. (2006). Roles of Pif1-like helicases in the maintenance of genomic stability. *Nucleic Acids Res., 34*, 4147-4153.

[39] Sung, P., Prakash, L., Matson, S. W., and Prakash, S. (1987). RAD3 protein of *Saccharomyces cerevisiae* is a DNA helicase. *Proc. Natl. Acad. Sci. U S A, 84*, 8951-8955.

[40] Durr, H., Flaus, A., Owen-Hughes, T., and Hopfner, K. P. (2006). Snf2 family ATPases and DExx box helicases: differences and unifying concepts from high-resolution crystal structures. *Nucleic Acids Res., 34*, 4160-4167.

[41] Yamada, M., Hayatsu, N., Matsuura, A., and Ishikawa, F. (1998). Y'-Help1, a DNA helicase encoded by the yeast subtelomeric Y' element, is induced in survivors defective for telomerase. *J. Biol. Chem., 273*, 33360-33366.

[42] Kapitonov, V. V., and Jurka, J. (2001). Rolling-circle transposons in eukaryotes. *Proc. Natl. Acad. Sci. U S A, 98*, 8714-8719.

[43] Kuznicki, K. A., Smith, P. A., Leung-Chiu, W. M., Estevez, A. O., Scott, H. C., and Bennett, K. L. (2000). Combinatorial RNA interference indicates GLH-4 can compensate for GLH-1; these two P granule components are critical for fertility in *C. elegans*. *Development, 127*, 2907-2916.

[44] von Zelewsky, T., Palladino, F., Brunschwig, K., Tobler, H., Hajnal, A., and Muller, F. (2000). The *C. elegans* Mi-2 chromatin-remodelling proteins function in vulval cell fate determination. *Development, 127*, 5277-5284.

[45] Wicky, C., Alpi, A., Passannante, M., Rose, A., Gartner, A., and Muller, F. (2004). Multiple genetic pathways involving the *Caenorhabditis elegans* Bloom's syndrome genes *him-6*, *rad-51*, and *top-3* are needed to maintain genome stability in the germ line. *Mol. Cell Biol, 24*, 5016-5027.

[46] Jeong, Y. S., Kang, Y., Lim, K. H., Lee, M. H., Lee, J., and Koo, H. S. (2003). Deficiency of *Caenorhabditis elegans* RecQ5 homologue reduces life span and increases sensitivity to ionizing radiation. *DNA Repair (Amst), 2*, 1309-1319.

[47] Lee, S. J., Yook, J. S., Han, S. M., and Koo, H. S. (2004). A Werner syndrome protein homolog affects *C. elegans* development, growth rate, life span and sensitivity to DNA damage by acting at a DNA damage checkpoint. *Development, 131*, 2565-2575.

[48] Cheung, I., Schertzer, M., Rose, A., and Lansdorp, P. M. (2002). Disruption of *dog-1* in *Caenorhabditis elegans* triggers deletions upstream of guanine-rich DNA. *Nat. Genet., 31*, 405-409.

[49] Barber, L. J., Youds, J. L., Ward, J. D., McIlwraith, M. J., O'Neil, N. J., Petalcorin, M. I., Martin, J. S. *et al.*, (2008). RTEL1 maintains genomic stability by suppressing homologous recombination. *Cell, 135*, 261-271.

[50] Duchaine, T. F., Wohlschlegel, J. A., Kennedy, S., Bei, Y., Conte, D., Jr., Pang, K., Brownell, D. R., *et al.* (2006). Functional proteomics reveals the biochemical niche of *C. elegans* DCR-1 in multiple small-RNA-mediated pathways. *Cell, 124*, 343-354.

[51] Tabara, H., Yigit, E., Siomi, H., and Mello, C. C. (2002). The dsRNA binding protein RDE-4 interacts with RDE-1, DCR-1, and a DExH-box helicase to direct RNAi in *C. elegans*. *Cell, 109*, 861-871.

[52] Steinmetz, L. M., Scharfe, C., Deutschbauer, A. M., Mokranjac, D., Herman, Z. S., Jones, T., Chu, A. M., *et al.* (2002). Systematic screen for human disease genes in yeast. *Nat. Genet., 31*, 400-404.

[53] van Haaften, G., Romeijn, R., Pothof, J., Koole, W., Mullenders, L. H., Pastink, A., Plasterk, R. H., *et al.* (2006). Identification of conserved pathways of DNA-damage response and radiation protection by genome-wide RNAi. *Curr. Biol., 16*, 1344-1350.

[54] Nakamura, M., Ando, R., Nakazawa, T., Yudazono, T., Tsutsumi, N., Hatanaka, N., Ohgake, T., *et al.* (2007). Dicer-related *drh-3* gene functions in germ-line development by maintenance of chromosomal integrity in *Caenorhabditis elegans*. *Genes Cells, 12*, 997-1010.

[55] Yoneyama, M., Kikuchi, M., Matsumoto, K., Imaizumi, T., Miyagishi, M., Taira, K., Foy, E., *et al.* (2005). Shared and unique functions of the DExD/H-box helicases RIG-I, MDA5, and LGP2 in antiviral innate immunity. *J. Immunol., 175*, 2851-2858.

Index

A

Aβ, 311, 312
AAA, 4, 412, 413, 430
AAC, 15
AAT, 312
ABC, 133, 199, 217, 367, 369, 381
abdominal cramps, 151
absorption, 58, 77
ACC, 27, 33, 34, 35, 39, 43, 392
acceleration, 83, 310
accessibility, 190
accuracy, 44, 163, 164
ACE, 81
acetaminophen, 82
acetate, 74, 221, 310, 360
acetylation, x, 86, 116, 187, 190, 191, 192, 193, 194, 196, 197, 198, 204, 207, 211, 212, 217, 220
acidic, 70, 74, 102, 138, 356
actin, 136, 202, 208
Actinobacteria, vii, 3, 21, 25, 27, 28, 29
activation, 66, 131, 132, 188, 190, 196, 198, 199, 202, 203, 204, 206, 207, 210, 211, 212, 216, 220, 222, 345, 355, 367, 369, 379, 381, 383
activators, 110, 189, 206
active oxygen, 53
active site, xi, 225, 226, 229, 259, 260, 315, 316, 331, 333, 334, 335, 336, 337, 339, 340, 341, 342, 346, 347, 348, 349, 357
acute, 138, 146, 156, 169, 173, 175, 177, 179, 180, 183, 184, 185, 215, 216, 217, 218, 219, 266, 267
acute infection, 266
acute leukemia, 218
acute myeloid leukemia, 215, 219

acute promyelocytic leukemia, 216, 219
Adams, 183, 209, 210
adaptation, 14, 16, 45, 174
adduction, 84
adducts, 54, 55, 56, 58, 60, 61, 62, 64, 65, 68, 69, 78, 80, 83, 88
adefovir, 271, 272, 273, 274, 275, 277, 278, 279, 280, 281, 282
adenine, 4, 74, 376
adenocarcinoma, 218, 220
adenosine, 271
adenovirus, 138, 164, 199, 203, 218, 220, 221, 222, 344, 349
adenoviruses, 170
adhesion, 223
administration, 66, 193, 198
ADP, 285, 286, 288, 292, 293, 359, 363, 364, 365, 366, 367, 384, 385
adults, 147, 266
advanced glycation end products (AGEs), 54, 55, 56, 60, 66, 67, 82, 84
adverse event, 74, 272
aerobic, 54, 60
aerosols, 139
AFM, 94, 95, 100, 104
Africa, 157, 183
Ag, 70, 264, 267
agar, 57, 59
AGC, 12, 27, 37
age, 82, 88, 139, 146, 147, 266
ageing, 86, 209
agent, x, 53, 138, 146, 187, 190, 201, 205, 218, 263, 272, 277, 303
agents, 138, 177, 189, 193, 198, 199, 214, 268, 269, 274, 276

aggregation, 108
aging, xiv, 54, 82, 85, 87, 88, 156, 209, 211, 411
aging process, 87
aid, 209, 310, 354
air, 360
alanine, 266
alanine aminotransferase, 266
albuminuria, 83, 86, 87
alcohols, 55
alkaline, 122, 133, 135, 200, 345
alkaline phosphatase, 133, 200
allele, 375
allograft, 88
allosteric, 298, 362, 406
alpha, 85, 88, 110, 217, 261, 279, 281, 346, 347, 350, 382, 390, 405
alpha interferon, 279, 281
ALT, 206, 266, 267, 271, 275, 276, 277
alternative, 5, 28, 29, 73, 132, 162, 184, 206, 222, 274, 276, 314
alters, 110, 190
Amadori, 53, 54, 56, 65, 66, 75, 78, 80, 82, 84, 85, 88, 219
amelioration, 14
amine, viii, 52, 54, 55, 76, 80, 84, 213, 333
amines, 54, 55, 68, 70, 73, 75, 85
amino, xii, xiv, 19, 20, 24, 25, 34, 38, 40, 42, 47, 53, 54, 57, 68, 70, 72, 76, 80, 84, 88, 135, 142, 152, 220, 277, 306, 329, 331, 332, 333, 334, 340, 341, 346, 347, 357, 359, 375, 376, 377, 388, 391, 392, 411, 413, 425, 426, 429
amino acid, xii, xiv, 19, 20, 24, 25, 34, 38, 40, 42, 53, 57, 70, 76, 84, 88, 135, 142, 152, 306, 329, 331, 332, 333, 334, 340, 341, 346, 347, 357, 359, 375, 376, 377, 388, 391, 392, 411, 413, 425, 426, 429
amino acids, 24, 25, 34, 53, 57, 70, 76, 88, 152, 332, 334, 377
amino groups, 54, 72, 76
amorphous, 360
amplitude, 397, 399, 400, 402
Amsterdam, 141
Amyloid, 86
amyloidosis, 86
anaerobic, 54, 132
analog, 271, 272, 362, 364
androgen, 214
angiogenesis, 199, 200, 201, 203, 219
angiogenic, 199, 201, 203, 219
animal models, 68, 272

animal waste, 170
animals, 55, 139, 147, 156, 172, 274
anisotropy, 363
annealing, 59, 284, 390, 393, 395, 397, 399, 400, 401, 402, 403, 404, 407
annotation, 47, 117
anorexia, 151, 156
antagonistic, 212
antagonists, 217
antiangiogenic, 199, 201, 203, 219
anti-angiogenic, 219
antiangiogenic therapy, 219
anti-apoptotic, 196, 199, 203
antibiotic, 62, 214
antibiotics, 323, 378
antibodies, 142
antibody, 41, 58, 60, 61, 64, 147, 194, 197, 199, 200, 202, 208
anticancer, 190, 195, 198, 199, 214, 223
anti-cancer, x, 187, 193, 201
anticancer drug, 190, 195
anticodon, 6, 12, 13, 15, 31, 44, 45, 46, 48
antigen, viii, 51, 147, 148, 151, 153, 175, 264, 278, 280, 281, 357, 379, 382
antigenicity, 176
anti-HIV, 271
antisense, 98, 189, 202, 208
antitumor, 201, 214, 219
antiviral, 264, 267, 268, 269, 270, 271, 272, 273, 274, 276, 277, 278, 279, 405, 432
antiviral agents, 269, 274, 276
antiviral drugs, 276, 277
antiviral therapy, 268, 270, 275
apoptosis, 188, 193, 194, 195, 196, 198, 202, 207, 209, 215, 216, 217, 218, 219, 220, 222
apoptotic, 194, 195, 196, 198, 202, 216, 217
application, 14, 17, 67, 73, 151, 174, 176, 183, 184, 267, 269, 272, 274, 404
archaea, vii, ix, 3, 21, 25, 27, 28, 29, 48, 91, 92, 93, 100, 107, 116, 305, 324, 326, 344
Argentina, 158
arginine, 12, 17, 44, 57, 59, 64, 65, 66, 69, 133, 191, 212, 356, 407
arrest, 115, 188, 193, 196, 202, 221, 222, 372, 386, 387, 425
Arrhenius equation, 292
arsenic, 217
arsenic trioxide, 217
ascites, 266
ascorbic, 87

Index

ascorbic acid, 87
Asia, xi, 157, 263
Asian, 169
asparagines, 15
aspartate, 86, 266, 333, 336
aspirin, 64, 81, 82, 84, 86, 87
assault, 53
assessment, 266, 269, 281
assets, 277
assumptions, 70, 229
asymmetry, 382
asymptomatic, 146, 148, 172, 266
asymptotically, 32
ataxia, 323
ATC, 27
atomic force, 110
Atomic Force Microscopy, 110, 111
atoms, 74
ATPase, xi, xiii, 84, 93, 140, 283, 284, 288, 290, 292, 293, 294, 295, 296, 298, 353, 354, 355, 357, 359, 366, 367, 369, 371, 374, 376, 377, 378, 380, 391, 430
attachment, 78, 111, 378
Australia, 138, 158, 173
autolysis, 133
autoradiography, 345
availability, 47, 49, 163

B

B. subtilis, ix, 102, 113, 114, 119, 120, 121, 122, 123, 124, 125, 126, 127, 128, 129, 132, 133, 134, 136, 304, 357, 372, 377
babies, 264
Bacillus, v, ix, 14, 47, 76, 88, 93, 110, 111, 112, 115, 119, 120, 126, 135, 136, 260, 285, 309, 311, 323, 347, 360, 369, 372, 375, 377, 382, 383, 385, 386, 387, 388
Bacillus subtilis, v, ix, 14, 47, 76, 88, 93, 110, 112, 115, 119, 120, 135, 136, 285, 323, 347, 375, 377, 382, 383, 386, 388
bacteria, vii, viii, x, xiii, 27, 45, 46, 47, 48, 51, 53, 62, 65, 68, 70, 73, 75, 78, 83, 92, 98, 102, 104, 106, 107, 112, 113, 114, 115, 120, 121, 138, 297, 319, 340, 354, 355, 357, 360
bacterial, vii, ix, 3, 4, 14, 15, 21, 23, 25, 28, 30, 45, 46, 47, 48, 56, 59, 60, 61, 64, 65, 67, 70, 75, 76, 91, 92, 93, 99, 101, 102, 106, 107, 108, 109, 110, 111, 112, 113, 114, 128, 133, 134, 350, 354, 357, 375, 383, 386, 412
bacterial cells, 70, 101, 106, 133

bacterial strains, 128, 134
bacteriophage, 14, 46, 48, 226, 227, 256, 259, 260, 261, 262, 284, 285, 286, 297, 304, 324, 340, 344, 347, 348, 349, 351, 354, 355, 358, 359, 362, 367, 368, 379, 380, 381, 382, 383, 384, 385, 387
bacteriophages, 326, 341, 344, 354, 355
bacterium, 14, 18, 120, 135, 324, 325, 377
Baraclude, 268
barrier, 55, 233, 292, 339
barriers, 94
base pair, 13, 64, 74, 146, 168, 227, 230, 232, 233, 234, 235, 241, 245, 246, 249, 252, 253, 256, 257, 258, 259, 286, 287, 290, 292, 295, 336, 338, 343, 344, 376
basic research, xiii, 330
Bax, 196
BCIP, 200
bcl-2, 196, 216
Bcl-2, 196, 203, 216
beads-on-a-string, 99, 106
behavior, 120
Beijing, 225, 283
bending, 94, 108, 116, 230
beneficial effect, ix, 119, 121
benefits, 67, 274, 276
benign, viii, 51
BH3, 202
bias, 13, 15, 16, 38, 40, 43, 45, 46, 48, 49
bile, 159, 170
Bim, 203
bioavailability, 66
biochemistry, 300, 320
biogenesis, xiv, 411, 412, 413, 414, 415, 416, 417, 426, 430
bioinformatics, 412
biological processes, 319
biological systems, 86, 134
biomarker, 56, 80
biomarkers, 81, 83
biopsy, 266
biosynthesis, 15, 46, 53, 133
biotechnology, 300, 320
biotic, 11
biotin, 133, 192, 193
birth, 156, 172, 320
bivalve, 179
black hole, 395
bladder, 196, 218
bladder cancer, 218
bleeding, 266

BLM, xiv, 411
blocks, 83, 223, 270, 326, 367, 395
blood, 55, 56, 75, 88, 156, 178, 199, 264, 266, 267, 390
blood glucose, 55
blood transfusion, 156
blot, 162, 165, 184, 193, 194, 202
Bolivia, 112
bonding, 74, 247, 260, 290, 298
bootstrap, 145
bovine, 58, 81, 147, 149, 178, 181, 182
Bradyrhizobium, 92
Brazil, 171
Brazilian, 168
breakdown, 55, 73
breast cancer, 191, 196, 213, 214, 216
Brownian motion, 262
bubble, 367, 372
budding, xiv, 113, 411, 412, 426, 427
buffer, 58, 60, 87, 303, 304, 306, 308, 310, 311, 312, 318, 319, 393, 394
building blocks, 270
Bulgaria, 3, 51, 57
Bundling, 94
bypass, 260

C

Caenorhabditis elegans, xiv, 411, 429, 430, 431, 432
calcitonin, 84
calf, 304
CAM, 158
Cambodia, 157, 158, 169
cAMP, 109
Canada, 49, 89, 137, 138, 158, 164
cancer, xiv, 188, 189, 190, 191, 192, 193, 196, 198, 199, 201, 202, 203, 204, 205, 206, 209, 210, 212, 213, 214, 215, 217, 218, 219, 220, 221, 222, 411
cancer cells, 188, 190, 192, 193, 196, 198, 199, 201, 202, 203, 204, 205, 206, 210, 213, 214, 215, 216, 217, 220, 221, 222
cancerous cells, 190, 203
Candida, 47
candidates, x, xi, 65, 101, 187, 196, 263, 428
capillary, 200, 201
CAR, 203, 221
carbohydrate, 36, 70, 88
carbohydrate metabolism, 36
carbohydrates, 81, 85
carbon, 53, 55, 57, 84, 128, 129, 132, 427
carbon dioxide, 132
carboxylates, 333, 337
carboxylic, 74, 333
carcinogenesis, 88, 188
carcinogenic, 56
carcinoma, xi, 199, 201, 205, 216, 217, 220, 223, 263, 266, 269, 272
carcinomas, 217
carrier, 265, 274, 279, 300
caspase, 196, 198, 202, 217, 220
caspase-dependent, 202, 217, 220
caspases, 196, 198, 217
CAT, 39, 40, 41, 42, 43, 344, 392
catabolism, 61
catalysis, 211, 231, 241, 333, 334, 338, 342, 347
catalyst, 320
catalytic effect, 72
cataract, 82
cataracts, 83
cation, 391
cattle, 142, 178
C-C, 183
CCC, 4, 27, 28, 39, 40, 42
CD95, 216
CDC, 275, 276
CDK, 202
cDNA, 48, 163, 168, 179, 189, 406
cell culture, 59, 153, 174, 272, 391
cell cycle, 113, 114, 115, 193, 195, 196, 207, 216, 221, 222, 223
cell death, 216, 217, 220, 330
cell differentiation, 195, 219, 223
cell division, 93, 188, 264
cell fate, 209, 431
cell growth, 115, 120, 121, 124, 188, 195, 216, 222, 223, 426, 427
cell invasion, 110
cell line, 156, 177, 188, 189, 190, 192, 194, 196, 198, 199, 203, 204, 205, 206, 207, 208, 215, 216, 217, 218, 219, 221, 222, 223, 272
cell lines, 188, 189, 190, 194, 196, 198, 199, 203, 204, 205, 206, 207, 208, 216, 217, 218, 221, 222, 272
cellulose, 129
cervical cancer, 210
CGC, 25, 28, 40, 392
CGT, 25, 28
Chad, 158
chain termination, 48
channels, 284
chaperones, 404, 407

chemical reactions, 234
chemical reactivity, 72
chemical stability, viii, 51
chemicals, viii, 51, 64
chemotaxis, 107, 133
chemotherapeutic agent, 192, 215
chemotherapy, 187, 209
chickens, 157, 174, 213
childhood, 168
children, 146, 147, 180, 181
chimpanzee, 273
China, 147, 158, 173, 174, 184, 185, 225, 265, 283, 296
chloramphenicol resistance, 125
chloroplast, 47, 92, 107
chloroplasts, 14
chromatid, 113
chromatin, x, xiv, 56, 99, 101, 102, 106, 107, 108, 109, 113, 114, 115, 116, 117, 187, 190, 191, 192, 195, 203, 204, 206, 209, 211, 212, 218, 222, 223, 411, 412, 413, 419, 427, 429, 430, 431
chromosome, vii, viii, xiv, 18, 52, 56, 60, 61, 62, 80, 93, 100, 101, 108, 109, 111, 112, 113, 114, 115, 117, 127, 135, 188, 211, 326, 330, 361, 372, 381, 384, 385, 386, 388, 411, 428
chromosomes, vii, 18, 92, 99, 107, 111, 112, 113, 117, 211, 330, 344, 368, 386
chronic lymphocytic leukemia, 202, 215, 220
cirrhosis, xi, 263, 266, 269, 279, 390
clams, 173
classes, vii, xiv, 3, 11, 18, 21, 26, 46, 196, 326, 411, 413, 426
classical, 55, 60, 211, 266
classification, 115, 140, 141, 142, 145, 151, 153, 158, 167, 170, 178, 185, 413, 430
cleavage, 70, 75, 80, 102, 111, 147, 196, 198, 215, 311, 320, 357, 361
clinical assessment, 266
clinical trial, x, 68, 187, 195, 201, 220, 391
clinical trials, x, 68, 187, 195, 201, 220, 391
clone, 406
cloning, 83, 183, 213, 342, 387, 392
clustering, 426
clusters, 17, 39, 127, 140, 142, 145, 413, 426
CML, 52, 58, 60, 61, 64
c-myc, x, 187, 189, 196, 201, 206, 207, 208, 210, 216, 218, 222, 223
codes, 5, 13, 108, 140, 376, 425
coding, vii, ix, 3, 5, 12, 13, 15, 17, 18, 21, 23, 30, 31, 32, 33, 34, 36, 39, 42, 43, 44, 49, 57, 98, 109, 110, 119, 120, 121, 122, 123, 128, 130, 143, 145, 146, 168, 213, 275, 320, 324, 349, 425
cofactors, 298, 365, 376, 384
cohesion, 113
cohort, 172
coil, 93, 108, 259, 323, 325
collaboration, 134
colon, 199, 215, 218
colon cancer, 199, 215
Colorado, 138, 170
colorectal cancer, 191, 196, 217
colors, 21, 24, 29
combination therapy, 276, 279
combined effect, 228, 257
communication, 140, 221
community, 138, 142, 172, 178
compaction, 93, 101, 104, 108, 112, 114, 116
compatibility, 17
competence, 129, 131, 136, 275
competency, 320
competition, 60, 402
complement, 395
complementary DNA, 316, 390
complexity, 165, 425
complications, 54, 67, 81, 84, 85, 151, 164, 265
components, vii, ix, 37, 91, 92, 93, 99, 101, 106, 382, 430, 431
composites, 399
composition, 14, 17, 39, 47, 49, 57, 83, 92, 107, 114, 163
compounds, viii, 11, 52, 54, 56, 63, 64, 67, 78, 81, 84, 192, 195, 219, 391
condensation, ix, 92, 93, 98, 102, 104, 105, 108, 112, 113, 115, 117
configuration, 295
confocal laser scanning microscope, 197
conformational analysis, 367
conformity, 63
confusion, 266
consensus, 29, 40, 76, 95, 141, 142, 145, 148, 157, 189, 206, 276, 355, 413
conservation, 145, 297, 377, 426, 427
constraints, 172
construction, viii, ix, 91, 93, 106, 119, 121, 331
consumption, ix, 119, 120, 128, 132, 138, 139, 151, 169, 171
contaminant, 300, 302, 305
contaminated food, x, 137, 151
contamination, x, 137, 139, 157, 162, 163, 305

control, ix, 43, 59, 64, 65, 68, 69, 73, 84, 92, 94, 98, 105, 110, 116, 122, 128, 129, 132, 133, 152, 156, 163, 164, 167, 171, 181, 184, 189, 194, 199, 200, 201, 202, 205, 208, 211, 212, 277, 370, 401, 417, 419
conversion, 6, 13, 25, 82, 88, 102, 284, 365
copolymer, 304, 319, 324
copolymers, xii, 299, 301, 323
correlation, vii, 3, 12, 15, 20, 29, 30, 39, 44, 46, 62, 67, 155, 164, 175, 210, 218, 266
correlation analysis, 29
correlations, 21
costs, 138
couples, 109
coupling, 58, 260, 284, 331, 339, 377
covalent, 54
Cp, 344, 349
CpG islands, 191
crab, 302
crosslinking, 367
cross-linking, 82, 84, 223
CRP, 95, 96, 97
crystal structure, 114, 228, 242, 243, 244, 249, 258, 259, 270, 272, 284, 285, 289, 333, 349, 360, 372, 382, 383, 403, 405, 407, 431
crystal structures, 249, 259, 372, 431
crystals, 360, 383
CTA, 27, 28, 392
CTD, 360
C-terminal, 34, 40, 42, 126, 284, 331, 332, 335, 340, 357, 358, 359, 360, 361, 369, 375, 387, 391, 392, 405
C-terminus, 351, 369, 375
cues, 114
cultivation, 59, 127, 134
culture, ix, 58, 59, 65, 120, 121, 122, 127, 128, 129, 132, 153, 156, 174, 183, 195, 198, 199, 200, 201, 202, 272, 391, 405, 427
culture conditions, 427
cutaneous T-cell lymphoma, 201, 215
Cyanobacteria, vii, 3, 21, 25, 27, 28
cycles, 59, 189, 202, 208, 309, 318, 406
cyclin D1, 201, 202, 215, 217, 218
cyclin-dependent kinase inhibitor, 196
cycling, 292, 395
cycloheximide, 204, 205
cysteine, 152
cytochemistry, 98
cytochrome, 46, 132, 196, 198
cytoplasm, viii, 13, 52, 55

cytoplasmic membrane, 112
cytosine, 4, 74, 191, 301
cytoskeleton, 156
cytosol, 56, 198
cytosolic, 133
cytotoxic, 201
cytotoxicity, 193, 203, 273

D

Darwinian evolution, 45
data set, 20
database, 18, 21, 30, 39, 117, 120, 325, 413, 425
dating, 14
DBP, 345
de novo, xii, 188, 204, 299, 300, 301, 302, 303, 304, 305, 306, 307, 308, 309, 310, 311, 312, 313, 314, 315, 316, 318, 319, 320, 321, 322, 324, 326, 329, 330
death, 82, 151, 178, 195, 196, 202, 215, 216, 217, 220, 330
deaths, 138, 146
decay, 45, 74, 413, 418
decoding, 12, 13, 14, 39, 43, 44
decomposition, 75
defects, 86, 427
defense, 428
deficiency, 62, 87
definition, 33, 34, 358
degenerate, 46
degradation, 54, 67, 81, 88, 188, 350, 413, 417, 431
degrading, 302, 399
dehydration, 54, 100
delivery, 135, 379
denaturation, 59, 372
Denmark, 197
density, 66, 101, 194
deoxynucleotide, 322
deoxyribonucleic acid, 322, 323
deoxyribose, 70, 338
dephosphorylation, 66, 132, 202
derivatives, 55, 60, 62, 65, 66, 68, 69, 75, 125
destruction, 111
detection, x, 56, 87, 108, 137, 138, 139, 140, 142, 143, 145, 148, 153, 159, 162, 163, 164, 165, 166, 167, 168, 169, 170, 171, 172, 173, 174, 175, 176, 177, 178, 179, 180, 181, 183, 184, 185, 195, 314, 320
detoxification, 66, 87
developed countries, 146
developing countries, 155

developmental change, 213
deviation, 13, 14, 17, 19, 23, 25, 32
diabetes, 54, 68, 82, 83, 85, 86, 87, 88, 89
diabetes mellitus, 85, 88
diabetic nephropathy, 67, 68, 82
diabetic neuropathy, 87, 89
diabetic patients, 54
diabetic retinopathy, 66, 83
dialysis, 86
diamonds, 243, 244
diarrhea, 180
diarrhoea, 146, 168
dietary, 83
dietary fat, 83
diets, 56
differentiation, 145, 167, 181, 193, 195, 207, 211, 214, 215, 216, 219, 222, 223
diffusion, 316
digestion, 100, 309, 310, 311, 316, 320, 361
dimer, 93, 316, 359, 361, 363, 364, 374
dimeric, 227, 357, 363, 370, 373, 374
dimerization, 373, 407
Discovery, vi, 322, 388, 411
discrimination, 145, 164, 330, 344
disease gene, 432
disease progression, 269
diseases, xiv, 88, 266, 270, 276, 300, 316, 411, 413
displacement, xii, 317, 324, 329, 331, 336, 338, 339, 340, 344, 346, 348, 349, 350, 365
dissociation, 231, 234, 318, 331, 339, 342, 362, 363, 364, 365, 367, 370, 397
distilled water, 58, 60
distortions, 74
distribution, viii, 4, 17, 21, 24, 30, 32, 39, 107, 109, 114, 127, 264
divergence, 47, 175
diversification, 106, 320, 326
diversity, 45, 80, 135, 152, 153, 157, 173, 177, 212, 261, 264, 280, 357
division, 93, 188, 264, 378
DNA damage, viii, 51, 62, 63, 73, 78, 80, 87, 88, 431
DNA lesions, 67, 260
DNA repair, viii, 51, 52, 56, 61, 283, 297, 412, 413, 419, 422, 423, 424, 427, 428
DNA sequencing, vii, 3, 4, 14, 78, 87
DNase, 204, 306, 313
domain structure, 111, 387
donor, 14, 74, 178, 267
dosage, 89

dosing, 274
double helix, 4, 355, 374, 392, 397
double-blind trial, 279
down-regulation, 96, 216, 220
drinking, 176
drinking water, 176
Drosophila, 44, 49, 117, 212, 213
drug exposure, 193, 201
drug resistance, xi, 124, 198, 260, 263, 276, 277
drug treatment, 194, 271, 274, 275, 276
drug-resistant, 222, 274, 276
drugs, xi, 214, 218, 263, 264, 268, 269, 273, 274, 275, 276, 277
DS-1, 149, 178
duplication, 104, 383
duration, 170, 275, 276, 277
dyslipidemia, 83
dysregulation, 190

E

E.coli, 33, 34, 35, 39, 44, 45, 70, 96, 302, 322, 354, 364
ears, 145
earth, 320, 324
East Asia, xi, 263
edema, 266
effluent, 176
effluents, 150
elasticity, 296
election, 48
electron, 112, 114, 139, 142, 148, 155, 162, 177, 359, 360, 374, 380, 382
electron microscopy, 112, 114, 139, 142, 148, 155, 162, 177, 359, 380
electrophoresis, 37, 40, 41, 70, 74, 150, 162, 189, 208, 345, 395, 396
electroporation, 120
electrostatic force, 234, 253
electrostatic interactions, 391
ELISA, x, 40, 52, 58, 61, 137, 139, 142, 148
elongation, viii, xii, xiii, 4, 6, 14, 17, 46, 49, 100, 189, 202, 208, 216, 262, 269, 271, 299, 300, 304, 308, 309, 310, 313, 314, 315, 316, 317, 318, 319, 321, 326, 330, 340, 344, 349, 353, 354, 355, 365, 372, 377, 378
embryos, 223
emission, 58, 132, 393
encoding, xiv, 17, 46, 98, 106, 112, 115, 122, 124, 125, 126, 132, 133, 150, 196, 209, 412, 430
endocytosis, 264

endonuclease, xii, 63, 173, 299, 300, 305, 309, 311, 313, 315, 316, 321, 323, 325
endosymbionts, 49
endothelial cell, 66, 67, 86, 87, 199, 200, 203
endothelial cells, 66, 67, 87, 199, 200, 203
energy, ix, xiii, 53, 55, 119, 120, 233, 247, 249, 256, 258, 284, 288, 289, 290, 291, 292, 296, 320, 353, 359, 361, 362, 366, 372, 373, 376, 384, 390, 406
energy consumption, ix, 119, 120
energy transfer, 384, 390, 406
England, 135, 138, 167, 388
entecavir, 271, 273, 274, 275, 277, 278, 280, 281
enterovirus, 138, 177
enteroviruses, 180
environment, viii, x, 91, 107, 120, 137, 138, 180, 190, 209, 235, 358
environmental change, 106, 115
environmental conditions, 97, 101, 102, 120
environmental factors, 88
enzymatic, viii, 52, 54, 55, 69, 85, 88, 93, 100, 148, 165, 166, 188, 331, 357, 359, 362, 375, 378, 391, 405, 406
enzymatic activity, 188, 357, 359, 405
enzyme immunoassay, 159, 175
enzyme-linked immunosorbent assay, 139, 176, 179
enzymes, x, xii, xiii, 53, 75, 85, 86, 88, 120, 128, 133, 156, 187, 196, 238, 262, 284, 299, 300, 305, 311, 312, 313, 321, 353, 354, 356, 370, 375, 381, 382, 401, 405
Epi, 212
epidemics, 145, 147, 156, 157, 167, 173
epidemiology, 171, 175, 179, 183, 184, 280
epigenetic, 190, 191, 204, 212, 223
Epigenetic control, 211
epithelial cells, 216, 221
epithelium, 210
equilibrium, 250, 251, 258, 259, 288, 289, 311, 362, 363, 364, 365, 399, 400
equilibrium state, 399
erythrocytes, 66, 83, 84, 86
Escherichia coli, v, viii, ix, xiii, 4, 45, 46, 47, 48, 49, 51, 56, 83, 84, 85, 88, 89, 92, 107, 108, 109, 110, 111, 112, 113, 114, 115, 117, 120, 134, 253, 256, 260, 261, 284, 297, 298, 322, 323, 325, 346, 349, 353, 378, 379, 380, 381, 382, 383, 384, 385, 386, 387, 404
ester, 74
estimating, 40
estrogen, 189, 206, 210
ethanol, 57, 197, 200

etiologic factor, 191
euchromatin, 190, 191
eukaryote, 106, 107, 211, 413, 426
eukaryotes, viii, xiv, 5, 6, 14, 47, 55, 81, 83, 91, 92, 94, 99, 106, 107, 117, 190, 322, 330, 411, 413, 428, 431
eukaryotic cell, 116
Europe, 157, 177
evolution, viii, 11, 12, 13, 14, 18, 43, 44, 47, 48, 51, 55, 148, 150, 152, 153, 170, 180, 182, 300, 305, 319, 320, 321, 322, 326, 327
evolutional processes, 107
excision, 56, 61, 80, 85, 253, 254, 427
excitation, 393
execution, xiii, 353, 354
exonuclease, xi, xii, 62, 225, 226, 227, 229, 231, 232, 233, 234, 235, 239, 242, 245, 246, 247, 248, 249, 250, 251, 252, 253, 254, 255, 257, 258, 259, 261, 262, 300, 301, 302, 306, 314, 315, 316, 324, 329, 331, 332, 333, 334, 336, 337, 339, 340, 341, 346, 347, 348, 350
experimental condition, 189, 345, 399
exposure, 115, 168, 193, 194, 201, 204, 360
expressivity, 46
extinction, 58, 392
extraction, 139, 148, 162, 163, 164, 166, 167, 173, 189
extrusion, 326
eyes, 355

F

faecal, 155, 159, 162, 164, 167, 174, 175, 179
failure, 175, 300
false negative, 164
false positive, 162
family, xii, xiv, 25, 27, 62, 80, 131, 139, 146, 151, 155, 157, 189, 196, 199, 206, 209, 211, 222, 263, 300, 329, 330, 332, 333, 334, 335, 337, 340, 346, 348, 350, 351, 373, 412, 426, 428, 430, 431
family members, xiv, 412, 426, 428
farms, 181
Fas, 196, 216
FasL, 196
fat, 56
fatigue, 151, 156
fatty acids, 55, 192
FDA, xi, 263, 267, 268, 273, 274, 276, 280
FDA approval, 268
fermentation, 116, 201, 214
fermentation broth, 201

ferritin, 114
fertility, 431
fetal, 272
fever, 156
fiber, 99, 100, 101, 106, 116, 117
fibers, ix, 91, 93, 94, 98, 99, 100, 101, 106
fibroblast, 200
fibroblasts, 199, 200, 206, 209, 218
fibrosis, 266, 267, 269, 274, 390
fidelity, x, 53, 75, 153, 225, 226, 260, 261, 262, 300, 323, 331, 344, 351
filtration, 362, 375
Finland, 138
first aid, 209
FISH, 100
fission, 210
FITC, 197
fixation, 44, 63, 85, 100
flexibility, 106, 342, 375
flow, 226, 313, 321
fluctuations, 13
fluorescence, 58, 61, 358, 363, 364, 384, 390, 393, 395, 396, 397, 398, 399, 400, 403, 406
fluorometric, 406
fluorophores, 64
folding, ix, xiv, 39, 91, 93, 99, 100, 107, 111, 117, 361, 389, 407
food, x, 54, 137, 138, 139, 151, 153, 156, 162, 163, 164, 165, 166, 169, 172, 178, 181, 184
formaldehyde, 192, 193, 194, 195, 197, 223
fragmentation, 54, 55
France, 158, 170, 178
free energy, 233, 247, 249, 258, 290, 292, 318, 366
free rotation, 343
freezing, 169
frequency distribution, 44
fructose, 55, 76, 85, 86, 87, 133
FTC, 273, 278
FTIR, 86
fuel, 405
fulminant hepatitis, 151, 156, 184
fumarate, 280
functional analysis, 429
functional changes, 275
fungi, 340
fungus, 191
fusion, 427

G

G4, 147, 149, 327

G8, 149
gas, 128, 129
gastric, 216
gastroenteritis, 138, 139, 146, 147, 168, 172, 173, 177, 178, 179, 180, 183, 184, 185
gastrointestinal, 146, 218, 220, 224, 266
GCC, 27, 34, 35, 325, 392
gel, 37, 39, 40, 41, 70, 74, 76, 148, 150, 162, 174, 189, 205, 208, 305, 306, 310, 311, 312, 319, 345, 362, 375, 393, 394, 395, 397, 398
GenBank, 14, 44, 140, 144, 146, 152, 155, 157, 161, 162
gene expression, vii, xiv, 3, 13, 14, 15, 16, 17, 30, 36, 39, 40, 42, 44, 46, 47, 82, 98, 102, 104, 115, 116, 133, 136, 190, 191, 196, 203, 214, 216, 219, 222, 264, 281, 411, 415, 429
gene promoter, 210, 215, 221, 222, 325
gene silencing, 111, 191, 213, 420
gene therapy, 199, 203, 219
gene transfer, 14, 47, 48, 324
generation, vii, 3, 53, 54, 59, 66, 67, 85, 125, 133, 215, 298, 378, 399
genetic code, 4, 5, 6, 11, 12, 13, 14, 15, 18, 43, 44, 45, 47, 48
genetic disease, xiv, 411, 413
genetic diversity, 45, 152, 153, 177
genetic drift, 49
genetic information, xiv, 102, 104, 107, 264, 300, 305, 320, 321, 324, 411
genetic marker, 62
genetic mutations, xiv, 411
genetics, 212, 323, 391, 429
genome sequences, xiv, 123, 164, 411
genome sequencing, 45
genomes, vii, viii, xii, xiv, 3, 14, 15, 18, 21, 23, 29, 44, 46, 47, 49, 91, 92, 107, 113, 120, 128, 146, 172, 264, 272, 319, 322, 329, 340, 344, 346, 354, 412
genomic, vii, viii, x, 3, 7, 14, 15, 21, 23, 30, 59, 82, 86, 91, 100, 101, 102, 107, 114, 120, 121, 143, 145, 147, 153, 156, 162, 163, 164, 166, 174, 176, 177, 178, 185, 217, 263, 274, 320, 326, 377, 431, 432
genomic regions, 101, 143, 153, 164, 185
genomics, 18, 45, 47, 120, 128, 134, 326
genotype, 61, 142, 147, 149, 151, 152, 153, 156, 157, 158, 159, 161, 165, 169, 174, 175, 177, 180, 181, 182, 184, 392, 395, 406, 407

genotypes, 140, 141, 142, 147, 150, 152, 155, 156, 157, 162, 165, 171, 173, 176, 183, 185, 264, 280, 281, 406
germ line, 431
Germany, 174, 197
GFP, 101, 113
GGT, 27, 392
Gibbs, 277, 278
gift, 205
GL, 46, 210, 212, 222
glass, 165
GlaxoSmithKline, 268, 273
globulin, 267, 268
glucose, 53, 54, 55, 57, 58, 60, 66, 67, 68, 73, 76, 77, 78, 80, 81, 82, 83, 84, 85, 86, 87, 88, 124
glutamate, 124, 286
glutamic acid, 377
glutamine, 377
glutathione, 81, 84
glycation, viii, 52, 54, 55, 56, 57, 60, 61, 62, 63, 64, 65, 66, 67, 68, 69, 70, 73, 74, 75, 76, 78, 80, 81, 82, 83, 84, 85, 86, 87, 88
Glycation, 52, 54, 57, 60, 62, 64
glycerol, 393, 427
glycolysis, 55
glycoprotein, 147, 198, 203, 217, 221
glycosides, 84
glycosylation, viii, 52, 54, 81, 82, 84, 85, 89
goals, 269, 321
gold, 163
gold standard, 163
gram-negative bacteria, 357, 375
Gram-positive, 120
gram-positive bacteria, xiii, 354, 357
grants, 378
graph, 58
groundwater, 168
grouping, 14, 29
groups, 13, 15, 23, 25, 29, 32, 33, 43, 54, 64, 67, 70, 72, 74, 76, 139, 151, 167, 191, 217, 219, 269, 276, 277, 309, 333
growth, ix, 15, 45, 56, 59, 60, 63, 64, 65, 66, 67, 70, 92, 98, 102, 106, 109, 110, 114, 115, 119, 120, 121, 122, 123, 124, 125, 127, 128, 129, 132, 188, 194, 196, 200, 202, 213, 214, 215, 216, 217, 218, 219, 222, 223, 314, 327, 375, 378, 425, 426, 427, 431
growth factor, 200, 214, 216
growth inhibition, 194, 215, 217
growth rate, ix, 45, 114, 119, 121, 127, 431

guanine, 4, 60, 74, 327, 432
guidance, 277
guidelines, 276

H

H. pylori, 375
H1, 59, 99, 116, 117, 227, 256, 361, 377
H$_2$, 58, 361, 377, 414
HA, 45, 153, 213, 217, 219, 223
haemoglobin, 54, 86
Hawaii, 141
HBV, vi, xi, 263, 264, 266, 267, 268, 269, 270, 271, 272, 273, 274, 275, 276, 277, 279
HBV antigens, 267
HBV infection, 265, 266, 267, 269, 272, 274, 276, 277
health, xi, 81, 88, 155, 166, 171, 263, 275
heart, 115, 391
heat, 80, 92, 138, 168, 174
heating, 59, 393
height, 36
Helicobacter pylori, 375, 387
helix, xi, 4, 74, 136, 283, 284, 286, 289, 296, 341, 355, 361, 374, 391, 392, 397
Helix, 323
hematologic, 201
hematopoietic, 220
hematopoietic cells, 220
Hemiptera, 112
hemoglobin, 54, 82, 86
hepatic failure, 277
hepatic fibrosis, 274
hepatitis, x, xi, xiii, 137, 138, 151, 155, 156, 157, 167, 168, 169, 170, 171, 172, 173, 174, 175, 176, 177, 178, 179, 180, 181, 182, 183, 184, 185, 263, 264, 265, 266, 267, 277, 278, 279, 280, 281, 282, 389, 390, 391, 404, 405, 406, 407
hepatitis A, 138, 151, 152, 153, 168, 170, 172, 181, 183
hepatitis B, xi, 263, 264, 265, 266, 267, 277, 278, 279, 280, 281, 282
hepatitis C, xiii, 176, 266, 281, 389, 390, 391, 404, 405, 406, 407
Hepatitis C virus, 405, 406
hepatocellular, xi, 191, 263, 266, 269, 272
hepatocellular cancer, 191
hepatocellular carcinoma, xi, 263, 266, 269, 272
hepatocytes, 278
hepatoma, 272
herbs, 149, 169

herpes, 270
herpes simplex, 270
heterochromatic, 213
heterochromatin, 190, 191, 212, 213
heterodimer, xiii, 330, 340, 342, 343, 381
heterogeneity, 145, 152, 172
heterogeneous, 101
high temperature, 307, 318
high-level, 204
high-risk, 156
Hilbert, 136
histidine, 44, 131
histological, 266, 275, 276
histology, 266, 269, 271, 272
histone, x, 92, 99, 107, 109, 110, 115, 116, 117, 187, 188, 190, 191, 192, 193, 194, 195, 196, 197, 198, 204, 206, 207, 211, 212, 213, 214, 215, 216, 217, 218, 219, 220, 221, 222, 223, 430
Histone deacetylase (HDAC), 190, 211, 212, 215, 216, 217, 218, 219, 220, 221, 223
histone deacetylase inhibitors, 215, 217, 218, 220, 222
HIV, 228, 260, 262, 264, 269, 270, 271, 272, 273, 275, 277, 279
HIV-1, 228, 260, 262, 269
HO-1, 189
holoenzyme, xiii, 107, 135, 227, 260, 353, 354, 369, 371, 374, 378
homogeneity, 381
homolog, 93, 112, 114, 117, 213, 375, 413, 431
homologous proteins, 427
homology, 116, 121, 165, 270, 281, 357, 358, 375, 377, 404, 426
homopolymers, 301
hormonal control, 55
hormone, 212
hospital, 173
hospitalizations, 138
host, ix, 14, 17, 73, 92, 115, 119, 121, 124, 126, 148, 156, 256, 264, 267, 269, 273, 355, 368, 369, 379, 385, 428
hot spring, 305
household, 168
human genome, 67, 222, 319, 326, 348
human immunodeficiency virus, 172, 270
human leukemia cells, 215
human papillomavirus, 210
humans, x, 64, 75, 83, 137, 138, 139, 142, 147, 152, 172, 183, 263, 426
hunting, 428

hybrid, 323
hybridization, 162, 165, 167, 169, 176, 184, 244
hybrids, 110
hydrazine, 68
hydro, 12, 356
hydrogen, 74, 109, 247, 286, 290, 333, 334
hydrogen bonds, 74, 286, 334
hydrogen peroxide, 109
hydrolysis, xi, xiii, 53, 70, 74, 76, 283, 284, 286, 287, 288, 290, 296, 298, 353, 356, 359, 362, 364, 365, 371, 373, 374, 377, 379, 380, 382, 383, 391, 405
hydrolyzed, 60, 74, 290, 362, 364, 366, 367, 376
hydrophilic, 12, 356
hydrophobic, 12, 356, 358, 361, 363
hydrophobicity, 12
hydroxyapatite, 173
hydroxyl, 331, 333, 342
hyperactivity, 212
hyperglycaemia, 88
hypermethylation, 191
hypersensitive, 116
hypersensitivity, 428
hypertensive, 83
hypothesis, 6, 7, 12, 13, 14, 16, 17, 45, 46, 54, 70, 82, 209, 228, 285, 286, 339, 344, 355, 359, 370
hypoxia, 201

I

ibuprofen, 81, 82
ice, 212
icosahedral, 151, 155, 264
identification, x, 76, 86, 136, 137, 140, 155, 162, 164, 165, 174, 176, 182, 185, 210, 325, 386, 387, 405, 430
identity, 48, 151, 157, 180, 241, 357, 375, 377
IFN, 17, 268
IgG, 159, 197, 200
IL-2, 199, 219
IL-8, 199, 219
image analysis, 382
images, 94, 104, 200, 382
imaging, 98, 111
imaging techniques, 98
imitation, 431
immobilization, 100
immortal, 206, 209
immortality, 188, 209
immune globulin, 267, 268
immune response, 180

immunity, 151, 264, 267
immunization, 267, 281
immunoassays, 153, 159, 184
immunofluorescence, 196, 197
immunoglobulin, viii, 51, 175, 182
immunological, 147
immunomodulatory, 268
immunoprecipitation, 101, 114
immunoreactivity, 58, 60, 61, 64
immunotherapy, 267
Immunotherapy, 268
implementation, 148, 165
impurities, 314, 321
in situ, 80, 100, 101, 195
in vitro, x, 54, 56, 65, 66, 68, 73, 74, 78, 87, 88, 110, 116, 117, 134, 137, 139, 163, 166, 199, 200, 201, 203, 205, 218, 272, 273, 278, 319, 325, 340, 345, 349, 354, 355, 369, 372, 375, 379, 380, 381, 386, 406, 407
in vivo, 48, 55, 56, 64, 65, 66, 68, 69, 73, 74, 75, 78, 81, 85, 87, 110, 115, 116, 117, 201, 212, 218, 219, 223, 275, 278, 316, 319, 365, 372, 375, 406
inactivation, 57, 73, 83, 121, 267
inactive, viii, 80, 91, 190, 276, 334, 377
incidence, 139, 155, 281
incubation, 54, 55, 57, 71, 74, 189, 197, 200, 272, 301, 302, 306, 308, 309, 311, 323, 345
independence, 259
India, 136, 147, 158, 167, 168, 172, 181
Indian, 169
indication, 44, 166, 427
indigenous, 182
indirect effect, 204, 207
indole, 358
inducer, 73, 201
inducible enzyme, 78
inducible protein, 136
induction, 102, 104, 133, 195, 196, 197, 202, 204, 205, 206, 207, 215, 216, 218, 219, 221
industrial, ix, x, 119, 120, 121, 122, 124, 126, 128, 133, 134
industrial application, 120, 121, 124, 126, 128, 133, 134
industrial production, ix, 119, 120
industry, 272
infants, 146, 147, 180, 266
infection, xi, 139, 140, 147, 151, 156, 168, 170, 174, 178, 179, 182, 184, 199, 203, 220, 263, 265, 266, 267, 269, 272, 276, 277, 278, 279, 280, 281, 282, 390

infections, 138, 147, 148, 151, 152, 156, 159, 169, 172, 175, 264, 266, 274, 276
infectious, x, 137, 138, 139, 164, 166, 167, 172, 177, 264, 267, 270, 281, 405, 406
infectious hepatitis, 172, 405
inflammation, 269
influenza, 180
infrared, 74
infrared spectroscopy, 74
ingestion, 173, 178
inheritance, 12
inherited disorder, 326
inhibition, 69, 70, 73, 74, 81, 82, 83, 84, 85, 94, 101, 127, 132, 181, 192, 194, 195, 202, 204, 205, 206, 215, 216, 217, 218, 219, 220, 221, 365, 367, 370
inhibitory, 65, 68, 70, 199
inhibitory effect, 68, 199
initiation, xiii, 5, 7, 19, 28, 29, 31, 48, 95, 109, 145, 190, 211, 302, 314, 315, 316, 321, 330, 331, 340, 341, 344, 346, 347, 348, 349, 350, 351, 353, 354, 355, 367, 368, 369, 372, 377, 378, 379, 381, 384, 385, 386, 388, 400, 401, 402, 404, 415
injury, 199, 218
innate immunity, 432
inorganic, 320
insects, 55
insertion, 42, 43, 111, 330, 333, 335, 337, 338, 339, 340, 344, 347, 348, 351, 375
insight, 93
inspection, 361
instability, 67, 101, 123, 326
institutions, 276
insulin, 363, 383
integration, 92, 115, 136, 163, 164, 166, 184, 247
integrity, 81, 114, 116, 190, 375, 413, 427, 428, 432
interaction, xiii, xiv, 6, 12, 15, 34, 40, 70, 74, 75, 86, 110, 131, 190, 227, 233, 234, 247, 249, 253, 256, 258, 285, 287, 290, 330, 341, 342, 348, 350, 353, 357, 359, 361, 363, 364, 369, 370, 372, 375, 378, 380, 383, 384, 386, 406, 412, 429
interactions, viii, xiii, 6, 11, 14, 30, 31, 43, 49, 83, 91, 107, 108, 113, 190, 227, 257, 286, 288, 290, 337, 341, 353, 359, 362, 363, 365, 369, 370, 372, 378, 379, 384, 385, 391, 429
interface, 361, 366
interference, xiv, 58, 64, 66, 411, 412, 427, 431
interferon, 267, 268, 274, 277, 279, 281, 390
interleukin, 199
interleukin-2, 199
intermolecular, 30, 375

intermolecular interactions, 30
interphase, 117
intervention, 65, 74, 81, 270
intrinsic, 16, 316, 331, 338, 364, 384, 428
inversion, 92
ionic, 302, 306, 324, 363
ionization, 56
ionizing radiation, 431
ions, 67, 87, 88, 115, 303, 320, 331, 333, 337, 347, 351, 377
Ireland, 179
irradiation, 428
island, 211
isolation, 84, 86, 115, 214
isoleucine, 12
isothermal, 142, 148, 153, 162, 172, 175, 184, 185, 345
Italy, 158, 169, 185

J

Japan, 51, 84, 91, 119, 120, 134, 158, 177, 179, 180, 182, 184, 187, 194, 200, 205, 299, 322, 411
Japanese, 120, 180, 183
jaundice, 151, 156, 266
Jefferson, 353
joining, 144, 146, 155, 161, 369
Jordan, 405
jumping, 262, 344, 349
Jun, 187
Jung, 221, 327

K

K^+, 84
K-12, viii, 45, 51, 56, 61, 83, 89, 114, 117, 297
KAP, 213
kappa, 217
KB cells, 193, 194, 198, 204
kidney, 84
kinase, 76, 83, 86, 87, 131, 196, 202, 219, 222, 393
kinase activity, 202
kinases, 76, 382, 405
kinetic model, 226
kinetic studies, 82
kinetics, 165, 226, 242, 261, 298, 301, 304, 324, 404
King, 182, 279, 344, 349, 386
Korea, 167, 169, 177, 187, 273, 281
Kyrgyzstan, 158

L

L1, 73
lactose, 61, 73
Lafayette, 51
lambda, 261, 379, 380, 384, 385
lamivudine, 271, 272, 274, 275, 276, 277, 278, 279, 280, 281, 282
land, 47
Langmuir, 327
language, 212
large-scale, ix, 113, 115, 119, 122, 125, 133
larval, 425
laser, 86, 197
lattice, 102
law, 60, 249
lens, 81, 83, 84, 87
lesions, viii, 51, 53, 68, 70, 80
lettuce, 181
leucine, 5, 361
leukaemia, 217
leukemia, 196, 201, 202, 215, 216, 218, 219, 220
leukemia cells, 202, 215, 216, 218, 220
leukemic, 223
leukocytes, 56, 80
life cycle, 15
life span, 188, 278, 431
life-threatening, 201
ligands, 216, 331, 332, 334, 335, 337, 348, 359
light cycle, 179
limitation, 15, 140, 164, 272
limitations, xiv, 344, 412
linear, xii, 92, 188, 304, 329, 330, 331, 332, 338, 340, 344, 346, 349, 350, 393, 396
linear regression, 393
lipid, 55, 56, 83, 84, 88, 346
lipid peroxidation, 55, 56, 83, 84
lipoprotein, 98, 264
liquid chromatography, 52
liver, 151, 156, 170, 178, 184, 221, 263, 266, 267, 268, 271, 272, 274, 276, 277, 280, 390
liver cancer, 390
liver cirrhosis, 266
liver damage, 267, 269
liver disease, 263, 266, 268, 274, 276, 277, 280
liver enzymes, 156
liver failure, 151, 269
liver transplant, 267, 390
liver transplantation, 267, 390
livestock, 147

loading, 202, 375, 377, 381, 382
localization, 94, 107, 113, 140
location, 17, 114, 357, 361
locus, 47, 100, 213, 297
London, 135, 171, 182
long period, x, 137
low molecular weight, 93
low temperatures, 311, 324
LSM, 197
lung, 199, 202, 205, 215, 218, 220, 222
lung cancer, 202, 215, 220, 222
lymph, 170
lymph node, 170
lymphoma, 201, 214, 215, 216, 220
lysine, 4, 52, 55, 60, 76, 77, 78, 81, 83, 84, 85, 86, 190, 191, 193, 194, 212, 213, 214, 347, 351
lysis, 117

M

M.O., 170
M1, 158, 161
machinery, 63, 80, 148, 320, 345
machines, 429
Mackintosh, 405
macromolecules, 11
magnesium, 310, 405
Maillard reaction, viii, 52, 54, 55, 60, 65, 67, 68, 74, 75, 80, 81, 82, 85, 86, 88
maintenance, xiv, 93, 114, 129, 188, 190, 191, 206, 211, 411, 412, 413, 415, 421, 423, 424, 427, 428, 430, 431, 432
malignancy, 202
malignant, 220, 221
malignant cells, 220, 221
maltose, 128, 132
mammalian cell, 82, 182
mammalian cells, 82, 182
management, 47, 275, 276, 280, 281
manipulation, 108
manufacturer, 189, 194, 195, 199, 200, 208
mapping, 47, 114, 347, 429
maritime, 309
market, 268, 391
mass spectrometry, 52, 81
matrix, 21
maturation, 414, 415, 418
Mb, ix, 119, 120, 121, 123, 124, 126
MDA, 217
meanings, 4, 12, 13, 23, 25, 32, 300
measurement, 60, 61, 77, 261, 390, 404, 406

measures, 45
meat, 156, 170, 177
media, 59, 60, 65, 70
medications, 214
medicine, 74
meiosis, 417
MEK, 218
melanoma, 196, 216
melt, 372
melting, 252, 318
melting temperature, 318
melts, 249
membranes, 208
Merck, 267, 268
mesothelioma, 216
messenger RNA, 4
metabolic, ix, 120, 121, 127, 128, 129, 133
metabolism, 36, 123, 130, 133, 207, 222, 284, 317, 413, 426
metabolite, 53, 55, 120, 123, 124, 136, 219
metabolites, viii, ix, 52, 53, 119, 125
metabolomics, 136
metal ions, 67, 88, 115, 303, 320, 331, 333, 337, 351
metals, 336
metastatic, 201, 216
metazoan, 429
methane, 52
methionine, 5, 12
methyl groups, 80, 191
methylation, x, 116, 187, 190, 191, 205, 206, 211, 212, 213
Mexican, 158, 161
Mexico, 157, 158
Mg^{2+}, 304, 325, 377
$MgSO_4$, 306, 308, 310, 312, 319
mice, 142, 219, 326
microarray, x, 89, 115, 136, 138, 151, 165, 166, 169, 175, 180, 196
microarray technology, x, 138, 165, 166
microbial, 135, 137, 219, 377
microorganisms, vii, ix, 4, 30, 119, 120
microscope, 112, 139, 197, 360
microscopy, x, 100, 107, 110, 111, 112, 114, 116, 117, 137, 139, 142, 148, 155, 162, 177, 197, 359, 380, 382
migration, 295, 297, 298, 376
mimicking, 31, 339
mineral water, 172
minerals, 320
mining, 325

Ministry of Education, 81, 428
misinterpretation, 162
mitochondria, 7, 12, 13, 47, 92, 107, 196, 198, 430
mitochondrial, 12, 14, 44, 84, 198, 202, 220, 412, 415, 419, 421, 423, 424, 427, 430
mitochondrial DNA, 412, 415, 421, 424
mitochondrial membrane, 84
mitogen, 222
mitogen-activated protein kinase, 222
mitosis, 115
mitotic, 99, 117
mixing, 401
mobility, 94
model system, 84
modeling, 83, 278, 358
models, 68, 226, 270, 272, 274, 372, 390, 391
modulation, 14, 30, 36, 195, 220, 377, 379, 384, 385
MOG, 416
moieties, 60, 80
molar ratio, 74
molecular biology, 300
molecular dynamics, 227, 259
molecular markers, 220
molecular mechanisms, 195, 413
molecular weight, 87, 93, 102, 199, 203
molecules, 55, 74, 99, 101, 163, 203, 216, 271, 284, 296, 304, 305, 313, 314, 320, 321, 323, 327, 345, 363, 370, 373, 374, 390, 391, 392, 399, 406
monkeys, 174
monoclonal, 41, 197, 202, 208
monoclonal antibody, 41, 202, 208
monocytes, 84
monomer, 284, 356, 357, 360, 362, 370, 371, 382, 383, 403, 405
monomeric, 295, 297, 356, 357, 370
monomers, 284, 320, 356, 360, 362, 365, 367, 370, 371, 373, 403
monosaccharides, 54, 87
monotherapy, 276, 279
Monroe, 167, 172, 175, 177, 179, 180, 183, 185
Moon, 219, 278, 279
Morocco, 158, 161
morphogenesis, 413
morphological, 127
morphology, ix, 106, 116, 120, 121, 127
mortality, 146, 156
mortality rate, 146, 156
mother cell, 132
mothers, 264
motion, xi, 225, 227, 255, 259, 262, 318, 344

motors, 115, 262
mouse, 86, 197, 200, 202, 208, 223, 427
movement, 113, 227, 230, 231, 235, 236, 242, 245, 255, 258, 262, 284, 286, 295, 333, 334, 337, 343, 363, 370, 373, 374, 380, 391
mRNA, x, xiv, 4, 15, 17, 38, 39, 41, 42, 43, 45, 48, 49, 98, 111, 140, 148, 188, 189, 190, 201, 202, 203, 204, 205, 206, 207, 208, 221, 264, 269, 326, 411, 412, 413, 414, 415, 416, 417, 418, 426, 431
mtDNA, 430
mucin, 182
multicellular organisms, 46, 428
multidrug resistance, 203, 217, 221
multiple myeloma, 218, 220
multiplexing, 163
mutagen, 84
mutagenesis, viii, xii, 51, 52, 53, 56, 62, 63, 64, 65, 67, 68, 69, 70, 83, 85, 329, 331, 335, 340, 346, 350, 359, 364, 377, 388
mutagenic, 53, 56, 67, 69, 70, 72, 73, 78, 81
mutation, 13, 47, 48, 53, 56, 62, 63, 64, 65, 68, 69, 80, 106, 109, 115, 135, 274, 275, 277, 280, 300, 306, 323, 334, 377
mutation rate, 56, 62, 63, 65, 68, 80, 274, 275
mutations, viii, xiv, 11, 13, 15, 51, 52, 61, 62, 63, 64, 67, 68, 73, 75, 80, 84, 88, 136, 150, 274, 275, 277, 279, 320, 340, 411, 413
MYC, 210, 223
myeloid, 215, 218, 219, 222
myeloma, 218, 220
myosin, 382, 405

N

Na$^+$, 84
N-acety, 64
NaCl, 57
NAD, 66
nanotechnology, 321
nanowires, 327
NAS, 142, 143, 148, 150, 153, 154, 162, 166, 174, 176, 179
National Academy of Sciences, 339, 345
National Institutes of Health, 378, 404
natural, 40, 44, 53, 73, 199, 203, 220, 260, 270, 272, 301, 304, 306
natural science, 260
natural sciences, 260
natural selection, 44
nausea, 151, 156
neck, 199, 201, 217, 222

neglect, 249
nematode, xiv, 411, 412, 425, 426, 427, 428, 429
nematodes, 413, 425, 428
neoplasms, 215
Nepal, 156
nephropathy, 66, 81, 88
nerve, 66, 84
Netherlands, 138, 158, 183
network, ix, 96, 120, 121, 133, 134, 189, 206, 207, 222, 286, 341, 429
neuroblastoma, 196, 216, 218
neuromuscular diseases, 300, 316
neurotoxicity, 69
New Jersey, 353
New York, 49, 135, 211, 261, 298, 322, 349, 350, 351, 378, 389
New Zealand, 158, 174
Nigeria, 157, 158
NIH, 81
nitrogen, 4, 116
NMR, 361
noise, 231, 232, 233, 258
nondisjunction, 115
non-enzymatic, viii, 52, 54, 55, 85
non-infectious, 164
non-random, 30
non-uniform, 49
normal, viii, ix, 51, 53, 56, 62, 63, 86, 119, 120, 121, 125, 127, 188, 190, 191, 197, 203, 206, 209, 211, 264, 266, 276, 308, 333, 356, 427
normalization, 26, 66, 86, 135, 271, 277
NSC, 215, 221
N-terminal, 82, 140, 190, 193, 285, 331, 332, 333, 340, 341, 350, 357, 358, 359, 360, 361, 367, 375, 380, 381, 387, 391, 392
nuclear, 80, 199, 212, 213, 417
Nuclear factor, 217
nuclease, 102, 309, 313, 315, 320
nuclei, 116
nucleic acid, vii, x, xiv, 11, 82, 110, 137, 139, 142, 148, 153, 162, 163, 164, 166, 172, 173, 174, 176, 179, 180, 181, 285, 305, 312, 319, 320, 321, 381, 390, 391, 392, 403, 406, 407, 411, 413
nucleocapsids, 264, 269
nucleoplasm, 264
nucleoprotein, 108, 116, 380, 384, 385
nucleosides, 270, 272, 273, 275
nucleosome, ix, 56, 91, 99, 106, 113, 116, 211, 213
nucleotide sequence, 31, 32, 46, 49, 142, 144, 157, 168, 170, 179, 182, 184, 281, 323

nucleotides, viii, 4, 7, 11, 12, 52, 53, 55, 62, 75, 76, 77, 78, 79, 80, 144, 146, 155, 156, 157, 161, 226, 228, 270, 271, 272, 273, 275, 298, 308, 312, 318, 331, 334, 340, 364, 365, 376
nucleus, vii, ix, 13, 91, 92, 196, 197, 264, 270
nutrient, 65, 412
nutrients, 120

O

oat, 197, 264
observations, 4, 30, 68, 98, 100, 131, 133, 285, 374, 428
occlusion, 112
Oceania, 157
ODN, 315, 318, 319
Okazaki fragment, 369, 370, 378, 418
oligomer, 403
oligomeric, 324, 357
oligomeric structures, 357
oligomerization, 379
oligomers, 284, 305, 320, 324
oligonucleotides, xiii, 165, 175, 309, 313, 315, 316, 321, 326, 363, 372, 389, 391, 392, 393, 395, 399, 400, 402
oligopeptide, 133
oncogene, 210, 213, 216, 217, 222, 223
oncogenes, 207, 222, 223
oocytes, 428
operator, 113
operon, 13, 44, 73, 76, 101, 121, 131, 388
optical, 58, 59, 65
optical density, 58, 59, 65
optimization, 11, 163, 183
oral, 66, 146, 151, 155, 168, 189, 192, 193, 198, 201, 202, 204, 214, 216, 221, 268, 269, 271, 274, 276, 277
ORC, 384
organ, 380
organelles, 13, 92
organic, viii, 51, 53, 320
organism, vii, 3, 14, 15, 18, 24, 26, 29, 378, 412
orientation, 241, 286, 334, 341, 343
osteosarcoma, 202, 216
outpatient, 146
ovarian cancer, 205, 222
ovary, 210
oxalate, 133
oxidation, 54, 55, 56, 60, 62, 80, 81, 83
oxidative, ix, 53, 54, 56, 67, 75, 80, 87, 88, 92, 101, 102, 104, 114, 115, 218

oxidative damage, 67, 88
oxidative stress, ix, 87, 92, 101, 102, 104, 114, 115
oxygen, 53, 55, 67, 83, 84, 215
oysters, 150, 170, 171, 172, 182

P

p53, 201, 202, 215, 216, 217, 222
packaging, 285, 286, 298, 391
pairing, 6, 45, 46, 406
Pakistan, 158
palladium, 360
paradox, xii, 299, 300
parameter, 242, 250, 253, 258, 292, 293, 296
parasites, 12, 13, 138
Paris, 111
PARP, 196, 198
particles, x, 137, 139, 142, 162, 163, 164, 182, 264, 272
passive, 281
pathogenesis, 54, 280
pathogenic, x, 88, 137, 138, 179, 428
pathogenic agents, 138
pathogens, 138, 139, 167, 168, 172
pathology, 54, 323
pathways, x, 60, 82, 83, 188, 189, 220, 364, 429, 431, 432
patient care, xi, 263, 269, 272
patients, 54, 87, 88, 146, 156, 169, 175, 180, 182, 201, 214, 215, 220, 264, 266, 267, 268, 271, 272, 274, 275, 276, 277, 278, 279, 280, 281, 390
PCR, x, 40, 59, 83, 125, 137, 142, 143, 145, 148, 149, 150, 151, 153, 154, 159, 160, 162, 163, 164, 165, 166, 167, 168, 169, 170, 171, 172, 173, 174, 175, 176, 177, 178, 179, 180, 181, 183, 184, 185, 189, 201, 202, 205, 208, 305, 392
peptide, 48, 53, 129, 131, 132, 264, 391, 394, 405
peptides, 131, 136, 192, 391, 394
perinatal, 156
periodic, 139
peripheral nerve, 66, 84
permit, 104, 343, 345
peroxidation, 55, 56, 83, 84
peroxide, 116
person-to-person contact, 139, 151
perturbations, 304
P-glycoprotein, 198, 203, 217, 221
pH, 57, 58, 60, 70, 72, 73, 74, 75, 302, 303, 304, 306, 307, 310, 312, 319, 393, 406
phage, 14, 17, 92, 340, 344, 345, 346, 347, 348, 349, 350, 354, 369, 381, 384, 386

pharmaceutical, 67, 88, 272
pharmaceutical industry, 272
pharmacokinetic, 214
pharmacokinetics, 272
pharmacological, 206, 214
pharmacology, 272, 274
phenol, 57
phenolic, 338
phenotype, 62, 63, 65, 68, 69, 93, 106, 127, 206, 425, 427
phenotypes, ix, xiv, 119, 120, 121, 127, 412, 413, 414, 425, 426, 427
phenotypic, 93, 121, 426, 427
phenotypic analyses, 427
phenylalanine, 4
phenylbutyrate, 195, 214
Philadelphia, 168, 173, 353
phosphatases, 132
phosphate, 52, 55, 56, 60, 64, 66, 67, 68, 69, 70, 72, 74, 76, 77, 78, 80, 81, 82, 83, 85, 86, 87, 131, 132, 133, 200, 230, 231, 232, 234, 287, 303, 333, 336, 337, 338, 342
phosphates, 337
phosphodiesterase, 133
phosphoprotein, 156
phosphorylation, x, 76, 131, 132, 187, 190, 191
phylogenetic, ix, 91, 93, 144, 151, 154, 164, 166, 168, 183, 184
physical interaction, 369, 384
physical properties, viii, 91, 107, 381
physicians, 54, 276
physicochemical, 12
physicochemical properties, 12
physics, 260
physiological, viii, 51, 53, 54, 55, 56, 63, 64, 77, 78, 81, 87, 88, 98, 290, 356
physiology, iv, 54, 67, 109
pig, 156, 170, 172, 174, 181, 184
pilot studies, 66, 84
placebo, 156
plants, 55, 88, 340
plaque, 168
plasma, 54, 55, 66, 86, 267
plasma proteins, 55, 86
plasmid, 40, 60, 70, 71, 72, 73, 74, 75, 76, 107, 112, 122, 125, 348, 367, 369, 379, 380, 385, 386, 392, 427
plasmids, ix, 18, 41, 42, 119, 120, 121, 126, 127, 219, 340, 344, 354
platforms, 142, 166

play, viii, xii, xiv, 52, 53, 93, 104, 106, 152, 198, 203, 207, 216, 284, 299, 320, 354, 362, 365, 369, 370, 375, 389, 404, 411, 412, 426, 428
PLC, 156
PLP, 52, 64, 66, 69, 70, 71, 72, 73, 74
point mutation, 64, 150
polarity, 345, 372, 374, 376
poliovirus, 164, 170
polyacrylamide, 37, 39, 40, 41, 148, 174, 393, 394
polymer, 301, 304, 315, 322, 324, 325, 383, 405
polymerase chain reaction, 142, 167, 168, 172, 173, 175, 179, 180, 183, 305
polymerization, xi, xii, 53, 59, 225, 226, 227, 229, 231, 232, 234, 238, 239, 241, 243, 256, 258, 262, 311, 314, 316, 320, 322, 325, 329, 330, 331, 332, 333, 334, 335, 336, 337, 338, 339, 340, 341, 344, 346, 348
polymers, xii, 299, 301, 318, 322, 325, 400
polymorphism, 44, 213
polymorphisms, 175
polynucleotide, 4, 393
polynucleotide kinase, 393
polypeptide, 4, 6, 49, 63, 76, 320, 323, 331, 355, 357, 359, 376, 377
polypeptides, 15, 86, 112, 184, 357, 359
polyproline, 4
polyunsaturated fat, 55
polyunsaturated fatty acid, 55
polyunsaturated fatty acids, 55
poor, 47, 123, 127, 145, 155, 269, 338, 396
population, 13, 20, 138, 147, 156, 175, 265, 266, 280
pore, 337
post-transcriptional regulation, 110
post-translational modifications, x, 187, 190
potassium, 124, 303, 310
power, viii, xi, 32, 51, 58, 165, 183, 283, 284, 286, 287, 288, 289, 290, 291, 292, 293, 295, 296, 371, 378
powers, xiii, 353
preclinical, xi, 68, 263, 273, 274, 277
prediction, 17, 39
preference, xi, 14, 16, 17, 27, 28, 44, 283
pregnant women, 156
press, 113
pressure, 44, 138, 274
prevention, x, 66, 68, 137, 179, 267
preventive, 276
primates, 273
priming, xii, 136, 244, 269, 270, 271, 279, 299, 315, 319, 320, 331, 336, 338, 340, 341, 342, 343, 344, 349, 350, 354, 365, 369, 370, 371, 372, 380, 381, 383, 386
probability, 229, 236
probe, 117, 145, 162, 163, 165, 166, 174, 179, 390, 395
process control, 164, 166, 176, 178
production, ix, 106, 110, 119, 120, 121, 122, 124, 126, 127, 128, 131, 132, 133, 134, 165, 215, 274, 319
productivity, ix, 119, 120, 121, 135
progeny, 264, 270
prognosis, 266
program, 18, 21, 31, 109, 132
prokaryotes, vii, viii, 3, 5, 6, 14, 21, 24, 25, 28, 29, 47, 83, 91, 107, 117, 350, 354, 369, 375
prokaryotic, vii, 3, 7, 15, 18, 27, 29, 108, 111, 322, 333, 334, 340, 346, 354, 355, 370, 378
proliferation, 104, 194, 195, 201, 202, 207, 211, 213, 216, 222
promoter, 40, 76, 110, 122, 131, 135, 189, 190, 191, 203, 204, 206, 210, 211, 212, 215, 221, 222, 223, 266, 274, 325, 355, 361, 364, 371, 378
promoter region, 191, 211, 222, 274, 325
promyelocytic, 196, 216, 217, 219
propagation, 120, 417
property, 78, 331, 376
prophylactic, 267
prostate, 205, 217, 221
prostate cancer, 205, 217, 221
proteases, 391
Proteasome, 220
protection, 67, 84, 115, 351, 428, 430, 432
protective role, 104, 114
protein binding, 351
protein family, 102, 378, 413, 428, 431
protein folding, 39
protein function, 354, 355
protein kinase C, 86
protein sequence, 358
protein structure, 313
protein synthesis, 4, 45, 48, 204, 205, 207
protein-coding sequences, 31
protein-protein interactions, 113, 341, 378, 384
proteobacteria, vii, ix, 3, 21, 25, 27, 92, 93, 104
proteolysis, 350, 382
proteome, xiv, 411
proteomics, 432
protocol, 145, 165, 166, 167, 194, 195, 199
protocols, 143, 145
protooncogene, 223

prototype, xiii, 148, 149, 152, 157, 353, 354
PSA, 419
pseudo, 363, 390
pseudogene, 425
Pseudomonas, 369
Pseudomonas aeruginosa, 369
PSI, 218
public, xi, xiv, 145, 155, 166, 171, 263, 412
public health, xi, 155, 166, 171, 263
purification, 87, 166, 323, 340, 346, 381, 392
pyridoxal, 56, 64, 66, 67, 68, 69, 72, 81, 83, 86, 87
pyridoxamine, 64, 65, 66, 67, 81, 82, 83, 84, 85, 87, 88
pyridoxine, 64, 67, 68
pyrimidine, 13
pyrophosphate, 65, 82, 338

Q

quality control, 164, 166, 181
quantum, 358
Quebec, 137
quorum, 136

R

radiation, 199, 218, 220, 428, 431, 432
radioisotope, 301
radius, 236, 249
Ramadan, 308, 317, 324
Raman, 58, 74, 86
random, 14, 30, 244, 274, 320, 323, 345
range, vii, 64, 163, 236, 247, 249, 273, 290, 311, 312, 356, 369, 379, 385, 393
ras, 219, 220
raspberries, 181
rat, 81, 83, 87, 216, 221
rats, 66, 68, 82, 83, 84, 85, 86, 87, 88
reactants, 54, 60
reaction mechanism, 348
reaction rate, 238, 395, 397
reaction temperature, 306, 311, 313
reaction time, 311, 313, 319
reactive oxygen, 67, 83, 215
reactive oxygen species, 67, 83, 215
reactivity, viii, 51, 60, 72, 157, 164
reading, 13, 30, 31, 32, 48, 140, 152, 172, 182, 184, 194, 212, 320, 391, 430
reagent, 77, 189
reagents, 165

recognition, xii, 6, 12, 15, 39, 48, 80, 107, 113, 206, 222, 259, 299, 310, 311, 312, 313, 321, 340, 343, 344, 348, 383
recombination, xiv, 61, 63, 125, 139, 140, 152, 169, 283, 284, 297, 327, 354, 377, 392, 403, 411, 419, 423, 428, 432
recovery, 163, 267, 344
recruiting, 190, 367
recycling, 394, 397
red blood cell, 88
red blood cells, 88
redox, 67, 88
reducing sugars, 82, 87
reductases, 66
redundancy, 426
reflection, 93
refractory, 214, 215, 281
regeneration, 349
regression, 216, 297, 393
regulation, ix, x, xiv, 13, 14, 15, 30, 47, 92, 93, 95, 96, 97, 98, 108, 110, 114, 116, 131, 136, 187, 190, 191, 207, 209, 211, 212, 216, 219, 220, 221, 223, 317, 362, 368, 370, 387, 388, 411, 423
regulators, 95, 130, 133, 212, 222, 362
relapse, 277
relationship, vii, 3, 20, 36, 39, 43, 49, 222, 269, 305, 309, 396
relationships, 133, 260, 346, 347, 349, 429, 430
relatives, 136
relaxation, 70, 71, 72, 74, 364
relevance, 81, 85, 279
remission, 268, 276
remodeling, xiv, 106, 109, 116, 190, 195, 213, 222, 372, 411, 412, 419, 423, 427, 429, 430, 431
remodelling, 430
renal, 66, 83, 220, 266, 271
renal cell carcinoma, 220
renal disease, 83
repair, viii, xi, xiv, 45, 51, 52, 53, 56, 61, 62, 63, 68, 75, 78, 80, 83, 85, 152, 283, 284, 297, 354, 377, 411, 412, 413, 419, 420, 422, 423, 424, 427, 428
reparation, 15, 166
repetitions, 344
repression, 105, 111, 190, 203, 206, 207, 211, 222, 223, 326
repressor, 95, 96, 113, 132, 210, 213, 326
reproduction, 162
Research and Development (R&D), 137, 272
residuals, 396

resistance, xi, 58, 62, 63, 84, 123, 124, 125, 135, 198, 203, 217, 220, 221, 260, 263, 271, 272, 275, 276, 277, 278, 281
resolution, xii, 259, 260, 329, 331, 334, 336, 338, 341, 342, 348, 431
respiration, 132, 427
restriction enzyme, xii, 299, 300, 310, 311, 312, 313, 321, 392
retardation, 71
retina, 66, 84
retinoic acid, 199, 217, 218
retinopathy, 87
returns, 20, 364
reverse transcriptase, 163, 167, 173, 179, 183, 188, 189, 210, 211, 221, 222, 228, 239, 260, 261, 262, 269, 270, 274, 275, 279
RFS, 273
rhombohedral, 383
ribonucleic acid, 324
ribose, 84
ribosomal, vii, 3, 14, 17, 18, 21, 30, 36, 39, 45, 48, 98, 430
ribosome, xiv, 28, 30, 31, 39, 40, 42, 43, 76, 411, 412, 413, 415, 426
ribosomes, 4, 42, 48, 49, 101
rings, 382
risk, 78, 156, 163, 165, 172, 179, 213, 266, 269, 271, 275, 276, 277, 279
risk factors, 179, 277
risks, 167
rivers, 170
RNAi, 273, 412, 414, 416, 418, 425, 428, 429, 430, 432
rolling, 305, 348
room temperature, xii, 290, 299, 393
rotations, 295
rotavirus, x, 137, 138, 149, 150, 167, 168, 170, 172, 173, 174, 175, 176, 177, 178, 179, 180, 181, 182, 183
Russia, 273

S

Saccharomyces cerevisiae, xiv, 115, 116, 411, 429, 430, 431
safety, 81, 120, 271, 272, 273
SAHA, 196, 199, 215, 217, 218
saliva, 167
salmon, 77
Salmonella, 14, 110, 356
salt, 57, 59, 98, 341, 363
salts, 74, 124
sample, 21, 100, 139, 157, 162, 163, 164, 166, 176, 178, 226
sample mean, 21
sanitation, 155, 169
SAS, 189, 201, 202, 204, 205, 207, 208
satellite, 326
scaffold, 378
scaffolding, 117
scavenger, 81
Schiff, 54, 55, 60, 66, 70, 72, 87, 279
Schiff base, 54, 55, 60, 66, 70, 72, 87
Schmid, 212, 223
school, 177, 178
scientific community, 142
scores, 144, 145, 146, 155, 161, 425
SDS, 40, 41, 52, 57, 76, 393
search, vii, 3, 64, 92, 404
searches, 18
searching, 72
seawater, 169
secrete, 120, 272
secretion, 98, 121, 127
sediment, 177
sedimentation, 301
sediments, 149
seeds, xii, 299, 309, 310, 314, 315, 316, 318, 320
segregation, 112, 113, 115, 422
selecting, 163, 275, 338
selectivity, 107, 212, 273
self-assembly, 176
senescence, 87, 188, 209
sensing, 428
sensitivity, x, 137, 139, 142, 145, 148, 159, 162, 163, 166, 194, 199, 204, 217, 220, 275, 330, 428, 431
separation, xiv, 113, 148, 167, 288, 297, 302, 389, 390, 395, 398, 399, 403, 406
sequencing, vii, xiv, 3, 14, 45, 48, 60, 84, 151, 153, 162, 166, 183, 412
series, ix, 40, 99, 119, 121, 123, 124, 125, 126, 127, 165
serine, 12, 132, 391
serology, 142
serum, 58, 84, 86, 159, 167, 170, 197, 266, 267, 268, 271, 276, 277
serum albumin, 58, 86
services, iv, 18
severity, 156
sewage, 176, 177

sex, 266
SGD, xiv, 412, 429
Shanghai, 46, 173
shape, 94, 226, 264, 334
shaping, 152
shares, 367, 377
sharing, 320
shellfish, 156, 169, 170, 176, 177
SHM, 392
shock, 102
short period, 288
shortage, 276
shunts, 66
side effects, 274, 390
sign, 33, 34
signal transduction, 132, 223
signaling, 116, 129, 131, 216, 220, 222, 223, 429
signaling pathway, 116, 220, 222, 429
signaling pathways, 220, 429
signals, 196
silencers, 211
silver, 174
similarity, xii, 93, 329, 330, 331, 357, 361, 375, 377, 413
simulations, 227, 259
SIS, 212
skin, ix, 119, 121, 125
Slovenia, 138, 182
sodium, 58, 60, 74, 192, 199, 214, 216, 217, 218, 221, 223
sodium butyrate, 192, 199, 216, 217, 218, 221, 223
solid tumors, 214
solubility, 77
solvent, 58, 333
somatic cell, 188
somatic cells, 188
sorbitol, 86
South America, 157, 172
Southampton, 141, 177
Spain, 51, 138, 158, 170, 329
spatial, 12, 30, 332, 335, 342, 374
species, vii, ix, 3, 13, 14, 15, 18, 19, 21, 22, 23, 25, 27, 28, 30, 39, 53, 54, 67, 83, 91, 92, 93, 100, 102, 106, 107, 157, 178, 215, 226, 263, 273, 309, 311, 369, 413, 425, 426, 427, 428
specificity, xiii, 14, 15, 77, 79, 87, 107, 162, 166, 284, 298, 304, 330, 341, 350, 356, 385, 386
spectrophotometric, 301
spectroscopy, 61, 74, 86, 327
spectrum, 18

speculation, 45
speed, 106, 313, 372, 378
sperm, 77
spheres, 287, 334, 337
spindle, 346
splenomegaly, 174
sporadic, 156, 179, 184
spore, 102, 120, 126, 132
squamous cell, 199, 201, 217
squamous cell carcinoma, 199, 201, 217
SSB, 345
stability, viii, 15, 51, 110, 156, 171, 257, 280, 302, 324, 331, 333, 339, 342, 415, 418, 430, 431, 432
stabilization, 335, 347, 349
stabilize, 94, 338, 375
stages, xii, xiii, 14, 54, 67, 75, 274, 299, 314, 353, 354, 365, 377
standard deviation, 19, 20, 22, 23, 59, 399, 401, 402
standardization, 166, 170
standards, 140, 145, 155
staphylococcal, 113
Staphylococcus, ix, 92, 112, 114, 375, 377, 387, 388
Staphylococcus aureus, ix, 92, 112, 114, 375, 377, 387, 388
starvation, 104, 110, 136
state regulators, 130
statistical analysis, 30
steric, 231, 260, 298, 338, 340
steroid, 212
steroidogenesis, 326
stochastic, 260
stock, 59
stoichiometry, 363, 365, 386
storage, vii, 169
strategies, 47, 182, 190, 275, 367
strawberries, 180
strength, 286, 302, 303, 306, 324, 363
Streptomyces, 340, 344
stress, viii, ix, 51, 55, 87, 92, 101, 102, 104, 114, 115, 295
stretching, 260
stroke, xi, 283, 284, 286, 287, 288, 289, 290, 291, 292, 293, 295, 296
strokes, 286
stromal, 221
stromal cells, 221
strong interaction, 228, 288, 378
structural characteristics, 93
structural gene, 386

structural protein, ix, 91, 92, 140, 147, 148, 156, 172, 264, 391
structural transitions, 116
structuring, 92
subdomains, 331, 332, 335, 336, 337, 338, 339, 341, 350
subgroups, 175
subjective, 315
substitution, 42, 64, 100, 175
substrates, xiv, 77, 78, 79, 80, 217, 221, 226, 303, 314, 331, 332, 335, 350, 367, 389, 393, 397, 402, 403, 404
subtraction, 196
sucrose, 55
sugar, 53, 76, 80, 84, 129, 230, 231, 232, 234, 287
sugars, 55, 61, 62, 66, 73, 75, 82, 83, 84, 87, 135
Sun, 112, 174, 212, 348
superoxide, 84
superposition, 342
supplements, 64
supply, 168, 315, 320
suppression, 42, 47, 63, 64, 81, 87, 205, 275, 276
suppressor, 6, 48, 64, 93, 191, 212, 219
surface water, 181
surveillance, 138, 162, 166, 168, 182
survival, viii, 51, 106, 194, 274, 390, 426
survival rate, 194
survivors, 431
suspensions, x, 137, 139
sustainability, 120
SV40, 116, 354, 357, 379, 381, 382
swine strain, 157
switching, xi, 225, 234, 258, 262, 276, 317
Switzerland, 168
symbiosis, 116
symbiotic, 107
symmetry, 94, 359, 361, 363
symptom, 156
symptoms, 68, 146, 151, 156, 266, 390
syndrome, xiv, 174, 411, 431
synergistic, ix, 120, 121, 127, 133, 199
synergistic effect, ix, 120, 121, 127, 133
systems, ix, 53, 54, 55, 69, 84, 86, 89, 92, 119, 120, 121, 129, 134, 136, 140, 143, 145, 148, 153, 155, 163, 164, 165, 166, 173, 174, 211, 341, 344, 354, 355, 369

T

Taiwan, 158, 174
tandem mass spectrometry, 81

tandem repeats, 188, 213, 312, 319
targets, x, 111, 145, 188, 220, 276
tau, 380
taxa, 18, 25, 29, 30
taxonomic, 29
taxonomy, vii, 3, 14, 18, 21, 24, 26, 29, 30, 171, 281
TCC, 27, 28, 392
T-cell, 199, 201, 220
technical assistance, 404
technology, x, 134, 138, 145, 148, 163, 165, 166, 272
Telbivudine, 268, 272, 275, 278, 279
telomerase, x, 188, 190, 203, 205, 206, 207, 209, 210, 211, 221, 222, 223, 224, 330, 431
telomere, 188, 209, 211, 222, 423
telomeres, 188, 206, 209, 326, 351, 421
TEM, 100
temperature, xii, 15, 138, 156, 290, 299, 304, 311, 312, 318, 324, 325, 355, 363, 377, 387, 388, 393
temporal, 113, 355
tension, xi, 225, 226, 241, 242, 243, 244, 252, 253, 254, 255, 256, 259, 260, 261
termination codon, 6, 13, 19, 28, 43
ternary complex, xii, 229, 230, 261, 329, 336, 337, 339, 342, 367, 369
tetracycline, 123, 125, 135
textbooks, 300
TGA, 19, 24, 25, 28
therapeutic interventions, 267
therapeutic targets, 220
therapeutics, 88, 267, 268
therapy, xi, 199, 201, 203, 212, 219, 220, 263, 268, 269, 271, 274, 276, 277, 278, 279, 280, 281
thermal stability, 156
thermodynamics, 259
Thermophilic, 304, 305, 306, 308, 310, 312
thermostability, 260
theta, 425
Thiamine, 65, 67, 82
thioredoxin, 227, 256, 259, 260, 261
Thomson, 405
threat, 179
threatening, 201
three-dimensional, 117
threonine, 35, 120
thymine, 4, 74
thymus, 304
timing, 45, 131, 276
TIMP, 219
tissue, 87, 88, 405

Index

Tokyo, 86, 117, 219, 411
Topoisomerase, 94
Topoisomerases, 93
topological, 94, 95, 104, 105, 109, 331
topology, ix, 92, 94, 95, 104, 105, 109, 113, 384
torque, 288, 289, 293, 294, 295
total costs, 138
toxic, 67, 69, 70, 272
toxic effect, 272
toxicities, 201
toxicity, 64
toxicology, 274
trace elements, 124
traits, ix, 119, 121, 124, 126
trans, 217, 218
transcript, 189, 190, 319
transcriptase, 163, 167, 173, 179, 183, 188, 189, 210, 211, 221, 222, 228, 239, 260, 261, 262, 269, 270, 274, 275, 279
transcription factor, 92, 97, 102, 106, 132, 191, 199, 206, 207, 213
transcription factors, 92, 97, 106, 206
transcriptional, 95, 101, 110, 116, 132, 133, 134, 136, 189, 190, 196, 206, 211, 212, 216, 222, 223, 275, 326, 423
transcriptomics, 136
transcripts, 140, 264, 326
transduction, 199, 357
transfection, 168
transfer, xi, 4, 14, 46, 47, 48, 173, 225, 226, 227, 233, 249, 250, 251, 252, 256, 257, 258, 259, 261, 262, 324, 384, 390, 406
transfer RNA, 4, 46, 47
transformation, 69, 73, 131, 135, 207, 210
transformations, 54, 55
transforming growth factor, 214
transfusion, 178
transgene, 199, 203, 218, 220, 221
transition, ix, xi, xiii, 13, 64, 120, 121, 132, 133, 225, 226, 227, 228, 229, 230, 231, 232, 233, 234, 241, 242, 245, 246, 251, 256, 258, 262, 292, 293, 326, 330, 331, 340, 342, 349
transition rate, 227, 241, 242, 245, 256, 292, 293
transitions, 53, 64, 109, 234, 241, 245, 256, 292, 323, 359
translation, viii, 4, 7, 13, 14, 15, 16, 17, 18, 28, 29, 30, 36, 37, 39, 41, 42, 43, 44, 45, 47, 48, 49, 57, 95, 98, 100, 101, 110, 148, 184, 274, 391, 413, 417, 431

translational, x, 12, 44, 46, 48, 59, 98, 101, 111, 187, 190
translocation, xi, xiii, 283, 284, 285, 286, 287, 288, 290, 292, 295, 296, 297, 298, 330, 338, 344, 346, 353, 372, 376, 383, 386, 387, 406
translocations, 295
transmission, x, 137, 138, 139, 147, 156, 169, 175, 177, 178, 179, 180, 182, 183, 266, 281
transplant, 390
transplantation, 267, 390
transport, 414, 417
transposons, 431
traps, 390, 391, 395, 397, 398, 399, 400
trial, 82, 201, 215, 220, 273, 278, 281
trichostatin, 192, 211, 216, 217, 219, 221
trichostatin A, 192, 211, 216, 217, 219, 221
triggers, 132, 190, 337, 432
trimer, 316, 363, 367, 370
Trp, 28, 44, 364, 392
trypsin, 357, 361
tryptophan, 12, 13, 82, 364, 384
TSA, 192, 193, 194, 195, 196, 197, 198, 199, 200, 201, 203, 204, 205, 207, 217
tubular, 264
tumor, 191, 201, 202, 206, 209, 210, 216, 218, 219, 222
tumor cells, 202, 206, 209, 210, 218
tumor growth, 219
tumorigenesis, 207
tumors, 214, 219, 222
turnover, 261, 393, 395, 396, 397, 400, 401, 403, 404, 414, 431
two-dimensional, 37, 39
type 2 diabetes, 88
tyrosine, 269, 333, 347
Tyrosine, 348

U

ubiquitin, 210
underlying mechanisms, 203
uniform, 49, 327
United Kingdom, 138, 175
United States, 138, 146, 156, 174, 176, 178, 180, 281
uridine, 362
urine, 151, 156
USSR, 302
UV, 102, 106, 115, 189, 208, 377
UV exposure, 115
uveal melanoma, 196, 216

V

vaccination, 264
vaccine, 156, 168, 182, 265, 267, 268, 280
Valdez, 217
validation, 163, 166, 175, 373
validity, 18
valine, 5
valproic acid, 221
values, 16, 19, 20, 21, 23, 26, 29, 32, 34, 35, 36, 39, 44, 64, 237, 238, 239, 240, 241, 242, 243, 244, 245, 249, 250, 255, 258, 290, 293, 296, 330, 393, 425
van der Waals, 290
variability, 150, 165, 167, 170, 180
variation, 83, 107, 139, 140, 142, 145, 166, 181, 238, 255, 302
variegation, 212
vascular endothelial growth factor, 200
vascular endothelial growth factor (VEGF), 200, 201
veal, 216
vector, 182, 218
vegetables, 149, 150, 169, 183
vein, 67, 200
velocity, 396
vertebrates, viii, 51
Vibrio cholerae, 112
Vietnam, 180
viral diseases, 270
viral gastroenteritis, 138, 177
viral hepatitis, 171, 183, 278
viral infection, xi, 172, 263, 267
viremia, 264
virological, 156, 266
virology, 163, 178, 179
virus infection, xi, 170, 178, 179, 203, 263, 279, 280, 282
virus replication, 277, 280, 281
viruses, vii, x, 137, 138, 139, 142, 145, 147, 153, 162, 163, 164, 165, 166, 167, 168, 169, 170, 172, 173, 176, 177, 178, 179, 180, 181, 183, 184, 263, 264, 266, 268, 273, 274, 280, 340, 344, 350, 391, 428

viscera, 156
viscosity, 236, 249
visible, 25, 76, 80
visualization, 111, 113
vitamin B1, 63, 64, 65, 68
vitamin B6, 64, 68, 69, 84
vitamin B6 deficiency, 68
vitamin D, 199
vitamins, 64, 67, 73
vitreous, 100, 112
vomiting, 156

W

Wales, 138
waste treatment, 170
water, x, 53, 58, 60, 137, 138, 139, 150, 151, 156, 157, 162, 164, 166, 168, 169, 171, 172, 176, 177, 181, 183, 205, 333, 334
water supplies, 157
wavelengths, 393
weak interaction, 74, 245, 256, 257, 286, 370
Weinberg, 210, 211
wells, 168, 194, 250
western blot, 202
white blood cells, 56
wild type, 38, 40, 42, 57, 61, 62, 166
wine, 169, 180
Wisconsin, 168
women, 156
workers, 56, 65, 66, 76, 77, 80
Wyoming, 167

X

xenografts, 218

Y

yeast, xiv, 46, 108, 113, 114, 115, 210, 261, 411, 412, 414, 416, 418, 419, 420, 421, 422, 423, 424, 425, 426, 427, 428, 429, 430, 431, 432
yield, 17, 38, 39, 42, 43, 44, 140, 358

Z

zinc, 213, 356, 383
zoonosis, 177
zoonotic, x, 137, 139, 147, 157, 170, 178